Preface

This volume is the third edition of the book entitled *"Principles of Environmental Science and Technology"*. The revised title reflects the enormous changes in the abatement of pollution which have taken place from 1988 to 2000. Environmental technology was the dominant tool in pollution abatement in the seventies and most of the eighties. Environmental science and environmental technology worked hand in hand to determine a concentration, translate it into an effect, discuss the acceptable effect level and find a suitable environmental technology which could provide this lower level by treatment of the point sources of the pollutant.

Three major developments have changed this strategy. Firstly, non-point sources were shown about 20 years ago to be as important as point sources. Another technology, ecotechnology, is therefore needed to cope with this problem, as the end-of-the-pipeline technology obviously cannot solve a problem associated with diffuse pollution. Secondly, it has been acknowledged that it is often less costly to change the production system to reduce emissions than to solve the problem by environmental technology, after the emissions have taken place. This development has been caused by the introduction of green tax and the growing importance of green products and production methods also for subcontractors. Thirdly, the procedure for a stepwise clear quantification of the pollution problem has found a concrete form as "environmental risk assessment" (ERA) along the same lines as environmental impact assessment which was introduced more than 20 years ago. Due to our increased knowledge about the environmentally important properties of more and more chemical compounds, and to a very clear ERA procedure, it is - although with some uncertainty - possible today to classify most compounds in a black (high risk), grey (maybe there is a risk) or white (no risk) zone.

These new tools, ecotechnology, cleaner technology, life cycle analysis and environmental risk assessment have changed environmental management strategy radically. Obviously, it is much more difficult to find a proper solution as the spectrum of possibilities is much wider. A more careful and detailed analysis is therefore required before a decision can be taken. Environmental management is getting even more multidisciplinary, as reflected in this volume, which touches on the above mentioned 4-5 different technologies, in addition to the use of basic scientific principles, ecology, ecotoxicology and environmental chemistry.

The second edition has more than 600 pages. It has therefore not been possible just to add the number of pages covering the new approaches. The environmental technology chapters have been reduced in size to give room for the new topics. Environmental technology has not changed radically during the last two decades - no new basic methods have been developed, but as a growing tree environmental technology has got new branches, i.e., many more modifications of the same basic methods. These new variants should not be forgotten as they sometimes offer new and promising solutions to several pollution problems. It has been attempted, to a certain extent, to mention some of these new variants of environmental technology, but it is recommended to examine all possible variants of environmental technology carefully in case these methods should be included in the overall environmental management solution. This is particularly the case when solutions to industrial waste water are discussed. Industrial waste water varies from case to case which makes it almost impossible to indicate a general solution.

This volume attempts to give an overview of *all* tools available today to abate the pollution problems and to give the basic considerations to be applied in selection of the most appropriate environmental management strategy. It is a very ambitious goal due to the high complexity of the problem and the wide spectrum of possibilities. It is my hope that I have been able to meet these objectives or at least give the readers an idea about the complexity of the problem and the number of possible solution models.

Ib Johnsen, Claus Lindegaard and Lars Kamp Nielsen have all given very useful inputs to the volume, as they have provided me with text covering important topics in air pollution and limnology.

The present work reflects the input of many colleagues, Bent Halling Sørensen, Søren Nors Nielsen, John Kryger and Hans Schrøder, to whom I express my most sincere thanks. Students from the Danish Agriculture University in Copenhagen contributed to the book by raising interesting questions and pointing out weaknesses in the first preliminary version of this text. The assistance of my son, Morten Vejlgaard Jørgensen, was of great value in producing many of the drawings in computer format.

Sven Erik Jørgensen
Copenhagen 31. January 2000

PRINCIPLES OF POLLUTION ABATEMENT
Pollution Abatement for the 21st Century

PRINCIPLES OF POLLUTION ABATEMENT
Pollution Abatement for the 21st Century

S.E. JØRGENSEN
Langkaer Vaenge 9, 3500 Vaerløse, Copenhagen, Denmark

2000

ELSEVIER
Amsterdam – Lausanne – New York – Oxford – Shannon – Singapore – Tokyo

ELSEVIER SCIENCE LTD
The Boulevard, Langford Lane
Kidlington, Oxford OX5 1GB, UK

First edition 2000

Library of Congress Cataloging in Publication Data
A catalog record from the Library of Congress has been applied for.

British Library Cataloging in Publication Data
A catalogue record from the British Library has been applied for.

ISBN: 0 08 043626 9 (Hardbound Edition)
0 08 043625 0 (Softbound Edition)

∞ The paper used in this publication meets the requirements of ANSI/NISO Z39.48-1992 (Permanence of Paper).
Printed in The Netherlands.

Table of Contents

1. Introduction

1.1 ENVIRONMENTAL MANAGEMENT

When the first green wave appeared in the mid and late sixties, it was considered a feasible task to solve the pollution problems. The visible problems were mostly limited to point sources and a comprehensive "end of the pipe technology" (= environmental technology) was available. It was even seriously discussed in the U.S. that what was called "zero discharge" could be attained by 1985.

It became clear in the early seventies that zero discharge would be too expensive, and that we should also rely on the self purification ability of ecosystems. That called for the development of environmental and ecological models to assess the self purification capacity of ecosystems and to set up emission standards considering the relationship between impacts and effects in the ecosystems. This idea is illustrated in Figure 1.1. A model is used to relate an emission to its effect on the ecosystem and its components. The relationship is applied to select a good solution to the environmental problems by application of environmental technology.

Meanwhile, it has been disclosed that what we could call the environmental crisis is much more complex than we initially thought. We could for instance remove heavy metals from waste water, but where should we dispose the sludge containing the heavy metals? Resource management pointed towards recycling to replace removal. Non-point sources of toxic substances and nutrients, chiefly originating from agriculture, emerged as new threatening environmental problems in the late seventies. The focus on global environmental problems such as the greenhouse effect and the decomposition of the ozone layer added to the complexity. It was revealed that we use as much as about 100,000 chemicals which may threaten the environment due to their more or less toxic effects on plants, animals, humans and entire ecosystems. In most industrialised countries comprehensive environmental legislation was introduced to regulate the wide spectrum of different pollution sources. Trillions of dollars have been invested in pollution abatement on a global scale, but it seems that two or more new problems emerge for each problem that we solve. Our society seems not geared to environmental problems or is there perhaps another explanation?

Recently, standards for environmental management in industries and green accounting have been introduced. The most widely applied standards today for industrial environmental management are the ISO 14000-series. These initiatives attempt to analyse our production systems to find new ways and methods to make our production more environmentally friendly. More than 100 countries have backed up the international standards for effective management of environmental impacts.

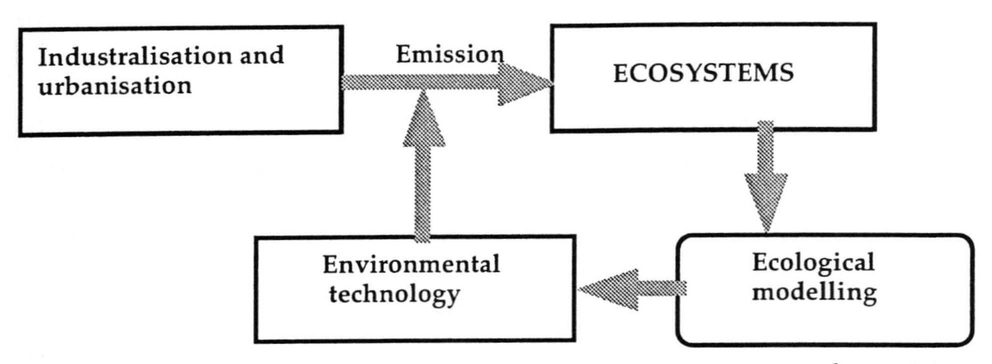

Figure 1.1. The strategy applied in environmental management in the early seventies is illustrated. An ecological model is used to relate an emission to its effect on the ecosystem and its components. The relationship is applied to select a good solution to the environmental problems by application of environmental technology.

The goal of ISO 14000 is to evolve a series of generic standards that provide management with a structured mechanism to measure and manage environmental risks and impacts. Standards have been or are being developed for:

1. Environmental management systems (ISO 14001-14004)
2. Environmental auditing (ISO 14010-14013)
3. Internal reviews (ISO 14014)
4. Environmental site assessments (ISO 14015)
5. Evaluation of environmental performance (ISO 14031)
6. Product-oriented standards such as environmental labelling, terms and definitions for self-declaration environmental claims (ISO 14020-14024).
7. Life-cycle assessment, LCA (ISO 14040-14043)

In Chapter 9 we will touch on all 7 items. The basic idea is that if we know the flows of matter and energy, we can also identify where we could recycle and where we could reduce loss of matter and energy. Point 7 will also be discussed briefly. LCA maps the total pollution originating from a product in its life time in order to develop products that to a greater extent can be recycled, have a longer life time or cause less pollution during their entire life time.

In many cases our production methods have not been charged the actual costs in relation to impacts on the environment and the depletion of our resources. A new tax, named green tax, attempts to charge the real environmental cost to our production. Due to these additional production costs which are often significant, the industries are forced to find new and more environmentally friendly production methods.

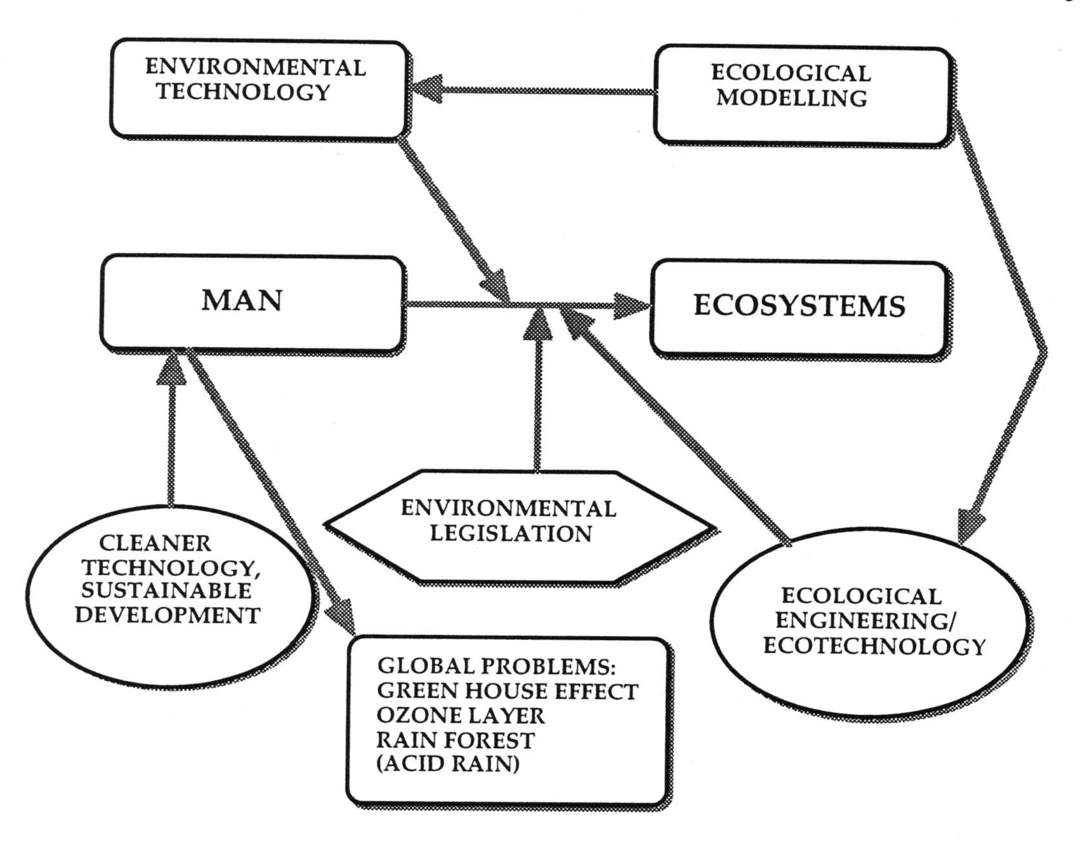

Figure 1.2 The use of environmental models in environmental management, which, today, is very complex and must apply environmental technology, cleaner technology and ecotechnology. Models are used to select the right environmental management strategy. In addition, the global environmental problems, which also require an the use of models as a synthesising tool, play an increasing role.

Figures 1.2 and 1.3 illustrate how complex environmental management is today. The first figure shows that a simultaneous application of environmental technology, ecotechnology, cleaner technology and environmental legislation is needed in environmental management. Figure 1.3 shows the flows of material (and energy) in the history of a product, from raw materials to final disposal as waste. P indicates emission of point pollutants and NP covers non-point pollution.

The number of products in the modern technological society is not known exactly, but it is probably in the order of 10^7-10^8. All these products emit pollutants to the environment during production, their transportation from producer to use, during their application and in their final disposal as waste. The core problem in environmental management is: how can we control these pollutants properly? The answer is that we have use on a wide spectrum of methods. Figure 1.3 illustrates where we can use various technologies to control and reduce pollution.

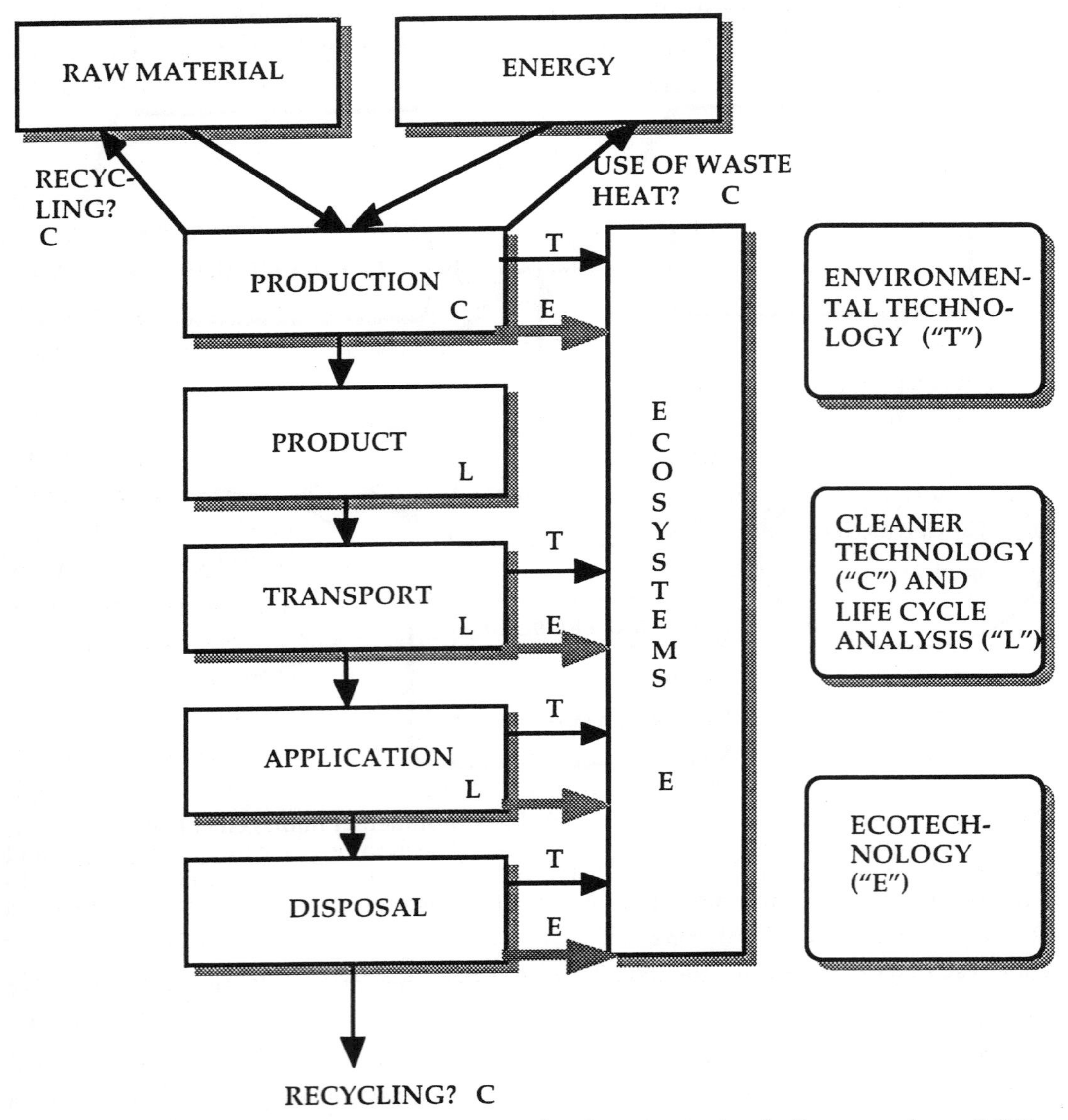

Figure 1.3. The arrows cover mass flows and the thin arrows symbolise control possibilities. Point pollution is indicated as a thin black arrow to the ecosystems, while non-point pollution is a thick grey arrow. Production includes both industrial and agricultural production, the latter being mostly responsible for non-point pollution. Environmental technology is applied on point pollution, while ecotechnology is needed for the solution of non-point pollution problems, but can also be applied to restore ecosystems. Cleaner technology explores the possibilities of changing the present production methods to obtain a reduction in pollution, either by the use of recycling of byproducts and waste products or by a more or less radical change of the production technology. Life cycle analyses are used to find where in the history of a product the pollution actually takes place with the scope to change the product, its transport and/or its application.

Environmental technology offers a wide spectrum of methods that are able to remove pollutants from water, air and soil. These methods are particularly applicable to cope with point sources. Chapters 5-7 will present the spectrum of solutions based on environmental technology.

Cleaner technology explores the possibilities of recycling byproducts or the final waste products or attempting to change the entire production technology to obtain a reduced emission. It attempts to answer the pertinent question: couldn't we produce our product by a more environmentally friendly method. It will to a great extent be based on environmental risk assessment, LCA and environmental auditing. The ISO 14000 series and risk reduction techniques are among the most important tools in the application of cleaner technology. This is covered in Chapter 9.

Ecotechnology covers the use of ecosystems to solve pollution problems, including the erection of artificial ecosystems. It also encompasses the technology that is applicable to restore more or less deteriorated ecosystems. It is presented in Chapter 8. The mentioned classes of technologies cover a wide spectrum of methods. We have for instance many environmental technological methods to cope with different waste water problems, and to select the right method or most often the right combination of methods, a profound knowledge of the applicability of the methods and of the processes and characteristics of the ecosystem receiving the emission is necessary.

Environmental legislation and green taxes may be used in addition to the classes of technology. They are not included in Fig. 1.3, as they may in principle be used as regulating instruments in every step of the flow from raw materials and energy to final waste disposal of the used product.

From this short introduction to the environmental management of today and the wide spectrum of methods which can be implemented to solve the environmental problems, we can conclude that environmental management is a very complex issue. A focal environmental problem may be solved by selection of another raw material or energy source, by changing partially or completely the method of production, by increasing use of recycling, by selection of the right combination of technological methods taken from any of the four classes of technologies mentioned above, by changing the properties of the product slightly, by a combination of environmental technology with recovery of the touched ecosystem and so on. The number of possible solutions is enormous, and yet, the environmental management strategy should attempt to find the optimum solution from an economic-ecological point of view.

There are of course no methods available that with high certainty can give the optimum solution to an environmental management problem, but it is clear that the probability of finding a close to optimum solution is higher the more knowledge the management team has about the problems and the wide spectrum of solution methods.

1.2 SUSTAINABILITY

Sustainability has become one of the buzz-words of our time. It is used again and again in the environmental debate - sometimes in a wrong context. It is therefore important to give a clear definition in the introduction to avoid misunderstandings later in the text. The Brundtland report (World Commission on Environment and Development, 1987) produced the following definition: sustainable development is development that meets the needs of the present without compromising the ability of future generations to meet their own needs. This definition has been widely accepted as authoritative (Willers, 1994). Note, however, that this definition includes no reference to environmental quality, biological integrity, ecosystem health or biodiversity. We will touch on all four issues in this volume. An economic treatment of the related concept sustainable accounting is made by Chichilnisky et al. (1998).

Conservation philosophy has been divided into two schools: resourcism and preservationism. They are understood respectively as seeking maximum sustained yield of renewable resources and excluding human inhabitation and economic exploitation from remaining areas of undeveloped nature. These two philosophies of conservation are mutually incompatible. They are both reductive and ignore non resources, and seem not to give an answer to the core question: how to achieve sustainable development, although preservationism has been retooled and adapted to conservation biology.

Lemons et al. (1998) are able to give a more down-to-earth solution by formulation of the following rules:

A. Output rule: waste emission from a project should be within the assimilative capacity of the local environment to absorb without unacceptable degradation of its future waste absorptive capacity or other important services.

B. Input rule: harvest rates of renewable resources inputs should be within the regenerative capacity of the natural system than generates them and depletion rates of nonrenewable resource inputs should be equal to the rate at which renewable substitutes are developed by human invention and investment.

Klostermann and Tukker (1998) discuss sustainability based on product innovation and introduce the concept of eco-efficiency, i.e., the reciprocal of the weighted sum of the environmental claims including ecological impacts, and draw on renewable and non-renewable resources. The concept of sustainable development will be applied several times during this volume with reference to the definition given above and with reference to the basic ideas presented in this section.

1.3 ENVIRONMENTAL SCIENCE

The past three decades have created a new interdisciplinary field as denoted environmental science which is concerned with our environment and the interaction between the environment and man. Understanding environmental processes and the influence man has on these processes requires knowledge of a wide spectrum of natural sciences. Obviously biology, chemistry and physics are basic disciplines for understanding the biological/chemical/physical processes in the environment. But environmental science draws also upon geology for an understanding of soil processes and the transport of material between the hydrosphere and lithosphere, on hydrodynamics for an understanding of the transport processes in the hydrosphere, and upon meteorology for an explanation of the transport processes in the atmosphere, just to mention a few of the many disciplines applied in environmental science.

Some also believe that general political decisions are of importance in environmental management and that sociological conditions influence man's impact on the environment. It is true that a relationship exists between all these factors and that a complete treatment of environmental problems requires the inclusion of political science and sociology in the family of environmental sciences.

At present, however, it seems unrealistic to teach all these disciplines simultaneously or to presume that one person has all the background knowledge required for a complete environmental solution taking all aspects into consideration at the same time. The right solution to environmental problems can be found only by cooperation between several scientists, and it is therefore advantageous if all the members of such a multidisciplinary team know each other's language.

Traditionally, scientists have worked to discover more and more about less and less. In environmental science, however, it is necessary to know more and more about more and more to be able to solve the problems. Environmental science has therefore caused a shift in scientific thinking, by demonstrating that although so much detailed knowledge and so many independent data have been collected, such details cannot be used by man to improve the conditions for life on earth unless they can all be considered together. It is the aim of environmental science to interconnect knowledge from all sciences - knowledge that is required to solve environmental problems.

It is the scope of this book to demonstrate how our present knowledge of natural sciences can be used and interconnected to understand how man influences life on earth. The role of socioeconomic disciplines will not be mentioned The author feels that it might over-complicate the issue to include these disciplines in the present treatment of environmental science.

The past three decades have seen an unprecedented accumulation of knowledge about the environment and man's impact upon it. Unfortunately, the very mass of this information explosion has created problems for those concerned with the application of this material in research and teaching. University courses structured to meet the growing demand for multidisciplinary treatment of environmental problems became a struggle for teacher and students. The teacher, inevitably a specialist in only one of the several disciplines, had to gather information from areas remote from his own field. Even after relevant information was obtained the rational organisation of the material for meaningful presentation in an environmental course became a massive problem.

Many excellent volumes about environmental science have been published during the last decade, but although some contain a pure interdisciplinary treatment of the subject, most concentrate on one particular aspect. Of course this raises a series of questions. How could one person write on so many diverse topics and still retain a sound factual base? How could so much material be organised to maintain continuity for the reader? How could such a book avoid containing only series of independent facts and problems? and so on. This book attempts to solve these problems by focusing on the principles of pollution abatement used to solve environmental problems within the framework of natural sciences. The multidisciplinary field of environmental science is growing rapidly: other sciences are bringing new knowledge into the field with an ever increasing rate, and new problems or new connections between existing problems are continually appearing. Consequently a book dealing with facts and problems will quickly become out of date and the knowledge learned by the students useless. By focusing on the principles instead of events, these obstacles can be overcome, as the same principles are equally valid in the solution of different and new environmental problems.

1.4 THE ENVIRONMENTAL CRISIS

Three pronounced developments have caused the environmental crisis we are now facing: the growth in population, industrialisation and urbanisation. Fig. 1.4 illustrates world population growth, past and projected. From the graph it can be seen that population growth has experienced decreasing doubling time, which implies that growth is more than exponential (exponential growth corresponds to a constant doubling time). Fig. 1.4 shows that growth from one billion to two billions took about 100 years, while the next doubling in population took only 45 years.

The net birth rate at present is about 370,000 people per day, while the death rate is 150,000 per day. The population growth is determined by the differences between the two: population increase = birth rate - death rate.

This implies that the world's population is increasing by more than 200,000 per day, or about 1.5 million per week corresponding to more than 80 million per year.

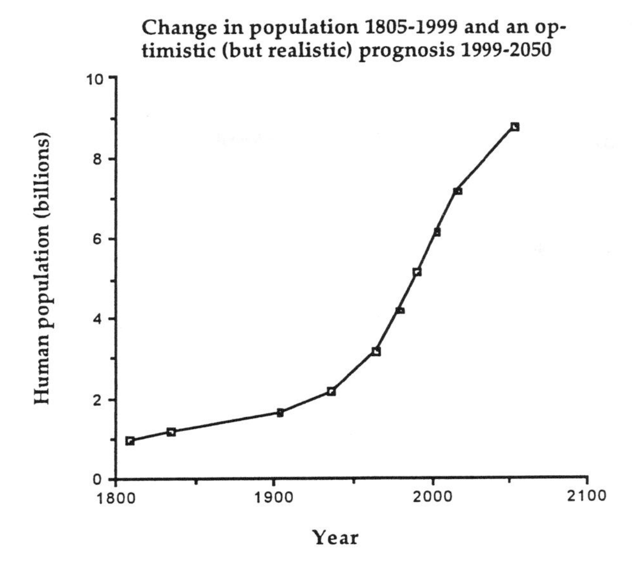

Figure 1.4 The graph shows the change in population 1805-1999. For the period 1999-2050 an optimistic prognosis is applied which predicts that the population growth will level off after the year 2015. Notice that the doubling time 2->4 is shorter than the doubling time 1->2. In the period 1805 ->1975 the growth therefore has been more than exponential.

The need to limit population growth is now more clearly perceived than before. Every living organism requires energy and material resources from its environment. Resources can be classified as renewable and non-renewable. Renewable resources are those that can maintain themselves or be continuously replenished if managed wisely. Food, crops, animals, wildlife, air, water, forest, etc. belong to this class. Land and open space can also be considered as renewable, but they shrink as the population increases. Although we cannot run out of these resources, we can use them faster than they can be regenerated or by using them unwisely we can affect the environment (Meadows et al., 1972).

Other resources, such as fossil fuels and minerals, are non-renewable resources, whose finite supplies can be depleted. Theoretically some of these resources are renewable, but only over hundreds of millions of years, while the time scale of concern to man is hundreds of years only. When we talk about finite supplies of resources we should qualify this by discussing finite supplies of substances presently considered to be resources. Often our most important consideration is whether the pollution costs from extraction and use of a resource outweigh its benefit as population or per capita consumption increase.

During the last few decades we have observed a distinct increase in pollution. Many examples illustrate these observations. The concentrations of carbon dioxide, sulphur dioxide and other gaseous pollutants have increased drastically. The concentrations of many toxic substances have increased in soil and water, and the ecological balance has been changed in our ecosystems. In many major river systems oxygen depletion has been recorded, and many recreational lakes are suffering from eutrophication (high concentrations of nutrients - mainly nitrogen and phosphorus).

What has caused this sudden increase in pollution? The answer is not simple, but the growth in population is obviously one of the factors that influences our environment. Other factors include man's rate of consumption and the type and amount of waste he produces.

Many now recognise two basic causes of the environmental crisis. The first, which occurs in the developing countries, is overpopulation relative to the food supply and the ability to purchase food even if it is available. The second occurs in the technologically advanced countries, in North America, Australia, Japan and Europe. These countries use 80-90% of the world's natural resources, although they only account for about 25% of the world's population. As a result the average consumer in these countries causes 25-50 times as great an impact on our life-support system as a peasant in a developing country (Davis, 1970).

The debate centres on the relation between pollution - or environmental impact - and population, consumption and technology. We can use the crude but useful model proposed by Ehrlich and Holdren (1971). They obtain the environmental impact, I, by multiplying three factors - the number of persons, P, the units of consumption per capita, C, and the environmental impact per unit of consumption, E,:

$$I = P \cdot C \cdot E \tag{1.1}$$

All three factors are equally important. We can illustrate the importance of all three factors by considering development in the U.S.A. between 1950 and 1970. The population increased 35% during these two decades, while per capita consumption increased about 51%. In the same period the production of environmentally harmful material has increased by between 40 and 1900%.

Indeed, during the same two decades we learnt to reduce the discharge into the environment. However, the total environmental impact increased considerably in the U.S.A. during the two decades from 1950 to 1970. The two first factors in Equation (1.1) are easy to find from the facts mentioned above, but in such a simple model the third factor will always be a subjective judgment. If we take the rapid growth in production of harmful substances (40-1900%) and also consider the measured increased concentration of such substances in major ecosystems, a reasonable value for the third factor would be around 2. It would give us the following increase in the total impact on the environment:

$$I = 1.35 \cdot 1.51 \cdot 2 = 4 \text{ times} \tag{1.2}$$

This is a very crude simplification, but an increase in environmental impact of four times during such a short period as 20 years must give us reason to worry about man's future on earth, unless we can use all our efforts to manage the problem.

If P increases as expected by a more pessimistic prognosis than the one presented in Figure 1.4, to twice the present population within year 2050, and if we allow C to increase by a factor of let us say 5 during the next about 50 years, then we have to decrease I ten times to maintain the same impact on the environment, which is a prerequisite for a sustainable development. How can we do that? The answer is by increasing the eco-efficiency ten times. It means to emit ten times less to the environment and use ten times less resources for one unit of production. It has been possible to increase the eco-efficiency in a number of industries by a factor of 4 (Klostermann and Tukker, 1998). So, why shouldn't we be able to increase the eco-efficiency a factor 10 during the coming 50 years? This is the message presented by Klostermann and Tusker.

Two major forces can lead to apathy: naive technological optimism - the idea that some technological wonder will always save us regardless of what we do - and the gloom and doom pessimism - the idea that nothing will work and our destruction is assured. The idea behind such a book as this is, of course, that something can be done, but the problem is very complex and difficult to solve. The best starting point must be an understanding of the nature of the problems and the principles and methods that can be used to solve them. It is hoped that by working through this book the reader will be able to grasp these basic concepts.

1.5 FOCUS ON PRINCIPLES

An understanding of environmental problems requires only the application of a few principles which must be coupled with environmental data. The principles are discussed in the next three chapters.

All life on this planet is dependent on the presence of a number of elements in the right form and concentration. Other elements not used in life-building processes should not be present in the biosphere or present only in very low concentrations. If these conditions are not fulfilled, life is either not possible or will be damaged. It is therefore of major importance in understanding environmental deterioration to keep a record of elements and compounds to ascertain whether an abnormal concentration of one or more elements or compounds can explain the environmental problem considered. Therefore the law of mass conservation should be widely used in understanding ecological reactions to pollution. Chapter 2 is devoted to the application of mass balances in the environmental context and the translation of mass balance results (concentrations) - to environmental effects for non-toxic substances. Chapter 4 will treat the same problems for toxic substances. Effects will in both chapters be discussed for all levels of the ecological hierarchy, i.e., cells, organs, organisms, population, ecosystems, ecosphere. In the context of the last two levels biodiversity will be discussed.

Energy is also required to maintain life. In a thermodynamic sense the earth can be considered a closed, but not isolated, system, which implies that the earth exchanges energy, but not matter, with the universe. The same consideration is sometimes also valid for ecosystems, although the characteristic pollution situation is an input of pollutants, which changes the concentrations in the ecosystem.

The earth's input and output of energy are approximately balanced. Solar radiation is the basic energy requirement for all life on earth, but after this energy has been used to maintain the biological, chemical and physical processes, it is converted to long wave radiation from the earth out to the universe. The balance between input and output assures that a constant average temperature is maintained.

The rate at which biological processes take place is strongly dependent on temperature which will be discussed in chapter 2. The present life on earth is dependent on a certain temperature pattern. Changes in the present temperature pattern can therefore have enormous consequences for all life on earth. Not only is the global energy balance of importance, but also the energy balance within ecosystems can give important information about the conditions of life. Chapter 3 is concerned with energy problems related to these environmental issues.

Emphasis will be laid on quantification of environmental problems throughout chapters 2-4, because it is only through the application of environmental principles for quantification of a problem that we have the right basis for selection of a feasible solution.

The selection of the right technology for the solution of an environmental problem requires a profound understanding of the problem itself. Therefore environmental science is an essential basic field for selection of the right technological solution.

Also chapters 5-9, about the available technology, attempt to focus on the principles and their application to find solutions to environmental problems. Such a solution is not always readily available, but can be obtained by use of environmental legislation. For example considerable lead pollution originated from the use of lead as an additive in gasoline, but as a result of legislation and the setting of a maximum permitted lead concentration in gasoline, such pollution has now (partly) diminished. This part of the book is written for all concerned with the solution of environmental problems. Design criteria are not included. The reader is thus given a basic knowledge of environmental, cleaner and eco-technologies to be able to select the best solutions available today and to discuss them with the specialists who must design and build whatever is required to the solution.

It is important to obtain a critical view of technological solutions, by considering the environments rather than the economy of prime importance, but also to understand when and which technological solutions have clear advantages over other possibilities. Only a clear environmental analysis of the problem can reveal whether it is advantageous from an ecological point of view to apply a technological solution.

Environmental technology has developed very rapidly during the last two decades. Many new methods are available today, and for most environmental problems a wide spectrum of methods (processes) are applicable. As a result it is considerably more difficult today to find the very best solution, but at the same time there are better possibilities for an acceptable technological solution.

The problems have been classified in terms of: water pollution, air pollution and solid waste pollution. This is a reasonable classification of environmental technology, as the methods and processes are dependent on the state of the pollutants. Chapter 5 is devoted to water pollution problems and to water resources, including the technology applied to production of potable and process water from surface and ground water. Chapter 6 discusses solid waste problems and Chapter 7 deals with air pollution problems. Each chapter is divided into sections dealing with problems that are considered of importance today.

Ecotechnology, presented in Chapter 8, is used increasingly today due to the increased role of diffuse pollution during the last two decades. Ecological principles that are the basis for ecotechnology will be presented in this chapter together with design examples and the application of ecotechnology in agriculture.

The polluter must pay principle is used widely today which implies that the cost of pollution abatement must be covered by the production. This has provoked an answer to the question: couldn't we produce the product by a more environmentally friendly method? Or could we make a product which is more environmentally friendly? These issues are discussed in Chapter 9.

As already underlined, a proper environmental solution will probably require the use of a combination of the methods that are presented in chapters 5-9.

1.6 HOW TO SOLVE ENVIRONMENTAL PROBLEMS

Principles and quantifications are used as keywords in our search for solutions to environmental problems. Fig. 1.5 gives a flow chart of a procedure showing how to go from emission of mass and energy to a solution of the related environmental problems. The chapters devoted to the various steps are indicated on the figure. Emission is translated into imission and concentration. The effect or impact of a concentration of a compound or energy is found by considering all the chemical, physical and biological processes that take place in the ecosystem. Equations for the most important of these processes are included in chapters 2-4.

This evaluation leads us to an acceptable ecological (and economic) solution by use of ecological engineering, cleaner technology and/or environmental technology. These technologies attack the problem respectively in the ecosystem, in the production unit and at the emission.

The presented procedure requires the application of principles and knowledge of environmental processes. Furthermore, the problem must be well understood and quantified in order to find the right solution.

Selection of methods will be discussed with reference to the principles presented in the first four chapters. Environmental technology is often based upon the same processes as found in nature for elimination of pollutants; in this case, reference to the natural process already mentioned will be given.

Generally there are 6 principal methods for solving pollution problems:

A. **To reduce the amount (energy and/or material) discharged by use of cleaner technology.** It may often be necessary to enforce the use of cleaner technology by legislation. For example, reduction of sulphur in fossil fuel and lead in gasoline.

B. **To recycle or reuse waste products.** This method is very attractive from an environmental point of view, as a resource and an environmental problem are solved simultaneously.

For example, recovery of chromium from waste-water and production of animal feed from slaughterhouse waste. This method may be based on environmental technology or cleaner technology or a combination of these two types of technologies.

C. **To decompose the waste to harmless components**. For example, biological treatment of municipal waste-water. This method is based on environmental technology in most cases, but may also be covered by an ecotechnological method. An artificial wetland can be used to treat waste water, or in situ treatment of contaminated soil by adapted microorganisms.

D. **To remove the waste for harmless deposition at another location**. For example, use of domestic waste as a soil conditioner. As transport is expensive and often involves other pollution problems, it is important to utilise various methods that can yield a reduced volume corresponding to a higher concentration of the pollutant. Several methods based on environmental technology are available.

E. **Restoration of ecosystems by use of ecotechnology**. Several methods for example for restoration of lakes are available.

F. **To avoid the problem by a planning based upon ecological principles.** This principal method is based upon the ecological principles presented as the basis for ecotechnology, an environmental impact assessment before a project is implemented or an environmental risk assessment before a production of a chemical is initiated.

All six methods are not equally applicable to all environmental problems, but it is important in pollution abatement to be receptive to new and non-traditional solutions. As already mentioned the right (best) answer to a pollution problem may be a combination of methods. The relationship between environmental issues and the other serious problems mankind is facing - shortage of resources, the energy crisis and ever-increasing population growth - make it absolutely essential to seek new ways.

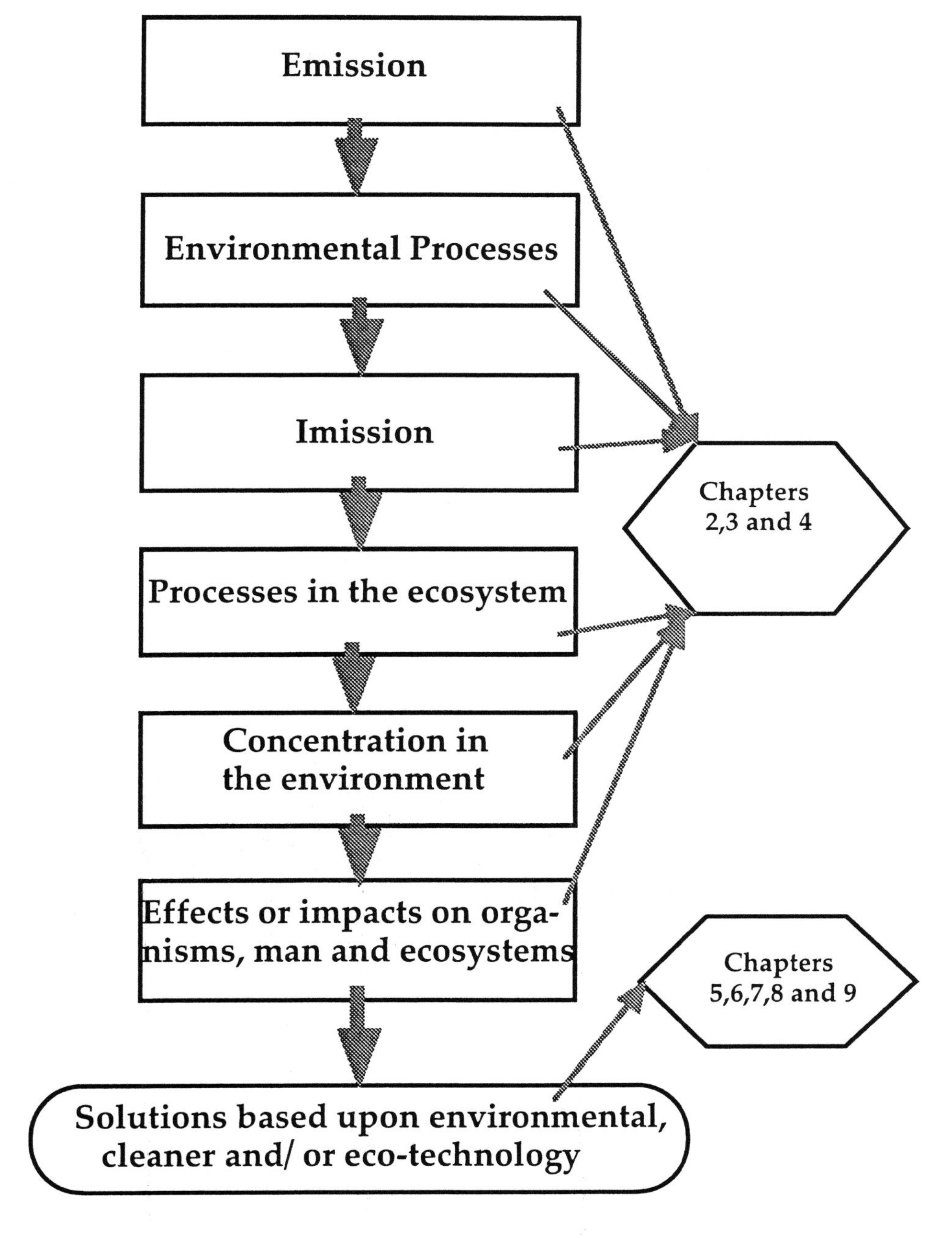

Fig. 1.5 Flowchart illustrating a procedure which can be used to go from emission to solution. The chapters devoted to the various steps are indicated on the figure.

2. Mass Conservation

2.1 EVERYTHING MUST GO SOMEWHERE

According to the law of mass conservation, mass can neither be created nor destroyed, but only transformed from one form to another. Thus everything must go somewhere. The notion of cleaning up the environment or pollution-free products is a scientific absurdity. We can never avoid pollution effects. Nobody - neither man nor nature - consumes anything; we only borrow some of the earth's resources for a while, extract them from the earth, transport them to another part of the planet, process them, use them, discard, reuse or reformulate them (Cloud, 1971).

The law of mass conservation assumes that no transformation of mass into energy takes place, which is formulated in accordance with Einstein as follows:

$$E = mc^2 \tag{2.1}$$

$$c = 3 \cdot 10^8 \text{ m sec}^{-1}$$

If we consider a system which exchanges mass with the environment, then the following equation is valid for an element, or a chemical compound,c:

$$dm_c / dt = \text{import - export} \pm \text{result of chemical reactions} \tag{2.2}$$

where m_c is the mass of c in the system. It is possible to compute concentrations, $c = m_c / V$, where V is the volume, in ecosystems by use of equation (2.2).

Mass is conserved; it can be neither created nor destroyed, but only transformed from one form to another. The form and location is of great importance for the effect of pollutants. We should always attempt to discharge our waste in such a way that the change in concentration of the most harmful forms becomes as low as possible. It is therefore noticeable that the four spheres have a completely different composition, as demonstrated in Appendix 1. Waste should generally be discharged in the sphere where the concentration would be changed least by the discharge. It is also possible to get a first rough idea about the impact by a comparison of the actual concentration in an ecosystem with the background concentration. If the difference between the two concentrations is minor, there is a high probability that the discharge is harmless, while a major difference could indicate that the discharge is harmful.

TABLE 2.1
Cu-concentrations (characteristic)

Item	Sphere represented	Concentration
Atmospheric particulates (unpolluted area)	Atmosphere	2 mg m^{-3}
Sea-water (unpolluted)	Hydrosphere	2 μg 1^{-1}
River water	Hydrosphere	10 μg 1^{-1}
Soil	Lithosphere	20 mg kg^{-1}
Fresh-water sediment	Lithosphere	40 mg kg^{-1}
Algae	Biosphere	20-200 μg 1^{-1}

Let us take a concrete example to illustrate these considerations: where should waste containing copper be deposited? To answer this question we need more information than that available in Table 2.1, although this table indicates that the highest concentration of copper is in the lithosphere. Therefore we can assume that the discharge of copper to the lithosphere will produce the smallest change in copper concentration of the four spheres, but we need to know something about the effect of copper in its different forms. This information is given in Table 2.2. It is seen that free copper ions are extremely toxic to some aquatic animals. Furthermore, it is demonstrated in Table 2.3 that free copper ions are bound to soil and sediments, which means that the most toxic form, which is copper ions, will often be present in the environment in low concentrations. The water deposition site is, in the first instance, selected in the sphere where the relative change in concentration is smallest. Furthermore, it is necessary to compare the form and the processes of the waste in the 4 spheres and consider this information for the deposition site specifically.

The values in Tables 2.1 - 2.3 are typical but must not be considered general, as they are strongly dependent on several factors not included in the tables. This does not imply that copper can be deposited in the lithosphere at any given concentration. It is only stated that nature has mainly deposited copper in the lithosphere and that the environmental effect here is smallest. It is necessary to control the concentration of free copper ions in the soil water and guarantee that this is far from reaching a toxic level.

TABLE 2.2

A. Lethal concentrations of copper ions (LC_{50}-values and lethal doses of copper (LD_{50}-values).

Species	Values
Asellus meridianus	LC_{50}^{*} = 1.7 - 1.9 mg l^{-1}
Daphnia magna	LC_{50}^{*} = 9.8 $\mu g\ l^{-1}$
Salmo gairdneri	LC_{50}^{*} = 0.1 - 0.3 mg l^{-1}
Rats	LD_{50} = 300 mg (kg body weight)$^{-1}$ (as sulphate)

* Dependent on pH, temperature, water hardness and other experimental conditions.

TABLE 2.3

Copper-binding capacities of soil and sediment samples

Sediment,Lake Glumsoe	28 mg g^{-1} dry matter at 5 mg l^{-1} in water. pH: 7.6, 20°C (free copper ions)
Humus-rich soil with high ion exchange capacity:	76 mg g^{-1} dry matter at 3.7 mg l^{-1} in soil water.

We can collect garbage and remove solid waste from sewage, but they must be either burned, which causes air pollution, dumped into rivers, lakes and oceans, which causes water pollution, or deposited on land, which will cause soil pollution. The management problem is not solved before the final deposit site for the waste is selected. Furthermore, environmental management requires that *all* consequences of environmental technology be considered. Eliminating one form of pollution can create a new form, as described above.

Finally, as the production of machinery and chemicals for environmental technology may also cause pollution, the entire mass balance must be considered in environmental management, including deposit of waste products and pollution from service industries. This problem can be illustrated by considering, as an example, the Lake Tahoe waste water plant, where municipal waste water is treated through several steps to produce a very high water quality. The conclusion is that it hardly pays, from an environmental point of view, to make such comprehensive waste water treatment. **A total solution to an environmental problem implies that all environmental consequences are considered by use of a total mass balance, including all wastes produced and the service industries.**

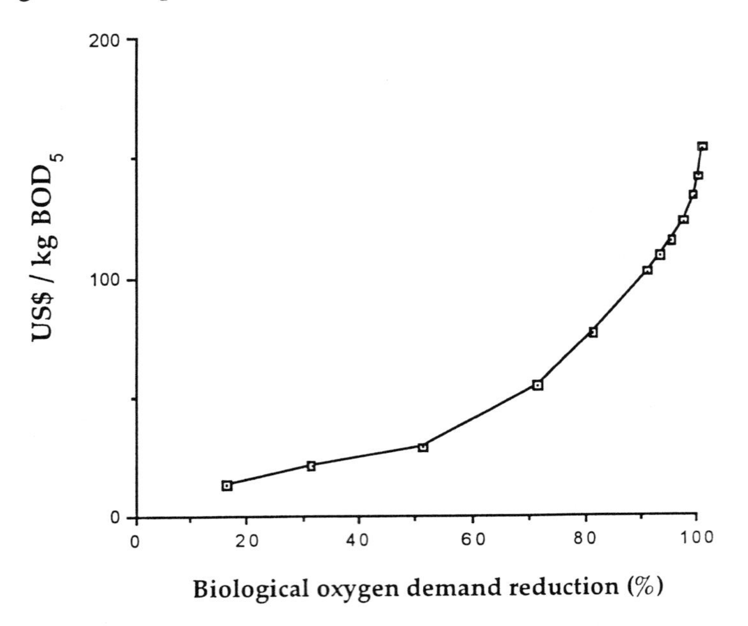

Figure 2.1. Relation between removal of BOD_5 (%) and the cost per kg BOD_5 removed from typical industrial water with a high concentration of biodegradable material. BOD_5 expresses the concentration of organic matter measured by the oxygen demand.

It is important to recognise that complete elimination of pollution is an unachievable goal. The task is to balance the cost of environmental degradation with the cost required to control that degradation. It is, however, generally much more difficult to identify and assess the costs associated with uncontrolled pollution or environmental degradation. Costs of pollution should include increased medical costs for sensitive people and loss of resources, but reduction in the quality of life and long term environmental effects are difficult to estimate. Typically, most of the pollution from a particular source may be controlled relatively inexpensive up to a certain efficiency. Figure 2.1 shows the relation between the degree of purification (in %) and the cost of treatment. Increasing the efficiency of the treatment toward 100% may produce exponential growth of the treatment costs. As the number of possible forms of by-product pollution often follows the trend of the cost, the same relation might exist between the environmental side-effects and the percent reduction of pollution. Consequently our problem is not the elimination of pollution, but its control. There exists, at least in principle, a minimum total cost for any activity as illustrated in Figure 2.2. A desirable goal for an environmental manager is to define that minimum cost.

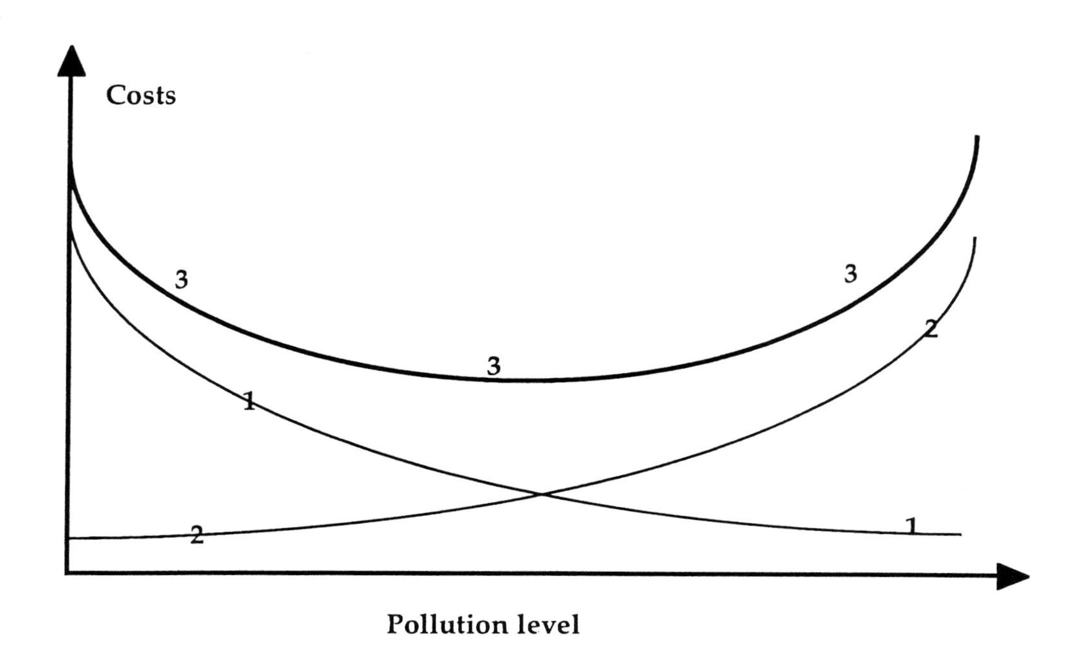

Figure 2.2. Cost of pollution versus pollution level. Curve 1 present the cost associated with pollution and environmental degradation. Curve 2 shows the cost of pollution control and curve 3 is the sum of the two costs.

Technology is essential in helping us to reduce pollution levels below a dangerous level, but in the long term pollution control must also include population control and control of the technology including its pattern of production and consumption. Wise use of existing technology can buy us some time to develop new methods, but the time we can buy is limited. The so-called energy crises are the best demonstration of the need for new and far more advanced technology. We do not know how much time we have - probably 20-50 years - so we had better get started now to make sure, we have the solution to man's many serious problems in time. The increasing cost of treatment with efficiency (Fig. 2.1) must be taken into account in urban planning. If it is decided to maintain environmental quality, increased urbanisation (which means increased amounts of wastes) will require higher treatment efficiency, which leads to a higher cost for waste treatment per inhabitant or per kg of waste. This fact renders the solution of environmental problems of metropolitan areas in many developing countries economically almost prohibitive.**The more waste it is required to treat in a given area, the higher the treatment efficiency that is needed and the higher the costs per kg of waste will be to maintain an acceptable environmental quality.**

2.2 THRESHOLD LEVELS

A pollutant can be defined as any material or set of conditions that created a stress or unfavourable alternation of an individual organism, population, community or ecosystem beyond the point that is found in normal environmental conditions (Cloud, 1971). The range of tolerance to stress varies considerably with the type of organism and the type of pollutant (Berry et al., 1974).

TABLE 2.4
Estimated environmental stress indices for pollutants now and in the future.

	Year 1999	Years 2000-2030
Heavy metals	80	110
Radioactive wastes	35	100
Carbon dioxide	100	175
Solid wastes	55	120
Water borne industrial wastes	65	100
Oil spills	50	90
Sulphur dioxide and sulphates	20	30
Waste heat	5	40
Nitrogen oxides	20	30
Litter	20	30
Pesticides	40	100
Hydrocarbons in air	10	20
Photochemical oxidants	15	20
Carbon monoxide	10	10
Organic sewage	15	20
Suspended particulates	10	20
Chemical fertilisers	70	100
Noise	10	25

To determine whether an effect is unfavourable may be a very difficult and often highly subjective process. A list of major pollutants can be found in Table 2.4, where the estimated environmental stress now and in the future is shown. Each environmental stress index is obtained by multiplying the weighted factors:

1. persistance (1 to 5),

2. geographical range (1 to 5; 1 means only of local interest, 5 that the pollutant is a global problem), and

3. complexity of interactions and effects (1 to 9). The highest possible index is therefore 225 (the data are taken from Howard, 1971, and modified by the author in 1999).

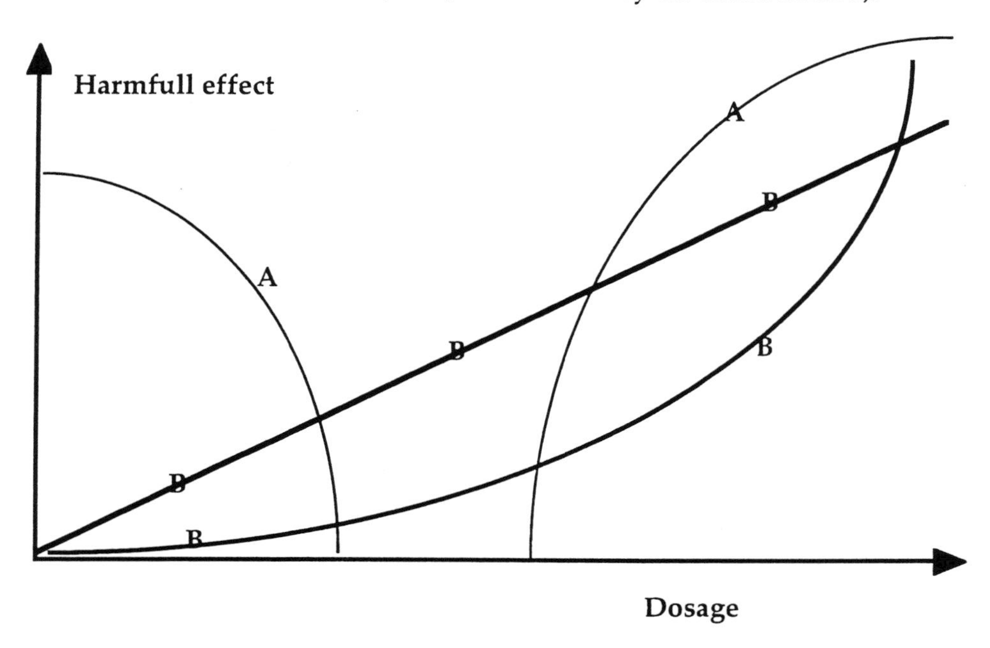

Figure 2.3 (A) Threshold agent. (B) Non-threshold or gradual agent. To have a threshold agent it is sufficient that one of the two A-plots is valid. The two B-plots represent two different dose-response curves.

It is important to recognise that there are both natural and man-generated pollutants. Of course, the fact that nature is polluting does not justify the extra addition of such pollutants by man, as this might result in the threshold level being reached.

In general we can classify pollutants into two groups:

1. non-threshold or gradual agents, which are potentially harmful in almost any amount, and

2. threshold agents, which have a harmful effect only above or below some concentration or threshold level.

This classification is illustrated in Fig. 2.3.

For the latter class we come closer to the limit of tolerance for each increase or decrease in concentration, until finally, like the last straw that broke the camel's back, the threshold is crossed. For non-threshold agents, which include several types of radiation, many man-made organic chemical, which do not exist in nature, and some heavy metals such as mercury, lead and cadmium, there is theoretically no safe level. In practice, however, the degree of damage at very low trace levels is considered negligible or worth the risk relative to the benefits obtained from using the products or processes.

Threshold agents include various nutrients, such as phosphorus, nitrogen, silica, carbon, vitamins and minerals (calcium, iron, zinc, etc.). When they are added or taken in excess the organism or the ecosystem can be overstimulated, and the ecological balance is damaged. Examples are the eutrophication of lakes, streams and estuaries from fertiliser run-off or municipal waste water. The threshold level and the type and extent of damage vary widely with different organisms and stresses. The thresholds for some pollutants may be quite high, while for others they may be as low as 1 part per million (1 ppm) or even 1 part per billion (1 ppb).

The threshold level is closely related to the concentration found in nature under normal environmental conditions.

Appendix 1 show the concentration of some important elements found in different parts of the ecosphere. Note that even uncontaminated parts of the ecosphere contain almost all elements, although some only in very low concentrations.

In addition, an organism's sensitivity to a particular pollutant varies at different times of its life cycle, e.g. threshold limits are often lower in the juvenile (where body defence mechanisms may not be fully developed) than in the adult stage. This is especially true for chlorinated hydrocarbons such as DDT and heavy metals - both of which represent some of the most harmful pollutants.

Pollutants can also be characterised by their longevity in process organisms and ecosystems. Degradable pollutants will be naturally broken down into more harmless components if the system is not overloaded. Non-degradable pollutants or persistent pollutants will, however, not be broken down or be broken down very slowly and the intermediate components are often as toxic as the pollutants. Knowledge of the degradability of the different components is naturally of great importance to environmental management, since an undesired concentration of pollutants in the environment is a function not only of the input and output, but also of the processes which take place in the environment. These processes must of course be considered whenever the concentration of pollutants is being computed.

Environmental impact assessment requires not only knowledge of the concentration and the form of a pollutant, but also of the processes which the pollutant might undergo in the environment.

2.3 BASIC CONCEPTS OF MASS BALANCE

The simplest case is an isolated system, where no processes take place, and therefore the concentration of all components is constant. An ecosystem is never an isolated system, but may be either an open system or a closed system. The former exchanges mass as well as energy with the environment. The input of energy to an ecosystem will cause cyclic processes (Morowitz, 1968), in which the important elements will play a part. Very few pollutants are completely chemically inert, most are converted to other components or degraded. The common degradation process including the degradation of persistent chemicals can be described by a first order reaction scheme:

$$dC / dt = k \cdot C \qquad (2.3)$$

where C is the concentration of the considered compound, t is the time and k a rate constant. k varies widely from the very easily biodegradable compounds, such as carbohydrates and proteins, to pesticides, such as DDT.

The so-called biological half-life time, $t_{1/2}$, is often used to express the degradability. $t_{1/2}$ is the time required to reduce the concentration to half the initial value. The relation between k and $t_{1/2}$ can easily be found:

$$\ln C_o / C(t) = kC$$

$$\ln 2 = k \cdot t_{1/2} \qquad (2.4)$$

Here C_o is the initial concentration.

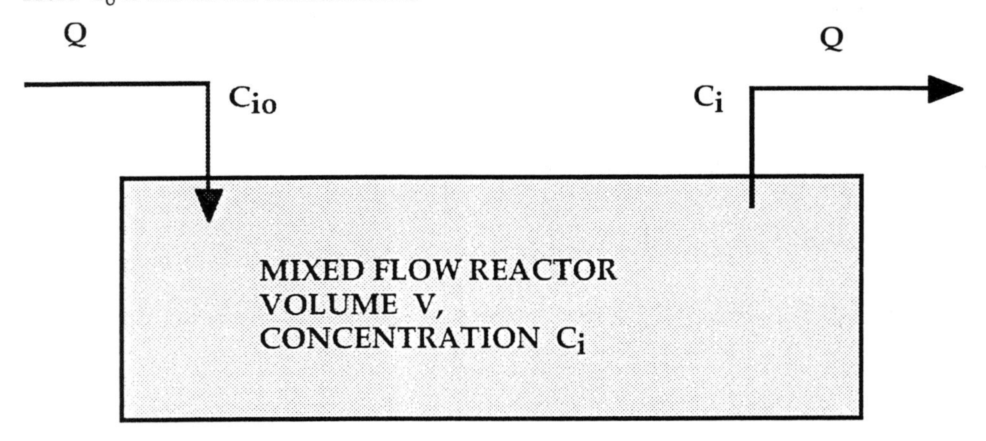

Figure 2.4 Principle of a mixed flow reactor.

Appendix 2 shows some characteristic rate constants for biodegradation and biological half-life times. k and $t_{1/2}$ are dependent on the reaction conditions, pH, temperature, ionic strength, etc. These conditions are indicated in the right-hand column of the tables. As seen, a wide spectrum of values is represented in these tables.

The continuous mixed flow reactor (abbreviated CMF) closely approximates the behaviour of many components in ecosystems. The CMF reactor is illustrated schematically in Fig. 2.4. The input concentration of component i is C_{io} and the flow-rate is Q. The tank (ecosystem) has the constant volume V and in the tank the concentration of i, denoted C_i is uniform. The effluent stream also has a flow-rate Q, and because the tank is considered to be perfectly mixed the concentration in the effluent will be C_i.

The principles of mass conservation can be used to set up the following simple differential equation (no reactions take place in the tank):

$$V dC_i / dt = QC_{io} - QC_i \tag{2.5}$$

In a steady state situation $dC_i = 0$, this means we have:

$$QC_{io} = QC_i \tag{2.6}$$

If a first order reaction has taken place in the tank (ecosystem) the equation will be changed to:

$$V dC_i / dt = QC_{io} - QC_i - kC_i \tag{2.7}$$

and in the steady state situation:

$$QC_{io} - QC_i = V * k * C_i \tag{2.8}$$

By dividing this equation with QC_i the following equation is obtained:

$$C_{io} / C_i - 1 = tr \times k \tag{2.9}$$

where $tr = V / Q$ is the retention time or the mean residence time in the tank (ecosystem). Rearrangement yields:

$$C_i / C_{io} = 1 / (1 + tr \times k) \tag{2.10}$$

Figure 2.5 shows this equation on a graph: $C_i = f(t)$ with $C_i = 2$ at $t = 0$. C_i will approach asymptotically the steady state concentration which can be found from equations (2.9) and (2.10). For a number of reaction tanks, m, in series, each with volume V, a similar set of equations can be set up:

$$C_i / C_{io} = 1 / (1 + tr \times k)^m \tag{2.11}$$

If the set of equations set up to describe the concentration is far more complicated, more processes must be taken into account, and it is necessary to use an ecological model (see Jørgensen, 1994). Space does not permit a detailed examination of more complicated models, but some processes of interest in an environmental context can be mentioned (in addition to **hydrophysical** and **meteorological** ones):

1. leaching of ions and organic compounds in soil,
2. evaporation of organic chemicals from soil and surface water,
3. atmospheric wash-out of organic chemicals,
4. sedimentation of heavy metals and organic chemicals in aquatic ecosystems,
5. hydrolysis of organic chemicals,
6. dry-deposition from the atmosphere,
7. chemical oxidation,
8. photochemical processes.

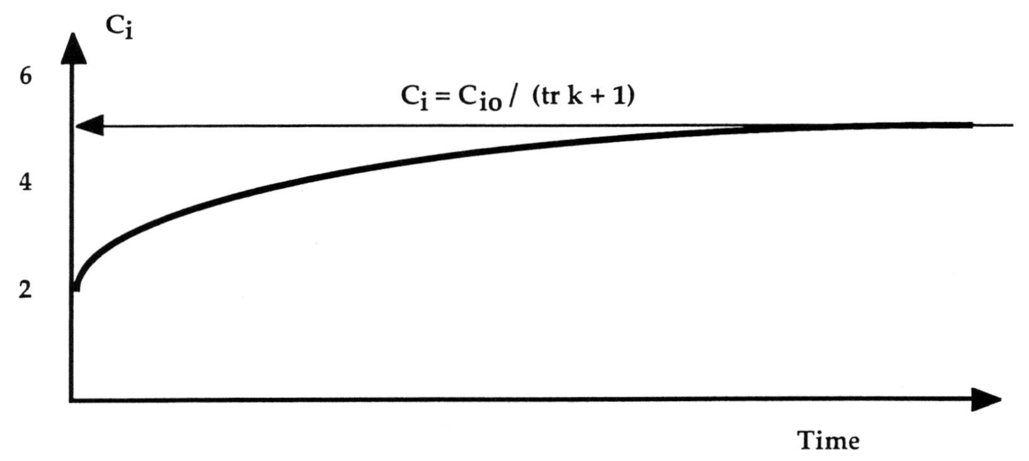

Figure 2.5. C_i = f(t) with C_i = 2 at t = 0. C_i will approach asymptotically the steady state concentration.

All these processes may be described as first order reactions (see equation 2.3) at least in some situations. Table 2.5 gives some typical examples of the 8 above-mentioned processes and an idea of where these processes are of importance. The list has not included **biotic** processes, which would further complicate the picture. These processes will be mentioned later in this chapter All chemical reactions proceed until equilibrium in accordance with an nth order reaction scheme, as presented above. At equilibrium no reaction takes place or rather the two opposite reactions have the same rate. If we consider a reaction:

$$aA + bB = cC + dD \qquad (2.12)$$

then the following equation is valid at equilibrium:

$$\frac{[C]^c\ [D]^d}{[A]^a\ [B]^b} = K \qquad (2.13)$$

where [ ] indicates concentration in M (moles per liter).

TABLE 2.5
Some chemicophysical processes of environmental interest

Process	Examples
Leaching of ions and organic compounds in soil	Nutrient run-off from agricultural areas to lake ecosystems
Evaporation of organic chemicals from soil and surface water	Evaporation of pesticides
Atmospheric wash-out of organic chemicals	Wash-out of pesticides
Sedimentation of heavy metals and organic chemicals in aquatic ecosystems	Most heavy metals have a low solubility in sea water and will therefore precipitate and settle
Hydrolysis of organic chemicals	Hydrolytic degradation of pesti cides in aquatic ecosystems
Dry deposition from the atmosphere	Dry deposition of heavy metals on land
Chemical oxidation	Sulphides are oxidised to sulphates, sulphur dioxide to sulphur trioxide, which forms sulphuric acid with water
Photochemical processes	Many pesticides are degraded photochemically

In a chemostate or in nature, equilibrium is rarely attained, as one of the reactants is continuously added to the system, However, many processes are rapid and equation (2.13) can be used as a good approximation if the time steps considered are significantly larger than the reaction time. This is often the case for the following environmental processes: adsorption, hydrolysis, chemical (but not biochemical) oxidations and acid-base reactions including neutralisation.

Concentrations of pollutants as a function of time can be found by use of the mass conservation principles. If many processes are involved simultaneously, the use of a computer model is needed. Equilibrium description might be used for rapid processes.

2.4 GROWTH AND LIFE CONDITIONS

This chapter will discuss and illustrate the effect of non-toxic substances on life, particularly on growth. Again, the concentration of the component is assumed to be available from computations such as those presented in section 2.3.

The life building processes require the presence of the elements characteristic to the biosphere (see Appendix 1). The composition of some organisms and their diets are given in Appendix 4 to illustrate these relationships.

The composition of the biosphere is closely related to the function of the elements.

The high concentrations of C, H and O are due to the composition of organic compounds. The nitrogen concentration results from the presence of proteins, including enzymes and polypeptides, and nucleotides. Phosphorus is used as matrix material in the form of calcium compounds, in phosphate esters and in ATP, which is involved in all energetically coupled reactions. (ATP is an abbreviation for adenosine triphosphate).

ATP is an energy-rich compound, because of its relatively large negative free energy of hydrolysis, denoted ΔG = approximately 42 kJ / mole:

$$H_2O + ATP = ADP + P + \Delta G_{high} \quad (2.14)$$
$$H_2O + ADP = AMP + P + \Delta G_{high}$$

AMP is not a high energy compound since:

$$AMP = A + P + \Delta G_{low} \quad (2.15)$$

These processes do not actually occur in living cells, but ATP participates directly or indirectly in group transfer reactions. A simple example will illustrate this point:

$$ATP + glucose = ADP + glucose\text{-}P \quad (2.16)$$

Whenever food components are decomposed by a living organism the energy released by the process is either used as heat or stored in the form of a number of ATP molecules. As with ATP it is possible to indicate a function for other elements needed by living organisms, which is related to the biochemistry of each organism.

The pattern of the biochemistry determines the relative need for a number of elements (in the order of 20-25), while other elements (the remaining 65-70 elements) are more or less toxic. Some elements needed by some species are toxic to others in all concentrations.

This does not imply that a particular species has a fixed composition of the required elements. The composition might vary within certain ranges

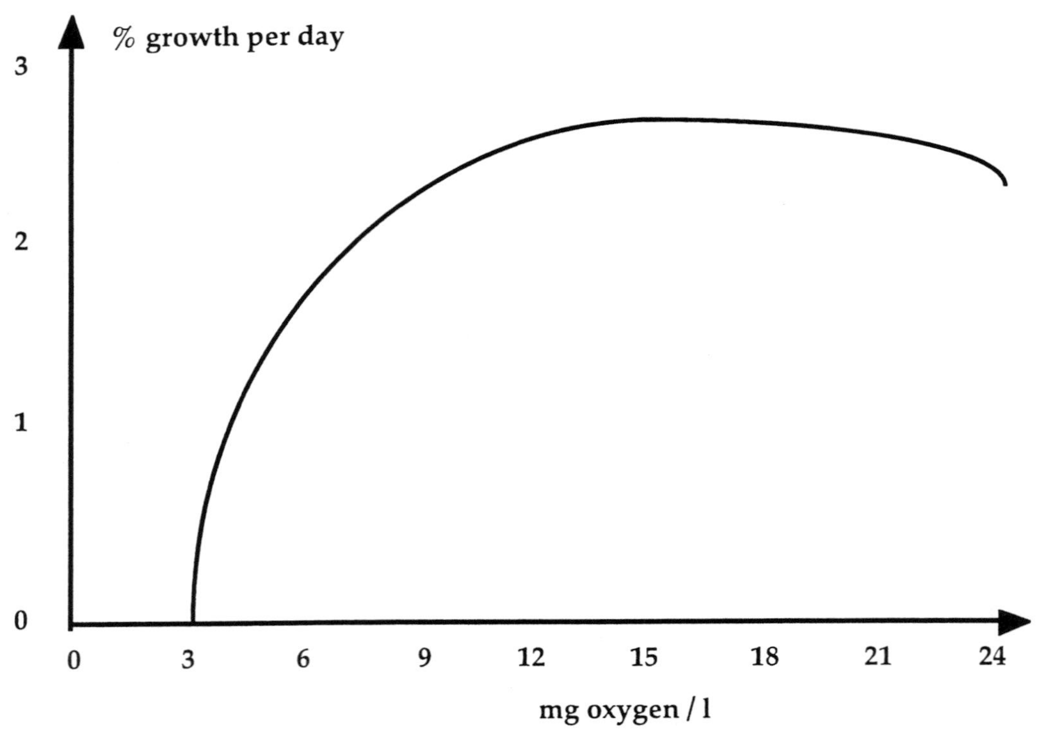

Fig. 2.6. Growth rate of Atlantic Salmon at unrestricted rations versus oxygen concentration. The growth rate is expressed for 50 g fish as per cent gain in dry weight per day (Davis et al., 1963).

It is known, for example, that algae species may have from 0.5 to 2.5 g phosphorus per 100 g of dry matter, with an average at about 1 g phosphorus per 100 g dry matter. High phosphorus concentrations are regarded as luxury uptake.

Animals require oxygen for respiration, and there is a pronounced relationship between oxygen concentration and growth rate and mortality (see Figure 2.6). Oxygen has a relatively low solubility in water (see Table 2.6). Low oxygen concentrations in aquatic ecosystems have often caused the destruction of fish due to discharge of otherwise harmless organic material which is biologically decomposed by microorganisms.

TABLE 2.6

Dissolved oxygen (ppm) in fresh, brackish and sea water at different temperatures and at different chlorinities (%). Values are amount of saturation.

C°	0%	0.2%	0.4%	0.6%	0.8%	1.0%	1.2%	1.4%	1.6%	1.8%	2.0%
1	14.24	13.87	13.54	13.22	12.91	12.58	12.29	11.99	11.70	11.42	11.15
2	13.74	13.50	13.18	12.88	12.56	12.26	11.98	11.69	11.40	11.13	10.86
3	13.45	13.14	12.84	12.55	12.25	11.96	11.68	11.39	11.12	10.85	10.59
4	13.09	12.79	12.51	12.22	11.93	11.65	11.38	11.10	10.83	10.59	10.34
5	12.75	12.45	12.17	11.91	11.63	11.36	11.09	10.83	10.57	10.33	10.10
6	12.44	12.15	11.86	11.60	11.33	11.07	10.82	10.56	10.32	10.09	9.86
7	12.13	11.85	11.58	11.32	11.06	10.82	10.56	10.32	10.07	9.84	9.63
8	11.85	11.56	11.29	11.05	10.80	10.56	10.32	10.07	9.84	9.61	9.40
9	11.56	11.29	11.02	10.77	10.54	10.30	10.08	9.84	9.61	9.40	9.20
10	11.29	11.03	10.77	10.53	10.30	10.07	9.84	9.61	9.40	9.20	9.00
11	11.05	10.77	10.53	10.29	10.06	9.84	9.63	9.41	9.20	9.00	8.80
12	10.80	10.53	10.29	10.06	9.84	9.63	9.41	9.21	9.00	8.80	8.61
13	10.56	10.30	10.07	9.84	9.63	9.41	9.21	9.01	8.81	8.61	8.42
14	10.33	10.07	9.86	9.63	9.41	9.21	9.01	8.81	8.62	8.44	8.25
15	10.10	9.86	9.64	9.43	9.23	9.03	8.83	8.64	8.44	8.27	8.09
16	9.89	9.66	9.44	9.24	9.03	8.84	8.64	8.47	8.28	8.11	7.94
17	9.67	9.46	9.26	9.05	8.85	8.65	8.47	8.30	8.11	7.94	7.78
18	9.47	9.27	9.07	8.87	8.67	8.48	8.31	8.14	7.97	7.79	7.64
19	9.28	9.08	8.88	8.68	8.50	8.31	8.15	7.98	7.80	7.65	7.49
20	9.11	8.90	8.70	8.51	8.32	8.15	7.99	7.84	7.66	7.51	7.36
21	8.93	8.72	8.54	8.35	8.17	7.99	7.84	7.69	7.52	7.38	7.23
22	8.75	8.55	8.38	8.19	8.02	7.85	7.69	7.54	7.39	7.25	7.11
23	8.60	8.40	8.22	8.04	7.87	7.71	7.55	7.41	7.26	7.12	6.99
24	8.44	8.25	8.07	7.89	7.72	7.56	7.42	7.28	7.13	6.99	6.86
25	8.27	8.09	7.92	7.75	7.58	7.44	7.29	7.15	7.01	6.88	6.85
26	8.12	7.94	7.78	7.62	7.45	7.31	7.16	7.03	6.89	6.86	6.63
27	7.98	7.79	7.64	7.49	7.32	7.18	7.03	6.91	6.78	6.65	6.52
28	7.84	7.65	7.51	7.36	7.19	7.06	6.92	6.79	6.66	6.53	6.40
29	7.69	7.52	7.38	7.23	7.08	6.95	6.82	6.68	6.55	6.42	6.29
30	7.56	7.39	7.25	7.12	6.96	6.83	6.70	6.58	6.45	6.32	6.19

Through the breakdown of the organic components oxygen is consumed, causing a critical low oxygen concentration in the water. This is especially dangerous at high temperatures where the biological degradation is fast and the solubility low (see Table 2.6). Section 2.6, which is devoted to oxygen balance in a river system, illustrates the interaction of these processes.

The relation between nutrient concentration and growth has long been known for crop yields. Fig. 2.7 illustrates what is called **Liebig's minimum law: if a nutrient is at a minimum relative to its use for growth, there is a linear relation between growth and the**

concentration of the nutrient. If the supply of other factors is at a minimum, further addition of the nutrient will, as shown, not influence growth.

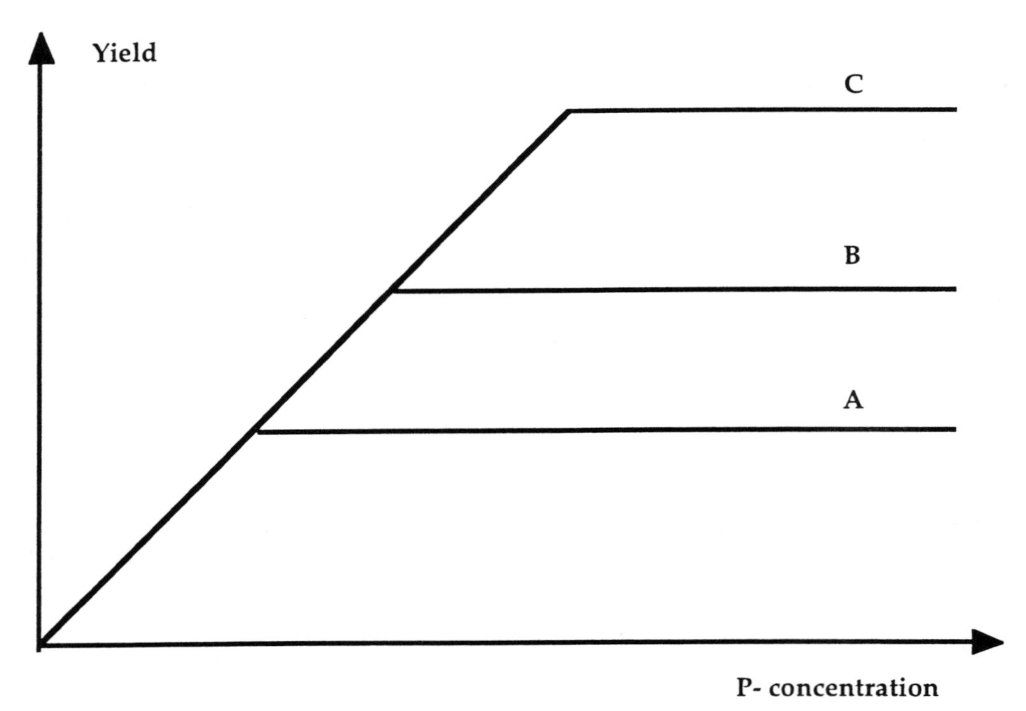

Fig. 2.7. Illustration of Liebig's law. The phosphorus concentration is plotted against the yield. At a certain concentration another component will be limiting, and a higher P-concentration will not increase the yield. The three levels, A, B and C correspond in this case study to three different potassium concentrations.

The relationship between growth, v = dw/dt, and nutrient concentration is described by Michaelis-Menten's equation, which is also used to give the relation between the concentration of a substrate and the rate of a biochemical reaction:

$$dw / dt = v = ks /(k_m + s) \tag{2.17}$$

where v = the rate (e.g. growth rate), k = a rate constant, s = the concentration of a substrate and k_m = the so-called half saturation constant. Notice that equation (2.17) corresponds to a first order reaction when s >> k_m. Growth follows a first order reaction, when the resources are abundant, while equation (2.17) describes the influence of limiting resources on the growth rate.

The situation in nature is often not so simple, as two or more nutrients (resources) may be limiting simultaneously. This can be described by:

$$v = K_r \times \left(\frac{N_1}{k_{m_1} + N_1} \times \frac{N_2}{k_{m_2} + N_2} \right) \tag{2.18}$$

N_1 and N_2 = the nutrient concentrations

K_r = a rate constant

k_{m_1} and k_{m_2} = half saturation constants related to N_1 and N_2

This equation will, however, often limit the growth too much and is in disagreement with many observations (Park et al., 1978). The following equation seems to overcome these difficulties:

$$v = K_r \times \min \left(\frac{N_1}{k_{m_1} + N_1}, \frac{N_2}{k_{m_2} + N_2} \right) \tag{2.19}$$

The relationship between nutrient discharge and its effect on an ecosystem is illustrated in section 2.7, where a nutrient balance for a lake is considered. If the nutrient concentration in a lake is increased the growth of algae will be enhanced. As a result, through photosynthesis, this will produce oxygen:

$$6CO_2 + 6H_2O \text{ -------> } C_6H_{12}O_6 + 6O_2 \tag{2.20}$$

However, the increased organic matter produced by the process will sooner or later decompose and thereby cause an oxygen deficit (for further details see section 2.7).

While nutrients are necessary for plant growth, they may produce a deterioration in life conditions for other forms of life. Ammonia is extremely toxic to fish, while ammonium, the ionised form, is harmless. The fish growth may be reduced at even relatively low concentrations of ammonia. The relation between ammonium and ammonia is dependent on pH:

$$NH_4^+ = NH_3 + H^+ \tag{2.21}$$

$$pH = pK + \log [NH_3] / [NH_4^+] \tag{2.22}$$

where $pK = -\log K$ and K = equilibrium for process (2.21). The pH value as well as the total concentration of ammonium and ammonia is important according to these equations. This implies that the situation is very critical in many hypereutrophic lakes during the summer, when photosynthesis is most pronounced, as the pH increases when the acidic component CO_2 is removed or reduced by this process. pK is about 9.3 in distilled water at 25°C, but increases with increasing salinity. At chlorinities of 2% pK is increased to about 9.7 at 25°C.

In arid climates primary production (the amount of solar energy stored by green plants) is strongly correlated with precipitation, as water is obviously "the limiting factor". Among many limiting factors, frequently the most important are various nutrients, water and temperature.

Temperature is also of great importance to life in ecosystems. The influence of the temperature on some biotic processes is given in Table 2.7.

TABLE 2.7

The influence of temperature on environmental processes:

A. $v(T) = v(20^oC) * k^{(T-20)}$

Process	k
Nitrification	1.07-1.10
NH_4^+ — NO_2^-	1.08
NO_2^- — NO_3^-	1.06
Org-N — NH_4^+	1.08
Benthic oxygen uptake	1.065
Degradation of organic matter in water	1.02-1.09
Reaeration of streams	1.008-1.026
Respiration of zooplankton	1.05

B. $v(T) = v(T-10^o) * k$

Process	k
Respiration of communities	1.7-3.5 (2.5)
Respiration of freshwater fish	2.18-3.28 (2.4)
Sulphate reduction in sediment	3.4-3.9
Excretion of NH_4^+ from zooplankton	2.0-4.3
Excretion of total nitrogen from zooplankton	1.5-2.5
Oxygen uptake of sediment	3.2
Respiration of zooplankton	1.77-3.28

TABLE 2.8

Average Annual Temperature (^oC).

Latitude	Year	January	July	Range
90°N	-22.7	-41.1	-1.1	40.0
80°N	-18.3	-32.2	2.0	34.2
70°N	-10.7	-26.3	7.3	33.6
60°N	-1.1	-16.1	14.1	30.2
50°N	5.8	-7.1	18.1	25.2
40°N	14.1	5.0	24.0	19.0
30°N	20.4	14.5	27.3	12.8
20°N	25.3	21.8	28.0	6.2
10°N	26.7	25.8	27.2	1.4
Equator	26.2	26.4	25.6	0.8
10°S	25.3	26.3	23.9	2.4
20°S	22.9	25.4	20.0	5.4
30°S	16.6	21.9	14.7	7.2
40°S	11.9	15.6	9.0	6.6
50°S	5.8	8.1	3.4	4.7
60°S	-3.4	2.1	-9.1	11.2
70°S	-13.6	-3.5	-23.0	19.5
80°S	-27.0	-10.8	-39.5	28.7
90°S	-33.1	-13.5	-47.8	34.3

Notice that the influence of the temperature on process rates is expressed by two different equations in Table 2.7. These two equations do not consider that many process rates will even decrease with increasing temperature above an optimum temperature. In this case a more complex equation should be used to describe the relationship between a process rate and the temperature. The equations presented in Table 2.7 may however, be used widely in practice. The amount of solar energy intercepting a unit of the Earth's surface varies markedly with latitude for two reasons. First, at high latitudes a beam of light hits the surface at an angle, and its light energy is spread out over a large surface area. Second, a beam that intercepts the atmosphere at an angle must penetrate a deeper blanket of air, and hence more solar energy is reflected by particles in the atmosphere and radiated back into space. A similar result of both these effects is that average annual temperatures tend to decrease with increasing latitude (Table 2.8). The poles are cold and the tropics are generally warm. In terrestrial ecosystems, climate is by far the most important determinant of the amount of solar energy plants are able to capture as chemical energy, or the gross primary productivity.

2.5 THE EQUILIBRIUM BETWEEN SPHERES

An increase or decrease in the concentration of components or elements in ecosystems are of vital interest, but: **The observation of trends in global changes of concentrations might be even more important as they may cause changes in the life conditions on earth.**

The concentrations in the four spheres: the atmosphere, the lithosphere, the hydrosphere and the biosphere are in this context of importance. They are determined by the transfer processes and the equilibrium concentrations among the four spheres. **The solubility of a gas at a given concentration in the atmosphere can be expressed by ɪ Henry's law which determines the distribution between the atmosphere and hydrosphere:**

$$p = H * x \qquad (2.23)$$

where
p = the partial pressure
H = Henry's constant
x = molar fraction in solution
H is dependent on temperature (see Table 2.9), and H is expressed in atmospheres. It may be converted to Pascals, as 1 atmosphere = 101,400 Pa. A dimensionless Henry's constant may also be applied. As $p = RT\,n/v = RTc_a$ and $x = c_h / (c_h + c_w)$, where c_a is the the molar concentration in the atmosphere of component h, expressed in mol /l and c_h is the concentration in the hydrosphere expressed also in mol /l and c_w is the mol/l of water (and other possible components). If we consider only two components in the hydrosphere: h and water, and that $c_h << c_w$, we can replace c_w with the concentration of water in water = 1000 /18 = 55.56 mol /l. We obtain according to these approximations the following equation:

$$c_a / c_h = H / (RxTx55,56) \qquad (2.24)$$

where H / (RxTx55,56) is the dimensionless Henry's constant.

The soil - water distribution may be expressed by one of the following two adsorption isotherms:

$$a = k\,c^b \qquad (2.25)$$

$$a = k'\,c / (c + b') \qquad (2.26)$$

where a is the concentration in soil, c is the concentration in water and k, k', b and b' are constants. Equation (2.25) corresponds to Freundlich adsorption isotherm and is a straight line with slope b in a log-log diagram, since log a = log k + b log c.

This is shown in Figure 2.8. Equation (2.26) The Langmuir adsorption isotherm, an expression similar to Michaelis-Menten's equation, see equation (2.17). If 1/a is plotted versus 1/c, see Figure 2.9, we obtain a straight line, the so-called Lineweaver-Burk's plot, as $1/a = 1/k' + b'/k'c$. When $1/a = 0$, $1/c = -1/b'$ and when $1/c = 0$, $1/a = 1/k'$.

b is often close to 1 and c is for most environmental problems small. This implies that the two adsorption isotherms get close to $a/c = k$, and k becomes a distribution coefficient. k for 100% organic carbon, usually denoted K_{OC}, may be estimated from K_{OW}, the octanol-water distribution coefficient which is the solubility in octanol divided by the solubility in water.

Several esitmation equations have been published in the literature; see for instance Jørgensen et al., 1997. The following log-log relationships between K_{OC} (100% organic carbon presumed) and K_{OW} are typical examples:

$$\log K_{OC} = -0.006 + 0.937 \log K_{OW} \quad \text{(Brown and Flagg, 1981)} \qquad (2.27a)$$
$$\log K_{OC} = -0.35 + 0.99 \log K_{OW} \quad \text{(Leeuwen and Hermens, 1995)} \qquad (2.27b)$$

TABLE 2.9

Henry's constant (atm) for gases as a function of temperature

Henry's constant (x 10^{-5})

Gas	Temperature (°C)						
	0	5	10	15	20	25	30
Acetylene	0.72	0.84	0.96	1.08	1.21	1.33	1.46
Air (atm)	0.43	0.49	0.55	0.61	0.66	0.72	0.77
Carbon dioxide	0.73	0.88	1.04	1.22	1.42	1.64	1.86
Carbon monoxide	0.35	0.40	0.44	0.49	0.54	0.58	0.62
Hydrogen	0.58	0.61	0.64	0.66	0.68	0.70	0.73
Ethane	0.13	0.16	0.19	0.23	0.26	0.30	0.34
Hydrogen sulphide	26.80	31.50	36.70	42.30	48.30	54.50	60.90
Methane	0.22	0.26	0.30	0.34	0.38	0.41	0.45
Nitrous oxide	0.17	0.19	0.22	0.24	0.26	0.29	0.30
Nitrogen	0.53	0.60	0.67	0.74	0.80	0.87	0.92
Nitric oxide	-	1.17	1.41	1.66	1.98	2.25	2.59
Oxygen	0.25	0.29	0.33	0.36	0.40	0.44	0.48

In the case that the carbon fraction of organic carbon in soil is f, the distribution coefficient, K_D, for the ratio of the concentration in soil and in water can be found as $K_D = K_{oc} x f$. If the solid is activated sludge (from a biological treatment plant) instead of soil, Matter-Müller et al. (1980) have found the following relationship:

$$\log FAS = 0.39 + 0.67 \log K_{ow} \quad (2.28)$$

where FAS is the ratio between the equilibrium concentrations in activated sludge and in water.

K_{ow} can be found for many compounds in the literature, but if the solubility in water is known it is possible to estimate the partition coefficient n-octanol- water at room temperature by use of a correlation between the water solubility in μmol /l and K_{ow}. A graph of this relationship is shown Figure 2.10.

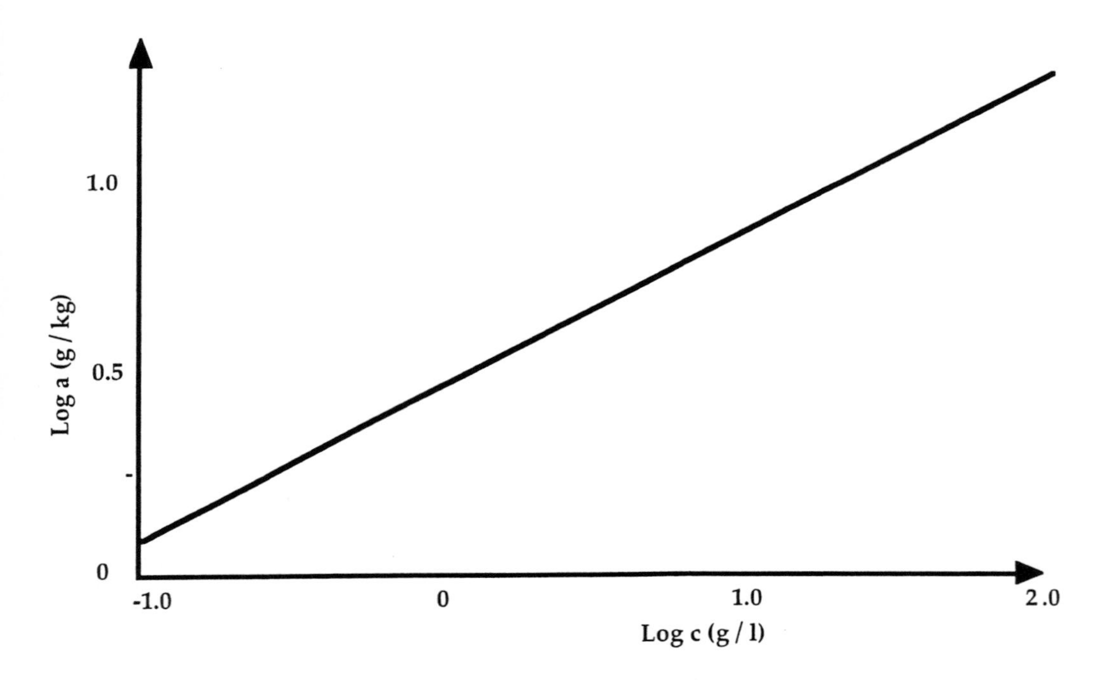

Figure 2.8. A log-log plot of the Freundlich adsorption isotherm is shown. The slope which is 1.15 / 3 = 0.383 represents b in equation (2.25) and log k = 0,48 which means that k = 3.1. The equation for the plot shown is therefore $a = 3.1\ c^{0.383}$

The distribution between the biosphere and the hydrosphere is also of importance. BCF is the ratio between the concentrations in an organism and in water. It can be found for many compounds and for some organisms in the literature. BCF may also be estimated, see Figure 2.11, where two log - log plots between BCF and K_{ow} are shown for mussels and fish (length 20-30 cm).

H, K_{OC}, K_D and BCF all express a ratio between two equilibrium concentrations in two different spheres. A transfer of a compound from one sphere to another will take place until the equilibrium concentrations have been attained. The rate of transfer will usually be proportional to the distance from equilibrium, and dependent on the diffusion coefficient of the compounds and of the resistance at the boundary layer between the two spheres. The reaeration of aquatic ecosystems follows this pattern. The rate of reaeration is proportional to the difference between the oxygen concentration at saturation and the actual oxygen concentration. The resistance at the boundary layer and the influence of the diffusion coefficient are usually covered by an empirical expression which is dependent on the temperature (the diffusion is strongly dependent on the temperature), the surface exposed to the atmosphere relative to the water volume and the rate of the water flow.

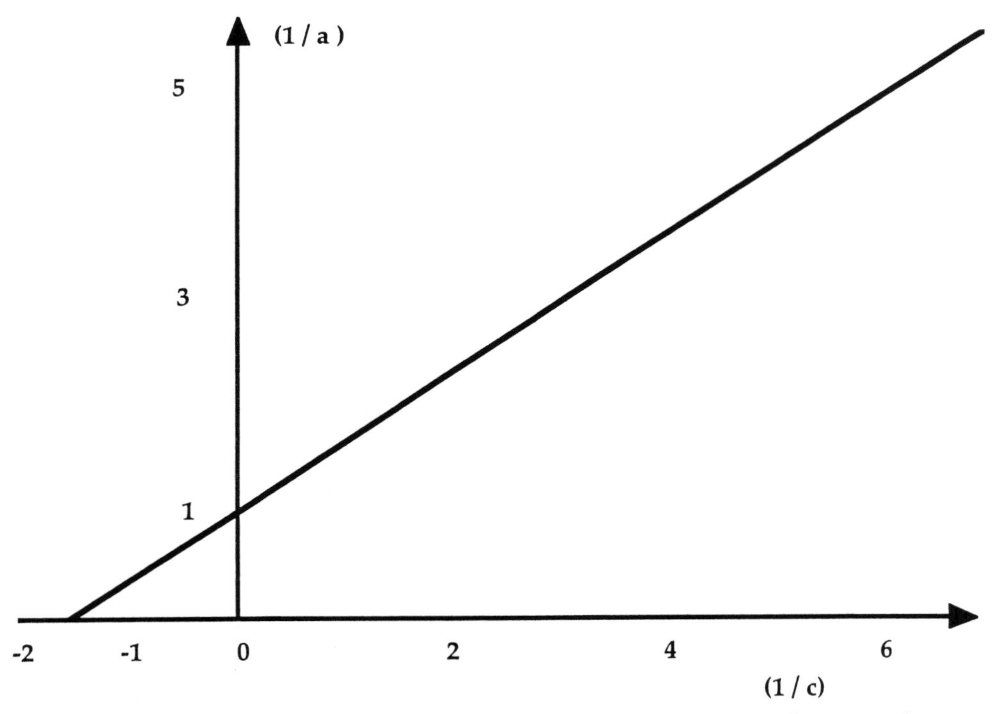

Figure 2.9 A Lineweaver-Burk's plot where 1/a is plotted versus 1/c. Form the plot, it is possible to read that $1/b' = -(-1.5)$, or $b' = 2/3$ and that $k' = 1$ ($1/k' = 1$). This means that it is a Langmuir adsorption isotherm: $a = c / (c + 2/3)$.

Example 2.1

From Table 2.6 we can see that the solubility of oxygen in fresh water 11.29 mg/l at $10^{\circ}C$. Show that it corresponds to the Henry's constant found in Table 2.9 for oxygen.

Solution

Henry Law is applied to find the molar fraction in solution,x (the partial pressure of oxygen is 0.21 atm, corresponding to 21% oxygen in the atmosphere):

$x = 0.21 \;/\; 33000 = 6.36 \times 10^{-6}$. This is translated to mg/l :

oxygen dissolved mg/l = $6.36 \times 10^{-6} \times 55.56 \times 32 \times 1000 =$ 11.31 mg/l.

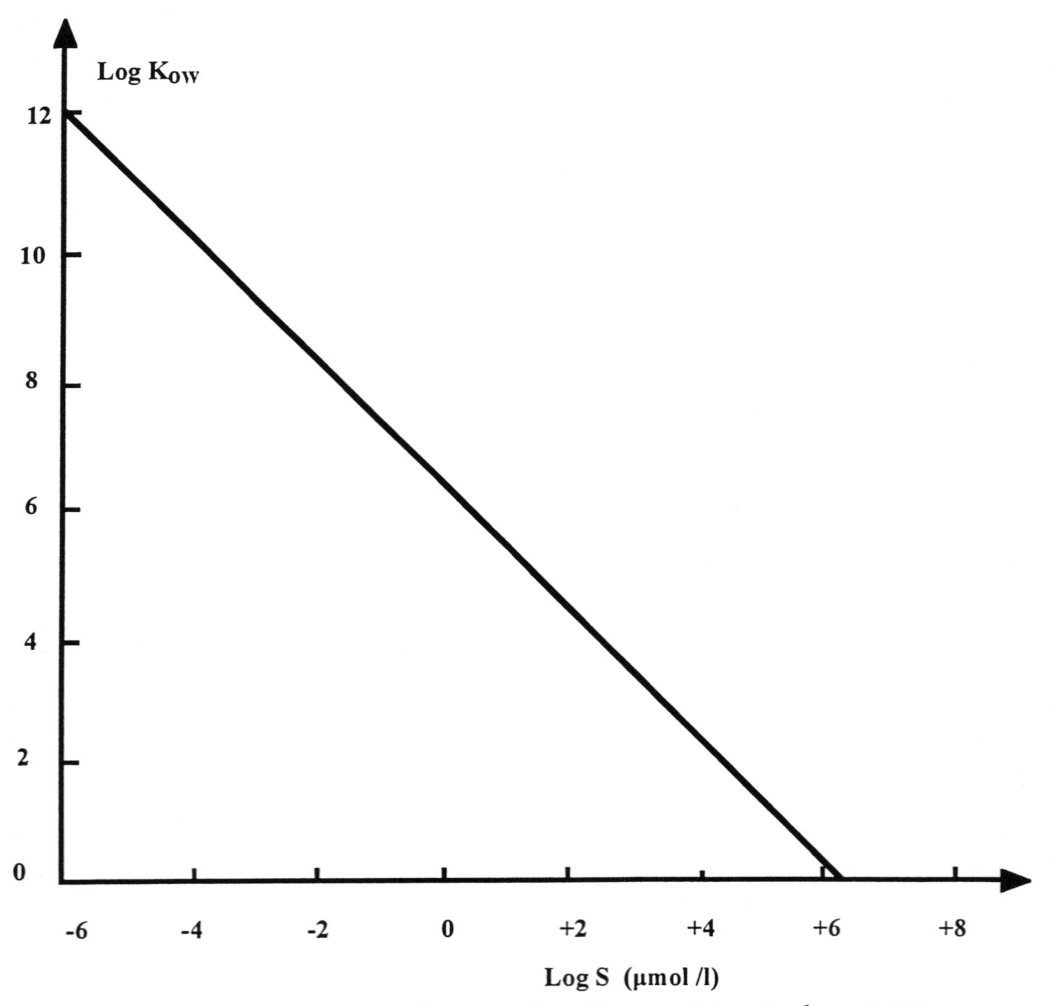

Figure 2.10 A linear regression between log K_{OW} and log S, the solubility in water expressed in μmol/l. The regression is based on a wide range of organic compounds (Chiou et al., 1977).

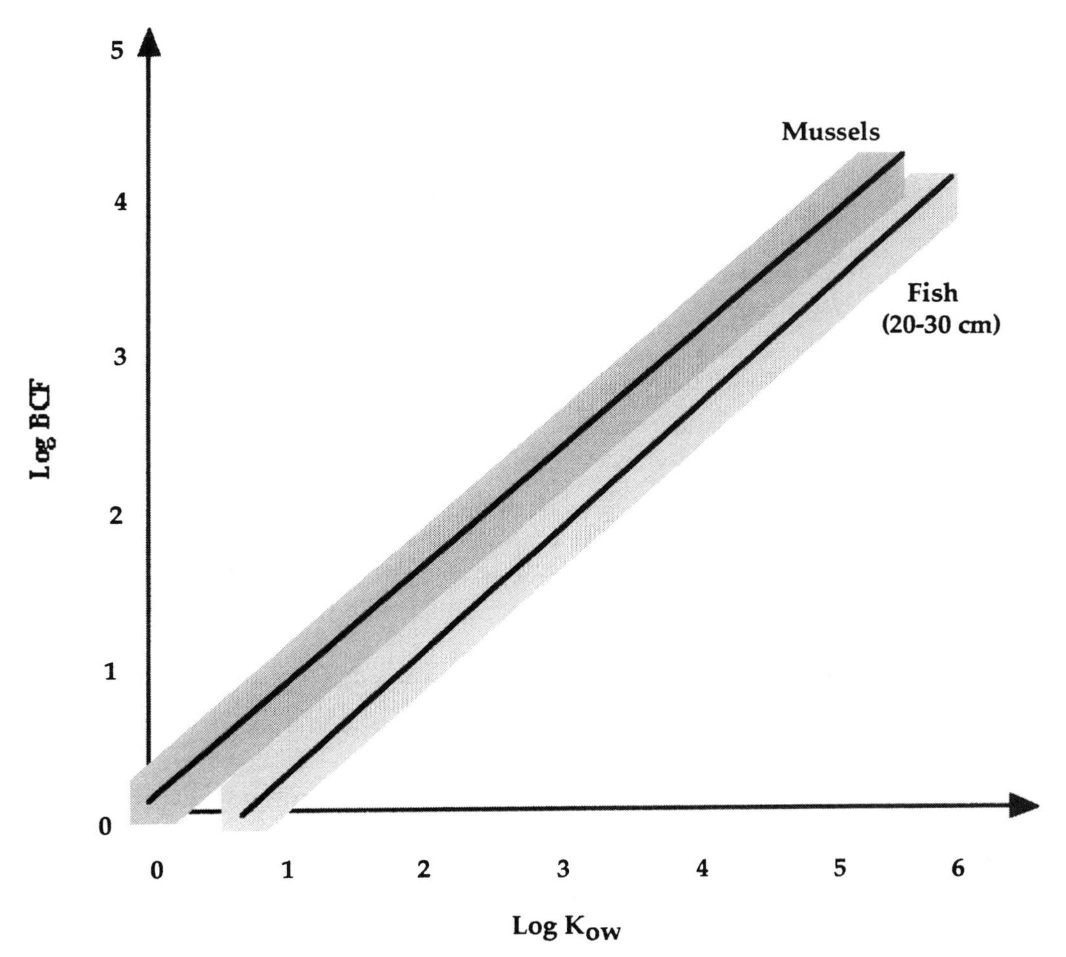

Figure 2.11 Relationships between log BCF (fish 20-30 cm long) or mussels and log K_{ow}. The two relationships are almost parallel. The grey zones shown around the straight lines indicate the bands corresponding to 95% of the observations on which the correlation is based. The correlations are based on 142 observations for the fish and 82 for the mussels. The observations are taken from several sources, among which the most important is Geyer et al., 1982.

The following empirical equations are often used (although many other empirical equations have been proposed in the literature, see Jørgensen and Gromiec, 1985):

$$\text{Reaeration} = K_a (C_s - C_a)$$

$$K_a \text{ at } 20^oC = 2.26 \text{ v / d} \qquad (2.29)$$

$$K_a \text{ at } t^oC = \text{Ka at } 20^oC \times e^{0.024x(t - 20)}$$

where K_a is the reaeration coefficient in the unit $(24h)^{-1}$, v is the flow rate of the water in m/s and d is the depth of the ecosystem (a stream) in m. The reaeration is found in mg/(1 24h).

2.6 THE GLOBAL ELEMENT CYCLES

Figure 2.15 illustrates the global carbon cycle. Carbon is used in photosynthesis in the form of carbon dioxide and hydrogen carbonate (HCO_3^-), and converted into organic matter (see equation 2.20). This is an energy-consuming process and the energy is supplied by solar radiation. The plants supply energy to herbivorous (plant eating) animals, which again are food for carnivorous (meat eating) animals. Every step in this food chain produces waste (dead animals and plants, faeces, excretion), which is the food for decomposers. Such organisms decompose (mineralise) organic components to inorganic matter, which can then be taken up by plants. By mineralisation and respiration carbon dioxide is produced. This recycling of matter is illustrated in Fig. 2.12. The system gets energy from an external source, the sun, which is the basis for the energy consumption of all life - plants, herbivorous and carnivorous animals and decomposers alike.

This natural recycling of carbon also occurs in the lithosphere and the hydrosphere (see Fig. 2.13). As carbon dioxide is a gas the food chain in the lithosphere is using the atmosphere to store carbon dioxide; hydrogen carbonate is stored in the hydrosphere. Carbon dioxide is exchanged between the atmosphere and the hydrosphere.

So far we have presented the biochemical part of the global carbon cycle. The rates of action in this cycle are relatively fast, while the process rates in the geochemical cycles are slow. Under the influence of pressure and temperature dead organic matter is slowly transformed to fossil fuel and to carbonate minerals.

These considerations raise the crucial question of whether human activity influences the carbon cycle. It does indeed. Through our use of fossil fuel, which produces carbon dioxide on burning, we remove carbon from its store in the lithosphere and transfer it to the atmosphere:

$$C_xH_{2y} + (x + y/2)O_2 \longrightarrow xCO_2 + yH_2O \tag{2.30}$$

Is this amount significant or negligible? As seen in Fig. 2.13 the input to the atmosphere is significant relative to its present concentration. The annual increase would be even higher if the hydrosphere was not taking about 60% of the carbon dioxide produced. However, this input into the hydrosphere is of minor importance for the carbon dioxide concentration in the hydrosphere (see Fig. 2.13).

In Fig. 2.14 the carbon dioxide concentration in the atmosphere is plotted against time for a prediction based on an increase in the consumption of fossil fuel of 2% per annum. Such predictions are only possible by means of ecological models, as many processes interact, as illustrated in Figure 2.13.

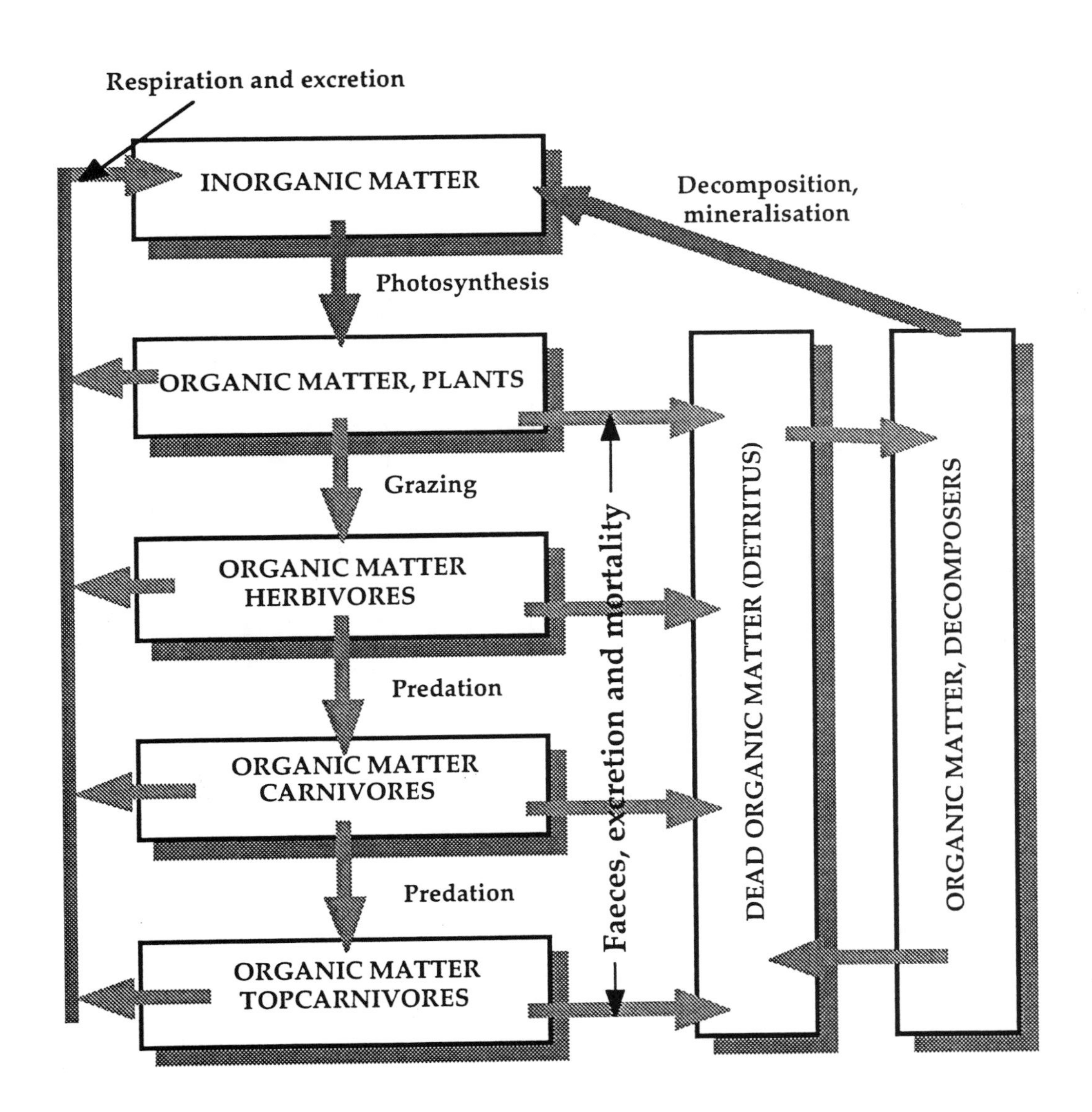

Fig. 2.12 Biochemical cycling of matter.

The diffusion of carbon dioxide to the deep sea must also be taken into account, as the oceans cannot be considered as mixed tank reactors. Furthermore the solubility of carbon dioxide varies according to pH which itself will change with increased uptake of the acidic component carbon dioxide:

$$CO_2 + OH^- \text{----->} HCO^- \qquad (2..31)$$

Concentration of air pollutants are mostly given as %, °/oo, or ppm on a volume/volume basis. The present concentration of CO_2 is about 0.0374% (1985) or 374 ppm. It is, however, easy to convert from percentage or ppm. to mass per volume.

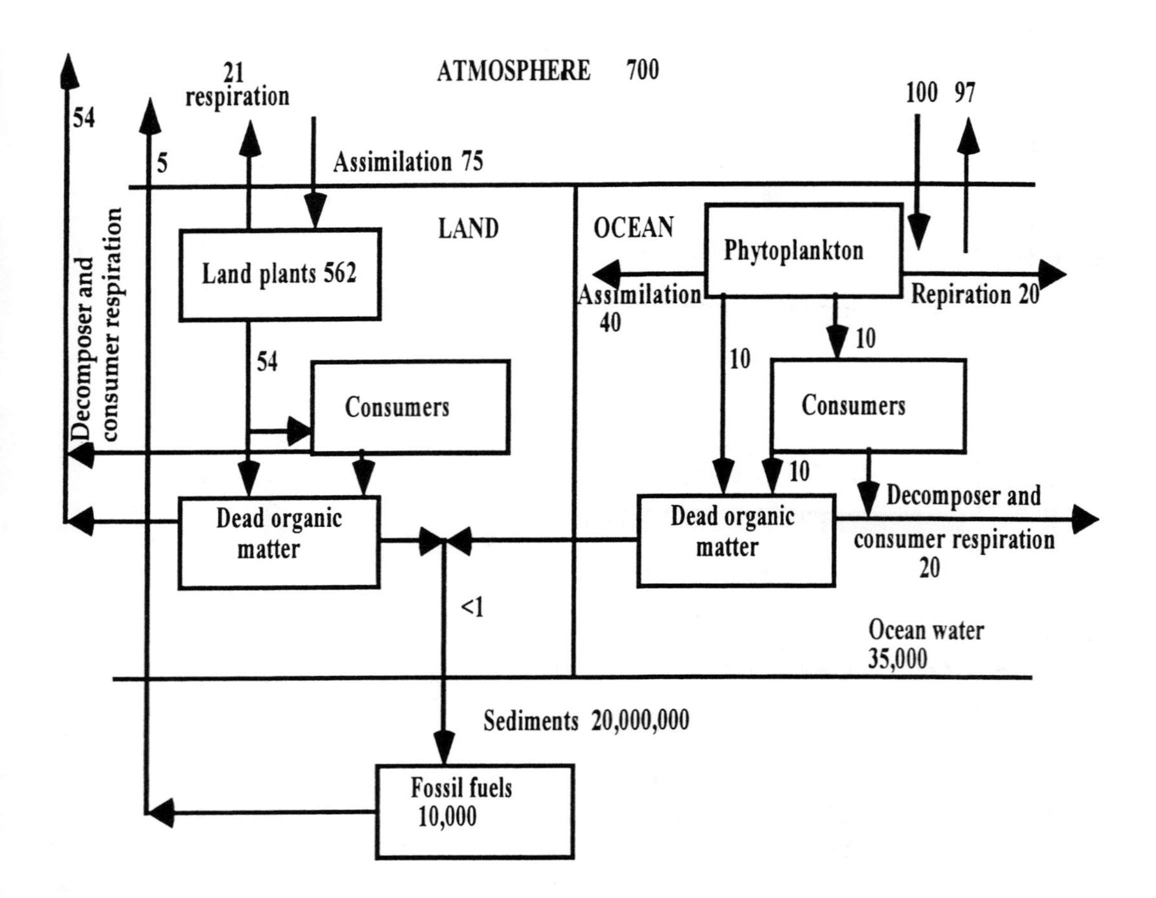

Fig. 2.13 Carbon cycle, global. Values in compartments are in 10^9 tons and in fluxes 10^9 tons/year.

Example 2.2

Convert 0.0374% CO_2 to mg/m^3 at 0°C and 1 atm.

Solution

Molecular mass is 12+2*16 = 44 g/mol

1 m^3 contains 0.034 * 1/100 m^3 = 0.000374 m^3 = 0.374 l

$p \times v = nRT \quad 1 \times 0.374 = n * 0.082 * 273$

$n = 1.67 \times 10^{-2}$ mol ≈ 0.675 g = 675 mg **675mg/m^3 at 0°C and 1 atm. ≈ 0.0374% CO_2**

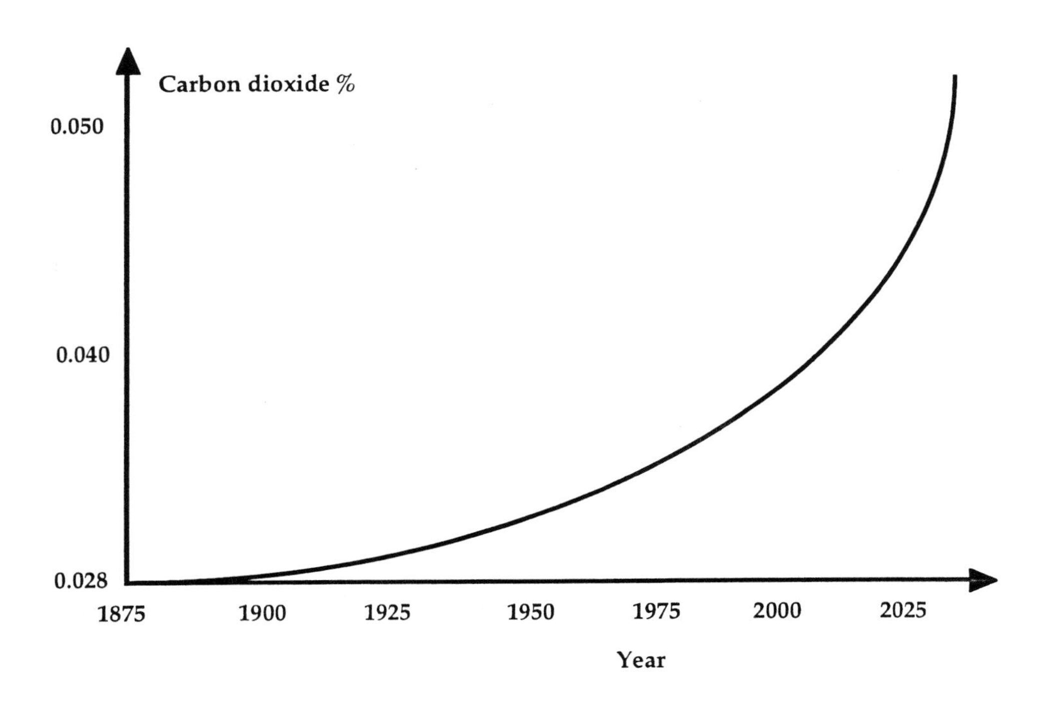

Fig. 2.14 The CO_2-concentration is plotted versus the year, taking into consideration that approximately 60% of the input to the atmosphere is dissolved in the sea.

Global cycles for all elements can be set up, although our knowledge of some elements is rather limited. However, in principle good examples exist of the cycles of the more important elements.

Fig. 2.15 shows the sulphur cycle, while Fig. 2.16 illustrates the global nitrogen cycle. Both cycles are out of equilibrium as a result of human activity. The production of nitrogen fertiliser has meant that gaseous nitrogen from the atmosphere is converted to ammonia or nitrate, which are deposited in the lithosphere, while the major part is washed out to the hydrosphere, where it may cause eutrophication problems on a local or regional scale.

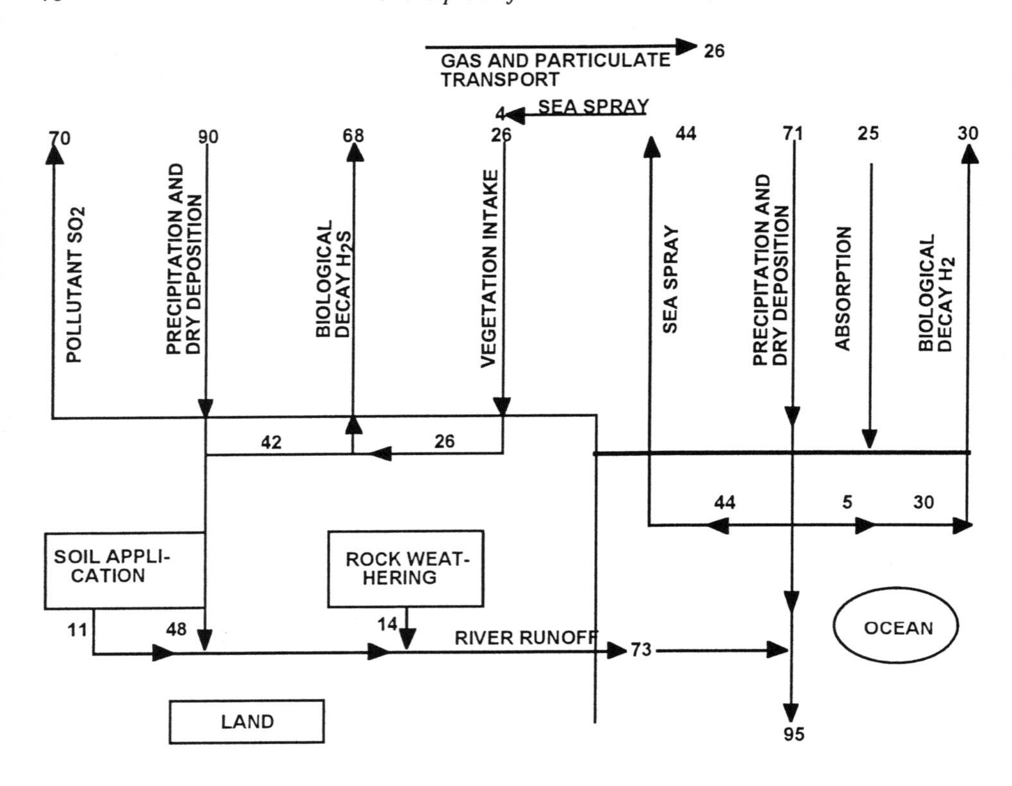

Fig. 2.15 The global sulphur cycle. The fluxes are indicated in the unit 10^9 kg S / y. Notice that sea spray and biological decay in the oceans balance river runoff. The lithosphere loses sulphur to the oceans but it is recycled back to the lithosphere by natural processes via the atmosphere. The most important anthropogenic flux is the sulphur dioxide pollution of the atmosphere. It is oxidised and hydrolysed to sulphuric acid and transferred back to the lithosphere in the form of acid rain.

There is a similar imbalance in the phosphorus cycle. Phosphorus minerals are also used in the production of fertilisers, which are applied in the lithosphere from where the major part is later transported to the hydrosphere. For a more comprehensive introduction to the subject of global element cycles, see Svensson and Söderlund (1976) and Bolin et al. (1979).

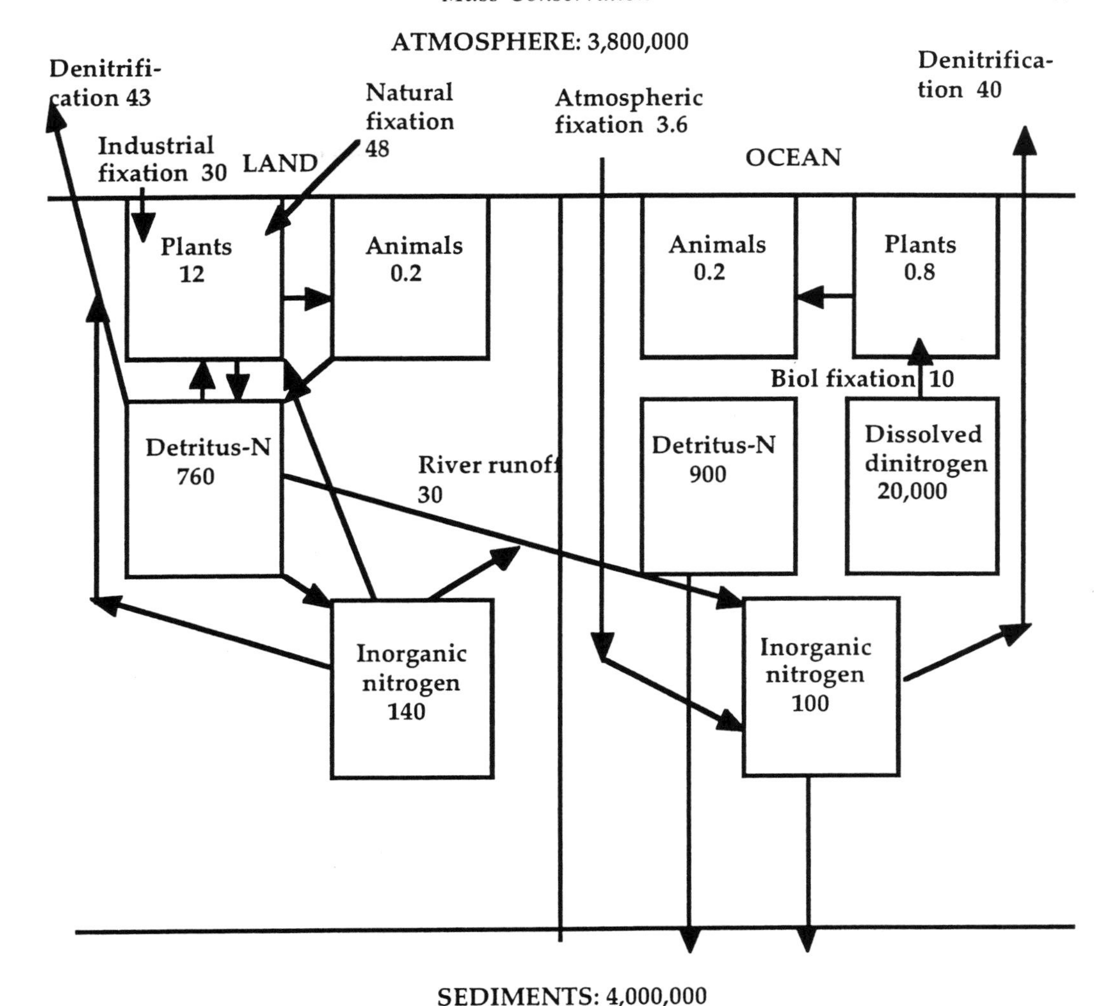

Fig. 2.16 The global nitrogen cycle. Values in compartments are in 10^9 tons N, and fluxes in 10^6 tons N /y. The global distribution of nitrogen and annual transfers can only be estimated within broad limits. The two quantities known with high confidence are the amount of nitrogen in the atmosphere and the industrial nitrogen fixation. Because of the extensive use of industrially fixed nitrogen, the amount of nitrogen available to land plants may significantly exceed the nitrogen returned to the atmosphere by denitrification. A significant amount of this excess fixed nitrogen is ultimately washed into the sea. This nitrogen doesn't effect the global nitrogen balance significantly, but may easily effect the coastal zones locally and cause an undesired eutrophication there.

2.7 OXYGEN BALANCE OF RIVERS AND STREAMS

Maintenance of a high oxygen concentration in aquatic ecosystems is crucial for survival of the higher life forms in aquatic ecosystems, as already discussed in Section 2.4. At least 5 mg l^{-1} is needed for many fish species. At 20-21 °C this corresponds to (see Table 2.6) 5/9 = 56% saturation. The oxygen concentration is influenced by several factors, of which the most important are:

1. **The decomposition of organic matter, often can be described by use of first order kinetic equation:** (see Section 2.3, equations 2.3 and 2.4)

$$L_t = L_o * e^{-K_1 t} \qquad (2.32)$$

where
L_t = concentration of organic matter at time t
L_o = initial concentration of organic matter
K_1 = rate constant

The concentration of organic matter can be expressed by means of the oxygen consumption in the decomposition process. This is usually BOD_5, which is the oxygen consumption measured over 5 days. The oxygen consumption after 5 days is therefore not included, but it will rarely be of importance, as all easily decomposed matter is broken down within 5 days.

BOD_5 of municipal waste water is usually 100-300 mg O_2 l^{-1} and for waste water after mechanical-biological treatment (concerning these treatment processes, see Chapter 5) 10-20 mg O_2 l^{-1}. Characteristic concentrations and rate constants are shown in Table 2.10.

2. **The nitrification of ammonia (ammonium)** in accordance with the following process:

$$NH_4^+ + 2O_2 \longrightarrow NO_3^- + H_2O + 2H^+ \qquad (2.33)$$

Ammonia is formed by decomposition of organic matter. Proteins and other nitrogenous organic matter are decomposed to simpler organic molecules such as amino acids, which again are decomposed to ammonia. Urea and uric acid, the waste products from animals, are also broken down to ammonia. Nitrifying microorganisms can use ammonia as an energy source, as the oxidation of ammonia is an energy-producing process. This decomposition chain is illustrated in Fig. 2.17, where it can be seen that the energy (chemicals) is decreased throughout the chain.

The nitrification process can be described by the following first order kinetic expression:

$$dN / dt = - K_N * t \qquad (2.34)$$

which after integration leads to the following expression:

$$N_t = N_o * e^{-K_N * t} \tag{2.35}$$

where

N_t = concentration of ammonium at time = t

N_o = concentration of ammonium at time = 0

K_N = rate constant, nitrification

N_t and N_o are here expressed *by the oxygen consumption* corresponding to the ammonium concentration in accordance with (2.35). Values for K_1, K_N, L_o and N_o are given for some characteristic cases in Table 2.10.

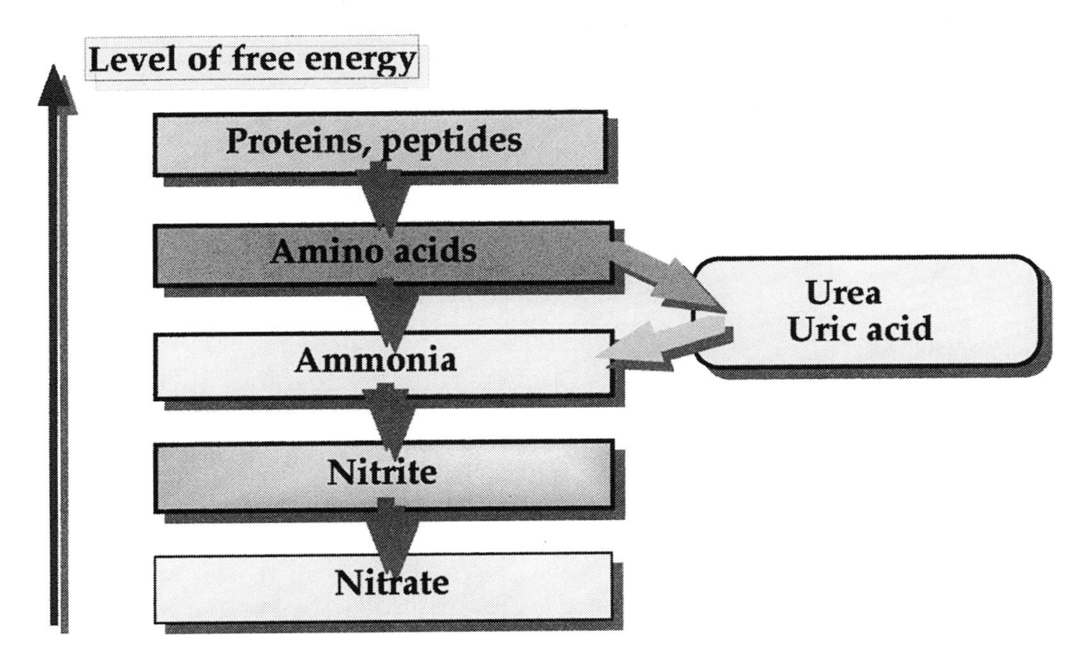

Fig. 2.17 Decomposition chain: protein to nitrate.

The relation between ammonium concentration and oxygen consumption in accordance with (2.33) is calculated to be (2 * 32)/14 = 4.6 mg O_2 per mg NH_4^+ - N, but due to bacterial assimilation of ammonia this ratio is reduced to 4.3 mg O_2 per mg NH_4^+ - N in practice.

K_1 and K_N are dependent on the temperature as illustrated in Table 2.7

TABLE 2.10

Characteristic values, K_1, K_N, L_o and N_o (20 °C)

	K_1(day^{-1})	K_N(day^{-1})	N_o	L_o
Municipal waste water	0.35-0.40	0.15-0.25	80-130	150-250
Mechanical treated municipal waste water	0.35	0.10-0.25	70-120	75-150
Biological treated municipal waste water	0.10-0.25	0.05-0.20	60-120	10-80
Potable water	0.05-0.10	0.05	0-1	0-1
River water	0.05-0.15	0.05-0.10	0-2	0-5
Diary waste water	0.4-0.5	0.12-0.2	200-300	1200-1600

3. **Dilution.** The dilution capacity of a stream can be calculated using the principles of mass balance (principles 2.1 and 2.7):

$$L_S * Q_S + L_W * Q_W = L_m * Q_m \tag{2.36}$$

where L represents the concentration (mass/volume) and Q the flow rate (volume/time). The subscripts S, W and m designate the stream, waste and mixture conditions, respectively.
4. **Settling or sedimentation**, nature's method of removing particles from a water body. Large solids will settle out readily, while colloidal particles can stay in suspension for a long period of time.
5. **Resuspension** of solids, common in aquatic ecosystems due to strong wind stress, flooding or heavy run-off.
6. **Photosynthesis**, which produces oxygen, see Section 2.4.
7. The use of oxygen for **respiration** by plants and animals.
8. **Oxidation of organic matter in the sediment**, which also causes consumption of oxygen and release of ammonia in accordance with the following equation:

$$\text{Organic matter} + O_2 \rightarrow CO_2 + H_2O + NH_4^+ \qquad (2.37)$$

9. If the oxygen concentration is below saturation (see Table 2.6), **reaeration** from the atmosphere will take place. The equilibrium between the atmosphere and the water can be expressed by use of Henry's law and the reaeration can be computed from equations (2.27) - (2.29).

The aeration is dependent on 1. the difference between the oxygen concentration at equilibrium and the present oxygen concentration, 2. the temperature, 3. the flow rate of water, and 4. the water depth; see equation (2.29)

The following alternative expression is satisfactory (Jørgensen and Gromiec, 1989):

$$K_a(20°C) = 649 * \frac{\sqrt{DI} * S^{1/4}}{d^{5/4}} \qquad (2.38)$$

where

DI = coefficient of molecular diffusion (liquid film m^2 day^{-1})

S = slope (m/m)

d = depth (m)

For Danish streams the following expression may be used:

$$K_a(20°C) = 37406\, v^{0.846}\, d^{-0.672}\, s^{1.154} \qquad (2.39)$$

where v is the average velocity (m/s), d is the average depth (m) and s is the slope expressed in m/m. $K_a(20°C)$ in Danish streams ranges from 1 to 10 1/24h. Many other alternative expressions can be found in the literature; see for instance the review by Gromiec (1983).

If all these processes are considered together the following equation for the oxygen concentration, C_t, is obtained:

$$dC_t / dt = P - R - K_1 * L_t - K_N * N_t + K_a (C_s - C_t) - S \qquad (2.40)$$

where

P = oxygen production by photosynthesis

R = oxygen consumption by respiration

S = oxygen consumption of sediment

Dilution can be considered by calculation of L_t and N_t. Sedimentation and resuspension may be included in the rate constants K_1 and K_N or by dividing L_t and N_t into the dissolved and suspended material. As P, R and S are dependent on the time, these computations lead to several coupled differential equations which can only be solved numerical by a computer. P is 0 for the time between sunset and sunrise. The lowest oxygen concentration is therefore often attained just before sunrise. It is of course of importance to measure or know the lowest oxygen concentration which is one of the most important factors for the water quality and the life of aquatic ecosystems.

More than seventy years ago Streeter and Phelps were using a simpler approach, as they only considered two processes to be of importance in the oxygen balance: biological decomposition and reaeration. They used the following differential equation:

$$dD / dt = K_1 * L_t - K_a * D \qquad (2.41)$$

where $D = C_s - C_t$. By use of equation (2.32), equation (2.40) can be reformulated:

$$dD / dt + K_a * D = K_1 * L_o * e^{-K_1 * t} \qquad (2.42)$$

This equation can be solved analytically. When $D = D_o$ at = 0, we get:

$$D = \frac{K_1 * L_o}{K_a - K_1} (e^{-K_1 * t} - e^{-K_a * t}) + D_o * e^{-K_a * t} \qquad (2.43)$$

For $K_1 - K_a = 0$, the solution is (equation 2.43) is of course not valid in this case):

$$D = (K_1 * t * L_o + D_o) * e^{-K_1 * t} \qquad (2.43a)$$

From equation (2.43) it is possible to plot D, C_t and the reaeration against time.

The point where the oxygen concentration is at a minimum is termed the critical point (see Fig. 2.18). It is from an environmental management point of view the most interesting concentration, as it represents the worst case.

This point can be found from:

$$dD / dt = 0; \; d^2D / dt^2 < 0 \text{ (maximum of D} \approx \text{minimum of oxygen)} \qquad (2.44)$$

$$t_c = \frac{1}{K_a - K_1} \ln \left(\frac{K_a}{K_1} - \left(\frac{K_a}{K_1} - 1 \right) \frac{D_o}{L_o} \right) \qquad (2.45)$$

$$D_c = (K_1 * L_o / K_a) * e^{-K_1 * t_c} \qquad (2.46)$$

If a nitrification term is added to the Streeter-Phelps equation, similar (but more complex) equations can be used to find an analytical solution.

In equations (2.40) to (2.46) time is used as independent variable, but if a river or stream is considered it might be useful to translate time intto distance in flow direction, x, by use of the flow velocity, v:

$$x = v * t \tag{2.47}$$

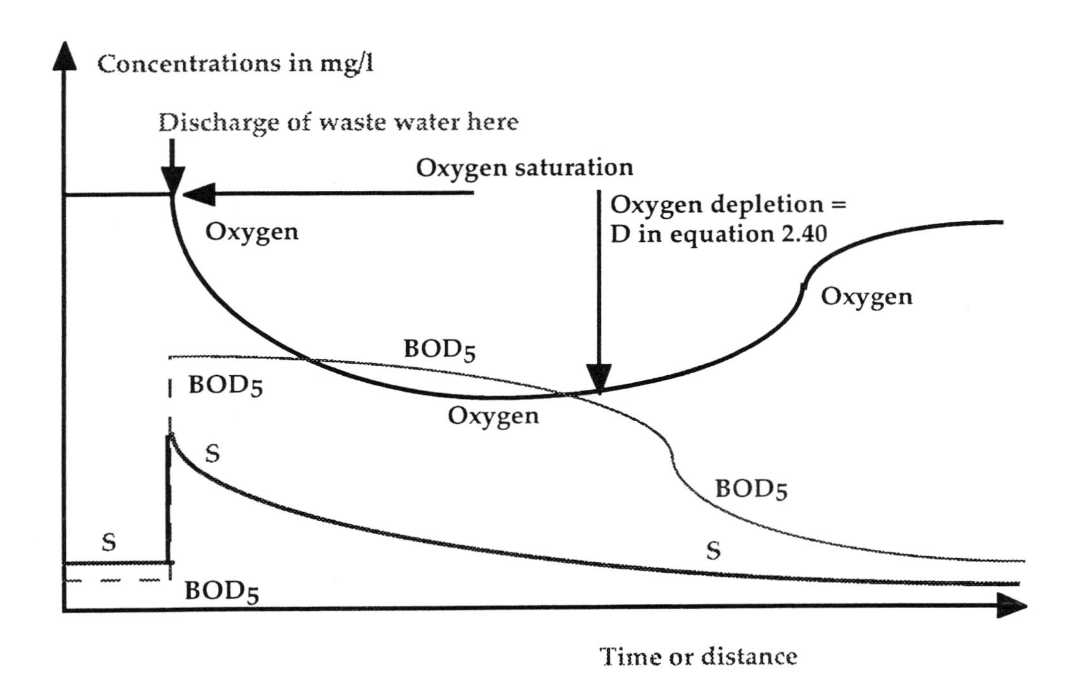

Fig. 2.18 Oxygen concentration, BOD_5, suspended matter, S, versus distance in running water.

The equations can be used to describe the oxygen profile in the flow direction, but if so, some hydraulic assumptions must be introduced: 1. that by discharge of polluted waste water into the river complete mixing with the flow will take place; 2. that the flow rate is the same throughout the entire cross-section of the river. These assumptions are hardly valid in practice, but may sometimes be used with good approximation. This means that the concentration, C, just after discharge of waste water can be calculated from the following simple expression:

$$C = \frac{Q_w * C_w + Q_r * C_r}{Q_w + Q_r} \tag{2.48}$$

where
Q_w = water flow, waste water ($1\ s^{-1}$)
C_w = concentration in waste water ($mg\ 1^{-1}$)
Q_r = water flow, river ($1\ s^{-1}$)
C_r = concentration in river water ($mg\ 1^{-1}$)

Example 2.2

A municipal waste-water plant discharges secondary effluent (after mechanical-biological treatment) to a surface stream. The worst conditions occur in the summer months, when stream flow is low and water temperature high. The waste water has a maximum flow of 12,000 m^3/day, a BOD_5 of 40 mg l^{-1} (at 20°C), a dissolved oxygen concentration of 2 mg l^{-1} and a temperature of 25°C. The stream has a minimum flow of 1900 $m^3\ h^{-1}$, a BOD_5 of 3 mg l^{-1} (at 20°C) and a dissolved oxygen concentration of 8 mg l^{-1}. Use K_1 = 0.15 (see Table 2.18 for the mixed flow). The temperature in the stream reaches a maximum of 22°C. Complete mixing is almost instantaneous. The average flow (after mixing) is 0.2 m sec^{-1} and the depth of the stream 2.5 m. Find the critical oxygen concentration.

Solution

$$K_a\ (20°C) = \frac{2.26 * 0.7}{2.5^{2/3}} = 0.25\ d^{-1} \qquad \text{(reaeration)}$$

Mixture: 500 m^3/h (ww) + 1900 m^3/h (stream) = in total 2400 m^3/h

$$BOD_5 \text{ of mixture : } \frac{500 * 40 + 1900 * 3}{2400} = 10.7\ mg\ l^{-1}$$

$L_5 = L_o * e^{-0.15*5}$
$BOD_5 = L_o - L_5 = 10.7$
$10.7 = L_o (1 - e^{-0.15*5})$
$L_o = 20.3\ mg\ l^{-1}$

Dissolved oxygen = $DO_{mixture}$ = (500*2+1900*8) / 2400 = 6.7 mg l^{-1}
$Temperature_{mixture}$ = (500*25+1900*22)/ 2400 = 22.6
K_1 at 22.6 = $0.15 * 1.05^{22.6-20}$ = 0.17 (temperature dependence)
K_a at 22.6 = $0.25 * e^{0.024(22.6-20)}$ = 0.27 (temperature dependence; see equations (2.27)-(2.29)
Initial oxygen deficit = D = 8.7 -6.7 = 2 mg l^{-1} (use Table 2.6)

$$\text{Critical location (time)} = \frac{1}{0.27-0.17} \ln\left(\frac{0.27}{0.17} - \left(\frac{0.27}{0.17} - 1\right)\frac{2}{20.3}\right) = 4.25\ d.$$

$$D_c = \frac{0.17 * 20.3}{0.27} e^{-0.17*4.25} = 6.2 \text{ mg l}^{-1}$$

This conditions occurs at 0.2 * 3600 * 24 * 4.25 m = 73440 m

<u>Oxygen concentration at critical point: 8.7 - 6.2 = **2.5 mg l^{-1}**</u>. This is a unacceptably low oxygen concentration which indicates that almonoid fish cannot live in the water.

Oxygen concentrations in river water can be determined by means of more complex ecological models, which take into account hydraulic components, growth of phytoplankton and zooplankton, oxygen consumption of the sediment, the spectrum of biodegradability of the components present in discharged waste water, and the presence of toxic matter affecting the biodegradability. For further information on these more complicated models, see Jørgensen (1980), Rinaldi et al. (1979), Orlob (1981) Jørgensen (1981), Jørgensen (1994) and Jørgensen et al. (1995a).

In section 2.4, we mentioned the importance of oxygen concentration for the ecological equilibrium and survival of aquatic ecosystems. Using a biological classification of running waters, based on the oxygen profile, it is easy to demonstrate how the oxygen concentration can be used to assess the ecological quality of the ecosystem. As the variations in oxygen concentrations only need to be small to significantly alter the ecology of running water, it could be concluded that an ecological examination of the aquatic ecosystem is a more sensitive instrument for assessing water quality than a chemical examination, although naturally a close relationship exists between the two results. Another advantage of the ecological examination is its ability to provide a picture of long-term conditions, while the chemical examination always gives a momentary picture. One of the most commonly used ecological examinations is the application of the saprobic system, which classifies running waters in 4 classes (Hynes, 1971). The following four degrees of pollution are identified (the often used American names are in brackets):

1. Oligosaprobic, I, (Clear Water Zone)
2. ß-mesosaprobic, II, (Recovery Zone)
3. °-mesasaprobic, III, (Active Decomposition Zone)
4. Polysaprobic, IV, (Degradation Zone)

The classification is based on an examination of plant and animal species present in the water at a number of locations. The species can be divided into four groups:

1. Organisms characteristic of unpolluted water
2. Species predominating in polluted water
3. Pollution indicators
4. Indifferent species

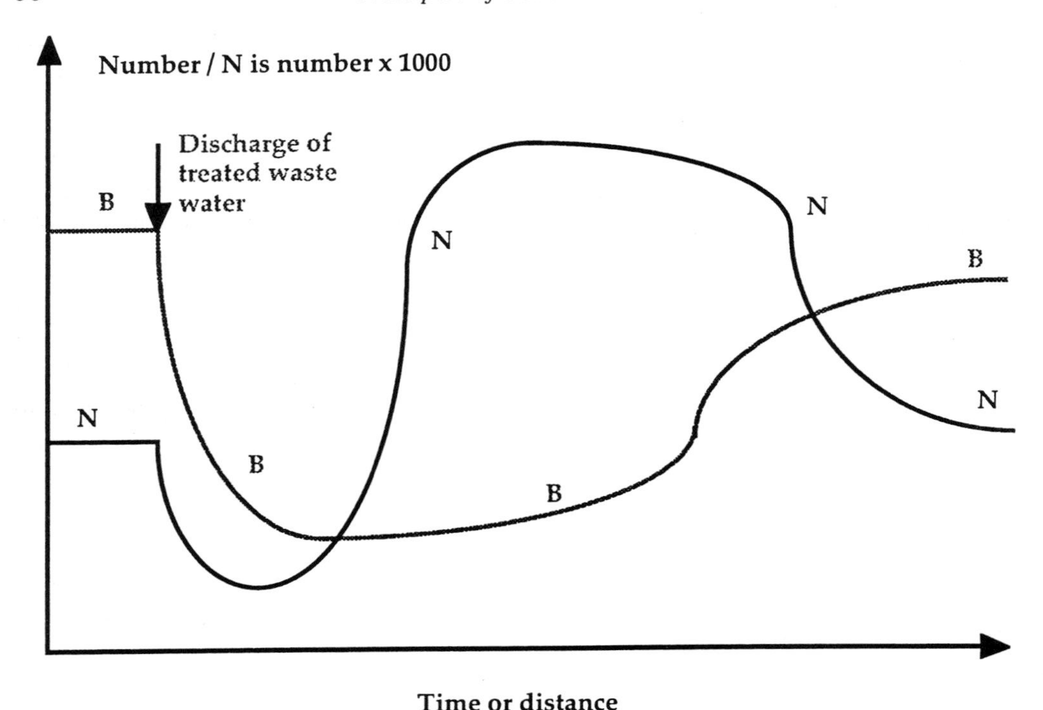

Fig. 2.19 N. Number of higher organisms, B. Number of species versus distance in running water. Classification with numbers in accordance with saprobic system.

The classification presented above includes only four classes. For more complicated systems, see Hynes (1971). Normally the assessment of the ecological conditions of running waters is a complicated procedure, and the application of the saprobic system requires some experience and is time consuming. Each species found is given an average saprobic value, s, by distributing 10 points over the four pollution degrees. Species with a broad distribution have their 10 points distributed evenly over all the degrees. The average value of s is found by multiplying the number of points from each degree by 1, 2, 3 and 4 respectively, and dividing the sum by 10. To allocate a higher weight to species with a narrow ecological distribution, each species is given an indicator value ranking from 1 to 5. Finally, the relative abundance of each species is calculated, and the saprobic value S for sampling the site can be estimated:

$$S = \Sigma\,(H \times s \times G) \,/\, \Sigma\,(H \times G) \qquad (2.49)$$

where H is the relaive number of individuals, s is the average saprobic value of each of the species, and G is the indicator value for the species. Examples of the G-value, the distribution of the 10 points and the resulting s-values are given in Table 2.11.

However, simple versions of the saprobic system are often used, based on only smell, appaerance and semi quantitative classification of macroinvertebrates as pollution indicators, pollution-dominant organisms or clean water organisms. The monitoring of streams based on biological indices such as saprobic index depends on the person performing the sampling and the accuracy of identification, and may thus vary considerably. The saprobic system is very sensitive to the current velocity, which means that at a given BOD-concentration a much lower degree is achieved at high current velocity; see Figure 2.20.

Although a direct translation of chemical parameters to ecological conditions is not possible, it is often advantageous to measure chemical parameters, because this is a much faster procedure. A relationship between the number of organisms and number of species on the one side and the oxygen conditions on the other side is often valid.; see Figure 2.19 and compare with Figure 2.18.

TABLE 2.11

Saprobic Index: Examples of the indicator value, G,the distribution of the 10 points over pollution degrees and the average value of the index, s. After Moth and Lindegaard, 1983.

		G	I	II	III	IV	s
Pollution indicators:	*Eristalis*	5				10	4.0
	Psychoda	5				10	4.0
Pollution dominants	*Tubificidiae*	4		1	3	6	3.5
	Chironomus	5		1	5	4	3.3
	Asellus	5		2	8		2.8
Clean-water animals	*Gammarus*	3	4	5	1		1.7
	Leuctra	2	3	5	2		1.9
	Protonemura	5	9	1			1.1
	Ephemera	3	5	4	1		1.6
Indifferent forms	*Corixa*	1	2	6	2		2.0
	Platambus	1	2	7	1		1.9

A tentative translation of the chemical analysis to the saprobic system is also shown in Table 2.12, but the table should be used very cautiously. As seen in this table it is not sufficient to measure the BOD_5 concentration alone: the ammonium and nitrate concentration must also be included in the examination to be able to obtain a full picture of the conditions. Furthermore, assessment of the diurnal variations is important, as it is the minimum oxygen concentration which is crucial. It can usually be recorded shortly before sunrise.

The saprobic system is used less today than 20 years ago. The focus today is, however, still on the oxygen conditions and the relationships between ammonia and nitrate. The problems associated with oxygen depletion are in most developed countries more or less solved today and in the streams where oxygen still is low, it is in most cases very clear what the environmental manager has to do.

TABLE 2.12

The saprobic-system and physio-chemical parameters

Chemical parameter	Zones: 4. Degradation	3. Active Decomposition	2. Recovery	1. Clear Water
Dissolved oxygen	0-3 mg l^{-1} <50% saturation	BOD_5 increasing >70% saturation BOD_5 decreasing <30% saturation	>60% saturation	>90% saturation
BOD_5	high	increasing <5 mg l^{-1} decreasing >20 mg l^{-1}	<5 mg l^{-1}	<3 mg l^{-1}
NH_4^+	0.5-2	0.3-1.2	<0.2	<0.1 mg N l^{-1}
NO_2^-	0-0.2	0-0.2	~0.2	<0.05 mg N l^{-1}
NO_3^-	very low	1-2 mg N l^{-1}	2-6 mg N l^{-1}	high
Turbidity	high	low	very low	very clear

A much faster method than the one based on the saprobic system is the **Trent Index,** Originally developed for the river Trent in England. It is based on identifiable species or systematic groups (genera or families) and it requires in principle only separation into different groups, not necessarily an identification of the groups. The groups are ranked, with

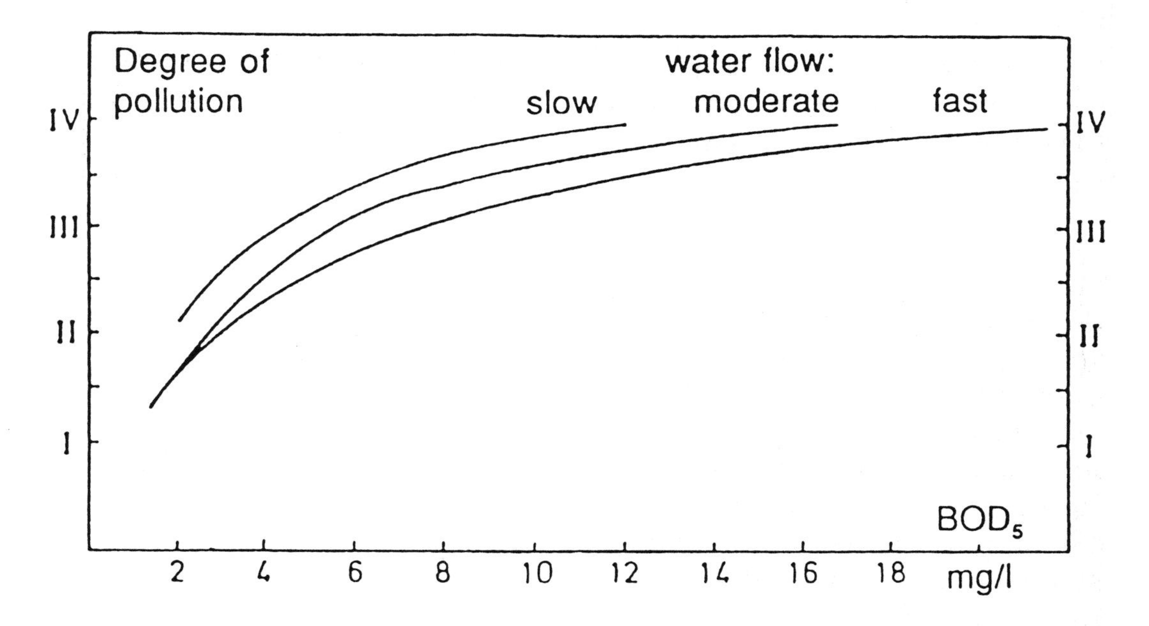

Fig. 2.20. The relationship between degree of pollution (ranging from I to IV) and average concentration of BOD5 in various water courses with slow, moderate and fast water flow.

the clean-water species such as stoneflies and mayflies at the upper end and the pollution-dominant species and pollution-indicator species such as tubificids and *Eristalis* at the lower end. The index value is simply found by moving down through the scheme (see Table 2.13). Variations of the Trent Index include **Chandler's Index**, assessing sensitivity to pollution and the relative abundance of each group on a scale from 1 to 100; and the **Danish Stream Fauna Index** (details see Miljøstyrelsen, 1998 and Lindegaard et al., 1999) which is specially adopted for Danish streams and uses negative values for some groups and only considers *Gammarus* if there are more than 10 individuals per sample.

Generally, pollution by biodegradable organic matter will cause stress on ecosystems, and the diversity will decrease. The diversity can be expressed simply by number of taxa or species, but the occurrence of a simgle individual will have too much importance and since the number of individuals is a function of sample size and sample number, a great number of samples has to be investigated to register all taxa and species. To overcome this problem the **Shannon Index** can be used; see Section 4.6, Table 4.5. Usually a significant decrease in the Shannon Index is observed at low oxygen concentrations and high BOD5 concentrations; compare also with Figure 2.19.

TABLE 2.13

The Trent Index represents a method of classification of watercourse pollution which is considerably simpler than the calculation of the saprobic index. After Moth and Lindegaard, 1983.

		Total number of groups present				
		0-1	2-5	6-10	11-15	>16
				Index		
Plecoptera	several groups	-	7	8	9	10
nymphs present	only 1 group	-	6	7	8	9
Ephemeroptera	several groups*)	-	6	7	8	9
nymphs present	only 1 group*)	-	5	6	7	8
Trichoptera	several groups§)	-	5	6	7	8
larvae present	only 1 group§)	4	4	5	6	7
Gammarus present	none of the above	3	4	5	6	7
Asellus present	none of the above	2	3	4	5	6
Tubificids and/or red midge larvae	none of the above	1	2	3	4	-
None of the above	possibly organisms such as *Eristalis*	0	1	2	-	-

*) Except for *Baetis rhodani*. §) Including *Baetis rhodani*

The following systematic entities are considered as groups:

Each genus of Tricladida (flatworms)
Oligochaeta (worms) except family Nuiididae
Each genus of Hirudinea (leeches)
Each genus of Mollusca (snails, mussels)
Each genus of Malacostraca
Each genus of Ephemeroptra (Mayflies) except *Baetis rhodani*
Each genus of Plecoptera (stoneflies)
Baetis rhodani (Mayfly)
Each family of Trichoptera (caddisfly)
Each genus of Neuroptera and Megaloptera (net-veined wings)
The family Chironomidae (midge larvae) except *Chironomus sp.*
Chironomus sp.
The family Simuliidae (black flies)
Each family of other Diptera (two winged) and each geneus of Elminterdae and Helodidae (beetles)
Each family of other Coleoptera (bugs)
Hydracarina (water mites)

Pyrite (FeS_2) and siderite ($FeCO_3$) are found in many soils. Pyrite is formed in anaerobic sediments where sulphate is reduced in the presence of iron, for instance, when seawater meets freshwater under anoxic conditions, as during the glacial eras. When the ground water level is lowered by draining or by water abstraction from lignite- and pits, or for irrigation and drinking purposes, the soil becomes oxidised and ochre, $Fe(OH)_3$ is formed:

$$FeS_2 + 3.5\, O_2 + H_2O \text{ ----> } Fe^{2+} + 2SO_4^{2-} + H^+ \qquad (2.50)$$

$$Fe^{2+} + 0.25O_2 + H^+ \text{ ----> } Fe^{3+} + 0.5\, H_2O \qquad (2.51)$$

The iron(II) ion is stable provided that pH < 3, but already at pH 3.5- 4.0 a microbial oxidation can take place. Above pH 4-5 a chemical oxidation occurs, and precipitation of ochre is observed:

$$Fe^{3+} + 3H_2O \text{ ----> } Fe(OH)_3 + 3H^+ \qquad (2.52)$$

For every mole of pyrite oxidised, four moles of hydrogen ions are formed and consequently the water becomes very acidic. pH-values as low as 1-2 can be found in drains from lignite pits. Ochre has a serious impact on the receiving streams. It precipitates on the respiratory organs of invertebrates and on the gills of fish. It inhibits the penetration of light, and especially the microbenthic algae suffer. The very fine ochre particles reduce the porosity of sediments and prevents the transport of oxygen down into the sediment, destroying bottom-living organisms and eggs from salmonid fishes. The direct toxic effect of iron(II) ions cannot be distinguished from the indirect effect of ochre precipitation. Diversity is likely to decrease.

It is possible to precipitate the iron with lime, calcium hydroxide, in basins before it reaches the stream, but the process is costly and only up to 95% of the iron(II) ions are oxidised.

The naturally meandering stream over a reach of 1 km typically has a denivellation of 4 m, or 0.4%. If the stream is regulated to a straight channel, the length may be reduced to 500 m, increasing the slope to 0.8%. If such a channel is not maintained, the erosional forces will bring the stream back to the original shape after some years. Reinforcement of the banks with stones or concrete will prevent re-meandering of the stream. The maintenance of running waters includes traditional macrophytes control by cutting in the stream and on the banks and by excavation of sedimented material. The removal of macrophytes and sediment is done mechanically. To reduce costs and the impact of the often damaging maintenance, more ecological methods have been introduced. The growth of the macrophytes can be reduced by planting shady trees and bushes along the river bank, and less frequent cutting the bank vegetation will reduce the admission of light to the stream. The use of herbivorous fish such as grass carp is implemented in some streams. These methods are based on ecotechnology which is presented in more detail in Chapter 8.

2.8 THE EUTROPHICATION PROBLEM

The word eutrophic generally means "nutrient rich". Naumann introduced in 1919 the concepts of oligotrophy and eutrophy. He distinguished between oligotrophic lakes containing little planktonic algae and eutrophic lakes containing much phytoplankton.

The eutrophication of lakes in Europe and North America has grown rapidly during the last decade due to the increased urbanisation and the increased discharge of nutrient per capita. The production of fertilisers has grown exponentially in this century and the concentration of phosphorus in many lakes reflects this growth (Ambühl, 1969). The word eutrophication is used increasingly in the sense of the artificial addition of nutrients, mainly nitrogen and phosphorus, to water. Eutrophication is generally considered to be undesirable, although it is not always so.

The green colour of eutrophic lakes makes swimming and boating more unsafe due to increased turbidity. Furthermore, from an aesthetic point of view the chlorophyll concentration should not exceed 100 mg m^{-3}. However, the most critical effect from an ecological viewpoint is the reduced oxygen content of the hypolimnion, caused by the decomposition of dead algae. Eutrophic lakes might show high oxygen concentrations at the surface during the summer, but low oxygen concentrations in the hypolimnion, which may cause fish kill, see Fig. 2.21 - the oxygen profile. The zones of deep lakes are shown in Fig. 2.21 with a typical oxygen profile.

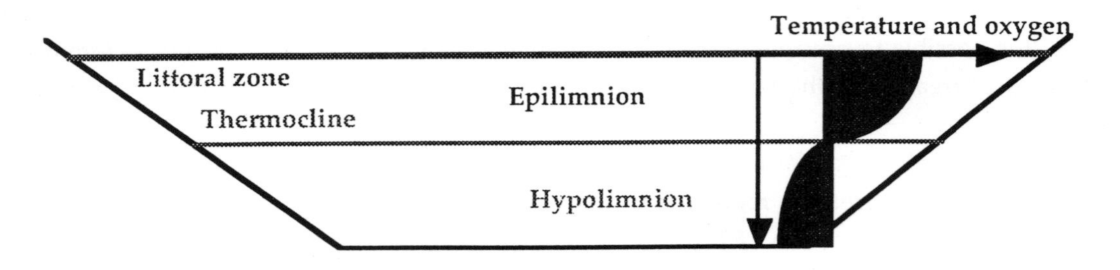

Fig. 2.21. Thermal stratification. Epilimnion, hypolimnion and littoral zone are shown. A plot of temperature and oxygen versus the depth in a typical summer situation for deep, temperate lakes.

On the other hand an increased nutrient concentration may be profitable for shallow ponds used for commercial fishing, as the algae directly or indirectly form food for the fish population. About 16-20 elements are necessary for the growth of freshwater plants, as demonstrated in Table 2.14, where the relative quantities of essential elements in plant tissue are shown. Compare with the table in appendix 4, where elements in *dry* plant tissues are given.

TABLE 2.14

Average fresh-water plant composition on *wet* basis

Element	Plant content (%)
Oxygen	80.5
Hydrogen	9.7
Carbon	6.5
Silicon	1.3
Nitrogen	0.7
Calcium	0.4
Potassium	0.3
Phosphorus	0.08
Magnesium	0.07
Sulphur	0.06
Chlorine	0.06
Sodium	0.04
Iron	0.02
Boron	0.001
Manganese	0.0007
Zinc	0.0003
Copper	0.0001
Molybdenum	0.00005
Cobalt	0.000002

The present concern about eutrophication relates to the rapidly increasing amount of phosphorus and nitrogen, which are normally present at relatively low concentrations. Of these two elements phosphorus is considered the major cause of eutrophication of lakes, as it was formerly the growth-limiting factor for algae in the majority of lakes but its usage has greatly increased during the last decades. Nitrogen is a limiting factor in number of East African lakes as a result of the nitrogen depletion of soils by intensive erosion in the past. However, today nitrogen may become limiting to growth in lakes as a result of the tremendous increase in the phosphorus concentration caused by discharge of waste water, which contains relatively more phosphorus than nitrogen. While algae use 4-10 times more nitrogen than phosphorus, waste water generally contains only 3 times as much nitrogen as phosphorus. Furthermore, nitrogen accumulates in lakes to a lesser extent than phosphorus and a considerable amount of nitrogen is lost by denitrification (nitrate to N_2).

The importance of phosphorus in the eutrophication of lakes is shown in Fig. 2.22, where the maximum algal concentration, expressed as μg chlorophyll-a per l, is plotted against the phosphorus concentration for several lakes. The correlation is obvious.

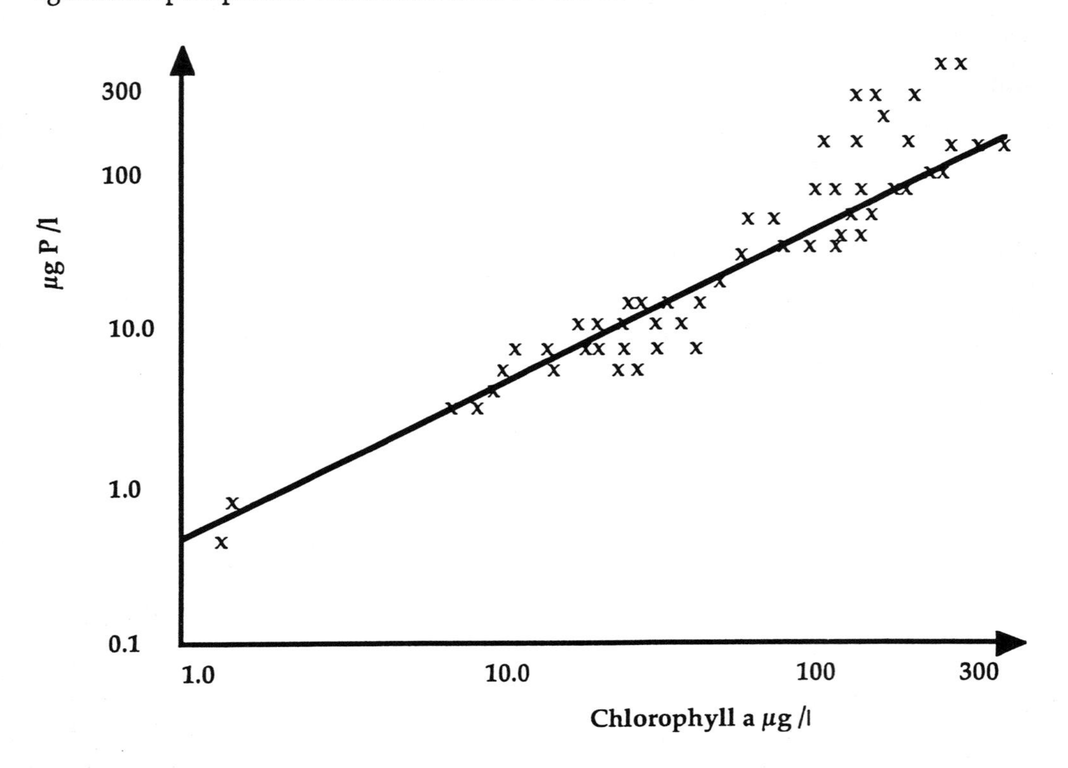

Fig. 2.22 Algae biomass (summer max) versus ortho-P for 56 lakes in England and Denmark.

From a thermodynamic point of view a lake can be considered an open system which exchanges material (waste water, evaporation, precipitation) and energy (evapora- tion, radiation) with the environment.

However, in many large lakes the input of material per year is unable to change the concentration significantly. In such cases the system can be considered as (almost) closed, which means that it is exchanging energy but not material with the environment. **The flow of energy through an ecosystem causes at least one cycle of material in the system** (see Morowitz, 1968).

The important elements all participate in process cycles. These cycles contain processes that determine eutrophication. The growth of phytoplankton is the key process eutrophication and it is therefore of great importance to understand the interaction processes regulating its growth which requires that the entire cycle is considered. A cycling of elements in lakes is illustrated for phosphorus and nitrogen in Figures 2.23 - 2.24 (reproduced from Jørgensen, 1994).

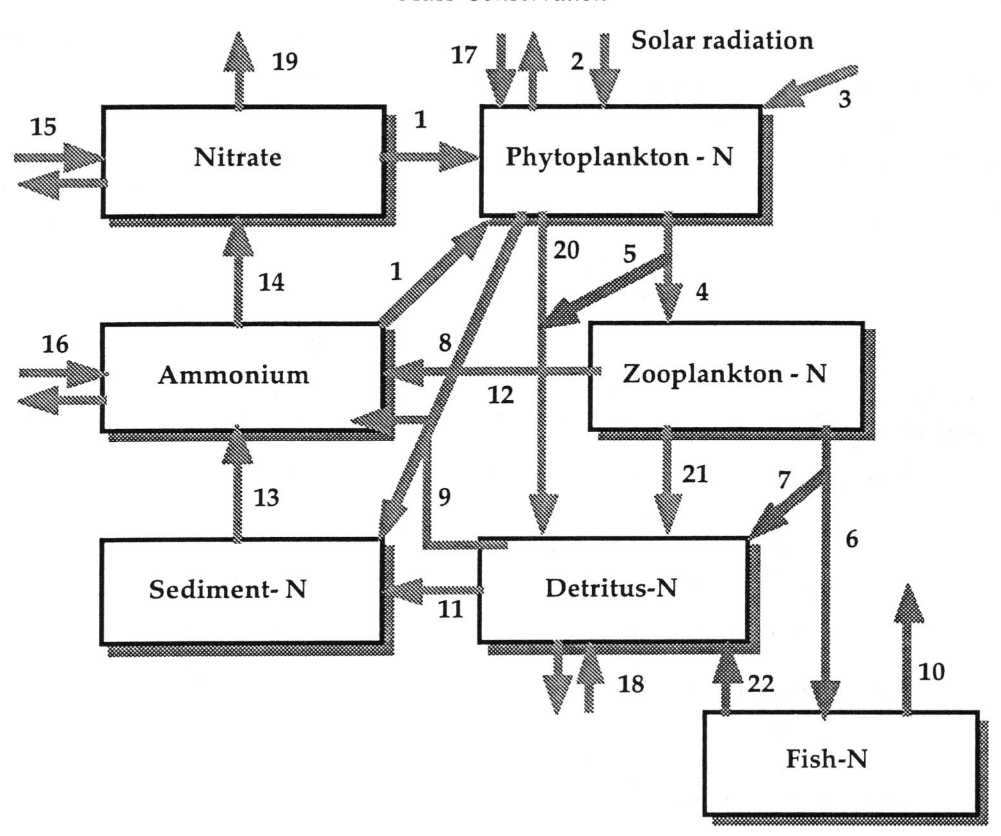

Fig. 2.23. The nitrogen cycle. The processes are: (1) Uptake of NO_3^- and NH_4^+ by algae, (2) Photosynthesis, (3) Nitrogen fixation, (4) Grazing with loss of undigested matter, (5), (6) is predation with loss of undigested material, (7), (8) Settling of phytoplankton (9) Mineralisation, (10) Fishery (11) Settling of detritus, (12) Excretion of ammonia by zooplankton (13) Release from sediment, (14) Nitrification, (15), (16), (17) and (18) are Inputs/outputs, (19) Denitrification. (20) (21) and (22) represent mortalities.

Primary production has been measured in great detail in many great lakes. This process represents the synthesis of organic matter, and can be summarised as follows:

$$\text{light} + 6CO_2 + 6H_2O \longleftrightarrow C_6H_{12}O_6 + 6O_2 \qquad (2.53)$$

This equation is necessarily an oversimplification of the complex metabolic pathway of photosynthesis, which is dependent on sunlight, temperature and the concentration of nutrient. The composition of phytoplankton is not constant (note that Table 2.12 only gives an average concentration), but to a certain extent reflects the chemical composition of the water. If, e.g. the phosphorus concentration is high, the phytoplankton will take up relatively more phosphorus - the luxury uptake.

Phytoplankton consists mainly of carbon, oxygen, hydrogen, nitrogen, phosphorus, and sulphur and without these elements no algae growth will take place. So each of these elements represents a limiting factor on algae growth. Another side of the problem is the consideration of nu- trient sources. It is important to set up mass balances for the most essential nutrients.

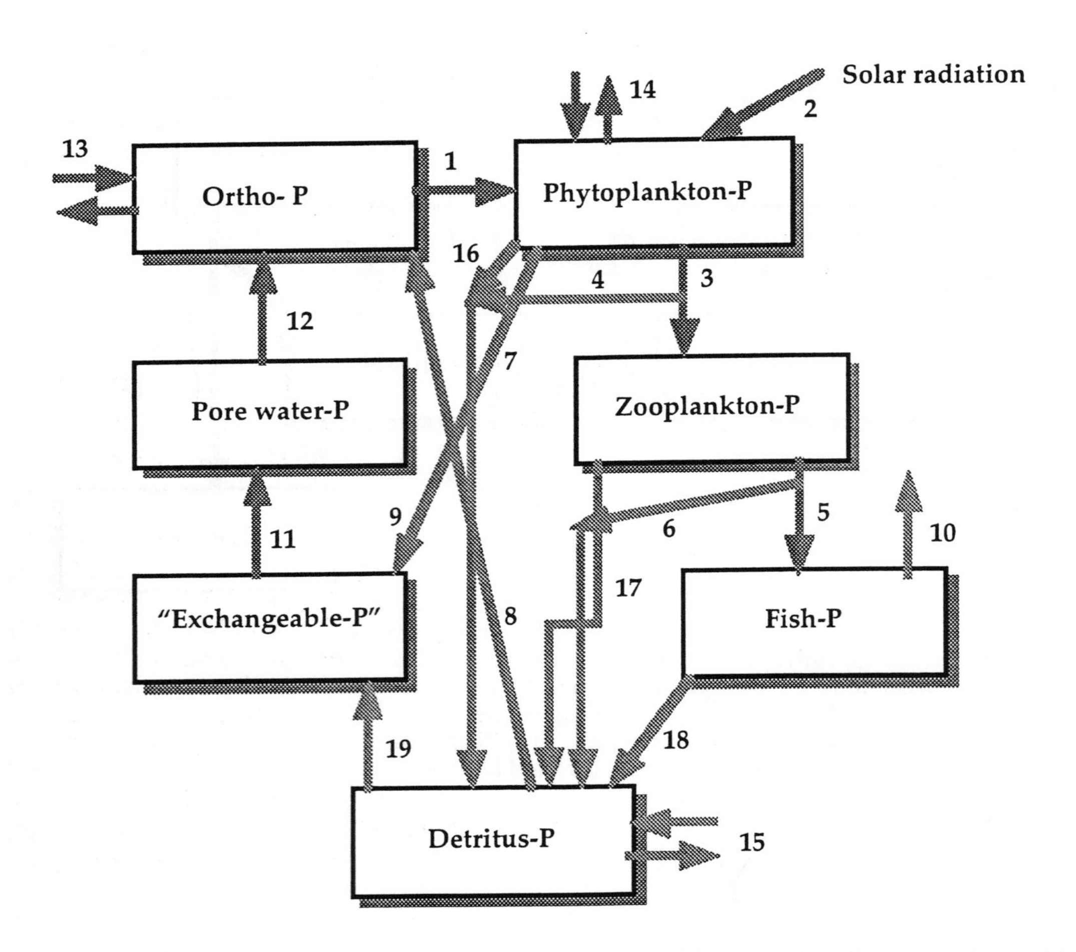

Fig. 2.24. The phosphorus cycle. The processes are: (1) Uptake of phosphorus by algae, (2) Photosynthesis, (3) Grazing with loss of undigested matter, (4), (5) is predation with loss of undigested material, (6), (7) and (9) Settling of phytoplankton (8) Mineralisation, (10) Fishery (11) Mineralisation of phosphorous organic compounds in the sediment, (12) Diffusion of pore water P (13) (14) and (15) are Inputs/outputs, (16), (17) and (18) represent mortalities and (19) is settling of detritus.

The sequence of events leading to eutrophication often occurs as follows. Oligotrophic waters often have a ratio of N:P greater than or equal to 10, which means that phosphorus is less abundant relative to the needs of phytoplankton than nitrogen. If sewage is discharged into the lake the ratio will decrease since, the N:P ratio for municipal waste water is about 3:1, and consequently nitrogen will be less abundant than phosphorus relative to the needs of phytoplankton. Municipal waste water contains typically 30 mg l^{-1} N and 10 mg l^{-1} P. In this situation, however, the best remedy for the excessive algal growth is not necessarily to remove nitrogen from the sewage, because the mass balance might show that nitrogen-fixing algae would produce an uncontrollable input of nitrogen into the lake.

It is necessary to set up a mass balance for the nutrients. This will often reveal, that the input of nitrogen from nitrogen-fixing blue green algae, precipitation and tributaries is already contributing too much to the mass balance for any effect to be produced by nitrogen removal from the sewage. On the other hand the mass balance may reveal that most of the phosphorus input (often more than 95%) comes from the sewage, and so demonstrates that it is better management to remove phosphorus from the sewage rather than nitrogen. It is, therefore **not important which nutrient is limiting, but which nutrient can most easily be made to limit the algal growth.**

These considerations have implied that the eutrophication process can be controlled by a reduction in the nutrient budget. For this purpose a number of eutrophication models have been developed, which take a number of processes into account. For details, see Jørgensen (1979), Jørgensen (1978), Jørgensen (1980), Orlob (1981), Jørgensen (1981), Jørgensen (1994) and Jørgensen et al. (1995). It will suffice here to consider the so-called Vollenweider plot (Vollenweider, 1969), which is much simpler to use than dynamic ecological models, but as it does not consider the dynamics of the phytoplankton population, the annual variation, the sediment and its interaction with the water body, it can only give a crude picture of the possible control mechanisms in existence. The plot is shown in Fig. 2.25, where the phosphorus loading in g/m^2 year is related to the depth of water. The diagram consists of three areas corres- ponding to oligotrophic, mesotrophic and eutrophic lakes.

A similar plot can be constructed for nitrogen and a comparison of the two diagrams can approximately show whether a possible reduction in the nitrogen loading would be a better management solution than a reduction in the phosphorus loading. Vollenweider has later (1975) improved these considerations by taking input, output and the net loss to the sediment into consideration and by using a correction factor for stratified lakes, but in cases where these improvements are required it is better to use a dynamic ecological model. Under all circumstances Vollenweider's plot should be used as a first approximation. Lakes can be classified in accordance with their primary production - the so-called oligotrophic-eutrophic series - which is shown in Table 2.15.

Typical oligotrophic lakes are deep, with the hypolimnion larger than the epilimnion. Littoral plants are scarce and the plankton density is low, although the number of species can be large. Due to the low productivity the hypolimnion does not suffer from oxygen depletion. The nutrient concentration is low and plankton blooms are rare, so the water is highly transparent.

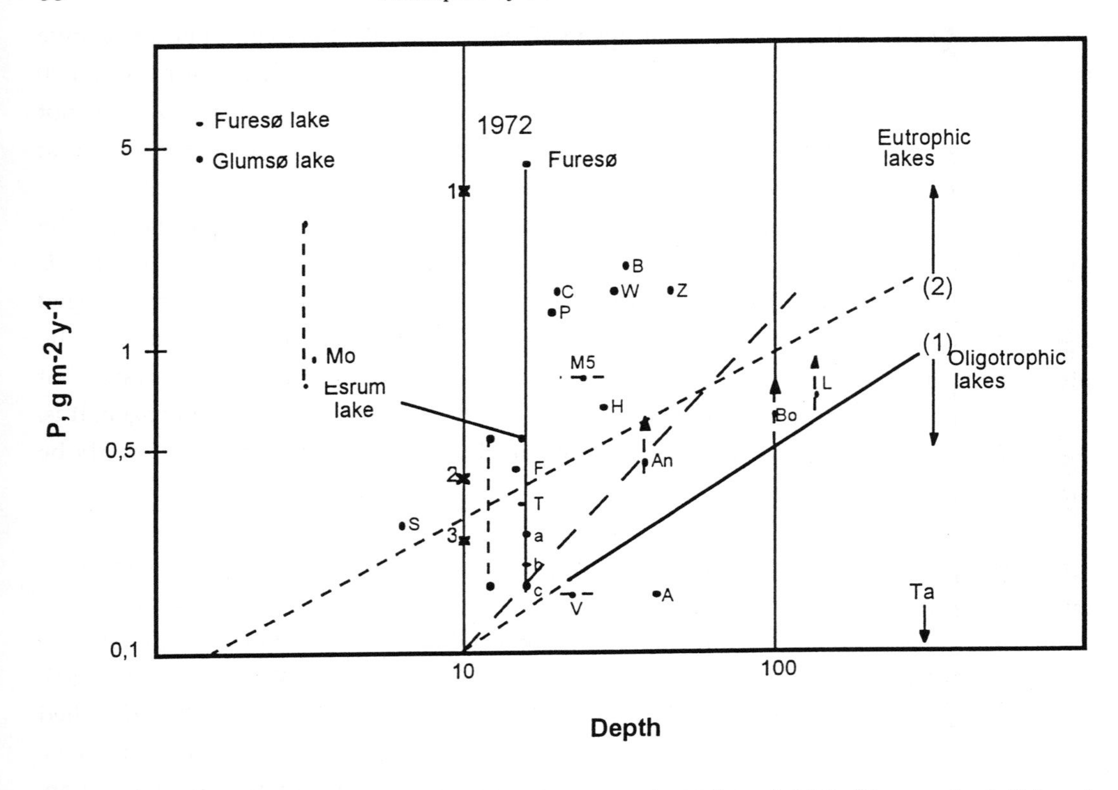

Fig. 2.25 a, b and c correspond to removal of 90%, 95% and 99% (Furesoe) of P input respectively. Glumsoe, 1972 (Jørgensen, Jacobsen and Høi, 1973):, Lyngby lake, 1972 (F.L. Smidth/MT, 1973); Esrom lake, 1972.

G	-	Greifensee	L	-	Lac Leman	An	-	Lac Annecy
P	-	Pfäffikersee	V	-	Vänern	Ta	-	Lake Tahoe
B	-	Baldeggersee	M	-	Lake Mendota	Bo	-	Bodensee
F	-	Furesoe (1954)	T	-	Türlersee Mo	-	Lake Moses	
H	-	Halwillersee	A	-	Aegerisee			
W	-	Lake Washington	Z	-	Zürichsee			

Example 2.3

The phosphorus input to Zürichsee (mid seventies; see Figure 2.26) was 90% due to waste water and 10% due to non-point sources. Which phosphorus removal efficiency is necessary to ensure that Zürichsee becomes mesotrophic?

Solution

The loading at that time is about 1.8 g/ m^2 year. It is necessary to reduce it to about 0.5 g/ m^2 to ensure mesotrophy. 0.18 g/ m^2 out of the total loading cannot be reduced, as it is coming from non-point sources. It means that is necessary to reduce 1.62 g/ m^2 to 0. 32 g/ m^2, which corresponds to an efficiency of 1.3x100 / 1.62 = 80.2%

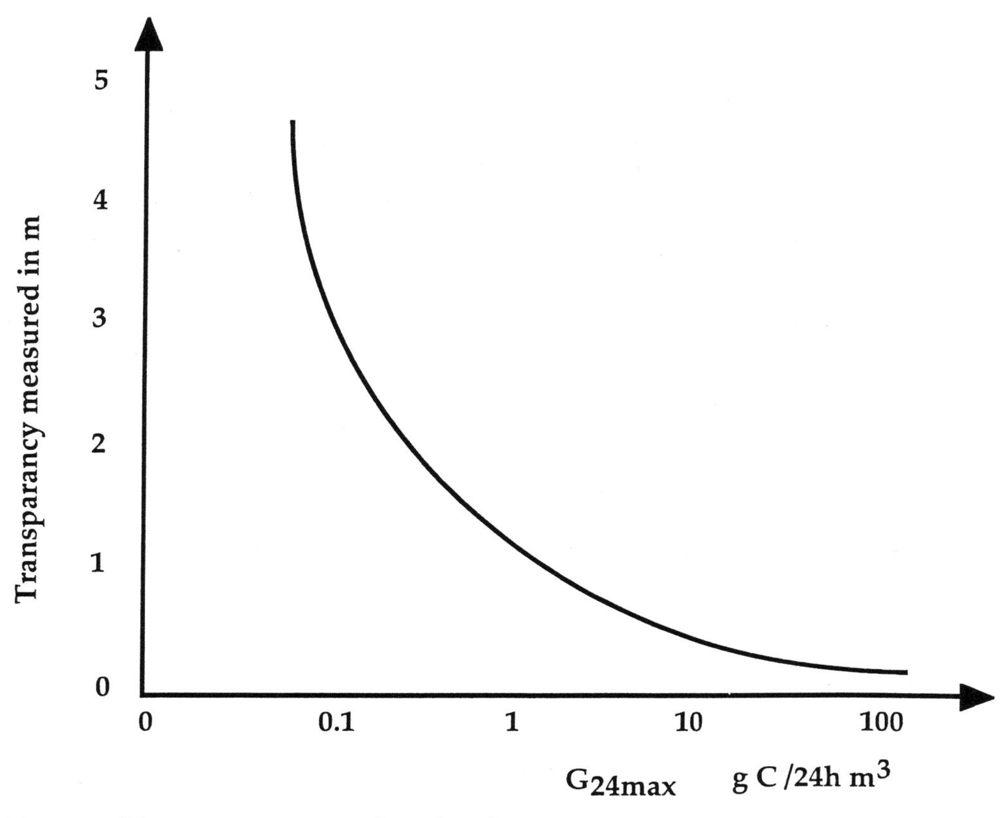

Fig. 2.26. The transparency, v, plotted to G_{24max}.

An approximate relationship between productivity and transparency is shown in Fig. 2.26. Here the transparency in m is plotted against the maximum G_{24}-value (the maximum production in mg per 24h and per m^3). This relationship does not always general apply, as the transparency is dependent not only on the phytoplankton concentration, but also on the concentration of inorganic suspended matter (e.g. clay) and the colour of the water (very humic rich lakes are brownish), but in most lakes the transparency is mainly determined by the phytoplankton. Eutrophic lakes are generally shallower and have a higher phytoplankton concentration, and thus a generally lower transparency. Littoral vegetation is abundant and summer and spring algal blooms characteristic. Another characteristic difference between oligotrophic and eutrophic lakes is the profile of the change in photosynthesis with depth (see Fig. 2.27).

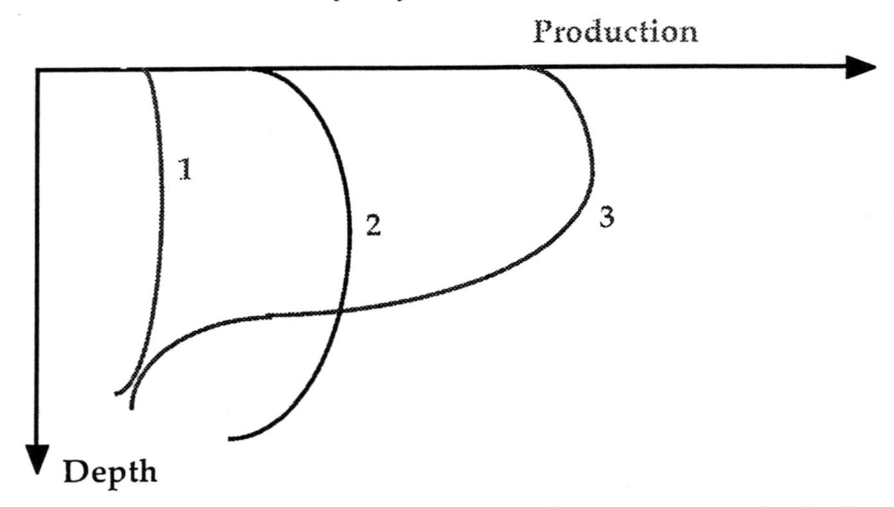

Figure 2.27. Productivity of phytoplankton per unit of volume verusus the depth in a series of lakes. (1) ologotrophic (2) mesotrophic and (3) eutrophic.

TABLE 2.15
Classification of lakes

Trophic type	Mean primary productivity (mg C/m^2/ day)	Phytoplankton density (cm^3/m^3)	Phytoplankton biomass (mg C/m^3)	Chlorophyll (mg/mg^3)	Dominant phytoplankton
Ultra-oligotrophic	< 50	< 1	< 50	0.01-0.5	
Oligotrophic	50-300		20-100	0.3-3	Chrysophyceae, Cryptophyceae,
Oligo-mesotrophic		1-3			Dinophyceae, Bacillario-phyceae
Mesotrophic	250-1000		100-300	2-15	
Meso-eutrophic		3-5			
Eutrophic	> 1000		> 300	10-50	Bacillario-phyceae Cyanophyceae,
Hypereutrophic		> 10			Chlorophyceae, Euglenophyceae
Dystrophic	< 50-500		< 50-200	0.1-10	

TABLE 2.15, continued

Trophic type	Light extinction coefficients (n/m)	Total organic carbon (mg/l)	Total P (µg/l)	Total N (µg/l)	Total inorganic solids (mg/l)
Ultra-oligotrophic	0.03-0.8		< 1-5	< 1-250	2-15
Oligotrophic	0.05-1.0	< 1-3			
Oligo-mesotrophic			5-10	250-600	10-200
Mesotrophic	0.1-2.0	< 1-5			
Meso-eutrophic			10-30	500-1100	100-500
Eutrophic	0.5-4.0	5-30			
Hypereutrophic			30- 5000	500- 15000	400-60000
Dystrophic	1.0-4.0	3-30	< 1-10	< 1-500	5-200

Much effort has been made to develop empirical models which can predict the retention of nitrogen and phosphorus in lakes. By using such a model we will be able to predict how the in-lake concentrations of an element are affected by changes in loading, in hydraulic regime and possibly also by changes in morphometry, provided that we have a loading estimation and a water balance. The nutrient loadings can be estimated by use of Table 5.16 (non-point sources), supplemented with the Danish estimations in Table 2.16, Table 5.4 (untreated waste water) combined with removal efficiencies for the treatment plants (Table 5.3 may be used) and Table 5.17 (rain water).

A simple steady state model can be derivated from equation (2.8):

$$QC_{io} = C_i (Q + Vk) \tag{2.54}$$

V is replaced by zA, where z is the depth and A the surface area of the lake. k represents the specific removal rate expressed in the unit 1/24h. This means that for phosphorus k represents the specific settling rate and for nitrogen the settling rate + the denitrification rate coefficient. By reorganisation of the equation, we get (tr = Q/V; see Section 2.3):

$$C_i (Q + Vk) / V = QC_{io} / V = C_i (tr + k) = L / z \tag{2.55}$$

$$\text{or } C_i = L / z(tr + k) \tag{2.56}$$

where L is the specific loading = QC_{io} / A. This is the sum of the loadings from different sources: river discharge, direct point sources, seepage and precipitation. This is the classical steady state model developed by Vollenweider. It is the basis for the considerations behind Figure 2.27. The hydraulic residence time, tr, is usually known, while the problem often is to determine k. The settling means, however, not total removal of the nutrients, as they may be partly released from the sediment again. This implies that the concentration of phosphorus may be expressed as a result of a longer retention time than corresponding to the hydraulic residence time, tr. It may be advantageous in this context to use R defined as the fraction of the loading that is retained in the lake and later washed out by the outlet: R = C_i V tr /L A = C_i z tr /L . We have:

$$C_i = L\,R\,/\,z\,tr \qquad (2.57)$$

TABLE 2.16
Range of run-off area coefficients and concentrations of nitrogen and phosphorus from various types of areas in Denmark

	Area coefficient kg /(ha*y)		Concentration weighted by water flow mg/l	
Nitrogen				
reference areas	1.9	2.9	1.6	1.9
cultivated areas				
without point sources	9.3	11.6	5.8	7.1
with point sources	12.8	14.1	7.2	2.9
with fish farms	12.6	13.4	3..4	3.4
Phosphorus				
reference areas	0.05	0.007	0.06	0.06
cultivated areas	0.21	0.27	0.14	0.16
fish-farm areas	0.76	0.78	0.17	0.18
waste water areas	0.44	1.46	0.33	1.12

For 94 Danish lakes covering 131 annual mass balances, 21 emperical retention models were tested. R was expressed as a function of hydraulic loading q = Q/A or hydraulic retention (residence) time, tr and of the average inlet concentration of phosphorus P. The lakes tested with the models represent a range of morphometry, hydrology and loading, as shown in Table 2.18.

Of the 21 models tested, four gave a better description than the remaining 17.

TABLE 2.17

Phosphorus load models tested with data from Danish lakes. All models express the retention coefficient, R, in terms of other parameter (for References and the model tests see Kristensen et al., 1990).

Number +Reference	Equation
1. Kirchner and Dillon (1975)	$R = 0.4088 \exp(-0.2899\ q) + 0.5912 \exp(-0.01019\ q)$
2. Ostrofsky (1978)	$R = 0.201 \exp(-0.0425q) + 0.574 \exp(-0.00949q)$
3. Vollenweider	$R = 10 / (10+q)$
4. Chapra (1975)	$R = 16 / (16+q)$
5. Chapra (1975)	$R = 12.4 / (12.4+q)$
6. Dillon and Kirchner (1975)	$R = 13.2 / (13.2+q)$
7. Ostrofsky (1978)	$R = 24/(30+q)$
8. Canfield and Bachmann (1981)	$R = 5.3 / (5.3+q)$
9. Nürnberg (1984)	$R = 15 / (18 + q)$
10. Reckhow (1979)	$R = (11.6 + 0.2q) / (11.6 + 1.2q)$
11. Canfield and Bachmann (1981)	$R = (2.99 + 1.7q) / (2.99+ 2.7q)$
12. Prairie (1988)	$R = (0.11+ 0.18q) / (1 + 0.18q)$
13. Prairie (1989)	$R = (0.25 + 0.18q) / (1 + 0.18q)$
14. Canfield and Bachmann (1981)	$R = 0.129tr^{0.451}P_{io}^{0.549}(1 + 0.129tr0.451P_{io}^{0.549})$
15. Rognerud et al. (1979)	$R = 1 - 0.63 \exp(-0.067tr)$
16. Berge (1987)	$R =1 - 0.436tr^{-0.16}$
17. Vollenweider (1976)	$R = 1 / (1 + tr^{-0.5})$
18. OECD (1982) final	$R= 1 - 1.55P_{io}^{-0.18} / (1 + tr^{0.5})^{0.82}$
19. OECD (1982) Nordic	$R= 1 - 1.12P_{io}^{-0.08} / (1 + tr^{0.5})^{0.92}$
20. OECD (1982) shallow	$R= 1 - 1.02P_{io}^{-0.12} / (1 + tr^{0.5})^{0.88}$
21. Frisk et al. (1980,1981)	$R = P_{io}tr / (30 + P_{io}tr)$

However, the medians of the percentage deviation between observed and evaluated R values were still large (33-41%). It appeared that in 24 % of the investigated lakes a negative R was measured and most of these lakes had an extremely low residence time (less than 19 days). In lakes with low residence time a negative R can be found because the steady state between loading and loss is never achieved, and a systematic error will occur if we use in-lake concentrations, the reason being that residence time, inlet concentrations and in-lake concentrations vary over the year.

Corresponding models for the relation between nitrogen loading and in-lake nitrogen concentrations are in principle more difficult to develop, since internal losses occur both from sedimentation and from denitrification, whereby nitrate is reduced to gaseous dinitrogen oxide and nitrogen escaping from the lake. For 69 Danish lakes (see Table 2.19), mass balances allowed concentrations to be measured of both absolute and relative loss from net sedimentation and by denitrification; see Table 2.20. The denitrification is calculated as the difference between total loss and sedimentary loss, calculated by multiplying the sedimentary phosphorus loss by the N/P ratio in the surface sediment:

$$N_{den} = N_{loss} - P_{loss} \times (N/P)_{surf.\ sed} \qquad (2.58)$$

Table 2.20 shows that on average 43% of the nitrogen loading was lost as sedimentation and denitrification. This is clearly a much higher percentage than the 25% loss of the phosphorus loading. The "extra" loss by denitrification amounted to 77% of the total loss.

Figure 2.28 shows the relation between the average inlet concentration of nitrogen and average in-lake concentration. Lakes with short residence times suffer the smallest loss.

TABLE 2.18
Statistical characteristics of the 131 Danish lakes applied for testing the models presented in Table 2.15.

Characteristics	average	std. dev.	min.	25%	median	75%	max.
lake area (km^2)	2.4	0.6	0.04	0.2	0.5	1.5	41
mean depth (m)	4.0	0.3	0.3	1.6	2.8	5.3	15.4
residence time, t_r (y)	1.1	0.2	0.004	0.05	0.26	1.35	20
q (m/y)	39.3	7.8	0.2	3.8	11.9	35.3	605
L (g P / (m^2y)	13.1	2.5	0.06	1.0	2.5	12.4	217
inlet conc. (μg P/l)	415	39	34	164	260	456	2396
conc. in lake (μg P/l)	301	35	16	99	162	398	3130

A test was made of three different models describing the relation:

$$N_{lake} = 0.45\ N_{in} \qquad (2.59)$$

$$N_{lake} = 0.42\ N_{in}\ tr^{-0.11} \quad (2.60)$$

$$N_{lake} = 0.42\ N_{in}\ tr^{-0.14}\ z^{0.17} \quad (2.61)$$

where N_{lake} is the annual mean of the in-lake concentration of total nitrogen, N_{in} is the annual mean in the incoming water, tr is the hydraulic residence time (years) and z is the average depth (m).

TABLE 2.19

Morphometrical data and total nitrogen concentrations (yearly basis) for 69 lakes, Kristensen et al., 1990.

Characteristics	average	median	std.dev.	min	max
lake area (km^2)	3.3	0.7	1.0	0.1	41
mean depth (m)	5.1	2.3	0.5	0.6	16
max.depth (m)	8.6	5.0	1.1	1.0	37
residence time, tr (y)	1.2	0.3	0.3	<0.1	14
inlet conc. (mg N/l)	5.6	5.0	0.8	0.6	15
conc. in lake (mg N/ l)	2.8	2.5	0.2	0.5	9

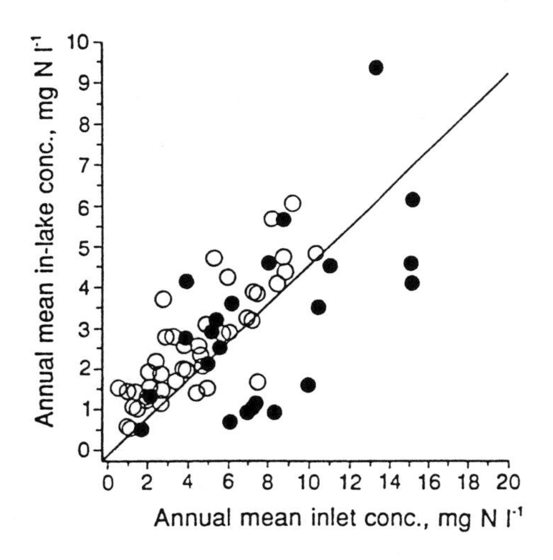

Fig. 2.28 The relation between in-lake and inlet concentration of nitrogen, both given annual averages. o: lakes with residence time less than one year •: lakes with residence time greater than one year.

TABLE 2.20

Nitrogen load, nitrogen loss and denitrification in the 69 lakes, cf. Table 4.17. Percent loss was calculated for each lake before the average was taken.

	average	median	n	std.dev.
nitrogen load (g N /(m^2 y)	142	52	69	35
nitrogen loss (g N /(m^2 y)	29	17	69	4
nitrogen loss (5)	43	41	69	4
denitrification (g N /(m^2 y)	23	16	58	3
denitrification (%)	33	30	58	3

TABLE 2.21

Six models of the relation between inlet and in-lake concentrations of nitrogen were tested with data from 98 Danish lakes. The table summarises the results and gives the percentage deviation in % of observed values. For additional references see Kristensen et al., 1990.

Model	mean	std.dev.	median	P(mean=0)
Equation (2.59)	-2.3	9.0	17.0	<0.257
Equation (2.60)	-4.1	6.5	7.0	<0.531
Equation (2.61)	-7.9	6.9	0.3	<0.802
Lijklema et al., 1989	-51	9.0	-25.0	<0.004
Bachmann, 1984	-153	118	94	<0.200
QECD, 1982	-17.1	8.9	-8.3	<0.001

Compared with the performance of other models (Table 2.21) the deviation between observed and calculated in-lake concentrations were reasonably small for the three models listed above. The reason shy the other models underestimated the nitrogen losses is probably that Danish lakes are rather uniform with respect to loading, residence time and depth, whereas the data used to calibrate the other models include data from lakes with smaller loading, longer residence time and greater depth.

The next step in the predictive modelling of a eutrophication response is the development of empirical relations between phytoplankton and nutrient level. A variety of such empirical models have been described in the literature and they are often quite different due to geographical differences. For example, lakes at higher latitudes have relatively more chlorophyll due to adaptation to lower insolation and different chlorophyll- extraction procedures may been used (Fig. 2.29). Therefore a direct relation between nutrient level and transparency depth might be more appropriate.

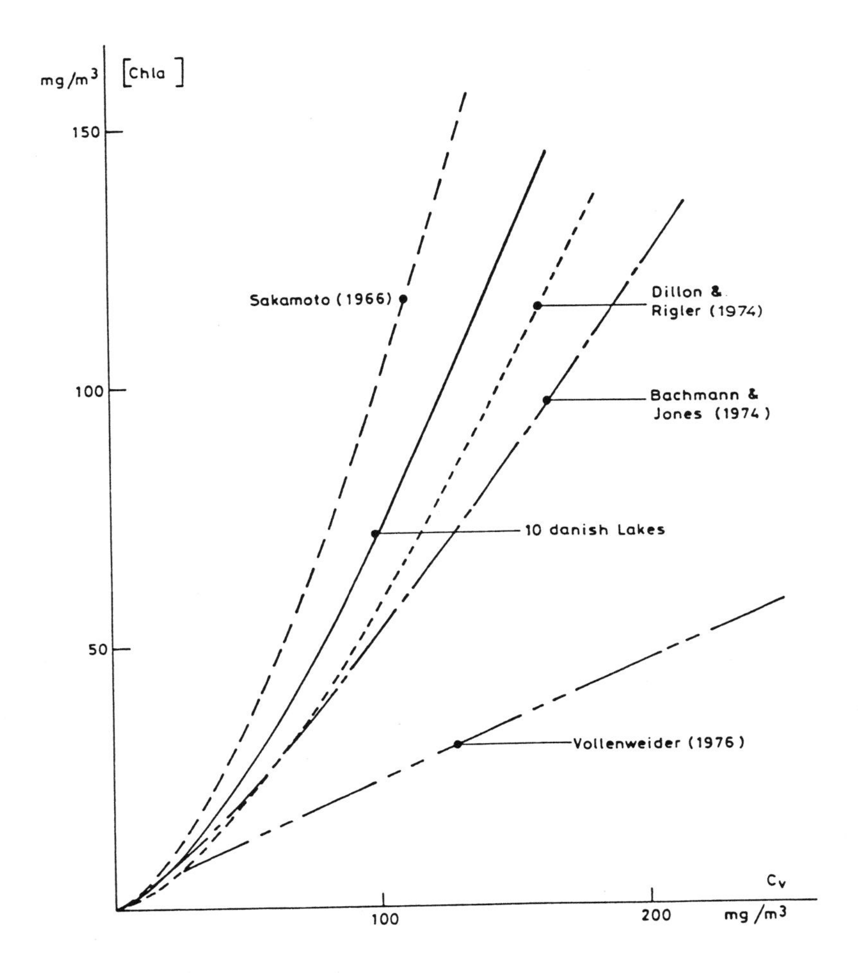

Fig. 2.29 Some emperical relations between summer chlorophyll concentration (chla) and annual average in lake phosphorus concentration; C_V, after Kamp Nielsen, 1986.

For 252 Danish lakes covering 498 measuring years (Table 2.22), the relation shown in Figure 2.30 was found. The relation is logarithmic because the transparency depth, z_{eu} (m) is related to the extinction coefficient n (1/m) by

$$n = 1.7 / z_{eu} \tag{2.62}$$

The extinction coefficient is found by measuring the light attenuation over depth:

$$I_z = I_o e^{-nz} \tag{2.63}$$

where I_z is the irradiance at depth z, and I_o is the irradiance at the surface. By taking logarithms and rearranging we can rewrite the equation as

$$nz = \ln I_o - \ln I_z \tag{2.64}$$

TABLE 2.22
Statistics of data from 252 Danish lakes, applied to find the relation in Figure 2.31. Asterisk means whole-year averages.

Characteristics	average	median	std. dev.	min.	max
lake area (km^2)	2.3	0.6	0.4	<0.01	41
mean depth (m)	4.0	2.6	0.2	0.5	15
max. depth (m)	8.7	5.5	0.9	1.0	34
residence time (y)	1.8	0.4	0.3	<0.01	21
mg total P/l	0.4	0.2	0.03	0.01	2
mg total P/l*	0.4	0.2	0.04	0.02	3
mg total N/l	2.3	2.0	0.09	0.2	9
mg total N/l*	2.9	2.3	0.11	0.2	10
transparency depth (m)	1.1	0.8	0.06	0.2	5
transparency depth (m)*	1.2	1.0	0.05	0.2	5
chlorophyll a (μg/l)	89	74	5.0	5.0	308
suspended matter (mg/l)	22	18	1.7	3.4	73

By inspection of the data material, an influence of the lake depth was found; shallow lakes had generally less transparency depth due to resuspension of sediments. A set of empirical models was developed, relating summer values, annual values, or combinations of the two (Table 2.23). In a typical Danish shallow lake there is often a distinct nitrogen limitation in lake summer and one would expect improved relationships if the lakes were screened for nitrogen limitation. However, separating the lakes according to the N/P ratio being above 10 (indicating P-limitation) or below 7 (indicating N-limitation) did not significantly change the relation between P and transparency depth (see the last two equations in Table 2.23).

Another indicator of a true P-limitation is the low P-concentration in the production period, but again no significant effect of this screening was found (Fig. 2.31). In another investigation, comparing a number of European lakes with respect to recovery, the screening for N-limitation by both N/P ratio and lower P concentrations in summer gave a significant difference in response direction (Fig. 2.32).

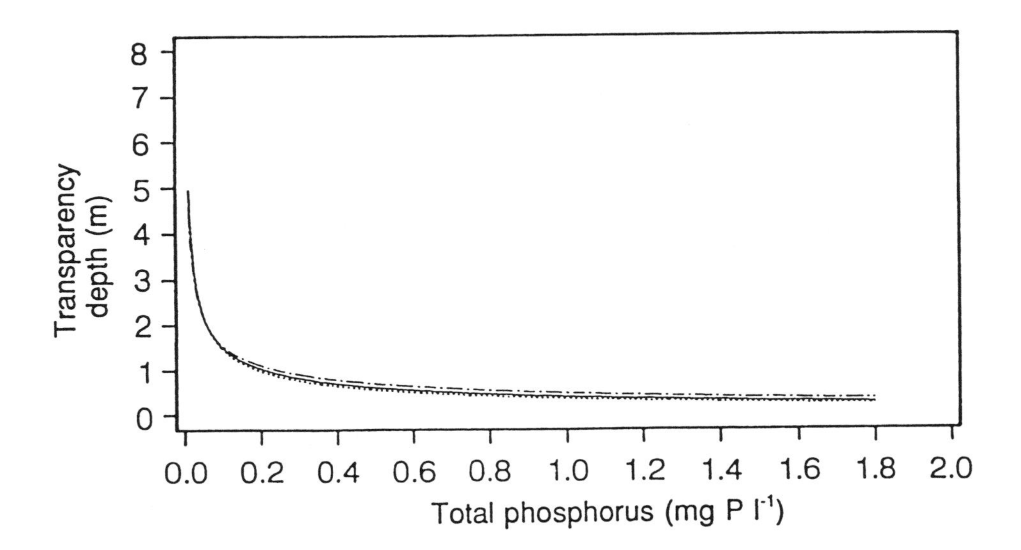

Fig. 2.30. The relationship between transparency depth, z_{eu}, and total P in a survey of 252 Danish lakes. Solid curve: both z_{eu} and P are summer averages; dotted curve: both z_{eu} and P are annual averages; dot-and-dash curve: z_{eu} is the summer average and P is the annual average.

Quantitative changes in phytoplankton biomass are accompanied by qualitative changes during the eutrophication process. Blue green algae and green algae become more and more dominant with increasing phosphorus level; see Fig. 2.35. Especially the blue greens are a nuisance since they float to the surface in competition for light. Here they are killed due to their high susceptibility to light inhibition; during their decomposition an unpleasant smell is developed and more seriously, some of the species are toxic to animals. Some filamentous blue greens are able to fix atmospheric nitrogen and reduce it to ammonia in specially developed cells (heterocysts) and it is generally believed that this nitrogen fixation is favoured by low nitrogen levels and low N/P ratios. But the material from Danish lakes show that the total phosphorus was more probably regulating the abundance of both non-N-fixing and N-fixing blue green algae, Figure 2.33. It is also remarkable that at very high P-concentrations the small green algae become dominant.

TABLE 2.23

Modelling the relationship between average transparency depth, z_{eu}, and total P in lake concentration.

Number	Equation*)
1	$z_{eu} = 0.44\ (+/-\ 0.038)\ P^{-0.54(+/-0.031)}$
2	$z_{eu} = 0.36(+/-\ 0.029)\ P^{-0.29(+/-0.028)}z^{0.51(+/-0.042)}$
3	$z_{eu} = 0.39(+/-\ 0.038)\ P^{-0.58(+/-0.034)}$
4	$z_{eu} = 0.34(+/-\ 0.028)\ P^{-0.29(+/-0.028)}z^{0.55(+/-0.040)}$
5	$z_{eu} = 0.52\ (+/-\ 0.042)\ P^{-0.48(+/-0.031)}$
6	$z_{eu} = 0.43\ (+/-\ 0.026)\ P^{-0.20(+/-0.022)}z^{0.55(+/-0.030)}$
7	$z_{eu} = 0.40\ (+/-\ 0.055)\ P^{-0.69(+/-0.064)}$
8	$z_{eu} = 0.34\ (+/-\ 0.0424\ P^{-0.60(+/-0.041)}$

*)Equations 1, 2, 3 and 4 are based on the summer mean and equations 5, 6, 7 and 8 on the annual mean of z_{eu}, while P represents the summer mean in equations 1 and 2 and the annual mean in the other equations for the total lake phosphorus concentration. For equations 7 and 8, the data were divided into two sets according to the N:P ratio being either > 10 or < 10 respectively. Kristensen et al., 1990.

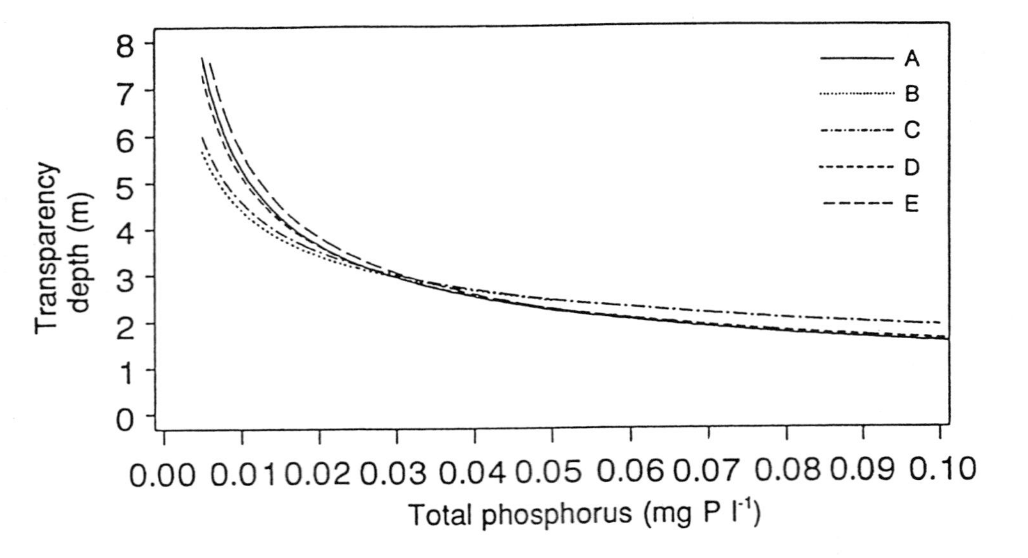

Fig. 2.31. The relationship between summer-average transparency depth, z_{eu}, and summer total phosphorus concentration, P, according to various maximum limits of P. A: all observations, B: P < 0.4 mg P/l, C: P< 0.2 mg P/l, D: P < 0.1 mg P/l and E: P<0.05 mg P/l. After Kristensen et al., 1990.

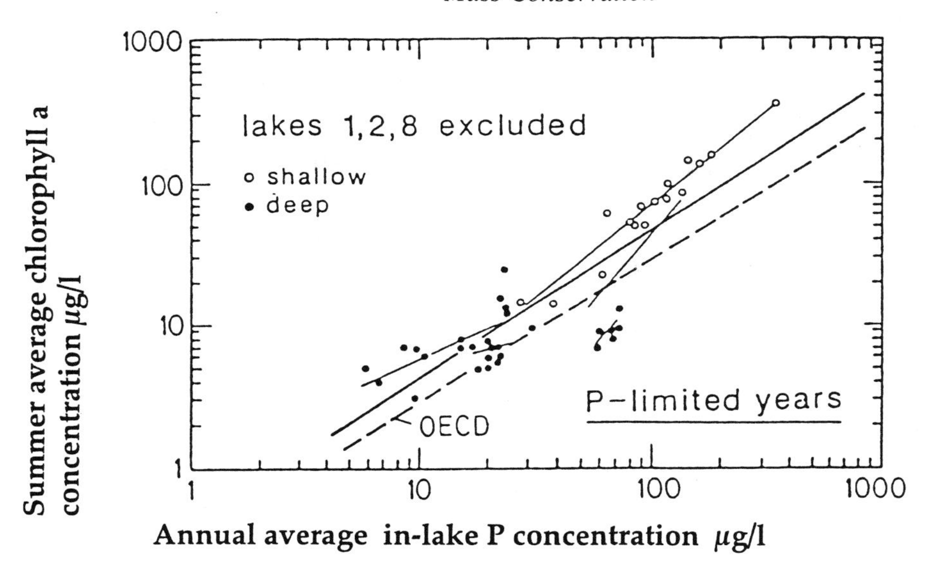

Fig. 2.32. Results of a regression analysis of the relationship between growth-seasonal chlorophyll-a and the in-lake annual average phosphorus concentration, performed on data screened for P-limitation to occur. Source: Kamp-Nielsen, 1989.

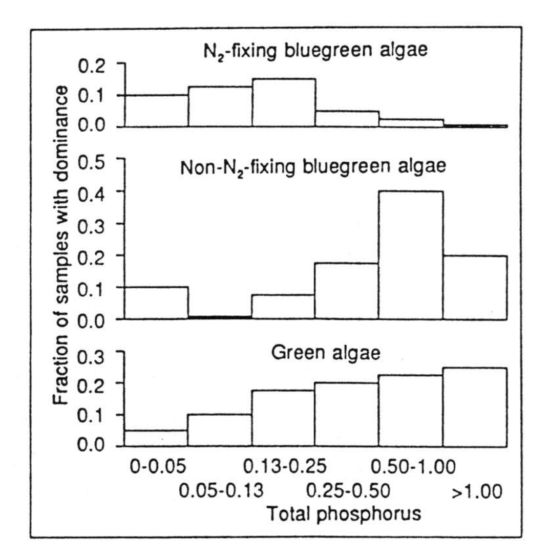

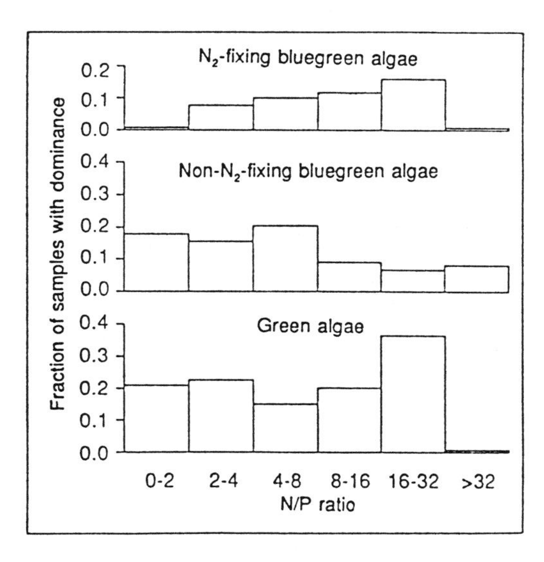

Fig. 2.33. Left: the fraction of samples with dominance of 1) blue green algae with nitrogen fixation 2) blue green algae without nitrogen fixation 3) green algae plotted versus average P-concentration, May-September. Right: the same plotted versus the N/P ratio. After Kristensen et al., 1990.

The competitive power of the blue greens is low at high light intensities and it has been suggested that a better description of blue-green algae abundance could be achieved by correcting the phosphorus dependence by the ratio z_{eu} / z_{mix}, the ratio between the photic depth (2.3xthe transparency depth) and the mixing depth (in non-stratified lakes >= the total depth). Figure 2.34 suggests that such a relationship holds and that a significant difference exists between deep lakes and shallow lakes, but there is no indication that their relative abundance is decreased at higher phosphorus levels as shown in the Danish material.

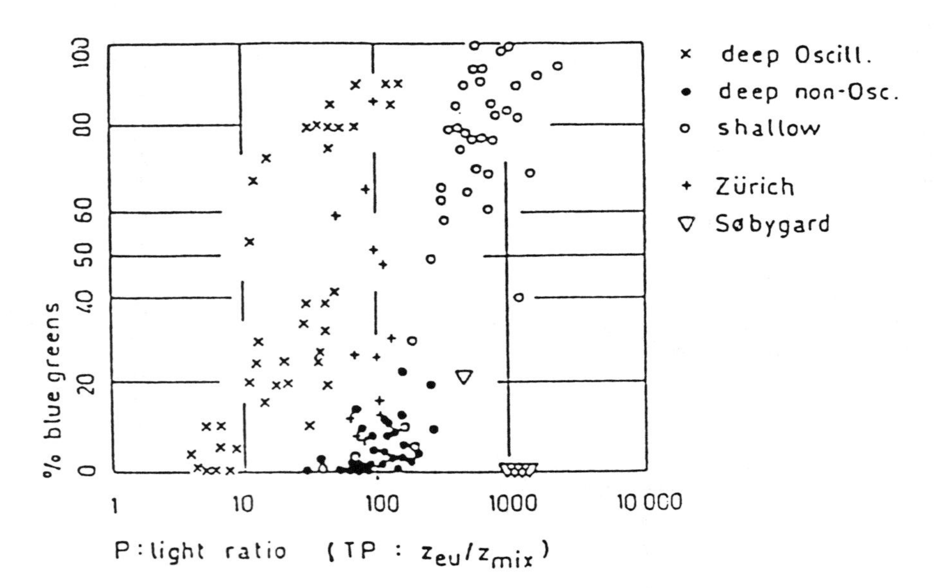

Fig. 2.34. Growth-seasonal average share of cyanobacteria in total algal biomass (arc sin transformed) as a function of growth seasonal P: light ratios a certain time lag before. Source Kamp Nielsen, 1989

The next level after phytoplankton in the limnic web is zooplankton. Its qualitative and quantitative composition is governed by both the availability of appropriate and sufficient food and by predation from fish. In oligotrophic lakes species preferring small particles dominate and in eutrophic lakes crustaceans such as Daphnia take over. The larger zooplankton species in the eutrophic lakes are excellent prey for the planktivorous fishes, and they increase their dominance relative to predatory fishes (Fig. 2.35).

An indirect effect of the increased phytoplankton biomass and the resulting decrease in transparency is the out shading of submerged macrophyte vegetation. Whereas emergent vegetation such as Phragmatis can form dense vegetation along lake shores and floating plants like water lilies can cover the near-shore surface, submerged plants greatly reduce their depth distribution. (Fig. 2.36).

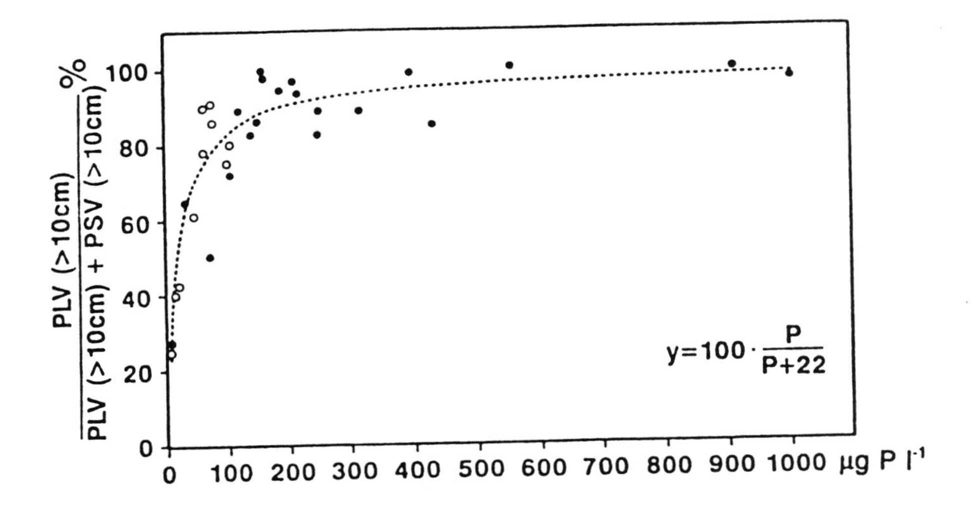

Fig. 2.35. The relationship between the ratio PLV /(PLV + PSV) and the average total P concentration (May-September). PLV = number of planktivorous fish. PSV = number of carnivorous fish, size < 10cm. o: lakes in Southern Sweden, • : Danish lakes. After Kristensen et al., 1990.

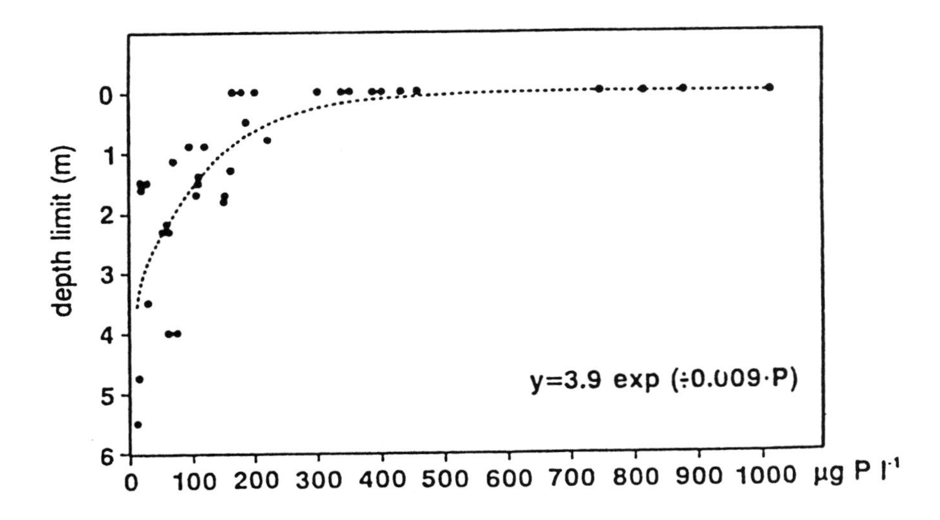

Fig. 2.36. The relationship between depth limit for the distribution of bottom vegetation and summer-average total P concentration in a number of Danish lakes. After Kristensen et al., 1990.

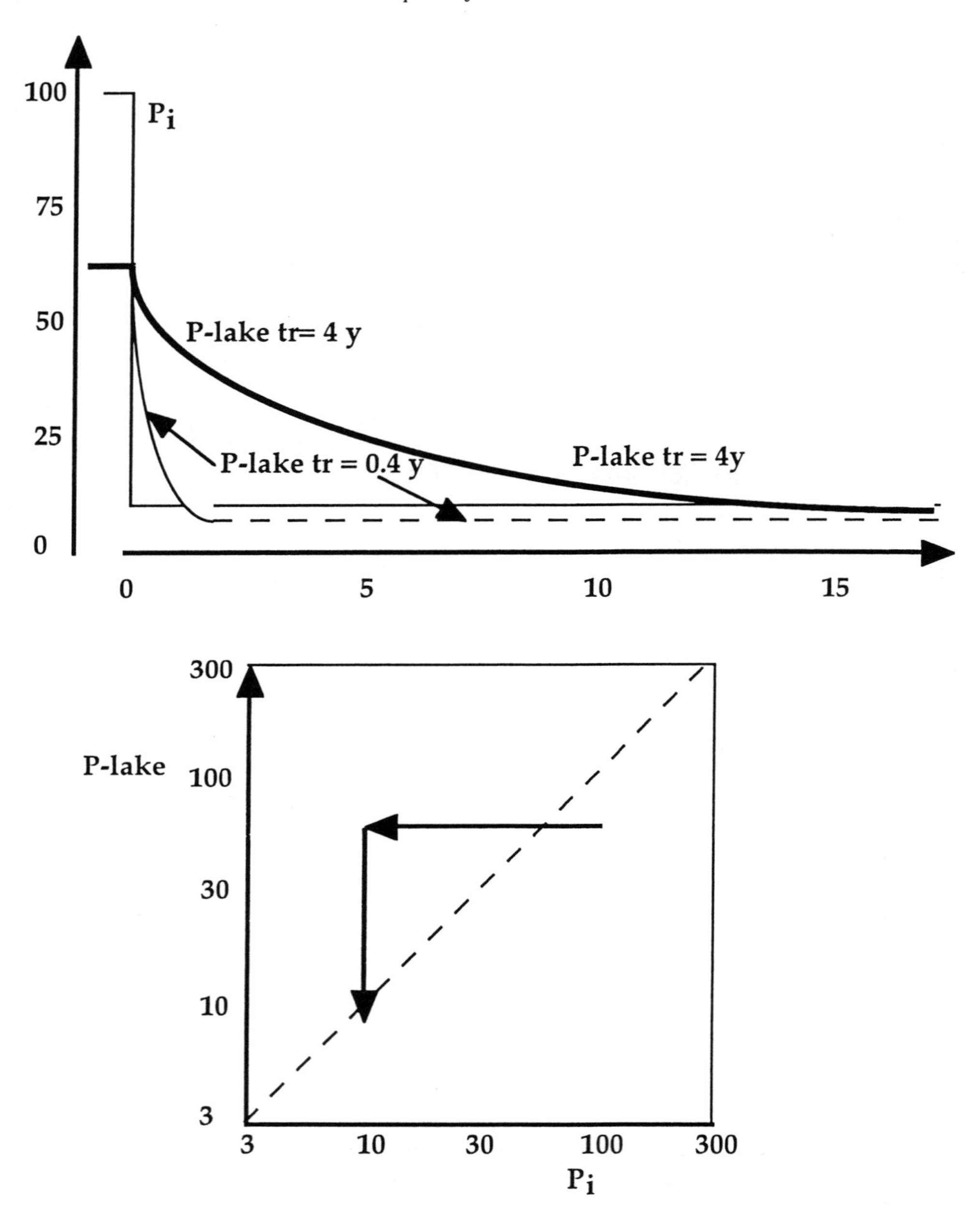

Fig. 2.37. The delaying effect of water dilution on the in-lake P concentration, P-lake following a load reduction, i.e., a drop in the inlet concentration, P_i. The upper figures shows the reduction in two hypothetical cases with different retention time, tr. The lower diagram shows a phase plot for tr = 4y.

The coincident disappearance of blue green algae and the very strong response in phytoplankton biomass at phosphorus concentrations below 100-120 μg P/l indicate that this concentration is the level not to be exceeded when reduction in phosphorus loading is considered. When the nutrient loading to a lake is suddenly reduced, the return to equilibrium between the new inlet concentration and the in-lake concentration is delayed due to the dilution effect which depends on hydraulic residence time and possibly also due to the 'sediment memory' or internal loading which is a continued release of nutrients from the sediment. The effect of dilution alone is shown in Fig. 2.37.

That dilution alone is part of the explanation for the delayed response is obvious from Fig. 2.38.. From the figure it appears that lakes with long residence times respond almost according to a dilution scheme or even faster, whereas lakes with short residence times have a long recovery period relative to the hydraulic residence time. The reason is that many lakes with short residence times are the lakes with the highest preceding loading (Fig. 2.39). and they have the biggest pools of sedimentary phosphorus. However, regardless of hydraulic residence time, most lakes will respond and recover within a few years.

The recovery pattern for some lakes are shown in Figure 2.40. For comparison, the dilution response alone is also shown. The lakes recover quickly, immediately after reduced loading, and a 50% recovery is achieved after 2-4 years, but total recovery may take ten years or more.

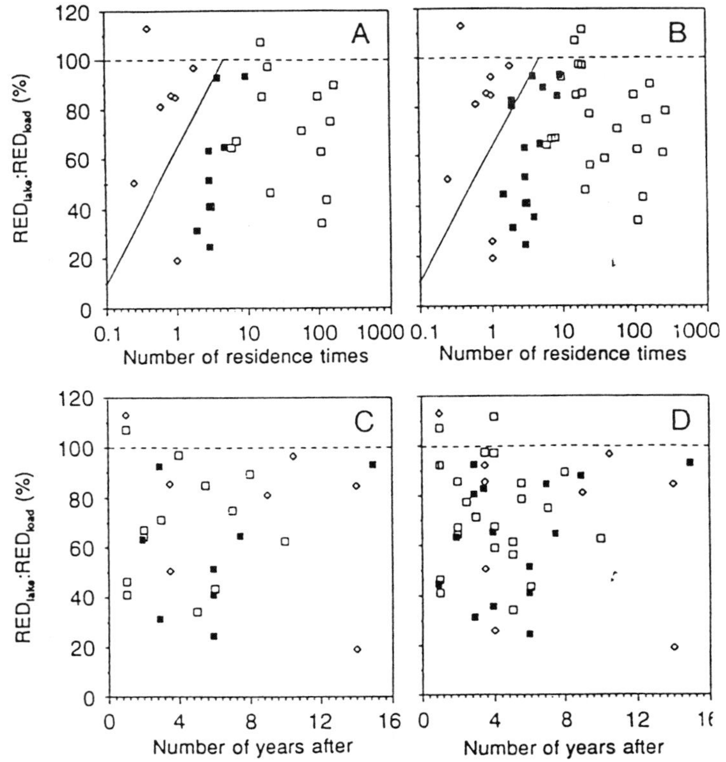

Fig. 2.38 The ratio (in %) between reduction in Plake and reduction in P-load plotted versus number of residence periods and number of years. A and C show Danish lakes, B and D show Danish lakes supplemented with data from Cullen and Forsberg, 1988. ❑ means that r < 0.5 year; ■ means that 0.5 < tr < 2.5 years and ◊ covers tr < 2.5 years.

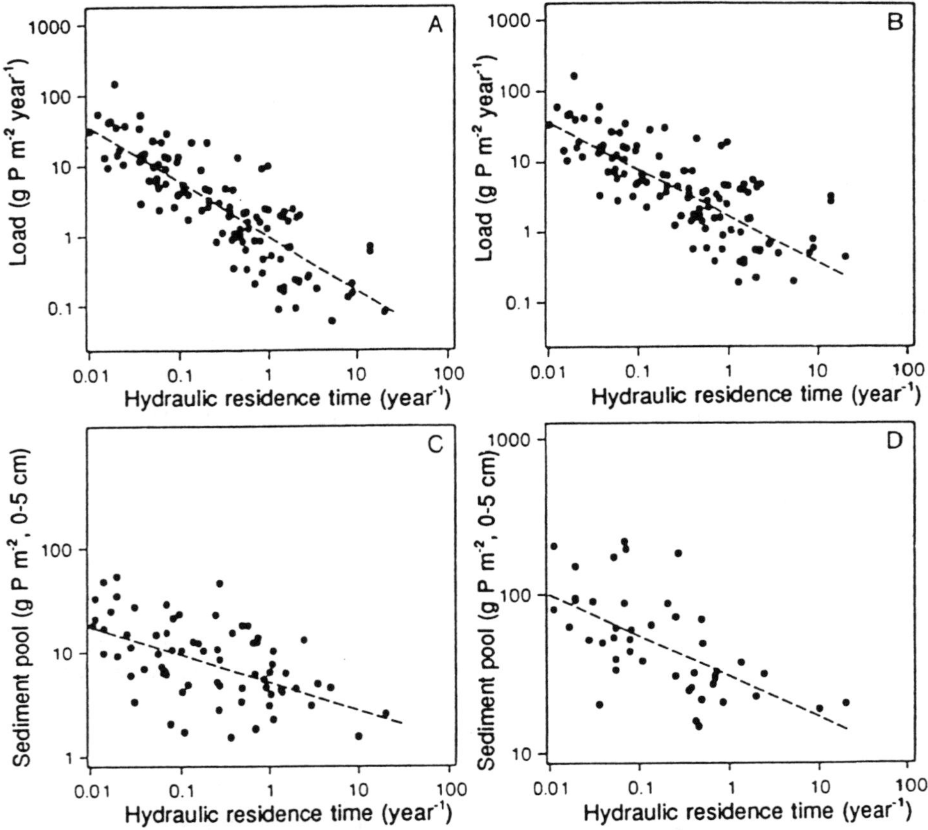

Fig. 2.39. Relationships between phosphorus load (A, B) and accumulated phosphorus pool in the sediment (C,D). and hydraulic residence time. A: P-load; B: P-load corrected for a difference in P-retention time in lakes with different tr values; C: P-pool 0-5 cm; D: P-pool 0-20 cm. After Kristensen et al., 1990.

Some of the resilience caused by release from sediments can be related to the accumulated pools of sedimentary phosphorus (Fig. 2.42), but there is also the phenomenon that in lakes with short residence time the released phosphorus is resedimented, because the residence time during the release period is relatively long. A big amplitude in hydraulic residence time is also a cause for delayed recovery.

In the investigation mentioned the response in phytoplankton biomass and transparency depth follows the reduction in phosphorus loading, but in some cases resilience in the biological structure occurrs; see Section 8.5 about restoration of lakes. In shallow lakes, where the submerged vegetation had disappeared, the response was often absent initially, because of the difference in the ecological structure of the entire food web. However, by biomanipulation it is possible to reestablish the old ecological structure, and thereby achieve a significant increase in transparency, provided that the phosphorus concentration is right (about 50-120 μg P /l). As seen from this example, reduction in the nutrient concentrations of waste water is not always sufficient to reestablish the habitat and the natural biological diversity. In such cases restoration measures can be implemented; see Section 8.5.

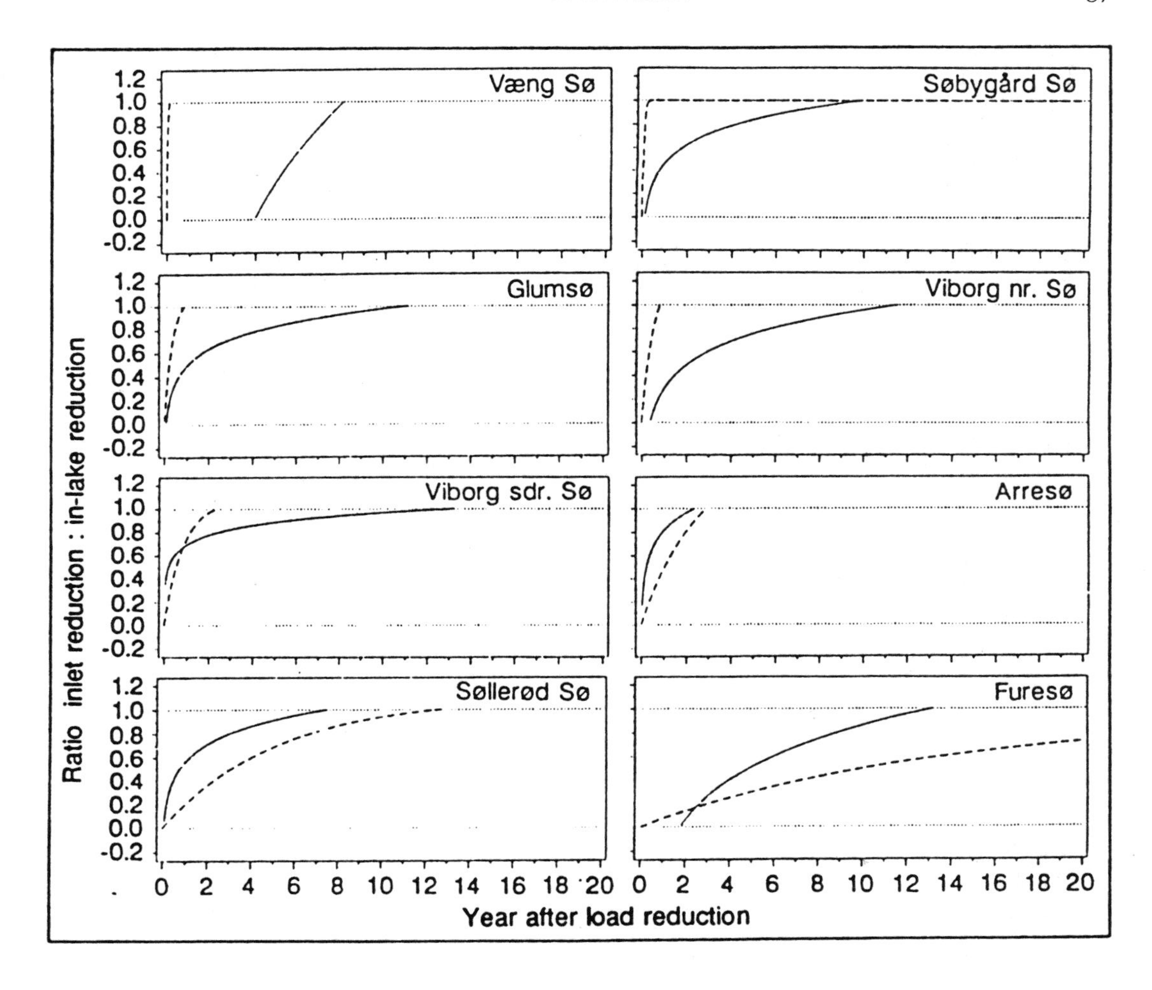

Fig. 2.40. The ratio inlet reduction to in-lake reduction, as observed in 8 Danish lakes. The solid curve represents the best fit with an exponential function, while the dotted curve shows the best fit with a simple dilution model. After Kristensen et al., 1990.

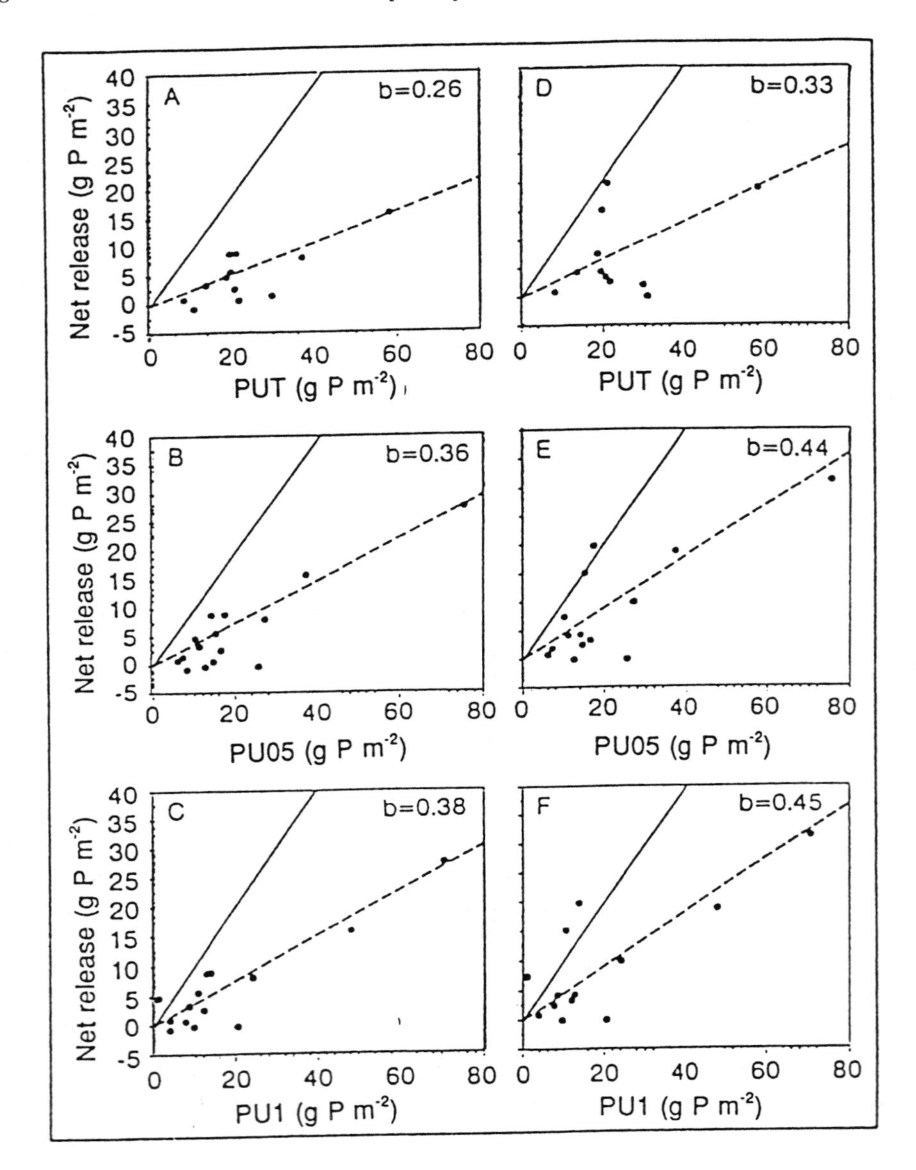

Fig. 2.41. The relationship between net release of total phosphorus (A, B, C: first five years after load reduction; D, E, F: entire known period since load reduction) and P-pool in the upper 10 cm-layer of the sediment. A, D: total P-pool; B, E: P-pool < 0.5 mg P / g D.W; C, F: P-pool < 1.0 mg P / g D.W. Solid line: release= pool; dotted line: linear regression of release as a function of pool; b indicates the slope of the regression line. Source: Kamp Nielsen, 1991.

2.9 MASS CONSERVATION IN A FOOD CHAIN

The mass flow through a food chain has been discussed in Section 2.2. The food taken in by one level in the food chain, I, is used in respiration (R), non-utilised (wasted) food (NUF), undigested food (faeces), F, excretion (urine), E, and growth and reproduction, G. If the growth and reproduction are considered as the net production, NP, we can state that:

$$NP = G = I - (NUF + F + R + E) \tag{2.65}$$

and we can call the ratio of the net production to the intake of food the efficiency (for further details about these considerations, see Chapter 3). The efficiency is dependent on several factors, but may be as low as 10% or even below 10%. **Any toxic matter in the food is unlikely to be lost through respiration and excretion, because it is usually much less bio-degradable than the normal components in the food.** This being so, the efficiency of toxic matter according to equation (2.65) is often higher than for normal food components, and as a result several chemicals, such as chlorinated hydrocarbons, including DDT and some heavy metals, can be magnified in the food chain. This phenomenon is denoted as biomagnification.

The assimilated food, A, is the food used for (growth + reproduction) + respiration + excretion:

$$A = I - (NUF + F) \tag{2.66}$$

Many organic toxic compounds are taken up (assimilated) by a high efficiency (more than 90%), i.e, the loss by faeces is low; see equation (2.66).

Heavy metals have fortunately a low assimilation efficiency. Approximately only 5-10% of their content in food is assimilated, but as they are excreted slowly and not removed by respiration, they still have a relatively high biomagnification.

Many organic compounds, included chlorinated hydrocarbons, have a particularly high biomagnification, **because they have 1) a high assimilation efficiency, 2) a very low biodegradability and 3) are only excreted from the body very slowly, because they are dissolved in fatty tissue.** This is illustrated for DDT in Table 2.24 and in Table 2.25 by a comparison of concentrations of DDT in various abiological and biological components . As man is on the last level of the food chain, relatively high DDT concentrations have been observed in human body fat. The concentration of a toxic component in an organism can be followed approximately by use of a simplified differential equation:

$$dTox / dt = \text{daily uptake via respiration and food} - k \times Tox \tag{2.67}$$

where the total daily uptake is found from the concentration in the ambient air of the toxic component times the efficiency of uptake via respiration plus the concentration in the food times the assimilation efficiency. It is assumed that the excretion follows a first order reaction which is approximately correct. k is the the first order excretion coefficient expressed for instance in the unit 1/24h.

TABLE 2.24
Biological magnification (Data after Woodwell et al., 1967)

Trophic level	Concentration of DDT (mg/kg dry matter)	Magnification
Water	0.000003	1
Phytoplankton	0.0005	160
Zooplankton	0.04	~ 13,000
Small fish	0.5	~ 167,000
Large fish	2	~ 667,000
Fish-eating birds	25	~ 8,500,000

TABLE 2.25
Concentration of DDT (mg per kg)

Atmosphere	0.000004	Rain water	0.0002
Atmospheric dust	0.04	Cultivated soil	2.0
Freshwater	0.00001	Seawater	0.000001
Grass	0.05	Phytoplankton	0.0003
Aquatic macrophytes	0.01	Freshwater fish	2.0
Invertebrates on land	4.1	Sea fish	0.5
Invertebrates in sea	0.001	Eagles, falcons	10.0
Herbivorous mammals	0.5	Swallows	2.0
Human food, plants	0.02	Human food, meat	0.2
Man	6.0		

TABLE 2.26
Lead in food

Food items	Typical lead concentration in mg/kg fresh weight England	Holland	Denmark
Milk	0.03	0.02	0.005
Cheese	0.10	0.12	0.05
Meat	0.05	<0.10	<0.10
Fish	0.27	0.18	0.10
Eggs	0.11	0.12	0.06
Butter	0.06	0.02	0.02
Oil	0.10	-	-
Corn	0.16	0.045	0.05
Potatoes	0.03	0.1	0.05
Vegetables	0.24	0.065	0.15
Fruits	0.12	0.085	0.05
Sugar	-	0.01	0.01
Soft drinks	0.12	0.13	-

The concentrations of lead in food items are shown in Table 2.26 to illustrate the presence of toxic substances in our food. The concentrations in the table are taken from the mid eighties, i.e., before the introduction of lead-free gasoline had shown any significant effect.

Equation (2.66) explains why the concentration of a toxic substance increases with increasing weight and age of the organism. This is exemplified in Figure 2.42 for fish with increasing weight. A steady state concentration, dTox / dt = 0, can be found from equation (2.66) as Tox = daily uptake via respiration and food / k. k is low for many toxic compounds including most toxic heavy metals such as mercury, lead and cadmium. This implies that the concentration of Tox becomes high and that it takes many years to reach a concentration close to the steady state situation.

It is, as already mentioned a few times, of great importance to quantify any imbalance in the concentrations of life-essential components, as well as toxic components, for the assessment of the life conditions in the ecosystem. This is of course valid on the global as well as on the regional and local level. **Many toxic substances are widely dispersed and a global increase in the concentration of heavy metals and pesticides has been recorded**, as exemplified for lead in Figure 2.43. The relationship between a global and a regional pollution problem and the role of dilution for this relationship are illustrated in Table 2.27, where the ratios of heavy metals concentrations in the river Rhine and in the North Sea are shown.

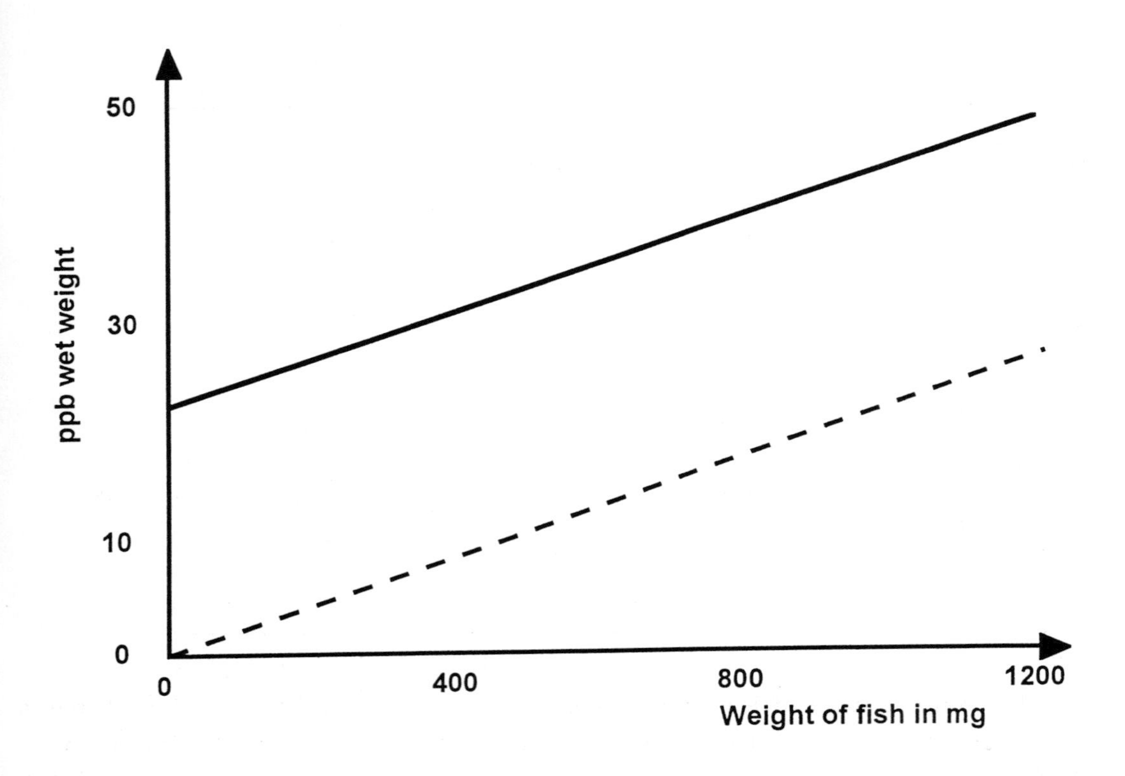

Fig. 2.42. Increase in pesticide residues in fish as weight of the fish increases. Top line (full) = total residues; bottom line (dottted) = DDE only. (After Cox, 1970).

TABLE 2.27

Heavy metal pollution in the river Rhine

	The river Rhine t/year	Ratio $\frac{\text{conc. in the Rhine}}{\text{conc. in the North Sea}}$
Cr	1000	20
Ni	2000	10
Zn	20,000	40
Cu	200	40
Hg	100	20
Pb	2000	700

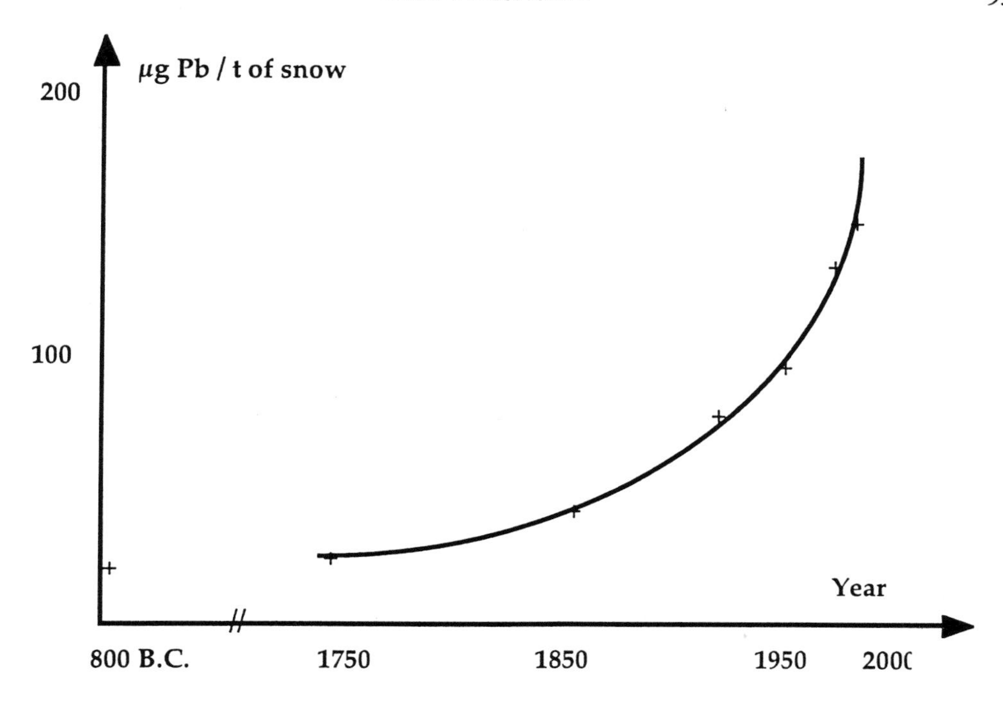

Fig. 2.43 Accumulation of lead in the Greenland ice pack from 800 B.C. to the present.

Independent of the level, the usual first step in a solution procedure for a toxic substance problem is to set up a list of pollution sources with information about the quantities emitted to the environment, the form in which the toxic substance is emitted and where in the environment (which spheres) the emission takes place. Based upon this information it is possible to assess the concentration of the toxic substance in various compartments of the environment. It is an essential step in an Environmental Risk Assessment, ERA, which will be presented in Chapter 4. It is often appropriate to start the solution of a pollution problem by setting up a mass balance for the considered component or element to clarify the sources of pollution and to state the most effective means of solving the problem.

2.10 THE HYDROLOGICAL CYCLE

Water is the most abundant chemical compound on earth (see Table 2.18), and has some unique properties. Its importance for all life on earth can be demonstrated as follows:

1. *Our body consists of 70% water* and we need at least 1.5 litres per day to survive. We can survive without food for perhaps 80 days, but only a few days without water.

2. *Water serves as a basic transport medium for lifegiving nutrients.*

3. *Water removes and dilutes many natural and man-made wastes.*

4. *Water has a great ability to store heat energy and to conduct heat, and has an extremely high vaporization temperature compared with its molecular weight.* These thermal properties are major factors influencing in the climatic pattern of the world, and in minimizing sharp changes in temperature on the earth.

5. *Water has its maximum density at 4 °C above its freezing point,* so solid water, ice, is less dense than liquid water. This is the reason why a water body freezes only on the top. If ice was denser than liquid water, lakes, rivers and oceans would freeze from the bottom up, killing most higher forms of aquatic life.

It is of course possible to quantify the global cycling of all compounds. It is because of the importance of water (see the 5 points above), of particular interest to obtain an overview of the global water cycle. **Water shows a physical cycling** (compare with the chemical cycling of the elements), **as demonstrated in Fig. 2.44. In this vast cycle, driven by solar energy, our supply of water is recyled again and again.** Water evaporates from the oceans, rivers, lakes and continents, and gravity pulls it back down as rain. Some of the water falls on the land sinks or percolates into the soil and ground to form ground water. The soil, can like a sponge, hold a certain amount of water, but if it rains faster than the rate at which the water percolates, water begins to collect in puddles and ditches and runs off into nearby streams, rivers and lakes. This run-off causes erosion. The water runs eventually into the ocean, which is the largest water storage tank. Because of this cycle, water is continually replaced, as indicated in Table 2.29.

Long-term average water-budget equations for extensive hydrological systems can be expressed as

$$P = E \tag{2.68}$$

where P is the precipitation inflow and E the evaporation outflow. The storage change, S, is zero. Water-budget equations for entire land and water masses must also contain the total water discharge from land to ocean, Q:

$$P + Q = E \quad \text{for oceans} \tag{2.69}$$

$$P = Q + E \quad \text{for land} \tag{2.70}$$

The numerical equality is illustrated in Table 2.30.

The short-term water-budget equation for a terrestrial ecosystem must include a storage term, S:

$$P = Q + E + S \tag{2.71}$$

TABLE 2.28

Water resources and annual water balance of the Continents of the World

Component	Europe	Asia	Africa	N. Amr	S. Amr	Australia	Total
Area (1E6 km^2)	9.8	45	30.3	20.7	17.8	8.7	132.3
Precipitation (km^3)	7165	32,690	20,780	13,910	29,355	6405	110,000
Total river runoff (km^3)	3110	13,190	4225	5960	10,380	1965	38,830
Underground runoff (km^3)	1065	3410	1465	1740	3740	465	11,885
Infiltration (km^3)	5120	22,910	18,020	9690	22,715	4905	83,360
Evaporation (km^3)	4055	19,500	16,555	7950	18,975	4440	71,475
% underground runoff of total	34	26	35	32	36	24	31

TABLE 2.29

The water cycle

Water in	Is replaced every
Human body	month
The air	12 days
A tree	one week
Rivers	a few days
Lakes	0.1 - 100 years
Oceans	3600 years
Polar ice	15,000 years

If subsurface flows are included, we have:

$$P + Q_i + L_i = E + Q_o + L_o + S \tag{2.72}$$

where Q_i and Q_o are surface inflow and outflow, respectively, and L_i and L_o are the corresponding subsurface flows.

TABLE 2.30

Mean annual water balance components for the earth

Item	Land	Ocean	Earth
Area (10^6 km^2)	148.9	361.1	510.0
Volume (10^3 km^3)			
Precipitation	111	385	496
Evaporation	- 71	- 425	- 496
Discharge	- 40	40	0
Mean depth (mm)			
Precipitation	745	1066	973
Evaporation	- 477	- 1177	- 973
Discharge	- 269	111	0

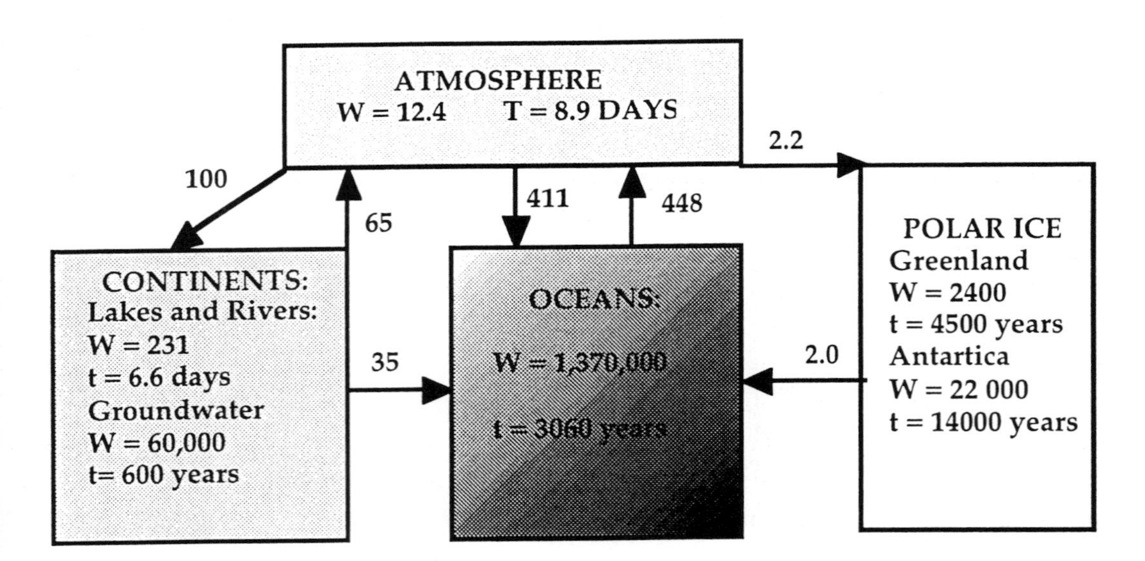

Fig. 2.44 The water cycle is shown. Water in compartment is indicated in 1000 km3 and the retention time is indicated in years. Number on fluxes represent 1000 km3 / y. The estimation of the ground water volume is to a depth of 5 km of the earth's crust; much of this water is not actively exchanged.

2. 11 GREEN ACCOUNTING

We have financial and production accounting, but we need also environmental accounting which is based on the mass- and energy accounting of a production or a distribution unit (system). Previously, we have attempted to eliminate the smoke and waste coming from these units by dilution and environmental technology (end-of-the-pipe technology), but obviously we need also to ask the following three pertinent questions:
1) how much smoke and waste is coming out of the unit?
2) why these amounts?
3) could we reduce these amounts of smoke and waste by changing the operation scheme of the unit?

Answers to these question will reveal that dilution and environmental technology do not suffice. It is necessary to manage the flows of energy and materials to optimise the unit. Otherwise, we may invest in (expensive) environmental technology which could be replaced by inexpensive adjustment of the energy and mass flows. Society could save large amounts of money by rational management through applying energy and materials accounts in an effort to attain the goal of "maximum environmental value for money".

A wave of certification schemes overwhelmed the society in the 1990s, of which the ISO 14000 family of standards and the EMAS registration are the most important. The certification schemes are based on standards, rules and regulations but do not ask the question "why?" (see point 2 above).

Unfortunately it is the general conception that green accounts are of the soft environmental reporting type with many well intentioned words, but green accounts should be based on rigorous mass and energy balances and flow diagrams with indication of measurable quantities and rates.

Environmental input-output analysis (Schrøder, 1999) begins with an examination of flow (of mass or energy) across a single system's or subsystem's boundaries. The number of interacting subsystems is initially small and is increased usually stepwise until there is a harmony between model complexity and data availability. Let us consider a system as shown in Figure 2.47. What is called the fund in the diagram represents elements that enter and come out of the system in the same amount. As an example, if a farmer sows one bag of barley and harvests 40 bags, 1 is the fund and 39 are the product.

We can now apply the conservation principle (see the symbols in Figure 2.45), assuming steady state (no accumulation):

$$I = P + W \tag{2.73}$$

The problem is to find the three unknown outputs, F, P and W, based on a given input, I. Equation (2.72) must therefore be supplied by two more equations. We may use the efficiency, ef, defined by the following equation:

$$ef = (P + F) / (I + F) \tag{2.74}$$

In addition, a farmer speaks often about $f = (P + F) / F$, an increase in yield on the seed fund.

If the production system consists of two interacting subsystems one can likewise apply the conservation principle for each subsystem. The situation is shown in Figure 2.48, where system 0 depicts the inputs and products, and system w the waste outputs. A matrix can be set up to overview the various flows (the applied notation is taken from Schrøder, 1999) :

To:	System 0	System 1	System 2	System w
From:				
System 0	0	x0,1	x02	0
System 1	0	0	x1,2	x1,w
System 2	x2,0	x2,1	x,2,2	x2,w
System w	0	0	0	0

Total inputs: x01 + x02
Total outputs: x1,w + x2,0 + x2,w

If we assume steady state, we can set up two balance equations for system 1 and system 2:

$$x0,1 + x2,1 = x1,2 + x1,w \tag{2.75}$$

$$x0,2 + x1,2 + x2,2 = x2,1 + x2,2 + x2,0 + x2,w \tag{2.76}$$

Since we have six outputs, we must define four more equations to be able to determine the four unknown outputs. We can set up the following efficiency equations which can also be used for guidance of realistic possibilities to manage the the amount of waste:

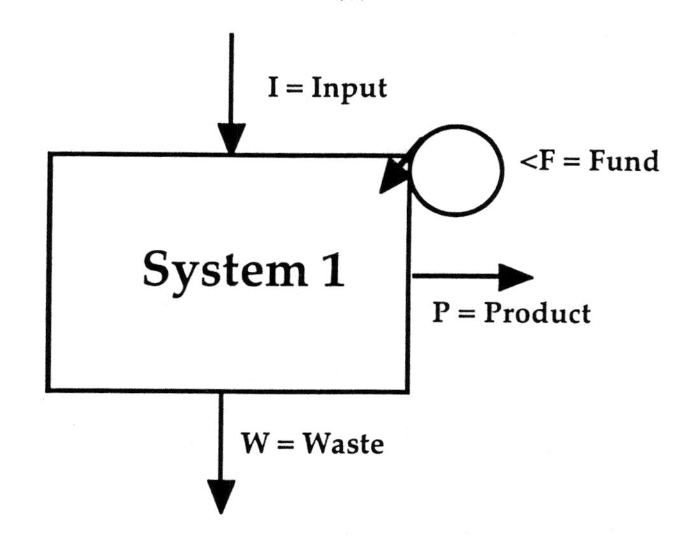

Figure 2.45 First level analysis of a system by green accounting. The symbols used in the text are indicated on the figure.

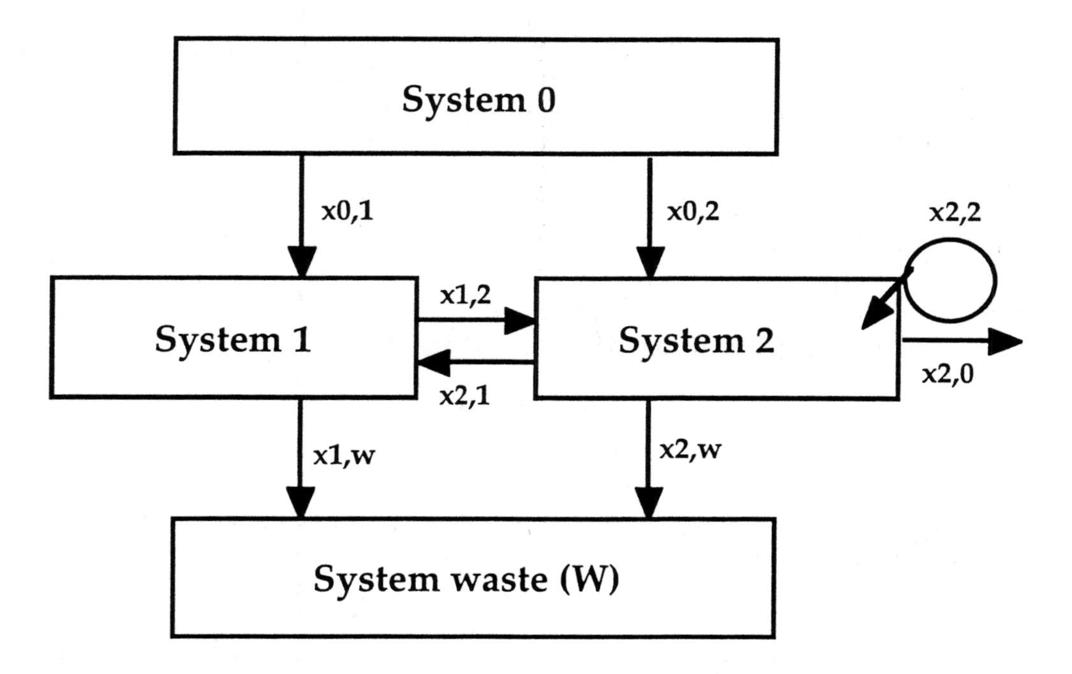

Figure 2.46 Two systems in interaction. A second level analysis.

$$ef1 = x1,2 \ / \ (x1,2 + x1,w) \tag{2.77}$$

$$ef2 = (x2,0 + x2,1 + x2,2) \ / \ (x2,w + x2,0 + x2,1 + x2,2) \tag{2.78}$$

$$f2 = (x2,0 + x2,1 + x2,2) \ / \ x2,2 \tag{2.79}$$

$$eftotal = x2,0 \ / \ (x0,1 + x0,2) \tag{2.80}$$

Other efficiency equations could of course be applied.

Schrøder (1999) has set up a green account, levels 1, 2, and 3 (four subsystems), for the use of nitrogen in Danish agriculture in the mid eighties, when the Danish Action Plant for the Aquatic Environment (Ministry of the Environment, Denmark, 1987) was launched. It was the goal to reduce the nitrogen emissions from Danish agriculture by 50% from about 479 M kg (see Figure 2.47) to about 440 Mkg. The level three analysis by Schrøder (1999) just before the implementation of the plan in the middle of the 1980s is shown in Figure 2.48.

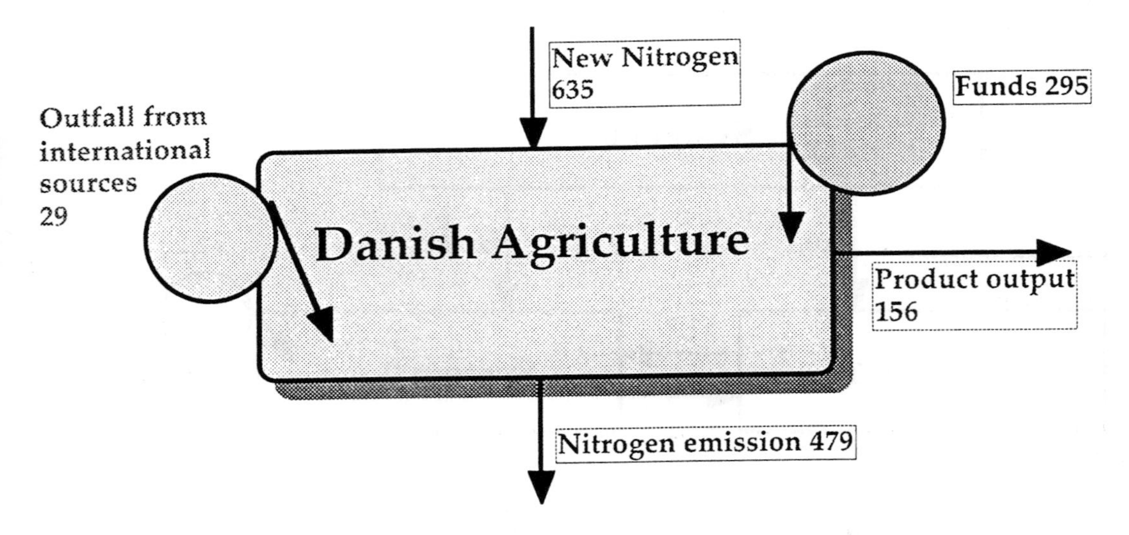

Figure 2.47 nitrogen flows and funds in Danish agriculture in the middle of the 1980s. A first level analysis of the system is applied. Unit: Mkg /y.

Table 2..31 shows the major nitrogen flows and efficiencies in the mid eighties and in the mid nineties, compared with the goals. As seen from these figures the plan has only been able to reduce the total nitrogen emission by about 20%, while the nitrate leaching has only been reduced by 15%. The bottleneck is the fertiliser efficiency. If we assume that the animal production is increased by 11% and the utilisation of nitrogen in manure can be further increased from 40% to 70%, the fertiliser efficiency must still be increased from 58% to 64% (see Table 2.31). This is very difficult, and it is hardly possible to achieve 50% of nitrogen emission without a reduction in the agriculture production which is probably not possible due to political obstacles. In addition, these calculations have been based on steady state conditions. It can not be excluded that a depletion of the stock of nitrogen may take place, maybe in the order of up to 100 Mkg per annum.

TABLE 2.31

Major Nitrogen Flows and Efficiencies in Danish Agriculture in the mid eighties and mid nineties and according to the Goals in the Environmental Action Plan

	Flow unit: M kg / y		
	Mid 1980s	**Mid 1990s**	**Goals**
Nitrogen inputs, total	635	564	414
Use of mineral fertiliser	393	290	115
Nitrogen in products	156	170	170
Nitrogen in animal products	81	91	91
Nitrogen in plant products	75	79	79
Nitrogen emission	479	394	244
Hereof leaching as nitrate	300	254	149
Fertiliser efficiency %	57	58	64
Fodder efficiency %	24	27	30
Utilisation efficiency of N in manure %	20	42	70

It has been shown that green accounts can be used to find where it is feasible to reduce waste outputs and thereby reduce the emission to the environment and save the enterprise, a region or a country for the corresponding consumption of material. The example of green accounting applied to Danish agriculture shows that the right environmental management plan cannot be made unless the situation is analysed by use of green accounting on a higher

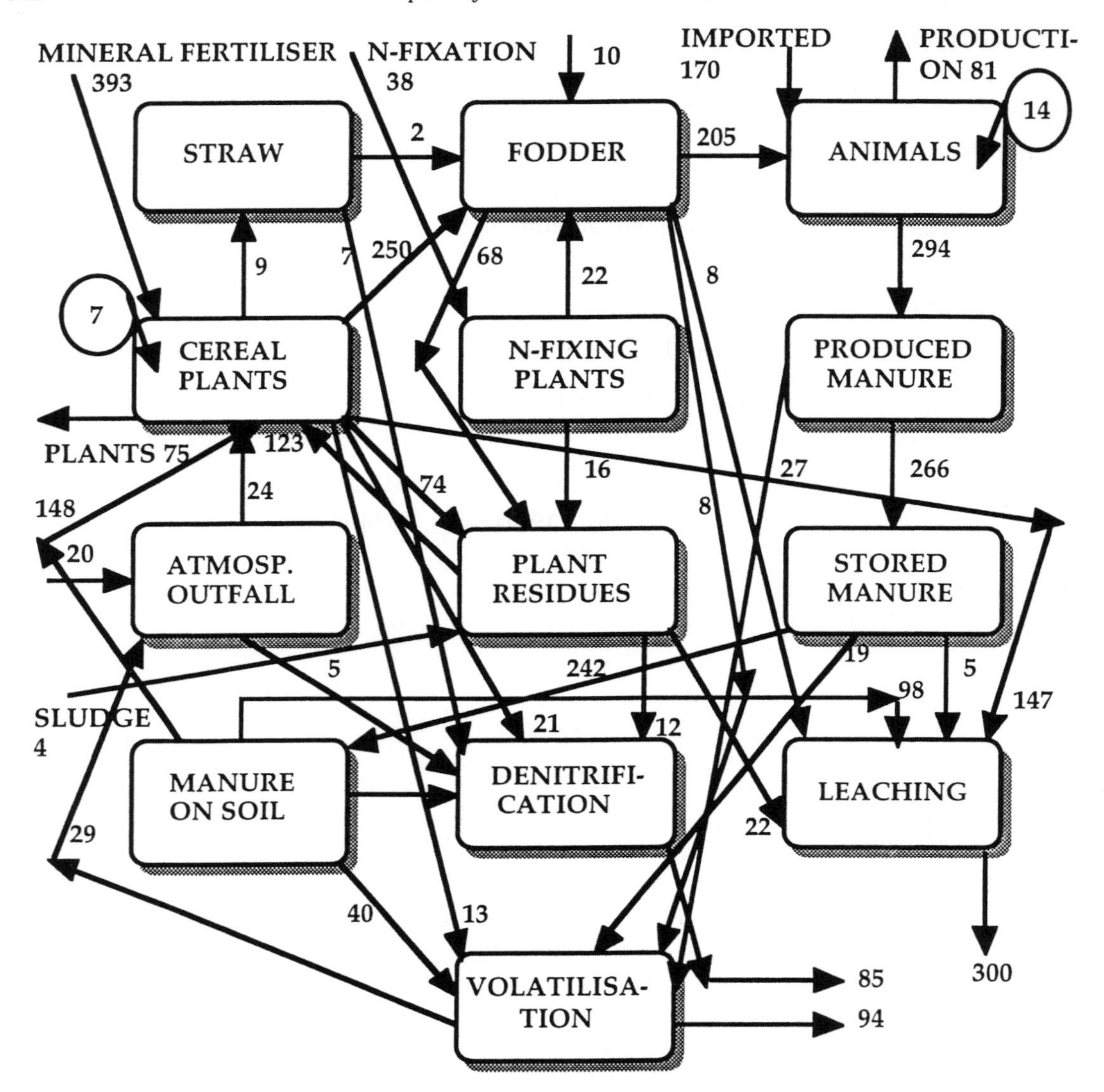

Figure 2.48 Nitrogen flows and funds in Danish agriculture in the mid-1980s in Mkg N /y. A third level system is applied. Clearly the low efficiencies (N in product, relatively to N in inputs or in waste) have to be improved to reduce the nitrogen emissions.

level which will uncover the processes which have a low efficiency and where waste reductions therefore can be obtained. It will also reveal whether an environmental management plan is realistic or not, as the increase in an efficiency needed to obtain a given waste reduction may be unrealistic under the accepted premises.

2. 12 QUESTIONS AND PROBLEMS

1. Indicate some compounds which could be discharged into a bay in the order of few mg per litre without harmful effect.

2. Consider a lake with a volume of 500,000 m^3. 2000 m^3 24 h^{-1} of waste water containing 0.5 mg l^{-1} methoxychlor is discharged into the lake. The natural flows to the lake corresponds to 4 months' retention time. Methoxychlor follows a first order decomposition rate, and the approximate half-life time in water can be found in Appendix 2 (no H_2O_2 is added). Find the equlibrium concentration in the lake water. Assume that precipitation and evaporation are balanced.

3. A person lives in an industrial area, where the Cd-concentration is as high as 5 μg per m^3. His daily respiration is 20 m^3 air. He smokes 20 cigarettes per day each containing 1 μg Cd. His food contains 75 μg Cd. The uptake efficiency of Cd is by lungs 25% and by digestion 7%. What is his daily intake? The excretion rate constant is 0.002 day^{-1}. Set up an equation to determine his Cd-concentration as a function of time. What will his maximum concentration be?

4. If iodine is the limiting factor for brown algae what is the approximate maximum concentration of brown algae in water with a concentration of 0.01 mg per litre iodine?

5. Find the critical point, by use of the Streeter-Phelps equation including nitrification, for the following case:

   100 l sec^{-1} waste water with BOD_5 = 30 mg l^{-1} and NH_4^+-concentration = 20 mg l^{-1} is discharged into a river with a flow of 1200 l sec^{-1}. The flow rate in the river is 0.5 m sec^{-1}, the depth is 4 m, and the temperature is 20°C.

   Is the saturation at the critical point acceptable?

   How would you classify the river at the critical point?

6. A lake has a phosphorus concentration of 80 μg l^{-1}. What is the expected maximum chlorophyll concentration?

7. A lake is 50 m deep and has an area of 10 km^2. The phosphorus input to the lake comes mainly from waste water. Approximately 50,000 m^3 per 24h of municipal waste is discharged into the lake. How would you classify the lake? The waste water is only treated by a mechanico-biological plant. What level of efficiency of phosphorus removal is necessary to improve the conditions of the lake to the mesotrophic and oligotrophic states, respectively?

8. The transparency in a lake is 0.8 m, estimate G_{24max}.

9. Write the following chemical reactions:
1) between acetic acid and nitrate under anaerobic conditions, and
2) between acetic acid and sulphate under anaerobic conditions.

10. Show that the conversion from ppm for a gaseous pollutant to $\mu g\ m^{-3}$ at 25°C and 760 mm Hg is given by:

10^3 (ppm) molecular weight of pollutant/24.5.

11. The average concentration of particulate lead is 6 $\mu g/m^3$, of which 75% is less than 1 μm in size. If a person respires 15 m^3 air daily and the uptake efficiency is 50% of the particles below 1 μm and negligible for those above this size, calculate how much lead is absorbed in the lungs each day.

12. The equilibrium: $2NO_2$ (g) = N_2O_4 (gas) is exothermic with the formation of about 60 kJ per mole. While nitrogen dioxide is brown, dinitrogen tetraoxide is colourless. Would you expect the colour of the photo-chemical smog to be deeper on a warm day or a cool day?

13. The distribution coefficient for DDT between olive oil and water is 925. The solubility or DDT in water is 1.2 $\mu g\ l^{-1}$. Calculate the solubility of DDT in olive oil in mg l^{-1} and molarity. The molecular weight of DDT is 354.5 g mol^{-1}.

14. The earth's rivers add to the oceans 30 * 10^{15} kg water per annum. Calculate, using the text, the residence time of K and Na, assuming a constant concentration of these elements in the oceans.

15. The average concentration of dissolved iron(III)ions in river water is 1 ppm, but in seawater it is only 8 ppb. What happens to iron(Fe^{3+})-bearing river water when it enters the sea?

16. Find the NH_3-concentration at pH = 7.6 when the total concentration of NH_3 + NH_4^+ has been found to be 4.6 mg/l.

17. Nitrogen gas in the atmosphere will dissolve in natural water bodies. Calculate the concentration that would be expected in pure water at room temperature by use of Henry's constant.

18. Explain biochemically why sulphur is a micronutrient in the growth of aquatic plants.

19. A person has a weight of 80 kg and is smoking 30 cigarettes a day. Each cigarette contains 1 μg cadmium. He respires 18 m^3 air / 24 h and lives in an industrial area, where the concentration of particulate cadmium is 5 μg cadmium / m3. His daily intake of food and water contains 100 μg cadmium. The uptake efficiency via the lungs is about 25% and via the intestines 7%. Excretion of cadmium follows a first order reaction scheme with a coefficient of 0.00045 $(24h)^{-1}$.

Write a differential equation to determine the cadmium content in the considered person as a function of the time.

Find the maximum concentration of cadmium, as a result of these processes. Comment on the result.

20. 2-chloro-4-nitrophenole has a water solubility of 2.35 mg/l. Estimate the octanol-water partition coefficient. Estimate the CF value of this compound for a fish of the length 25 cm.

21. 1 million m^3/ y drainage water from an agricultural area contains 5 mg nitrate-N /l and 0.1 mg P /l.. The water is discharged via a wetland directly to a lake with a depth of 10 meter and natural tributaries bringing 1 million m^3/ year of almost nutrient free water. The volume of the lake is 2,000,000 m^3. Denitrification takes place in the wetland which contains organic matter with 50% carbon and 0.25% phosphorus.

A. How much phosphorus is released by this process, if it is presumed that all nitrate is denitrified?

B. Use a Vollenweider plot to find what the phosphorus loading (coming from the drainage water and released by the denitrification) means for the eutrophication of the lake.

C. How much phosphorus must be removed to ensure that the lake can be designated mesotrophic?

22. How much nitrogen is dissolved at 20°C in equilibrium with the atmosphere? Answer this question by use of Henry's constant; see Table 2.9.

3. Energy Conservation

3.1 FUNDAMENTAL CONCEPTS RELATED TO ENERGY

Energy is defined as the ability to do work, and the behaviour of energy can be described by the first and second laws of thermodynamics. **The first law of thermodynamics states that energy may be transformed from one type to another but is never created or destroyed.** It can also be applied in a more ecological way as follows: you cannot get something for nothing - there is no such thing as a free lunch (Commoner, 1971).

Thus when a change of any kind occurs in a closed system (see 2.3 for definition) an increase or decrease in the internal energy occurs, heat is evolved or absorbed and work is done. Therefore:

$$\Delta E = Q + W \tag{3.1}$$

where

ΔE = change in internal energy

Q = heat absorbed

W = work done on the system

As mentioned in Section 2.1 a relationship exists between mass and energy, which dictates that energy is produced as a result of nuclear processes. Equation (3.1) assumes that such processes have not taken place. In environmental science, we are primarily concerned with the quantity of incident solar energy per unit area in an ecosystem and the efficiency with which this energy is converted by organisms into other forms. This situation is illustrated in Fig. 3.1, where the fate of solar radiation upon grass-herb vegetation of an old field community in Michigan is shown (Golley, 1960). The transformation of solar energy to chemical energy by plants conforms with the first law of thermodynamics:

Energy of solar radiation -> Solar energy assimilated by plants + solar energy reflected + energy used for evaporation and

Solar energy assimilated by plants -> chemical energy of plant growth + heat energy of respiration (3.2)

For the next level in the foodchain, the herbivorous animals, an energy balance can also be set up:

$$F = A + UD = G + R + UD \tag{3.3}$$

where

F = the food intake converted to energy (Joules)

A = the energy assimilated by the animals
UD = undigested food or the chemical energy of faeces
G = chemical energy of animal growth
R = the heat energy of respiration

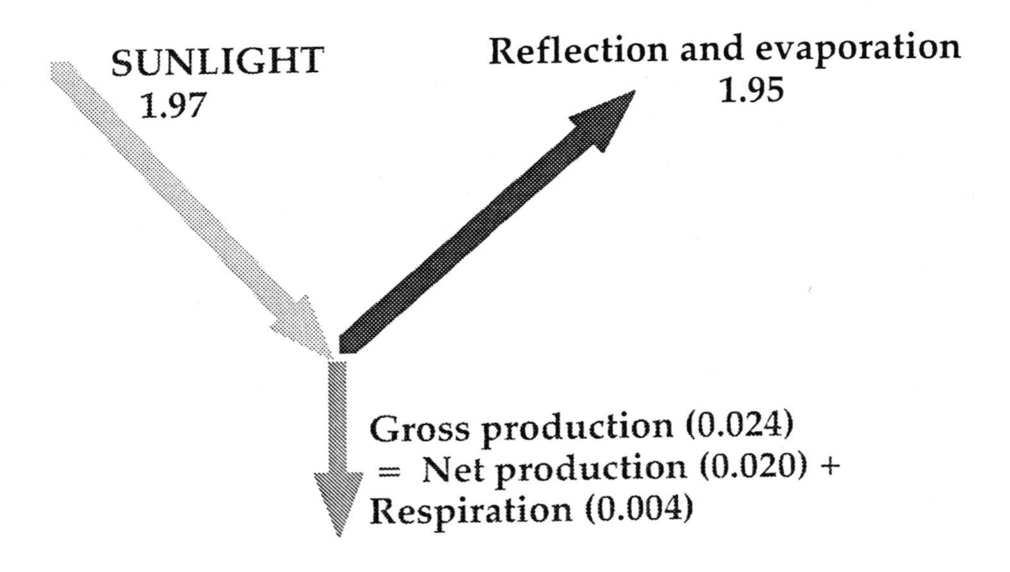

Fig. 3.1. Fate of solar energy incident upon the perennial grass-herb vegetation of an old field community in Michigan. All values in GJ m^{-2} y^{-1}.

These considerations are along the same lines as those mentioned in Chapter 2, where the mass conservation principle was applied. The conversion of biomass to chemical energy may be illustrated by the chemical energy content of carbohydrates which is about 17 kJ /g, of proteins which is about 20-21 kJ /g and of fat which is about 37-38 kJ /g.

In nature we can distinguish two processes: spontaneous processes, which occur naturally without an input of energy from outside, and non-spontaneous processes, which require an input of energy from outside. These facts are included in the second law of thermodynamics, which states that processes involving energy transformations will not occur spontaneously unless there is a degradation of energy from a non-random to a random form, or from a concentrated into a dispersed form. In other words all energy transformations will involve energy of high quality being degraded to energy of lower quality (e.g. potential energy to heat energy). The quality of energy is measured by means of the thermodynamic state variable entropy (high quality ~ low entropy). The second law of thermodynamics can be more precisely expressed as the existence of a state variable S, defined by:

$$dS = d_eS + d_iS \tag{3.4}$$

where $d_eS = dQ/T$ and $d_iS \geq 0$ (=0 for reversible processes), which will change during any process such that $dS \geq 0$.

Entropy has, however, the disadvantage that it has the unit J /K and is not defined far from thermodynamic equilibrium. It is therefore advantageous to apply another concept denoted as exergy.

Exergy is defined as the amount of work (= entropy-free energy) a system can perform when it is brought into thermodynamic equilibrium with its environment. Figure 3.2 illustrates the definition. The considered system is characterized by the extensive state variables S, U, V, N1, N2, N3..., where S is the entropy, U is the energy, V is the volume and N1, N2, N3 ..... are moles of various chemical compounds, and by the intensive state variables, T,p, μc1, μc2, μc3....The system is coupled to a reservoir , a reference state, by a shaft. The system and the reservoir are forming a closed system. The reservoir (the environment) is characterized by the intensive state variables To, po, μc1o,, μc2o, μc3o.....and as the system is small compared with the reservoir, the intensive state variables of the reservoir will not be changed by interactions between the system and the reservoir. The system develops toward thermodynamic equilibrium with the reservoir and is simultaneously able to release entropy-free energy to the reservoir. During this process the volume of the system is constant as the entropy-free energy must be transferred through the shaft only. The entropy is also constant as the process is an entropy-free energy transfer from the system to the reservoir, but the intensive state variables of the system become equal to the values for the reservoir. The total transfer of entropy-free energy in this case is the exergy of the system. It is seen from this definition that exergy is dependent on the state of the total system (= system + reservoir) and not dependent entirely on the state of the system. Exergy is therefore not a state variable. In accordance with the first law of thermodynamics, the increase of energy in the reservoir, ΔU, is:

$$\Delta U = U - Uo \tag{3.5}$$

where Uo is the energy content of the system after the transfer of work to the reservoir has taken place. According to the definition of exergy, Ex, we have:

$$Ex = \Delta U = U - Uo \tag{3.6}$$

As $$U = TS - pV + \sum_{c} \mu c\, Ni \tag{3.7}$$

(see any textbook on thermodynamics), and:

$$U_o = T_o S - p_o V + \sum_c \mu c_o N_i,$$

we get the following expression for exergy :

$$Ex = S(T - T_o) - V(p - p_o) + \sum_c (\mu c - \mu c_o) N_i \qquad (3.8)$$

Exergy covers, as demonstrated above, the high quality energy, i.e., the energy which can be used to do work. It is therefore a very useful concept in calculation of the efficiency of power plants, where the energy efficiency is 100% (energy is never lost), while the exergy efficiency is in the order of 35% -65% depending on the use of the waste heat.

As reservoir, reference state, we can for instance select the same system but at thermodynamic equilibrium, i.e., assume that all components are inorganic and at the highest oxidation state, if sufficient oxygen is present (nitrogen as nitrate, sulphur as sulphate and so on). The reference state will in this case correspond to the ecosystem without life forms and with all chemical energy utilized or as an "inorganic soup". Usually, it implies that we consider $T = T_o$, and $p = p_o$, which means that the exergy becomes equal to the Gibb's free energy of the system, or the chemical energy content + the thermodynamic information (see below) of the system. Notice that the above shown equation also emphasizes that exergy is dependent on the state of the environment (the reservoir = the reference state), as the exergy of the system is dependent on the intensive state variables of the reservoir.

Notice that exergy is not conserved - only if entropy-free energy is transferred, which implies that the transfer is reversible. All processes in reality are, however, irreversible, which means that exergy is lost (and entropy is produced). Loss of exergy and production of entropy are two different descriptions of the same reality, namely that all processes are irreversible, and we unfortunately always have some loss of energy forms which can do work to energy forms which cannot do work. So, the formulation of the second law of thermodynamics in terms of exergy is: All real processes are irreversible which implies that exergy is lost. Energy is of course conserved by all processes according to the first law of thermodynamics. It is therefore wrong to discuss, as already mentioned briefly, an energy efficiency of an energy transfer, because it will always be 100%, while the exergy efficiency is of interest, because it will express the ratio of useful energy to total energy which always is less than 100% for real processes. It is therefore of interest to set up for all environmental systems in addition to an energy balance also an exergy balance, too. Our concern is loss of exergy, because that means that "first class energy" which can do work is lost as "second class energy" which cannot do work.

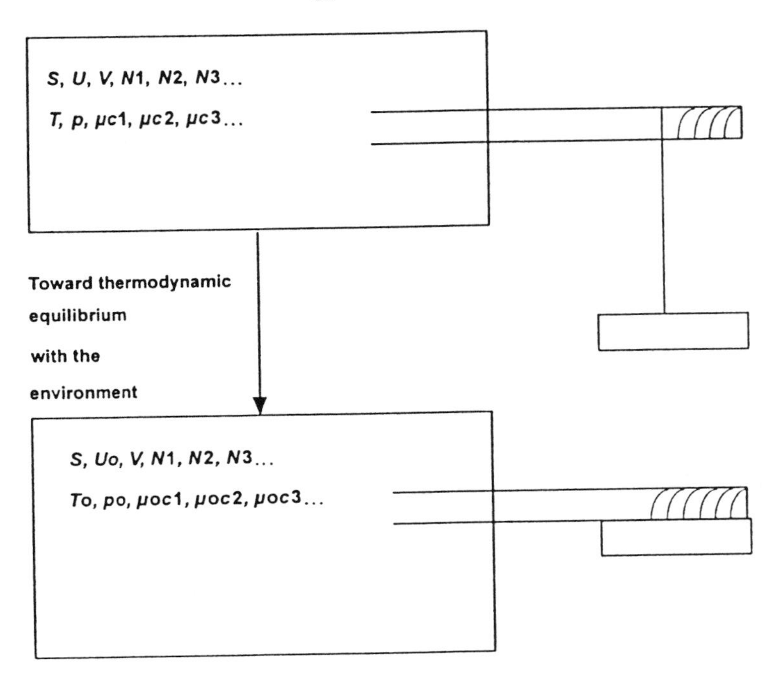

Fig. 3.2. Definition of exergy as the amount of the work the system can perform relatively to a reference state, for instance the environment.

Exergy seems more useful to apply than entropy to describe the irreversibility of real processes, as it has the same unit as energy and is an energy form, while the definition of entropy is more difficult to relate to concepts associated with our usual description of the reality. Finally it should be mentioned that the self organizing abilities of systems are strongly dependent on temperature, as it is discussed in Jørgensen et al. 1997. Exergy takes the temperature into consideration as the definition shows, while entropy doesn't.

Notice furthermore that information also contains exergy. Boltzmann (1905) showed that the free energy of the information that we actually possess (in contrast to the information we need to describe the system) is k*T* ln I, where I is the information we have about the state of the system, for instance that the configuration is 1 out of W possible (i.e., that W = I) and k is Boltzmann's constant = $1.3803 * 10^{-23}$ (J / molecules*deg). This implies that one bit of information has exergy equal to k T ln2. Transformation of information from one system to another is often almost an entropy-free energy transfer. If the two systems have different temperatures, the entropy lost by one system is not equal to the entropy gained by the other system, while the exergy lost by the first system is equal to the exergy transferred and equal to the exergy gained by the other system, provided that the transformation is not accompanied by any loss of exergy. In this case it is obviously more convenient to apply exergy than entropy.

Exergy is as seen closely related to information theory. A high local concentration of a chemical compound, for instance, with a biochemical function that is rare elsewhere, carries exergy and information. On more complex levels, information may still be strongly related to exergy but in more indirect ways. Information is also a convenient measure of physical structure. A certain structure is chosen out of all possible structures and defined within certain tolerance margins.

Biological structures maintain and reproduce themselves by transforming energy and information, from one form to another. Thus, the exergy of the radiation from the sun is used to build the highly ordered organic compounds. The information laid down in the genetic material is developed and transferred from one generation to the next. The chromosomes of one human cell have an information storage capacity corresponding to 2 billion K-bytes! When biological materials are used to the benefit of mankind it is in fact the organic structures and the information contained therein that are of advantage, for instance, when using wood. Information is of course of utmost importance for production systems, where the right management can be crucial for the efficiencies of the processes utilized in the system. Green accounting or environmental audit is a new method of environmental relevance to provide more information about the system; see Chapter 2. It can be shown, that exergy differences can be reduced to differences of other, better known, thermodynamic potential, (see Jørgensen, 1997) which may facilitate the computations of exergy in some relevant cases. As seen the exergy of the system measures the contrast - it is the difference in free energy if there is no difference in pressure, as may be assumed for an ecosystem or an environmental system and its environment - against the surrounding environment. If the system is in equilibrium with the surrounding environment the exergy is of course zero.

The second law of thermodynamics may be formulated by use of the concept of exergy as follows:**All real processes are irreversible which implies that exergy will be lost by all real processes. This means that high quality energy = exergy = energy which can do work will be lost and converted to heat energy without temperature differences which implies that it cannot be utilised to do work. ΔEx for any process < 0.**

Since the only way to move systems away from equilibrium is to perform work on them, and since the available work in a system is a measure of its exergy, we have to distinguish between the system and its environment or between the system and the same system at thermodynamic equilibrium alias for instance an inorganic soup. It is therefore reasonable to use the available work, i.e., the exergy, as a measure of the distance from thermodynamic equilibrium.

As we know that ecosystems due to the through-flow of energy have the tendency to develop away from thermodynamic equilibrium losing entropy or gaining exergy and information, we can put forward the following proposition of relevance for ecosystems: **Ecosystems attempt to develop toward a higher level of exergy.**

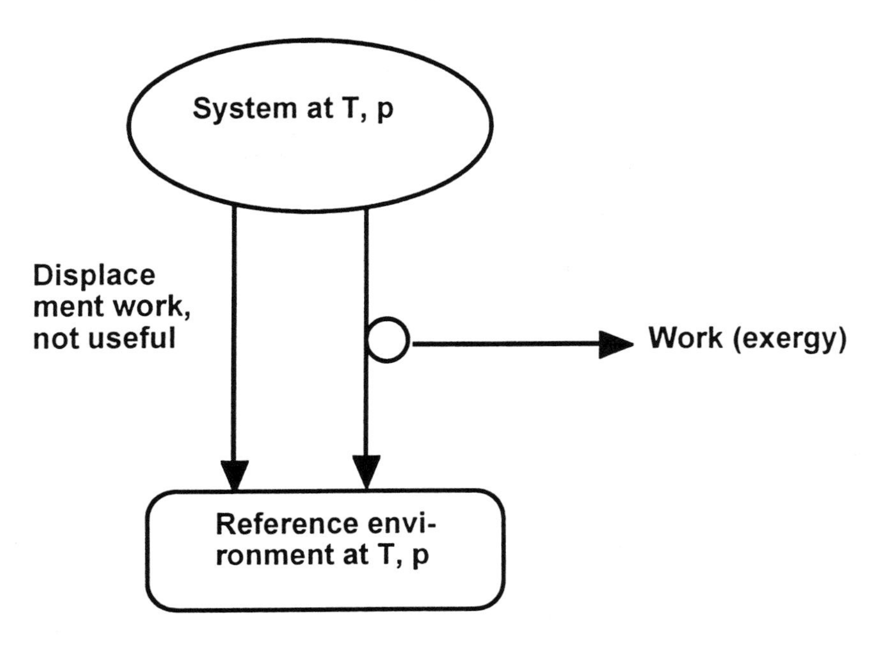

Fig. 3.3. The exergy content of the system is calculated in the text for the system relatively to a reference environment of the same system at the same temperature and pressure, but as an inorganic soup with no life, biological structure, information or organic molecules.

This description of exergy development in an ecosystem makes it pertinent to assess the exergy of ecosystems. It is not possible to measure exergy directly - but it is possible to compute it according to equation (3.8), if the composition of the ecosystem is known. If we presume a reference environment that represents the system (ecosystem) at thermodynamic equilibrium, which means that all the components are inorganic at the highest possible oxidation state if sufficient oxygen is present (as much free energy as possible is utilized to do work) and homogeneously distributed in the system (no gradients), the situation illustrated in Figure 3.3 is valid. As the chemical energy embodied in the organic components and the biological structure contributes by far the most to the exergy content of the system, there seems to be no reason to assume a (minor) temperature and pressure difference between the system and the reference environment. Under these circumstances we can calculate the exergy content of the system as coming entirely from the chemical energy: $\sum_{c}(\mu c - \mu co)$ Ni.

We can find by these calculations the exergy of the system compared with the same system at the same temperature and pressure but in the form of an inorganic soup without any life, biological structure, information or organic molecules. As $(\mu c - \mu co)$ can be found from the definition of the chemical potential replacing activities by concentrations, we get the following expression for the exergy:

$$Ex = RT \sum_{i=0}^{i=n} Ci \ln Ci / Ci,o \tag{3.9}$$

where R is the gas constant, T is the temperature of the environment (and the system; see Figure 3.3), while Ci is the concentration of the i'th component expressed in a suitable unit, e.g. for phytoplankton in a lake Ci could be expressed as mg /l or as mg /l of a focal nutrient. Ci,o is the concentration of the i'th component at thermodynamic equilibrium and n is the number of components. Ci,o is of course a very small concentration (except for $i = 0$, which is considered to cover the inorganic compounds), corresponding to a very low probability of forming complex organic compounds spontaneously in an inorganic soup at thermodynamic equilibrium. Ci,o is even lower for the various organisms, because the probability of forming the organisms is very low with their embodied information which implies that the genetic code should be correct.

The total exergy of an ecosystem *cannot* be calculated exactly, as we cannot measure the concentrations of all the components or determine all possible contributions to exergy in an ecosystem. If we calculate the exergy of a fox for instance, the above shown calculations will only give the contributions coming from the biomass and the information embodied in the genes, but what is the contribution from the blood pressure, the sexual hormones and so on? These properties are at least partially covered by the genes but is that the entire story? We can calculate the contributions from the dominant components, for instance by the use of a model or measurements that cover the most essential components for a focal problem.

Organisms, ecosystems and the entire ecosphere possess the essential thermodynamic characteristic of being able to create and maintain a high state of internal order or a condition of high exergy (or low entropy, which entropy can be said to measure disorder, lack of information about molecular details, or the amount of unavailable energy). Loss of exergy is a result of a continual dissipation of energy of high utility - light or food - to energy of low utility - heat. Order is maintained in the ecosystem by respiration, which continually produces disorder (heat). The second law of thermodynamics explains why ecosystems can maintain organisation or order. A system tends spontaneously toward increasing disorder (or randomness), and if we consider the system to consist of an ecosystem and its surroundings, we can understand that order (exergy) can be produced in the ecosystem if more disorder (loss of exergy) is created in its surroundings. In the ecosystem the ratio of total community respiration to the total community biomass (denoted as the R/B ratio) can be regarded as the maintenance to structure ratio or as a thermodynamic order function for homogeneous systems.

The behaviour of energy in an ecosystem can conveniently be termed the energy flow, because the energy transformation occurs only in one direction towards lower quality energy. However, as the chemical compounds carry chemical energy and the elements cycle, energy will also cycle. Solar radiation is the inflowing energy, which is used by phototrophic organisms to produce biomass. The chemical energy in the biological matter produced is of a non-random character, but at the same time heat is produced by respiration and becomes the outflowing energy. The chemical cycling of the elements is a result of this unidirectional flow of energy. Without solar radiation there would be no order in the ecosystem and no cycling of elements would take place. From a physical standpoint the environmental crisis is a crisis of lost exergy = production of entropy, as pollution makes disorder. An example of this is given in Fig. 2.30, which illustrates the accumulation of lead in the Greenland ice pack from 800 B.C. to the present. This steadily increased accumulation demonstrates that lead released to the atmosphere is distributed worldwide and that exergy is correspondingly decreased. That exergy is lost by distribution of pollutants can be demonstrated by a simple model consisting of two bulbs of equal volume, connected with a stop cock. If one chamber contains one mole of a pure gas and the second is empty, then opening the valve between the two chambers causes a loss in exergy :

$$\Delta Ex = -\int dQ = -Q = -RT * \ln V_2 / V_1 = RT * \ln 2 \qquad (3.10)$$

where ΔEx is the loss in exergy, V_2 is the volume occupied by the mole of gas after the valve was opened, while V_1 is the volume before the valve was opened. Q is the heat produced.

Thus paradoxically, the more we attempt to maintain order, the more energy we require and the greater stress we inevitably put on the environment.

3.2 ENERGY USE AND ENERGY RESOURCES

A human needs about 9,000 kJ per day to survive, but today the average inhabitant in most developed countries uses more than 900,000 kJ, 100 times more than the survival level. Most energy is consumed by the industrial nations. They have 28% of the world's population but consume 80% of the world's energy.

As already discussed, only 10-20% of the initial biomass can be used by the organisms on the next level in the foodchain. As biomass can be translated into energy (see Section 3.1), this is also true of energy transformation through a foodchain. This implies that the short foodchain of grain —> human should be prefered to the longer and more wasteful grain —> animal —> human. Of course the problem of food shortage cannot be solved so simply, since animals produce proteins with a more favourable amino acid composition for human food (lysine is missing in plant proteins) and eat plants that cannot all be used as human food at present. But to a certain extent food production can be increased by making the foodchain as short as possible. This concept also applies to the use of fossil fuel or other energy sources. Fig. 3.4 shows such an energy chain where the more links there are the greater the waste of energy.

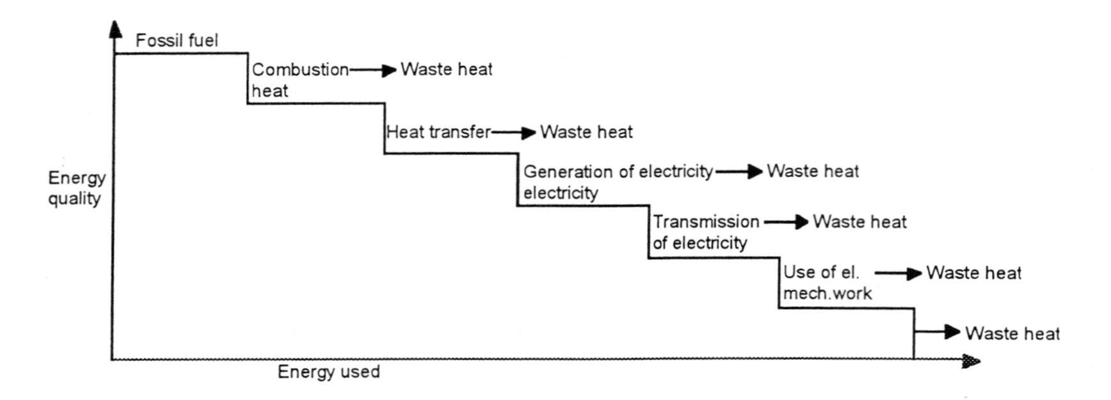

Fig. 3.4. Degradation of energy through an energy chain.

At each link in the chain energy is transferred from one form to another and heat is wasted as a consequence of the second law of thermodynamics. In Table 3.1 is listed the net energy efficiency for some of the most commonly used energy systems. We cannot escape the second law of thermodynamics, but we can minimize energy waste:

1. by keeping the energy chain *as short as possible,*
2. *by increasing the efficiency,* i.e. the ratio of useful energy output to the total energy input,
3. *by wasting as little heat to the surroundings as possible,* e.g. by *insulation,* and *using heat produced by energy transfer* (heat produced at power stations can be used for heating purposes).

The first law of thermodynamics can also be applied to the energy situation. The only free energy available is solar radiation, because the earth is an open system that gets energy from external sources; if we have to get energy from other sources it will always cost energy to provide energy. For example, at present only one third of the oil in an average reservoir is recovered, and the energy needed to produce the steam that is injected into the borehole to increase the degree of recovery may exceed the energy recovered as oil. As a result, although total energy production is declining total energy consumption is increasing.

Fig. 3.5 demonstrates how steeply energy consumption has increased in the twentieth century, and that the gap in per capita energy consumption between developed and developing nations is widening.

Based on the best possible information available today, it seems clear that the major supplies of oil and gas will probably run out somewhere between the years 2025 and 2050, while there is still sufficient coal for 200-400 years. Coal, however, cannot replace oil and gas completely because of pollution problems. Since it takes 25 to 50 years to develop a new energy system, we have no time to lose in seeking replacements for oil and gas.

TABLE 3.1
The exergy efficiencies of common individual systems

Conversion system	Percentage efficiency approximately
Processing of natural gas	97
Mining of nuclear fuel (uranium)	95
Processing of coal	92
Processing of oil	88
Home natural gas furnace	85
Electric car storage battery	80
Surface mining of coal	78
Extraction of natural gas	73
Propeller driven wind turbine	70
Home oil furnace	65
Fuel cell	60
Processing of nuclear fuel	57
Deep mining of coal	55
Fossil fuel power plant with proposed MHD topping cycle	50
Steam turbine engine	45
Diesel engine	42
Offshore oil well	40
Fossil fuel power plant and proposed nuclear breeder plant	38
Today's nuclear power plant	31
Internal combustion engine	30
Fluorescent lamp	28
Wankel engine	22
Advanced solar cells (if developed)	15
Today's solar cells	10
Incandescent lamp	5

We will not go further into energy problems, since this book is concerned with the principles of environmental science, but it is worth mentioning that the world's greatest energy problem today is food or maybe rather the distribution of food: about 40% of the world's population do not receive the basic minimum quantity or quality of energy needed to maintain good health. At the same time the population growth in the less developed countries is more than 2% with a doubling time of 30 years or less.

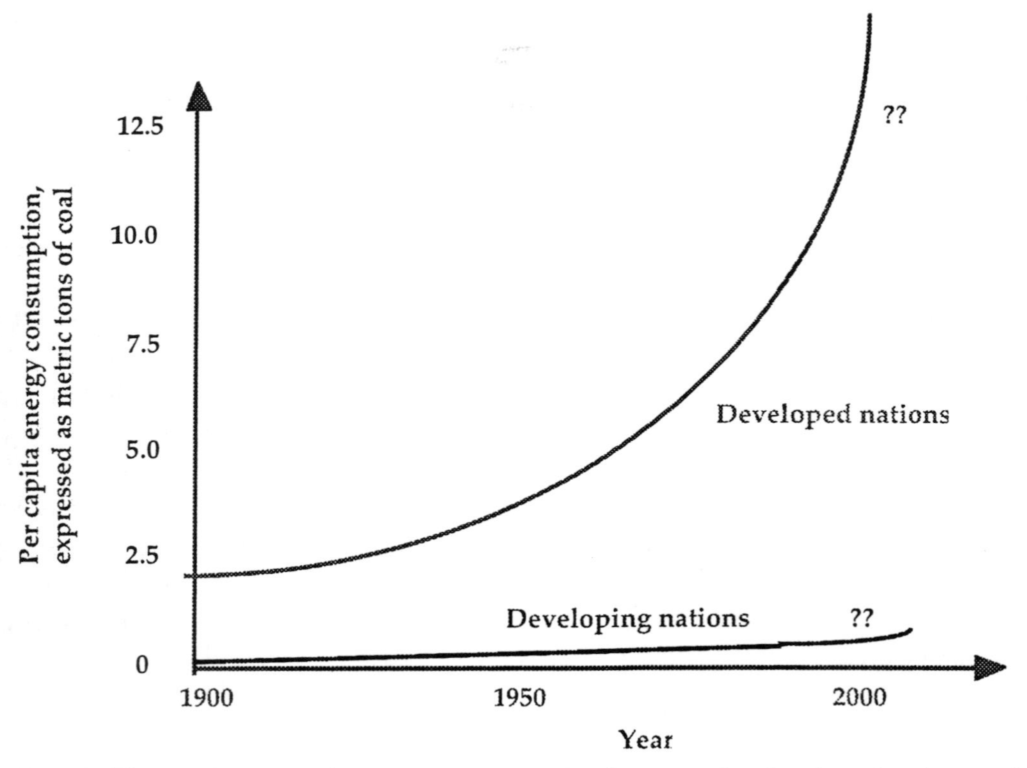

Fig. 3.5. The gap in per capita energy consumption between the developed nations and the underdeveloped nations is widening. (Source: United Nations)

3.3 PRODUCTIVITY

The *primary productivity* of an ecological system is defined as the rate at which radiant energy is stored by the photosynthetic and chemosynthetic activity of producer organisms, chiefly green plants, as chemical energy in the form of organic matter. An often used unit is used g (kcal, kJ) or g dry matter per m^2 or m^3 per 24h or per year. It is essential to distinguish between *gross primary productivity,* which is the total rate of synthesis including the organic matter used up in *respiration,* and *net primary productivity,* which is the rate of storage of organic matter. The rate of energy storage at consumer levels is referred to as *secondary productivity.*

The standing crop of biomass present at any given time must not be confused with net productivity, which is the change in biomass per unit of time:

$$P = dB / dt = f(B, \text{temp, environmental factors}) \qquad (3.11)$$

where P = productivity, B = biomass.
Table 3.2 gives some characteristic figures for gross and net primary productivity and Table 3.3 gives the estimated annual gross primary production of the biosphere and its distribution among major ecosystems.

Primary production in terms of food is summarized in Table 3.5 A+B, where the major food crops in developed and developing countries are compared. As can be seen, the developing countries have a lower per area unit production compared with the developed countries, which use fuel-subsidized agriculture. In 1999 there were an estimated $6*10^9$ people in the world each requiring $4.2*10^6$ kJ per year, amounting to a total of about $25*10^{15}$ kJoules per year, which represents only about 0.7% of the net primary production of the biosphere.

However, we must add to this the huge food consumption of all domestic animals. The standing crops of livestock in the world are at least 5 times that of humans in terms of equivalent food requirements. Thus man and his domestic animals consume at least 6.5% of net production of the whole biosphere or at least 12.5% of that produced on land. In addition, man also consumes huge quantities of primary production in the form of fibres, so his percentage consumption of net production on land is actually 15%. But what if the population doubles? Will there be enough food? Can we still continue to use and eat animals?

The situation is rapidly becoming critical, and our planning for the future must include the following points:

1. No more than about 25% of the land is truly arable. Irrigation of huge areas of dry land will require large expenditure of money (energy) and will have some severe ecological side-effects.
2. We are able to reduce the number of domestic animals, but man's need for animal proteins must not be underestimated.
3. Theoretically crop yields in the developing countries can be increased significantly, but these countries do not have the money to invest in agriculture and have growing populations. Even now it is almost impossible to maintain the existing per capita production of food, which is already too small.
4. High productivity can only be maintained by the use of energy subsidies and will be accompanied by environmental pollution due to the heavy use of machinery, herbicides, insecticides and fertilizers.

TABLE 3.2

Characteristic figures for gross and net primary producion

Ecosystem	Gross prim. production	Net prim. production
Agricultural land	650 g dry weight $m^{-2}y^{-1}$ (range 100-4000)	
Alfalfa field	24000 kcal $m^{-2}y^{-1}$	15200 kcal $m^{-2}y^{-1}$
Desert	200 kcal $m^{-2}y^{-1}$	150 kcal $m^{-2}y^{-1}$
Oak-Pine forest	11500 kcal $m^{-2}y^{-1}$	5000 kcal $m^{-2}y^{-1}$
Pine forest	12200 kcal $m^{-2}y^{-1}$	7500 kcal $m^{-2}y^{-1}$
Rain forest	45000 kcal $m^{-2}y^{-1}$	13000 kcal $m^{-2}y^{-1}$
Coal reef	18.2 g dry weight $m^{-2}day^{-1}$	
Grassland	1200 g dry weight $m^{-2}y^{-1}$ (range 250-2500)	500 g dry weight $m^{-2}y^{-1}$ (range 150-1500)
Eutrophic lakes	2.1 g dry weight $m^{-2}day^{-1}$	
Sargasso Sea	0.5 g dry weight $m^{-2}day^{-1}$	
Silver Spring	17.5-35 g dry weight $m^{-2}day^{-1}$	
Total biosphere	2000 kcal $m^{-2}y^{-1}$	
Temperate forest	8000 kcal $m^{-2}y^{-1}$	3600 kcal $m^{-2}y^{-1}$

TABLE 3.3

Estimated gross primary production (annual basis) of the biosphere and its distribution among major ecosystems

Ecosystem (10^6 km^2)	Total Area	Gross Primary Productivity (kcal/m^2/year)	Total Gross Production (10^{16} kcal/year)
Marine *)			
Open ocean	326.0	1,000	32.6
Coastal zones	34.0	3,000	6.8
Upwelling zones	0.4	6,000	0.2
Estuaries and reefs	2.0	20,000	4.0
Subtotal	362.4	-	43.6

TABLE 3.3 (continued)

Terrestrial **)			
Deserts and tundras	40.0	200	0.8
Grasslands and pastures	42.0	2,500	10.5
Dry forests	9.4	2,500	2.4
Boreal coniferous forests	10.0	3,000	3.0
Cultivated lands with little or no energy subsidy	10.0	3,000	3.0
Moist temperate forest	4.9	8,000	3.9
Fuel subsidized (mechanized agriculture	4.0	12,000	4.8
Wet tropical and subtropical (broadleaved evergreen forests)	14.7	20,000	29.0
Subtotal	135.0	-	57.4
Total for biosphere (not included ice caps)	500.0	2,000	100.0

*) Marine productivity estimated by multiplying net carbon production figures by 10 to get the number of kcal, then doubling these figures to estimate gross production and adding an estimate for estuaries (not included in his calculations).
**) Terrestrial productivity based on net production figure by Lieth and Whittaker, 1975, doubled for low biomass systems and tripled for high biomass systems (which have high respiration) as estimates of gross productivity. Tropical forests have been upgraded in light of recent studies, and the industrialized (fuel subsidized) agriculture of Europe, North America, and Japan has been separated from the subsistence agriculture characteristic of most of the world's cultivated lands.

These points illustrate well how the environmental crisis, the energy crisis and the food crisis are linked together. Only the solution of all three problems simultaneously can be considered a real solution.

3.4 ENERGY IN ECOSYSTEMS

The transfer of energy from its source in the plants through herbivorous animals (primary consumers) to carnivorous animals (secondary consumers) and further to the top carnivore (tertiary consumers) is called *the foodchain.* By using the first law of thermodynamics an energy balance can be set up for each level in the food chain as illustrated in Figure 3.6. In nature the food relationships are rarely as simple as a foodchain, but rather form a foodweb.

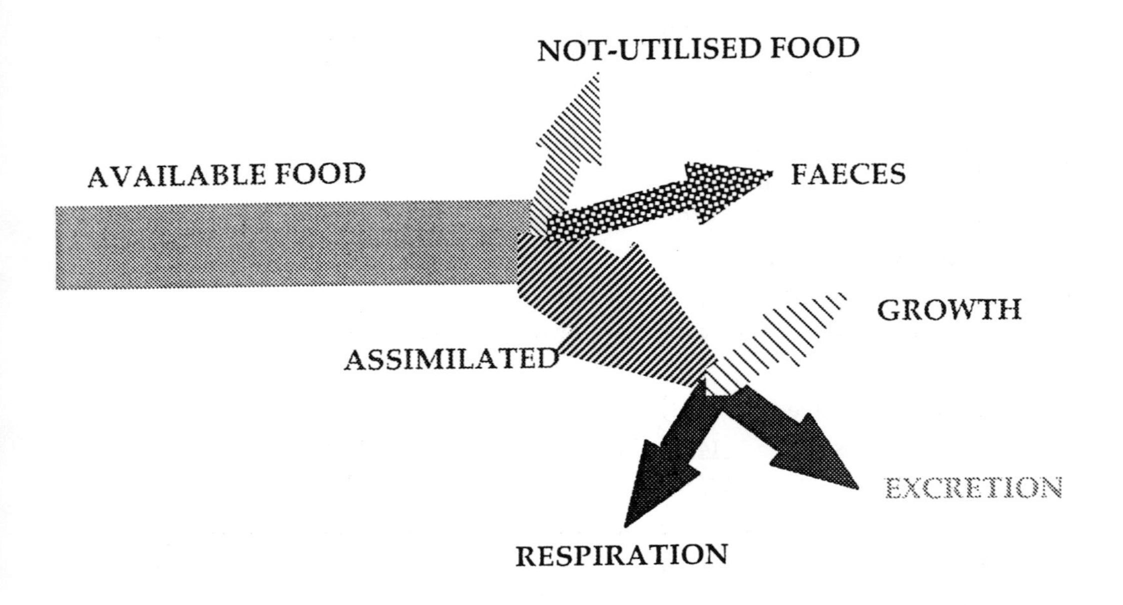

Fig. 3.6. An energy balance on the organism level. The available food is either not utilised, not digested (faeces) or assimilated. The assimilated food is utilised for growth, excretion and respiration.

The relationships can also be illustrated by means of so-called ecological pyramids, which can either represent the number of individuals, the biomass (or energy content) or the energy flow on each level in the foodchain or foodweb (see Fig. 3.7). Only the energy flow forms a true pyramid, in accordance with the first law of thermodynamics. The number pyramids are affected by variation in size and the biomass pyramids by the metabolic rates of individuals.

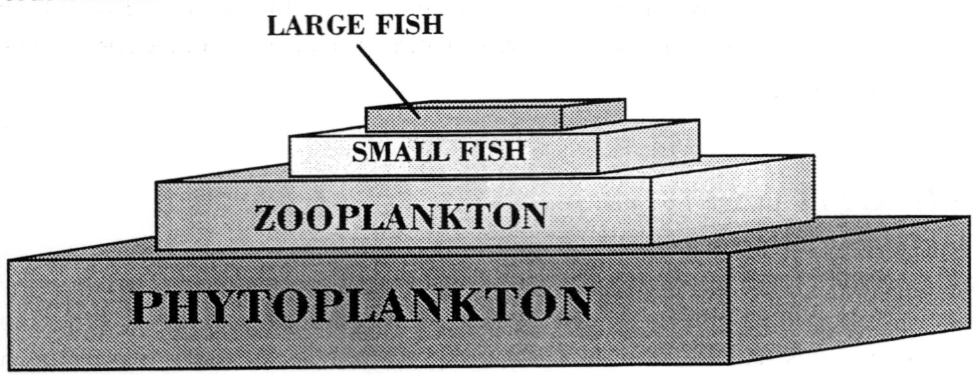

Fig. 3.7 An energy flow pyramid in the sea.

Ecological energy flows are of considerable environmental interest as calculation of biological magnifications are based on energy flows (see Chapter 2).

Ecological efficiency should also be mentioned here, see Table 3.4, where some useful definitions are listed. Some characteristic efficiencies are listed in Tables 3.5 - 3.7.

TABLE 3.4
Ecological efficiency

Concept	Definition *)
Lindeman's efficiency	Ratio of energy intake level n to n-1: I_n/I_{n-1}
Trophic level assimilation efficiency	A_n/A_{n-1}
Trophic level production efficiency	Net Production $_n$/NetProduction$_{n-1}$
Tissue growth efficiency	Net Production$_n$/A_n
Ecological growth efficiency	Net Production$_n$/I_n
Assimilation efficiency	A_n/I_n
Utilization efficiency	I_n/NetProduction $_{n-1}$

TABLE 3.5
Assimilation efficiency (A/I) for selected organisms (after various authors)

Taxa	A/I value
Internal parasites	
Entomophagous Hymenoptera *Ichneumon* sp.	0.90
Carnivores	
Amphibian (*Nectophrynoides occidentalis*)	0.83
Lizard (*Mabuya buettneri*)	0.80
Praying mantis	0.80
Spiders	0.80 to 0.90
Warm- and cold-blooded herbivores	
Deer (*Odocoileus* sp.)	0.80
Vole (*Microtus* sp.)	0.70
Foraging termite (*Trinervitermes* sp.)	0.70
Impala antelope	0.60
Domestic cattle	0.44
Elephant (*Loxodonta*)	0.30
Pulmonate mollusc (*Cepaea* sp.)	0.33
Tropical cricket (*Orthochtha brachycnemis*)	0.20
Detritus eaters	
Termite (*Macrotermes* sp.)	0.30
Wood louse (*Philoscia muscorum*)	0.19
Soil-eating organisms	
Tropical earthworm (*Millsonia anomala*)	0.07

TABLE 3.6
Tissue Growth Efficiency (NP/A) for selected organisms (after various authors)

Taxa	NP/A value
Immobile, cold-blooded internal parasites	
Ichneumon sp.	0.65
Cold-blooded, herbivorous and detritus-eating organisms	
Tropical cricket (*Orthochtha brachycnemis*)	0.42
Other crickets	0.16
Pulmonate mollusc (*Cepaea* sp.)	0.35
Termite (*Macrotermes* sp.)	0.30
Termite (*Trinervitermes* sp.)	0.20
Wood louse (*Philoscia muscorum*)	0.16
Cold-blooded, carnivorous vertebrates and invertebrates	
Amphibian (*Nectophrynoides occidentalis*)	0.21
Lizard (*Mabuya buettneri*)	0.14
Spiders	0.40
Warm-blooded birds and mammals	
Domestic cattle	0.057
Impala antelope	0.039
Vole (*Microtus* sp.)	0.028
Elephant (*Loxodonta*)	0.015
Deer (*Odocoileus* sp.)	0.014
Savanna sparrow (*Passerculus* sp.)	0.011
Shrews	Even lower values

TABLE 3.7
Ecological Growth Efficiency (NP/I) for selected organisms
(after various authors)

Taxa	NP/I value
Herbivoruous mammals	
Domestic cattle	0.026 (0.44 x 0.057)
Impala antelope	0.022 (0.59 x 0.039)
Vole (*Microtus* sp.)	0.020 (0.70 x 0.285)
Deer (*Odocoileus* sp.)	0.012 (0.80 x 0.014)
Elephant (*Loxodonta*)	0.005 (0.30 x 0.015)
Birds	
Savanna sparrow (*Passerculus* sp.)	0.010 (0.90 x 0.011)
Herbivorous invertebrates	
Termite (*Trinervitermes* sp.)	0.140 (0.70 x 0.20)
Tropical cricket (*Orthochtha brachycnemis*)	0.085 (0.20 x 0.42)
Other crickets (New Zealand taxa)	0.050 (0.31 x 0.16)
Pulmonate mollusc (*Cepaea* sp.)	0.130 (0.33 x 0.30)
Detritus-eating and soil-eating invertebrates	
Termite (*Macrotermes* sp.)	0.090 (0.30 x 0.30)
Wood lause (*Philoscia muscorum*)	0.030 (0.19 x 0.16)

TABLE 3.7 (continued)

Tropical earthworm (*Millsonia anomala*)	0.005 (0.076 x 0.06)
Carnivorous vertebrates	
Lizard (*Mabuya* sp.)	0.100 (0.80 x 0.14)
Amphibian (*Nectophrynoides occidentalis*)	0.180 (0.83 x 0.21)
Carnivorous invertebrate	
Spiders	0.350 (0.85 x 0.42)
Internal parasites	
Ichneumon sp.	0.580 (0.90 x 0.65)

There is a close relationship between energy flow rates and organism size. Some of these relationships are illustrated in Figs. 3.8-3.10. All these examples illustrate the fundamental relationship between size (surface) and the biochemical activity in the organisms. The surface determines the contact with the environment quantitatively, and thereby the possibilities of taking up food and excreting waste substances. The same relationship explains the graphs shown, where biochemical processes of hydrophilic toxic substances are involved. These figures are constructed from literature data and the excretion and uptake rates (for aquatic organisms) follow the same trends as the metabolic rate. This is of course not surprising, as the excretion is strongly dependent on the metabolism and the direct uptake is dependent on the surface.

The concentration factor, which indicates the ratio of concentration in the organism to the concentration in the medium, also follows the same lines; see Fig. 3.10. By equilibrium, the concentration factor can be expressed as the ratio between the uptake and excretion rates, as shown in Jørgensen (1979a). As most concentration factors are determined by equilibrium, the relationship found seems reasonable. Intervals for concentration factors are here indicated for some species, in accordance with the literature (see Jørgensen et al., 2000).

The principles illustrated in Figs. 3.8-3.10 can be applied generally on mainly hydrophilic compounds. In other words, it is possible to find the uptake and excretion rates and the concentration factors, provided that these parameters are available for the considered element or compound for one, but preferably several, species. With the parameter for one species, it is possible to draw a line with a slope of - 1 in a diagram similar to Figs. 3.8-3.10. With the parameters for two or more species, it is possible to control whether a slope of - 1 is correct. When a plot similar to Figs 3.8-3.10 is constructed, it is possible to read off the parameters for any organisms or species provided that the size of the organism or species is known. For lipohilic components, the lipid content of the organism is also of importance. This will be discussed further in Chapter 4.

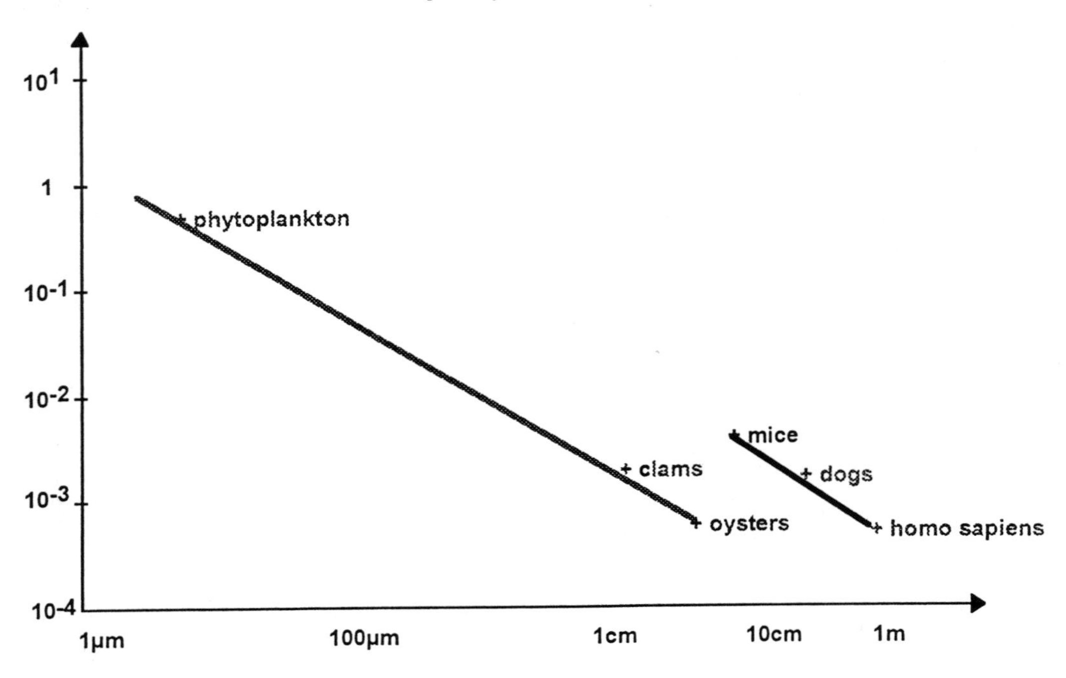

Fig. 3.8. Excretion of Cd ($24h^{-1}$) plotted against the length of various organisms.

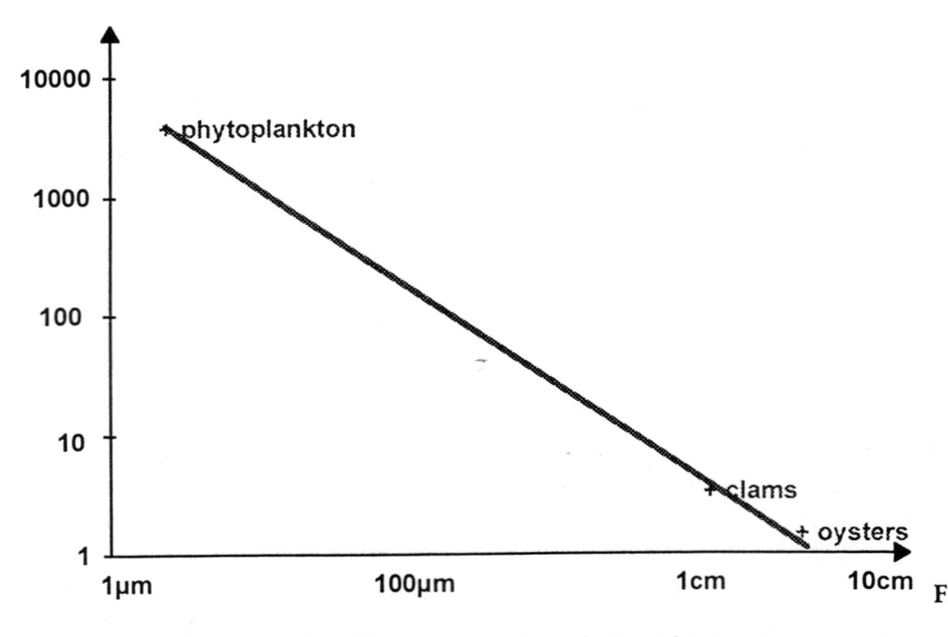

Fig

Fig. 3.9. Uptake rate for Cd (µg/g 24H) plotted against the length of various organisms.

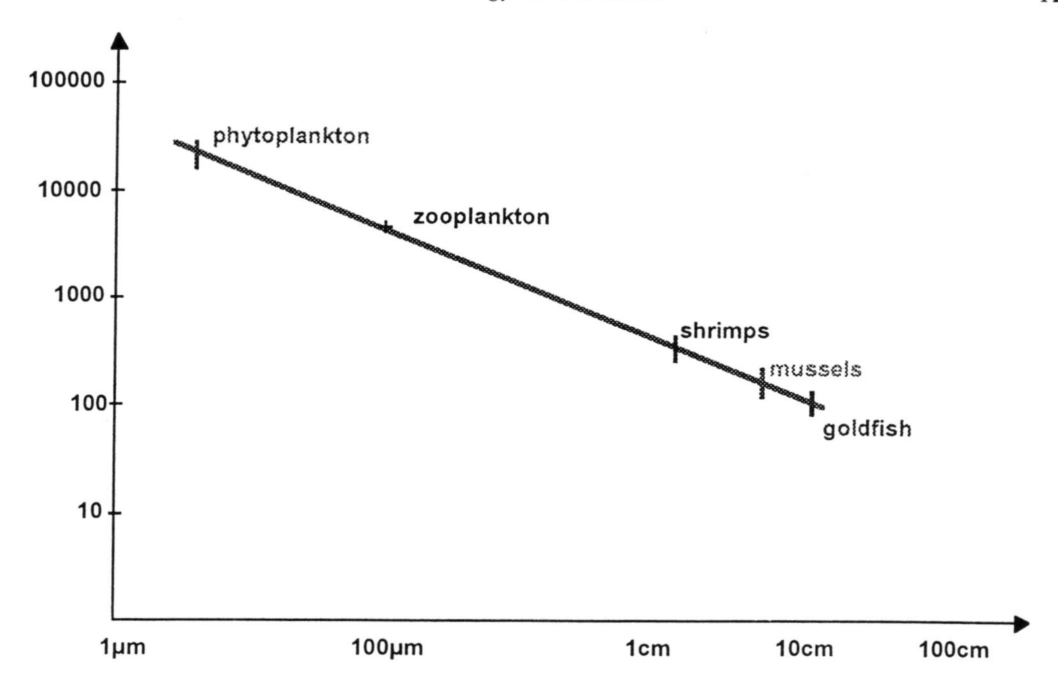

Fig. 3.10. CF for Cd versus the length of various organisms.

There is a fixed upper limit to the total biomass of any species that can be supported in a given habitat. This fixed limit is referred to as the *carrying capacity* of that particular habitat for a particular species. **The carrying capacity is related to the organism size in a similar manner to energy flow rates:**

$$C = w * p^{3/2} \tag{3.12}$$

where C is the carrying capacity, w is the mean weight of individuals and p is the population density (White and Herper, 1970).

This means, since C is fixed, that we can increase the mean weight of the plants in a population or the population density, but not both. Space itself can be considered an important resource, independently of any other resources. Increasing the space per individual may have beneficial effects, even where other resources remain constant. Any self-sustaining ecosystem will contain a wide spectrum of organisms ranging in size from tiny microbes to large animals and plants. The small organisms account in most cases for most of the respiration (energy turnover), whereas the larger organisms comprise most of the biomass.

These considerations are based on allometric principles (Peters, 1983 and Straskraba et al., 1999), which can be used to assess the relationship between the size of the units in the various hierarchical levels and the process rates, determining the need for the rate of energy supply. All levels in the entire hierarchy of an ecosystem are, therefore, due to the hierarchical organization, characterized by a rate which is ultimately determined by their size. Openness is proportional to the area available for exchange of energy and matter, relative to the volume = the inverse space scale (L^{-1}). It may also be expressed as the supply rate = k · gradient · area relative to the rate of needs, which is proportional to the volume or mass.

An ecosystem must, as previously mentioned, be open or at least non-isolated to be able to import the energy needed for its maintenance.

Table 3.8 illustrates the relationship between hierarchical level, openness, and the four scale hierarchical properties presented in Simon (1973).

The openness is here expressed as the ratio of area to volume. For the higher levels in the hierarchy approximate values aare used. As we move upwards in the hierarchy, the exchange of energy (and matter) becomes increasingly more difficult due to a decreasing openness.

It becomes increasingly more difficult to cover needs, which explains why energy density, time scale and dynamics decrease according to the inverse space scale or openness, or expressed differently as the rates are adjusted to make the possible supply of energy sufficient. These considerations are consistent with the relationship between size and time scale of levels in the hierarchy, as presented by O Neill et al. (1986) and Shugart and West (1981).

The energy received by ecosystems as solar radiation comes in small packages (quanta, hñ, where ñ is the frequency), which makes only utilization on the molecular level possible. The energy can be used on all hierarchical levels by an interactive coupling . The exchange of energy and matter on each level is dependent on openness, measured by the available area for exchange of energy and matter relative to the volume. Openness becomes the measure of the dynamics of the hierarchical level. Openness is inverse to hierarchical scale.

Exchange of matter and information with the environment of open systems is not absolutely necessary, as energy input (non-isolation) is sufficient (the system is non-isolated) to ensure maintenance far from equilibrium. However, it often gives the ecosystem some additional advantages, for instance by input of chemical compounds needed for certain biological processes or by immigration of species offering new possibilities for a better ordered structure of the system. The importance of the latter consequence of openness is clearly illustrated in the general relationship between species diversity, SD, of ecosystems on islands and the area of the islands, A:

$$SD = C^* A \text{ (number)}, \tag{3.13}$$

where C and z are constants. The perimeter relative to the area of an island determines how "open" the island is to immigration or dissipative emigration from or to other islands or the adjacent continent. The unit (L^{-1}) is the same as the above used area to volume ratio as a measure of openness.

TABLE 3.8

The relationship between hierarchical levels, the approximate magnitudes of their openness and approximate values of the typical four scale-hierarchical properties (energy portions / volume, space scale, time scale, and behavioural frequencies)

Hierarchical level	Openness (m^{-1})	Energy (kJ/m^3)	Space scale (m)	Time scale (sec.)	Dynamics (g/m^3sec)
Molecules	10^9	10^9	10^{-9}	$< 10^{-3}$	10^4-10^6
Cells	10^5	10^5	10^{-5}	10-10^3	1-10^2
Organ	10^2	10^2	10^{-2}	10^4-10^6	10^{-3}-0.1
Organism	1	1	1	10^6-10^8	10^{-5}-10^{-3}
Populations	10^{-2}	10^{-2}	10^2	10^8-10^{10}	10^{-7}-10^{-5}
Ecosystems	10^{-4}	10^{-4}	10^4	10^{10}-10^{12}	10^{-9}-10^{-7}

Different species have very different types of energy use to maintain their biomass. For example the blue whale uses most (97%) of the energy available for increasing the biomass for growth and only 3% for reproduction.

Whales are what we call K-strategists, defined as species having a stable habitat with a very small ratio between generation time and the length of time the habitat remains favourable. It means that they will evolve toward maintaining their population at its equilibrium level, close to the carrying capacity. K-strategists are in contrast to r-strategists which are strongly influenced by any environmental factor. Due to their high growth rate they can, however, utilise suddenly emergent favourable conditions and increase the population rapidly. Many fishes, insects and other invertebrates are r-strategists. The adult female reproduces every season she is alive and the proportion going into reproduction can be over 50%.

Ecological energy flows are very sensitive to man's impact on ecosystems. A detailed picture of the energy flow in an ecosystem is a very useful tool for understanding this influence.

The function of the ecosystem is closely related to the energy flow, and any change in the flow will mean a change in ecological function.

Fig. 3.11 shows the energy flow in a man-made ecosystem. Notice that it is characterised by an input of man-controlled energy.

By comparison with the energy flows in a natural ecosystem, it can be seen that the energy flow is simpler. Man-made ecosystems (chiefly agriculture systems) often have little ecological diversity, in other words the number of species is small. In many cases, however, this renders the ecosystem very vulnerable to any (unexpected) change. This is discussed further in Chapter 4. The stability of the ecosystem may not be changed due to a lower diversity, but the ability to absorb (buffer) unexpected changes may because of low diversity.

Man's influence on an ecosystem quite often means a simplification or decrease in its complexity, although the relationship between an ecosystem's stability and complexity is rather complicated .

A possible relationship between ecosystem's stability and biodiversity will be discussed in Section 4.5 . It will here be emphasised that there is no clear and simple relationship between ecosystem's stability and biodiversity, which was, however, the general opinion in ecology 25 years ago.

A total energy balance for a terrestrial ecosystem can be set up by use of the first law of thermodynamics :

$$R + C + H + ¥ * E = O \quad (3.14)$$

where R is the net radiation, C is the conductance in solids, H is the convection in moving fluids and ¥ * E is the latent heat exchange associated with evaporation. C, H and ¥ * E are negative.

Net radiation is the dominant term in the equation. It is the balance between incoming and outgoing fluxes of shortwave and longwave radiation. R can be found from:

$$R = (1-r)\,G + é(L - \partial T^4) \quad (3.15)$$

where G and L are incoming fluxes of shortwave and longwave radiation, respectively; ∂ is a constant (= $81.7 * 10^{-12}$ for R in 1y min^{-1}), r is the reflectivity coefficient, é the emissivity coefficient and T the absolute temperature. r, é and T are ecosystem surface characteristics.

Generally é ≈ 1 for natural mineral and organic surfaces, while r, expressed as a percentage (= albedo), varies widely with the surface composition.

Notice that the albedo is high for snow which implies that the greenhouse effect corresponding to higher temperature will decrease the albedo of the earth. From Table 3.12 we can read that 1% decrease relative to the present albedo will give an increase in the global temperature of 0.14 ^{o}C.

This relationship may cause a "runaway" effect:
higher temperature ----> lower albedo -----> higher temperature -------> and so on.

Table 3.10 shows the energy and water budget for land areas of the earth as a function of the latitude.

TABLE 3.9
Typical albedoes for some natural surfaces

Surface		Description; conditions	Albedo
Water			
	Liquid	Solar altitude: 60°	5
		30°	10
		20°	15
		10°	35
		5°	60
	Solid	Fresh snow	75
		Old snow	50
		Glacier ice	30
Ground			
	Soil	Dark organic	10
		Clay	20
		Light sandy	30
	Sand	Gray: wet	10
		dry	20
		White: wet	25

TABLE 3.9 (continued)

Typical albedoes for some natural surfaces

Surface		Description; conditions	Albedo
		dry	35
	Rock	Sandstone spoil, dry	20
		Black coal spoil, dry	5
Vegetation			
	Grass	Typical field: green	20
		dry	30
		Dry steppe	25
		Tundra, heather	15
	Crops	Cereals, tobacco	25
		Cotton, potato	20
		Sugar cane	15
	Trees	Rain forest	15
		Eucalyptus	20
		Red pine forest	10
		Hardwoods in leaf	18

TABLE 3.10

Mean values of the components of water and energy budgets for land areas of the earth.

Latitude	Water budget (mm y^{-1})			Energy budget (kly y^{-1})		
(degrees)	P	Q	E	lE	H	R
Polar (N)	176	- 106	- 70	- 4	4	0
70-60	428	-227	- 201	-12	- 8	20
60-50	577	-259	- 318	-19	-11	30
50-40	535	-155	- 380	- 22	-23	45
40-30	534	-122	- 412	-24	-36	60
30-20	611	-245	- 366	- 21	-48	69
Tropics (N)	1,292	- 436	- 856	-50	-22	72
Tropics (S)	1,576	- 546	-1,030	-60	-13	73
20-30	564	- 88	- 476	-28	-42	70
30-40	660	-165	- 495	-29	-33	62
40-50	1,302	-914	- 388	-23	-18	41
50-60	993	-605	- 388	-23	- 8	31
60-70	429	-369	- 60	- 4	- 9	13
Polar (S)	148	-120	- 28	- 2	6	- 4

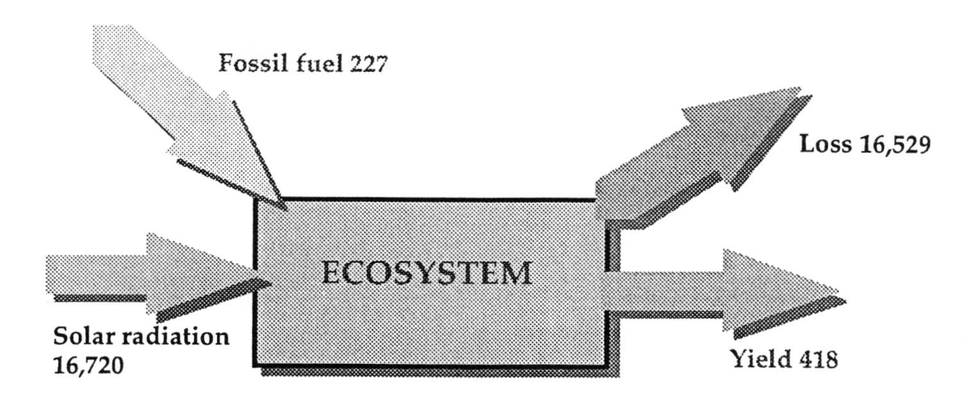

Fig. 3.11. Man-controlled ecosystem (agriculture). Numbers in kJ $m^{-2}day^{-1}$.

3.5 ENERGY CONSUMPTION AND THE GLOBAL ENERGY BALANCE

Recently the subjects of climate change and, particularly, the possible effects of CO_2 (carbon dioxide) on the climate are being widely discussed. Currently the question of interaction between atmospheric carbon dioxide and climate is a very critical one, that could greatly affect future energy strategies.

In 1975 the world consumption of fossil fuel corresponded to 4000 Tg C /y. Today, 25 years later the carbon emission has increased to more than 6000 Tg C/ y. Numerous estimations of future minimum energy demands have been made recently. A typical average scenario gives a total global energy demand of 35 TW in year 2030 or about 3 times the present energy consumption. The average energy consumption is predicted at 4.4 kW per capita in 2030, while only a few developed countries today have an energy consumption larger than 2.5 kW. The cumulated overall world energy demand by 2030 will be approximately 1000 TW for this reference scenario. The crucial questions are: can we meet the demand for such a high energy consumption? If so, how, and what will the environmental consequences be?

Another problem is of course the size of available fossil energy resources needed to supply this large energy demand. The economically recoverable reserves of fossil energy predict that in 2030 only a fraction of the 35 TW can be supplied in the form of conventional oil, gas and coal. The geological resources of fossil energy are much larger, but cannot be recovered under present and predicted technological and economic conditions.

A further constraint on the massive consumption of fossil fuel is represented by the CO_2 problem. And what we are considering here is: how such large scale use of fossil fuel would influence the global climate.

The use of fossil fuel means the oxidation of carbon or carbon hydrides, which produces a stoichiometric amount of carbon dioxide:

$$C + O_2 \rightarrow CO_2 \qquad (3.16)$$

$$C_xH_y + (x + y/4)O_2 \rightarrow 1/2\, y\, H_2O + xCO_2 \qquad (3.17)$$

Having determined that the burning of fossil fuels produces carbon dioxide, we can next consider how this carbon dioxide (and water vapour) produced by oxidation will influence the climate. This is a very complex question and any attempt to an answer requires a comprehensive knowledge of the global cycle. Another crucial issue related to the long-term prediction model is the estimation of the ultimate recovery of coal, oil and natural gas. Table 3.11 gives the most recent estimates of global fossil fuel resources and the corresponding amounts of carbon dioxide that would be produced, calculated from the following information (Ziemen et al., , 1977):

Coal 2.00 moles of carbon dioxide per MJ

Oil 1.68 moles of carbon dioxide per MJ
Gas 1.29 moles of carbon dioxide per MJ

TABLE 3.11
Recent estimates of global fossil fuel resources and the corresponding amount of carbon dioxide expressed as 10^{15} moles (a summary of various sources of information)

	Estimates of global fossil fuel resources (year 2000)	
	10^{21} J	10^{15} mol CO_2
Solid fuel	150 -300	300 - 600
Liquid fuel	40 -120	60 -200
Tars and shales	15 -30	25 -50
Natural gas	10 -75	12 -95

One part of the problem is the quantitative determination of the carbon pools and the fluxes between these pools, another is the relationship between the atmospheric carbon dioxide concentration and the climate. This issue is also very complex as different subsystems interact with each other. The global climatic system consists of 5 subsystems: the atmosphere, the oceans, the cryosphere (ice and snow), the land surface and the biosphere. These subsystems interact with each other through a wide variety of processes. The climatic state (Wisniewski and Lugo, 1992) is defined as the average and the variability of the complete set of atmospheric, hydrospheric and cryospheric variables over a specified time interval and in a specified domain of the earth-atmosphere system.

The climatic states are subject to fluctuations of a statistical origin in addition to those of a physical nature. These statistical fluctuations in the weather are unpredictable over the time scales of climatological interest and are therefore referred to as "noise" or the "inherent variability of the climatic system". With reference to the carbon dioxide problem, we are not concerned with the inherent variability of the climate, but with the natural variability caused by changes in external and/or internal conditions.

However, any estimate of a change in climate caused by an increase in carbon dioxide concentration must take into consideration the fact that climate fluctuates naturally with time, and therefore that it takes a long time to detect a significant change in climate. It should also be pointed out that problems could arise because of the time scales of natural variability, e.g. the time constant for heat storage in the oceans. At the edge of the third millenium, it is, however, possible to conclude that an increase in the global average temperature probably in the order of 0.5°C, has taken place.

A global average model can be used to compute changes in average climate for a given change in atmospheric carbon dioxide concentration. Such models have been available the last 20 years, although results are still uncertain due to lack of knowledge about feedback mechanisms. The question of regional climatic changes is, however, of more practical importance. In this context it should be stressed that **it is possible to have drastic changes in the regional climate without any substantial change in the global average situation.**

The global energy balance (see Fig. 3.12) is influenced by the carbon dioxide concentration in the atmosphere. The atmosphere is transparent for solar radiation, but prevents, to a certain extent, the loss of heat from the earth's surface (by longwave radiation). The radiation balance of the atmosphere is essentially determined by the presence of optically active minor constituents, such as water vapor, carbon dioxide, ozone, aerosols and others. An increased carbon dioxide concentration means increased absorption of longwave radiation and consequently an increased global average temperature. Carbon dioxide has what is called a greenhouse effect, but the atmosphere also contains a number of other minor constituents capable of causing changes in the global average temperature. Table 3.12 lists the calculated surface temperature change for a 1% relative increase in each of the parameters given (the table is based upon the following sources: Reck, 1978; Wang et al., 1976; Reck and Fry, 1977; Ramanathan, 1975; Hummel and Reck, 1978, Wisniewski and Lugo, 1992 and Koch and Mooney, 1996). However, from this table it can be seen that the concentration of carbon dioxide seems to be the most important agent in this respect, although the role of trace gases has not yet been investigated in detail.

The relationship between the atmospheric carbon dioxide concentration and the surface temperature of the earth has been calculated as the average of several model results. Changes in cloud cover can be taken into account by assuming either a fixed cloud top altitude (CTA) or a fixed cloud top temperature (CTT). If CTA is assumed a doubling of the concentration of carbon dioxide in the atmosphere from the concentration about 100 years ago of 280 ppm to 560 ppm will imply a 1.5^{o}C increase in the average global temperature, while assumption of CTT will predict a 4.5^{o}C temperature increase. If we use the figures in Table 3.12, a doubling will correspond to an increase in the global average surface temperature on 2°C, while the present observed increase during the last 100 years would give a temperature increase of 0.6^{o}C which is close to the observed increase of about 0.5^{o}C according to the metereological records.

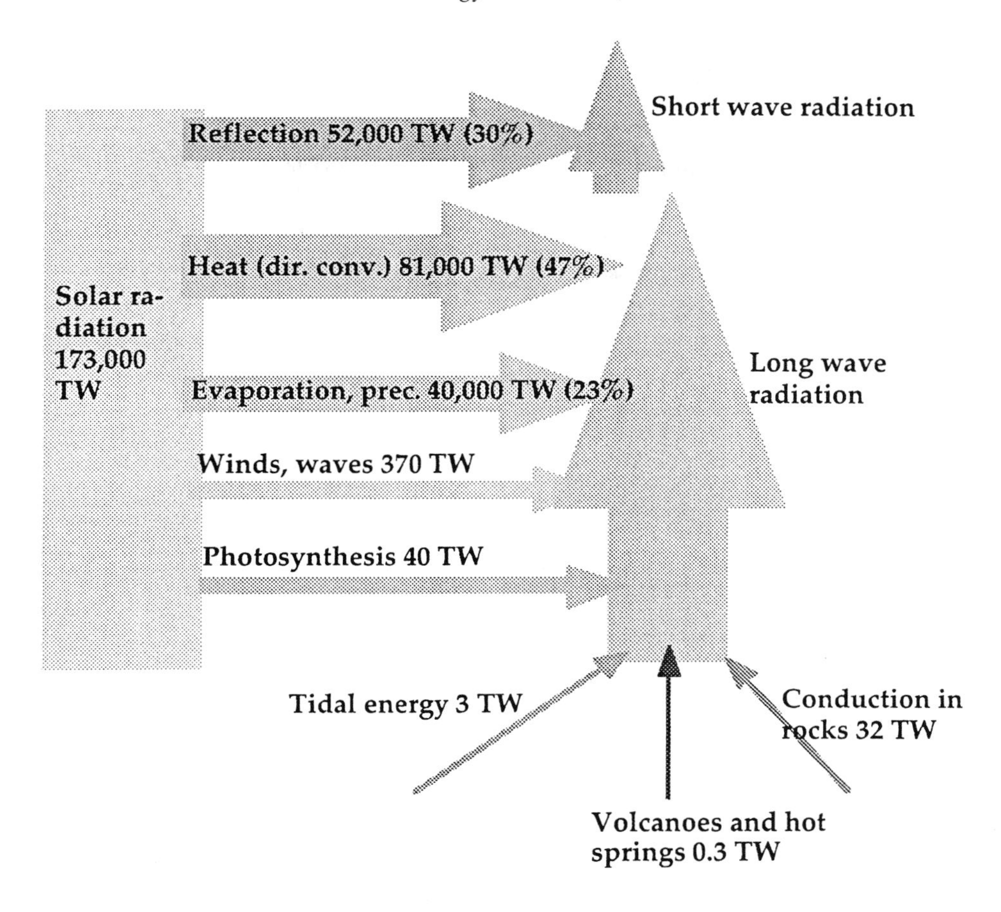

Fig. 3.12. The global energy balance. More than 99% of the input is solar radiation. Terrestrial energy (nuclear, thermal and gravitational energy) and fossil energy are negligible in relation to the global energy balance.

Even if the concentration of carbon dioxide could be accurately predicted, vast uncertainties about its effect on the climate would remain. However, there is strong evidence to suggest that additional atmospheric carbon dioxide will cause global warming in the range of an average 2 to 4 °C rise in temperature for a doubling in the carbon dioxide concentration which is likely to occur between year 2035 and year 2050, dependent to a great extent on the measures we take to reduce the consumption of fossil fuel.

As well as the global average surface temperature the atmospheric circulation pattern is also affected. The general atmospheric circulation is characterized by poleward transport of heat energy. The general temperature increase caused by a higher carbon dioxide concentration enhances the poleward transport of latent heat in middle latitudes, resulting in a general movement of agroclimatic zones. For example, summer temperatures may become too high for corn and soybean, so that the corn belt will be shifted north. Agricultural productivity is very much dependent upon climate, but the reponse is very complex, as each crop has a unique response to climatic variations (Thompson, 1975 and Rozema et al., 1992).

TABLE 3.12
Calculated surface temperature change for a 1% increase in each of the parameters given

	1975 value	Units	ΔT_s	Relative rank
Albedo	30	%	-0.14	1
Relative humidity	74	%	0.065	0.46
CO_2	330 (2000: 375)	ppmv	0.020	0.14
Airborne particles extraction coefficient	0.1	km^{-1}	-0.0063	0.057
N_2O	637	ppbv	0.0044	0.031
O_3	0.37	atm-cm	-0.0032	0.023
CH_4	1.6	ppm	0.002	0.014
H_2S	3.5	ppmv	0.0012	0.0086
$CHCL_3$	$<2 * 10^{-4}$	ppmv	0.0010	0.0071
NH_3	$6 * 10^{-3}$	ppmv	0.0009	0.006
HNO_3	10^{-3} - 10^{-2}	ppmv	0.00060	0.004
$CFCl_3$	$2.3 * 10^{-4}$	ppmv	0.00035	0.0025
SO_2	2	ppmv	0.00020	0.0014
CF_2Cl_2	$1.3 * 10^{-4}$	ppmv	0.00019	0.0014
C_2H_4	$2 * 10^{-4}$	ppmv	0.0001	0.0007
CH_3Cl	$5 * 10^{-4}$	ppmv	0.0001	0.0007
CCl_4	$1 * 10^{-4}$	ppmv	0.0001	0.0007

ΔT_S = surface temperature change (K) for 1% increase in the present average value

The response of snow and ice cover to climatic factors varies greatly in time. Typical extents of cover are 0.1-1.0 years for seasonal snow, 1.0-10 years for sea ice and 1000-100,000 years for ground ice and ice sheets. Of primary interest, of course, are the times of seasonal temperature transition across the 0°C threshold (-1.8 °C for seawater). Another important effect is the large albedo differences between snow cover (appr. 0.8) and snow-free ground (0.1-0.25) or between ice (0.65) and water (0.05-0.1). At present perennial ice covers 7% of the world's oceans and 11% of the land surface with a further percentage of land under seasonal snow cover. Barry (1978) has attempted to predict the effect on the cryosphere of a doubling in the atmospheric carbon dioxide concentration, assuming that it will result in a global warming of 1.5°C with a rise of 5-6°C in polar latitudes (Manabe and Wetherald, 1975 and Koch and Mooney, 1996).

It is concluded that:

1. there will be negligible changes for major ice sheets and ground ice on a time scale of centuries,
2. changes in seasonal snow cover may also be minimal. Any increase in winter snowfall would probably be offset by increased summer melt,
3. it is possible that greater melt of pack ice leading to more open water, would reduce the solar radiation input.

However, it is also concluded that major deficiencies exist in terms of data on snow cover extent and depth. Many crucial research problems remain, especially because our knowledge about the effects of a rise in temperature on the melting glaciers and ice sheets is rather limited.

Although much more research is needed before we can map the relationship between the consumption of fossil fuel and climatic conditions, it can be concluded from this brief review that we have quantitative knowledge of the interactions between the subsystems to justify constructing a model of this relationship. It is only through the use of such a model that we can assess in which direction we should conduct our further research to discover ways of setting up more reliable models.

The problem discussed in this chapter is a very complex one, because its solution requires knowledge of the interactions between a large number of subsystems. A break-down of the problem is attempted in Fig. 3.13: energy policy determines the energy demand and both determine the cosumption of fossil fuel, which in turn is related to carbon dioxide production. The concentration of carbon dioxide in the atmosphere is a function of both the global carbon cycle including all its processes, and carbon dioxide production. The carbon dioxide concentration and the global average surface temperature are related in a very complex way, and the latter will cause several other global changes, such as higher sea level, shifts in crop belts, etc - changes which it is very difficult to predict with our present knowledge. However, such predictions of global changes are needed before we can set up a realistic energy policy.

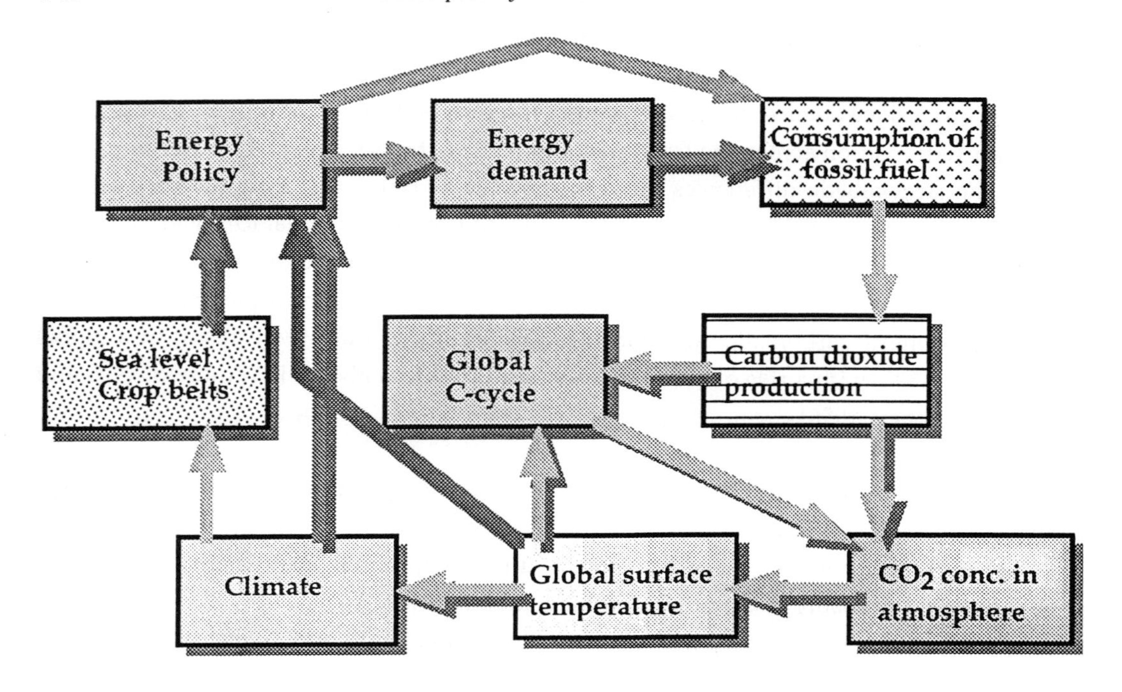

Fig. 3.13. A break-down of the CO_2-problem into subsystems. Interactions between the subsystems are shown.

Example 3.1.

Estimate the increase of the global surface temperature (year 2000) as a result of a decrease of the albedo from 30% (year 2000) to 29%, an increase of the present (year 2000) carbon dioxide concentration from 375 ppmv to 525 ppmv and the present methane concentration from 1.7 ppmv to 1.9 ppmv. It is realistic to expect these changes to happen between the year 2000 and year 2030.

Solution

We use Table 3.12 to make some preliminary estimations:

$$(-3.3)*(-0.14) + 40*0.02 + 0.002*11.8 = 1.3^{\circ}C$$

3.6 QUESTIONS AND PROBLEMS

1. Draw all possible parallels between a foodchain and an energy chain.

2. List at least 6 environmental factors which influence the primary productivity.

3. Find the CF for Cd for a 4 m long tuna fish.

4. How would an increasing humidity influence the global average temperature? What would be the temperature change for an increase from the present humidity of 74% to 80%, and how would such a change influence the corn yield and cryosphere?

5. What would be the expected increase in surface temperature if 10% of the earth's surface area which is now covered by fresh snow and ice becames tundra?

6 Compare the formation of carbon dioxide by combustion of ethane and of ethanol. The heat of combustion for ethane is 59 kJ / g and for ethanol 27.6 kJ / g. Which of the two fuels gives the smallest amount of carbon dioxide per unit of energy? Compare also with coal, oil and gas.

7. Estimate the increase in the global surface temperature as a consequence of doubling both the carbon dioxide and methane concentration in the atmosphere.

4. Risks and Effects

4.1 WHAT IS ENVIRONMENTAL RISK ASSESSMENT (ERA)?

Treatment of industrial waste water, solid waste and smoke is very expensive. Consequently, the industries attempt to change their products and production methods in a more environmentally friendly direction to reduce the treatment costs. Industries need, therefore, to know how much the different chemicals, components and processes are polluting our environment. Or expressed differently: what is the environmental risk of using a specific chemical compared with another alternative chemical ? If industries can reduce their pollution just by switching to another chemical or process, they will of course consider doing so to reduce their environmental costs. An assessment of the environmental risk associated with the use of a specific chemical and a specific process gives the industries possibility of making the right selection of chemicals and processes to the benefit for the economy of the enterprise and the quality of the environment.

Society similarly needs to know the environmental risks of all chemicals applied in society so as to phase out the most environmentally threatening chemicals and set standards for the use of all other chemicals. The standards should ensure that there is no serious risk in using the chemicals, provided that the standards are followed carefully.

Modern abatement of pollution therefore includes environmental risk assessment, ERA, which may be defined as the process of assigning magnitudes and probabilities to the adverse effects of human activities. The process involves identification of hazards such as the release of toxic chemicals to the environment by quantification of the relationship between an activity associated with an emission to the environment and its effects. The entire ecological hierarchy is considered in this context which implies that the effects on the cellular (biochemical) level, on the organism level, on the population level, on the ecosystem level and for the entire ecosphere should be considered.

The application of environmental risk assessment is rooted in the recognition that:

1) the cost of elimination of all environmental effects is impossibly high,

2) decisions in practical environmental management must always be made on the basis of incomplete information.

We use about 100,000 chemicals in such amounts that they may threaten the environment, but we know only about 1% of what we need to know to be able to make a proper and complete environmental risk assessment of these chemicals. Later in this chapter will be given a short introduction to available estimation methods which it is recommended we can apply if we cannot find information about properties of chemical compounds in the literature. A list of the relevant properties is also given in this context, and it is discussed what these properties mean for the environmental impact.

ERA is in the same family as environmental impact assessment, EIA, which attempts to assess the impact of a human activity. EIA is predictive, comparative and concerned with all possible effects on the environment, including secondary and tertiary (indirect) effects, while ERA attempts to assess the probability of a given (defined) adverse effect as result of a considered human activity.

Both ERA and EIA use models to find the expected environmental concentration (EEC) which is translated into impacts for EIA and to risks of specific effects for ERA. The development of models is not covered in this volume, although the applicability of different classes of models has been mentioned in Chapter 2 and will be further discussed in this chapter. Development of the ecotoxicological models that are applicable in assessment of environmental risks is treated in detail in Jørgensen (1994). An overview of ecotoxicological models is given in Jørgensen et al. (1995a).

Legislation and regulation of domestic and industrial chemicals with respect to the protection of the environment have been implemented in Europe and North America for decades. Both regions distinguish between existing chemicals and introduction of new substances. For existing chemicals the European Union requires for instance according to Council Regulation no. 793/93 an assessment of risk to man and environment of priority substances by principles given in the Commission Regulation no. 1488/94. An informal priority setting (IPS) is used for selecting chemicals among the 100,000 listed in the The European Inventory of Existing Commercial Chemical Substances. The purpose of IPS is to select chemicals for detailed risk assessment from among the EEC high production volume compounds, i.e., > 1000 t/y (about 2000 chemicals). Data necessary for the IPS and an initial hazard assessment are called Hedset and cover issues as environmental exposure, environmental effects, exposure to man and human health effects.

The risk assessment of new notified substances is in the EU based on data submitted according to Directive 67/548/EEC. The directive provides a scheme of step-wise procedure which is in both North-America and Europe approximately as presented below. Tests are often required to provide the data needed for the ERA.

At the UNCED meeting in Rio de Janeiro in 1992 on the Environment and Sustainable Development, it was decided to create an Intergovernmental Forum on Chemical Safety (IGFCS, Chapter 19 of Agenda 21). The primary task is to stimulate and coordinate global harmonisation in the field of chemical safety, covering the following principal themes: assessment of chemical risks, global harmonisation of classification and labelling, information exchange, risk reduction programs and capacity building in chemicals management (Karcher, 1998).

The uncertainty plays an important role in risk assessment (Suter, 1993). Risk is the probability that a specified harmful effect will occur or in the case of a graded effect, the relationship between the magnitude of the effect and its probability of occurrence.

Risk assessment has emphasised risks to human health and has to a certain extent ignored ecological effects. It has, however, been acknowledged that some chemicals that have no or only little risk to human health cause severe effects on for instance aquatic organisms. Examples are chlorine, ammonia and certain pesticides. **A up-to-date risk assessment comprises therefore considerations of the entire ecological hierarchy** which is the ecologist's view of the world in terms of levels of organisation. Organisms interact directly with the environment and it is organisms that are exposed to toxic chemicals. The species-sensitivity distribution is therefore more ecologically credible (Calow, 1998). The reproducing population is the smallest meaningful level in ecological sense. However, populations do not exist in vacuum, but require a community of other organisms of which the population is a part. The community occupies a physical environment with which it forms an ecosystem.

Moreover, both the various adverse effects and the ecological hierarchy have different scales in time and space which must be included in a proper environmental risk assessment; see Figure 4.1. For example, oil spills occur at a spatial scale similar to those of populations, but they are briefer than population processes. Therefore a risk assessment of a oil spill requires considerations of reproduction and recolonisation that occur on a longer time scale and that determine the magnitude of the population response and its significance to natural population variance.

Uncertainties in risk assessment are taken most commonly into account by application of safety factors. Uncertainties have three basic causes:

1) the inherent randomness of the world (stochasticity)

2) errors in execution of assessment

3) imperfect or incomplete knowledge

The inherent randomness refers to uncertainty that can be described and estimated but cannot be reduced because it is characteristic of the system. Meteorological factors such as rainfall, temperature and wind are effectively stochastic at levels of interest for risk assessment. Many biological processes such as colonisation, reproduction and mortality also need to be described stochastically.

Human errors are inevitably attributes of all human activities. This type of uncertainty includes incorrect measurements, data recording errors, computational errors and so on.

The uncertainty is considered by use of an assessment (safety) factor from 10 to 1000. The choice of assessment factor depends on the quantity and quality of toxicity data; see Table 4.1. The assessment or safety factor is used in step 3 of the environmental risk assessment procedure, presented below. Other relationships than the uncertainties originating from randomness, errors and lack of knowledge may be considered when the assessment factors are selected, for instance cost-benefit. This implies that the assessment factors for drugs and pesticides for instance may be given a lower value. Lack of knowledge results in undefined uncertainty that cannot be described or quantified. It is a result of practical constraints on our ability to accurately describe, count, measure or quantify everything that pertains to a risk estimate. Clear examples are the inability to test all toxicological responses of all species exposed to a pollutant and the simplifications needed in the model used to predict the expected environmental concentration.

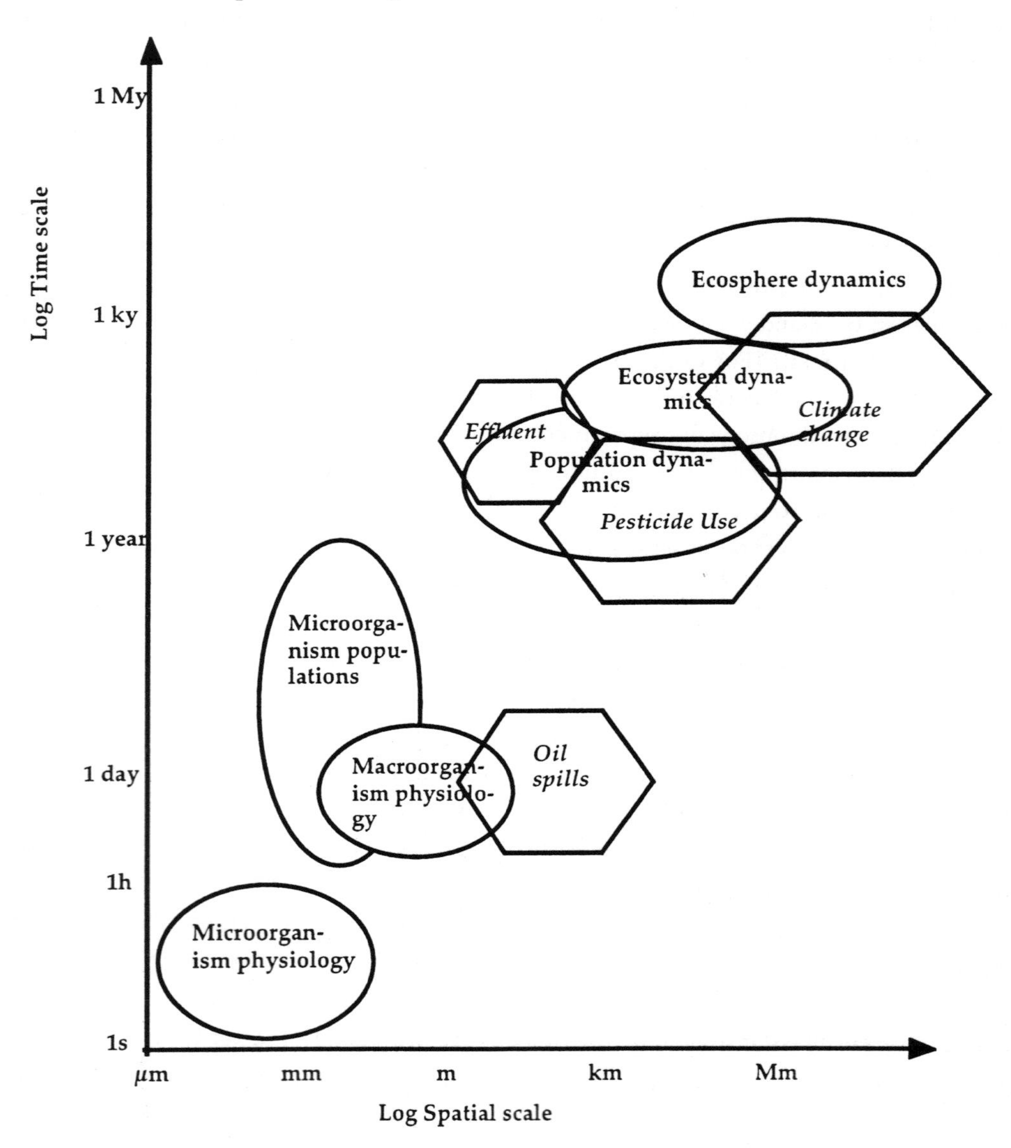

Fig. 4.1. The spatial and time scale for various hazards (hexagons, italic) and for the various levels of the ecological hierarchy (circles, non-italic).

The most important feature distinguishing risk assessment from impact assessment is the emphasis in risk assessment on characterising and quantifying uncertainty. It is therefore of particular interest in risk assessment to be able to analyse and estimate the analysable uncertainties. They are natural stochasticity, parameter errors and model errors. Statistical methods may provide direct estimates of uncertainties. They are widely used in model development; see for instance Jørgensen (1994).

The use of statistics to quantify uncertainty is complicated in practice by the needs to consider errors in both the dependent and independent variables and to combine errors when multiple extrapolations should be made. Monte Carlo analysis is often used to overcome these difficulties; see for instance Bartell et al. , 1992.

Model errors include inappropriate selection or aggregation of variables, incorrect functional forms and incorrect boundaries. The uncertainty associated with model errors is usually assessed by field measurements utilised for calibration ad validation of the model (Jørgensen, 1994).

TABLE 4. 1

Selection of assessment factors to derive PNEC (see also step 3 of the procedure presented below)

Data quantity and quality	Assessment factor
At least one short-term LC_{50} from each of the three trophic levels of the base set (fish, zooplankton and algae)	1000
One long-term NOEC (non-observed effect concentration, either for fish or Daphnia)	100
Two long-term NOECs from species representing two trophic levels	50
Long-term NOECs from at least three species (normally fish, Daphnia and algae) representing three trophic levels	10
Field data or model ecosystems	case by case

4.2 HOW TO PERFORM AN ERA?

Risk assessment of chemicals may be divided into nine steps, which are shown in Figure 4.2. The nine steps correspond to questions which the risk assessment attempts to answer to be able to quantify the risk associated with the use of a chemical. The nine steps are presented in detail below with reference to Fig. 4.2.

Step 1: Which hazards are associated with the application of the chemical? This involves gathering data on the types of hazards - possible environmental damage and human health effects. The health effects include congenital, neurological, mutagenic, endocrine disruption (so called oestrogen) and carcinogenic effects. It may also include characterisation of the behaviour of the chemical within the body (interactions with organs, cells or genetic material). What is the possible environmental damage includes lethal effects and sublethal effects on growth and reproduction of various populations.

As an attempt to quantify the potential danger posed by chemicals, a variety of toxicity tests have been devised. Some of the recommended tests involve experiments with subsets of natural systems for instance microcosms or with entire ecosystems. The majority of testing new chemicals for possible effects has, however, been confined to studies in the laboratory on a limited number of test species. Results from these laboratory assays provide useful information for quantification of the relative toxicity of different chemicals. They are used to forecast effects in natural systems, although their justification has been seriously questioned (Cairns et al., 1987).

Step 2: What is the relation between dose and responses of the type defined in step 1? It implies knowledge of NEC (non-effect concentration), LD_x- (the dose which is lethal to x% of the organisms considered), LC_y- (the concentration which is lethal to y% of the organisms considered) and EC_z-values (the concentration giving the indicated effect to z% of the considered organisms) where x, y and z express a probability of harm. Appendix 5 gives an overview of the concept used in this volume. The answer can be found by laboratory examination or we may use estimation methods. Based upon these answers a most probable level of no effect, NEL, is assessed. Data needed for steps 1 and 2 can be obtained directly from scientific libraries, but are increasingly found via on-line data searches in bibliographic and factual databases. Data gaps should be filled with estimated data. It is very difficult to get complete knowledge about the effect of a chemical on all levels from cells to ecosystem. Some effects are associated with very small concentrations, for instance the oestrogen effect. It is therefore far from sufficient to know NEC, LD_x-, LC_y- and EC_z-values. These problems will be touched on further in Sections 4.4. and 4.5.

Step 3: Which uncertainty (safety) factors reflect the amount of uncertainty that must be taken into account when experimental laboratory data or empirical estimations methods are extrapolated to real situations? Usually, safety factors of 10 - 10,000 are used. The choice is discussed above and will usually be in accordance with Table 4.1. If good knowledge about the chemical is available a safety factor of 10 may be applied. If on the other hand , it is estimated that the available information has a high uncertainty, a safety factor of 10,000 may be recommended. Most frequently, safety factors of 50-100 are applied. NEL (non-effect level) times the safety factor is named the predicted non-effect level, PNEL. The complexity of environmental risk assessment is often simplified by deriving the predicted no effect concentration, PNEC, for different environmental components (water, soil, air, biotas and sediment).

Step 4: What are the sources and quantities of emissions? The answer requires thorough knowledge of the production and use of the chemical compounds considered, including an assessment of how much of the chemical is wasted in the environment by production and use? The chemical may also be a waste product which makes it very difficult to determine the amounts involved. For instance the very toxic dioxins are waste products from incineration of organic waste. Chapter 2 has touched on the application of mass balances to assess the emissions and the resulting concentrations in different components of the environment.

Step 5: What is (are) the actual exposure concentration(s)? The answer to this question is named the predicted environmental concentration, PEC. Exposure can be assessed by measuring environmental concentrations. It may also be predicted by a model, when the emissions are known. The use of models (see Section 4.3) is necessary in most cases either because we are considering a new chemical, or because the assessment of environmental concentrations requires a very large number of measurements to determine the variations in concentrations in time and space. Furthermore, it always provides an additional certainty to compare model results with measurements, which implies that it is always recommended to both develop a model and make at least a few measurements of concentrations in the ecosystem components, where it is expected that the highest concentration will occur. Most models will demand an input of parameters, describing the properties of the chemicals and the organisms, which also will require extensive application of handbooks and a wide range of estimation methods. The development of an environmental, ecotoxicological model requires therefore extensive knowledge of the physical-chemical-biological properties of the chemical compound(s) considered. The selection of a proper model is discussed later in this chapter (Section 4.3).

Step 6: What is the ratio PEC / PNEC? This ratio is often called the risk quotient. It should not be considered an absolute assessment of risk but rather a relative ranking of risks. The ratio is usually found for a wide range of ecosystems for instance aquatic ecosystems, terrestrial ecosystems and ground water.

Steps 1-6 are shown in Figure 4.3 which is completely in accordance with Figure 4.2 and the information given above.

Step 7: How will you classify the risk? The valuation of risks is made in order to decide on risk reductions (step 9). Two risk levels are defined:

1) the upper limit, i.e., the maximum permissible level (MPL), and

2) the lower limit, i.e., the negligible level, NL. It may also be defined as a percentage of MPL, for instance 1% of MPL.

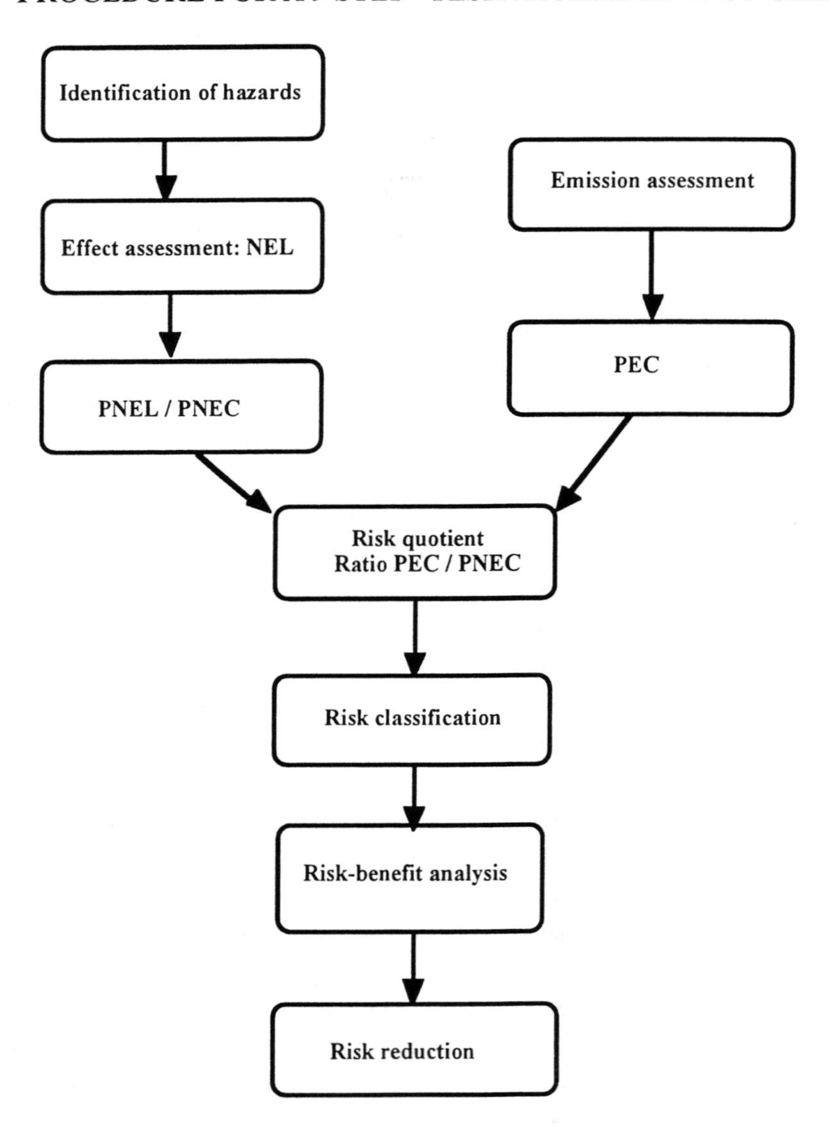

Fig. 4.2. The presented procedure in nine steps to assess the risk of chemical compounds. Steps 1-3 require extensive use of ecotoxicological handbooks and ecotoxicological estimation methods to assess the toxicological properties of the chemical compounds considered, while step 5 requires the selection of a proper ecotoxicological model.

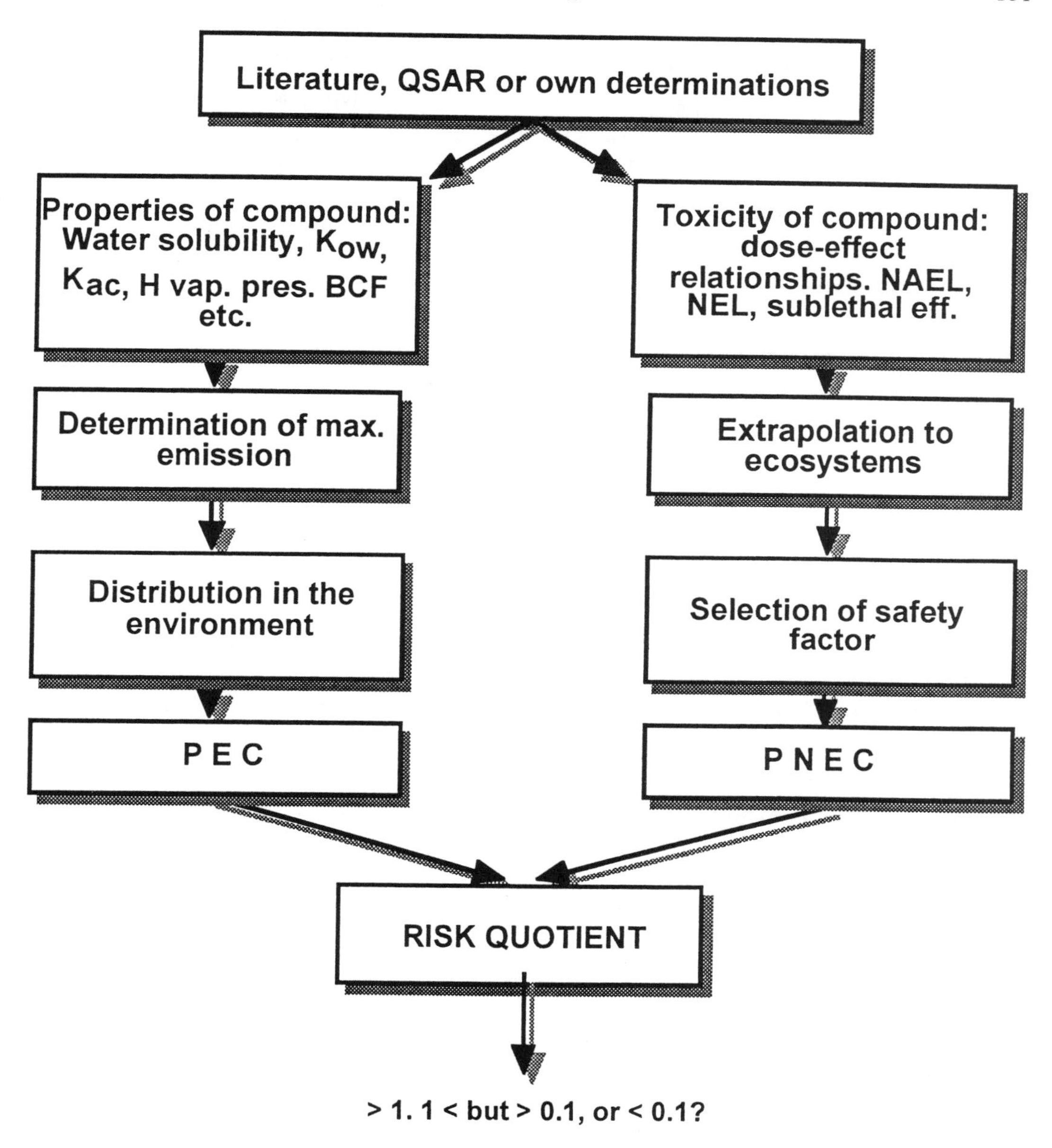

Fig. 4.3. The steps 1-6 are shown in more detail for practical applications. The result of these steps leads naturally also to assessment of the risk quotient.

The two risk limits create three zones: a black, unacceptable, high risk zone > MPL, a grey, medium risk level and a white, low risk level < NL. The risk of chemicals in the grey and black zones must be reduced. If the risk of the chemicals in the black zone cannot be reduced sufficiently it should be considered to out phase the use of these chemicals.

Step 8: What is the relation between risk and benefit? This analysis involves examination of socioeconomic, political and technical factors, which are beyond the scope of this volume. The cost-benefit analysis is difficult, because the costs and benefits are often of a different order.

Step 9: How can the risk be reduced to an acceptable level? The answer to this question requires deep technical, economic and legislative investigation. Assessment of alternatives is often an important aspect in risk reduction.

The steps 1, 2, 3 and 5 require knowledge of the properties of the focal chemical compounds, which again implies an extensive literature search and/or selection of the best feasible estimation procedure. In addition to Beilstein it can be recommended to have at hand the following very useful handbooks of environmental properties of chemicals and methods for estimation of these properties in case literature values are not available:

S.E. Jørgensen, S. Nors Nielsen and L.A. Jørgensen, 1991. Handbook of Ecological Parameters and Ecotoxicology, Elsevier, 1991.

P.H. Howard et al., 1991. Handbook of Environmental Degradation Rates. Lewis Publishers.

K. Verschueren, 1983. Handbook of Environmental Data on Organic Chemicals. Van Nostrand Reinhold.

P.H. Howard. Handbook of Environmental Fate and Exposure Data. Lewis Publishers.
Volume I. Large Production and Priority Pollutants. 1989
Volume II. Solvents. 1990.
Volume III. Pesticides. 1991.
Volume IV. Solvents 2. 1993.
Volume V. Solvents 3. 1998.

G.W.A. Milne, 1994. CRC Handbook of Pesticides. CRC.

W. J. Lyman, W.F. Reehl and D.H. Rosenblatt, 1990. Handbook of Chemical Property Estimation Methods. Environmental Behaviour of Organic Compounds. American Chemical Society.

D. Mackay, W.Y. Shiu and K.C.Ma. Illustrated Handbook of Physical-Chemical Properties and Environmental Fate for Organic Chemicals. Lewis Publishers.
Volume I. Mono-aromatic Hydrocarbons. Chloro-benzenes and PCBs. 1991.
Volume II. Polynuclear Aromatic Hydrocarbons, Polychlorinated Dioxins, and

Dibenzofurans. 1992.
Volume III. Volatile Organic Chemicals. 1992.
Jørgensen, S.E., Mahler, H. and Halling Sørensen, B., 1997. Handbook of Estimation Methods in Environmental Chemistry and Ecotoxicology. Lewis Publishers.

Steps 1-3 are sometimes denoted as effect assessment or effect analysis and steps 4-5 exposure assessment or effect analysis. Steps 1-6 may be called risk identification, while environmental risk assessment, ERA, encompasses all the 9 steps presented in Figure 4.2. Particularly step 9 is very demanding, as several possible steps in reduction of the risk should be considered, included treatment methods, cleaner technology and substitutes to replace the examined chemical.

During the last 5-6 years it has in North-America, Japan and EU been considered to treat medicinal products similarly to other chemical products, as there is in principle no difference between a medicinal product and other chemical products. This has, however, only resulted in the introduction after 1. January 1998 of the application of environmental risk assessment for new veterinary medicinal products. At present, technical directives for human medicinal products do not in the EU include any reference to ecotoxicology and the assessment of their potential risk (Jensen, Nielsen and Halling-Sørensen, 1998). However, a detailed technical draft guideline issued in 1994 indicates that the approach applicable for veterinary medicine also would apply to human medicinal products. Presumably, ERA will be applied to all medicinal products in the near future, when sufficient experience with veterinary medicinal products has been achieved. Veterinary medicinal products on the other hand are released in bigger amounts to the environment, as manure for instance in spite of its possible content of veterinary medicine is utilised as fertiliser on agricultural fields.

It is also possible to perform an environmental risk assessment where the human population is in focus. Ten steps corresponding to Figure 4.3 are shown in Figure 4.4, which is not significantly different from Figure 4.3. The principles for the two types of environmental risk assessment are the same. Figure 4.4 uses the non-adverse effect level (NAEL) and non-observed adverse effect level (NOAEL) to replace the predicted non-effect concentration and the predicted environmental concentration is replaced by the tolerable daily intake (TDI).

This type of environmental risk assessment has particular interest for veterinary medicine which may contaminate food products for human consumption. For instance, the use of antibiotics in pig feed has attracted a lot of attention, as they may be found as residue in pig meat or may contaminate the environment though the application of manure as natural fertiliser.

Selection of a proper ecotoxicological model is a first initial step in the development of an environmental exposure model, as required in step 5. It will be discussed in more detail in the next section.

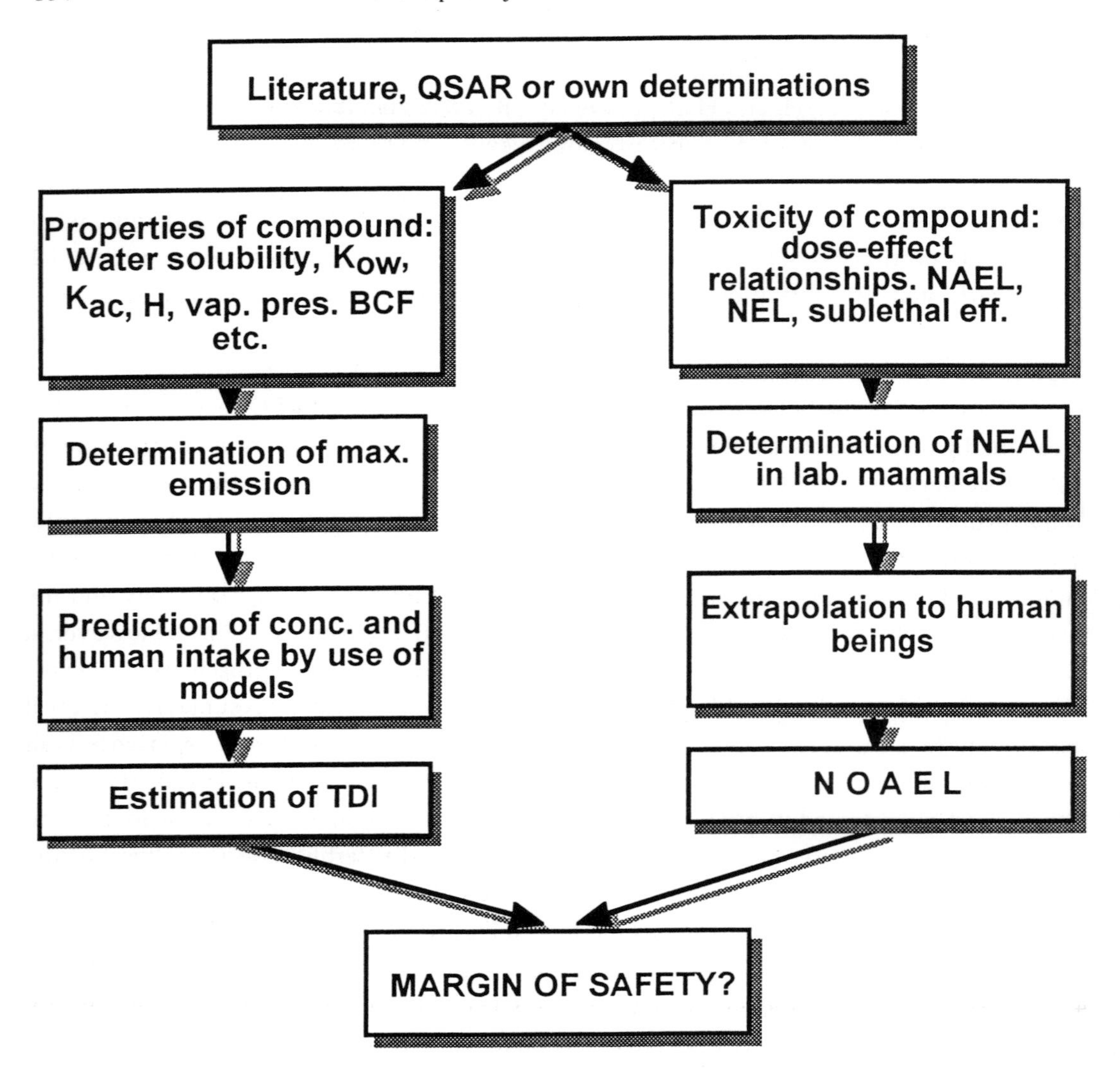

Fig. 4.4. Environmental risk assessment for human exposure. It leads to a margin of safety which corresponds to the risk quotient in Figures 4.2 and 4.3

4.3 SELECTION OF ECOTOXICOLOGICAL MODELS

This section will not present how an ecotoxicological model is developed. That would require a more complete presentation of the topic "environmental modelling" which cannot be made in less than 150-200 pages. Those interested in this topic should referred to Fundamentals of Ecological Modelling by S.E. Jørgensen (1994, a new and third edition is expected in 2001). We will, however, present in this section the characteristic features of toxic substance models and how to select the best applicable model among the spectrum of available toxic substance models.

Toxic substance models attempt to model the fate and effect of toxic substances in ecosystems. Toxic substance models differ from other environmental models in that:

1. The need for parameters to cover all possible toxic substance models is great, and general estimation methods are therefore used widely. See also Section 4.6.
2. The safety margin should be high, for instance, expressed as the ratio between the actual concentration and the concentration that gives undesired effects.
3. They require possible inclusion of an effect component, which relates the output concentration to its effect. It is easy to include an effect component in the model; it is, however, often a problem to find a well examined relationship to base it on.
4. They need to be simple models due to points 1 and 2, and our knowledge of process details, parameters, sublethal effects, antagonistic and synergistic effects is limited.

It may be an advantage to clarify several questions before developing a toxic substance model:

1. Obtain the best possible knowledge about the processes of the considered toxic substances in the ecosystem. As far as possible knowledge about the quantitative role of the processes should be obtained.
2. Attempt to get parameters from the literature and/or from one's own experiments (in situ or in the laboratory).
3. Estimate the parameters not found under 2. Estimation methods are shortly reviewed in Section 4.6.
4. Estimate which processes and state variables it would be feasible and relevant to include in the model. If there is the slightest doubt then include at this initial stage too many processes and state variables rather than too few.
5. Use a sensitivity analysis to evaluate the significance of the individual processes and state variables. This often may lead to further simplification.

In many problems it may be necessary to go into more detail on the effect on the organism level in order to answer the following relevant questions:

1. Does the toxic substance accumulate in the organism?
2. What will be the long term concentration in the organism when the uptake rate, excretion rate and biochemical decomposition rate are considered?
3. What is the chronic effect of this concentration?
4. Does the toxic substance accumulate in one or more organs?
5. What is the transfer between various parts of the organism?

6. Will decomposition products eventually cause additional effects?

A detailed answer to these questions may require a model of the processes that take place in the organism, and a translation of concentrations in various parts of the organism into effects. This implies, of course, that the intake = (uptake by the organism)*(efficiency of uptake) is known. Intake may either be from water or air, and may also be expressed by concentration factors, which are the ratios between the concentration in the organism and in the air or water.

But, if all the above mentioned processes were to be taken into consideration for just a few organisms, the model would easily become too complex, contain too many parameters to calibrate, and require more detailed knowledge than it is possible to provide. Often we even do not have all the relations needed for a detailed model, as toxicology and ecotoxicology are still in their infancy. Therefore, most models in this class will not consider too many details of the partition of the toxic substances in organisms and their corresponding effects, but rather will be limited to the simple accumulation in the organisms and their effects. Usually, accumulation is fairly easy to model and the following simple equation which was applied in Chapter 2 is often sufficiently accurate:

$$dC/dt = (ef*Cf*F + em*Cm*V)/W - Ex*C = (INT)/W - Ex*C \qquad (4.1)$$

where C is the concentration of the toxic substance in the organism; ef and em are the efficiencies for the uptake from the food and medium respectively (water or air); Cf and Cm are the concentration of the toxic substance in the food and medium respectively; F is the amount of food uptake per day; V is the volume of water or air taken up per day; W is the body weight as either dry or wet matter; and Ex is the excretion coefficient (1/day). As can be seen from the equation, INT covers the total intake of toxic substance per day.

This equation has a numerical solution:

$$C/C(max) = (INT*(1 - exp(Ex*t)))/(W*Ex) \qquad (4.2)$$

where C(max) is the steady state value of C:

$$C(max) = INT/(W*Ex) \qquad (4.3)$$

Synergistic and antagonistic effects have not been touched on so far. They are rarely considered in this type of model for the simple reason that we do not have much knowledge about these effects. If we have to model combined effects of two or more toxic substances, we can only assume additive effects, unless we can provide empirical relationships for the combined effect.

The applied models are in general sufficiently accurate to give a very applicable picture (overview) of the concentrations of toxic substances in the environment, due to the application of high safety factors. The application of the estimation methods renders it feasible to construct such models, even our knowledge of the parameters is limited. The estimation methods obviously have a high uncertainty, but the large safety factor helps in accepting this uncertainty. On the other hand our knowledge about the effects of toxic substances is very limited - particularly at the organism and organ level. It must not be expected, therefore, that models with effect components give more than a preliminary rough picture of what is known today in this area.

A classification of ecotoxicological models should be mentioned, because it illustrates very clearly the thoughts behind modelling in environmental chemistry. The classification uses three classes of models, as also presented in Chapter 2:

A. Fate models, included fugacity models.

B. Ecosystem specific models of toxic substances.

C. Models of toxic substances in a typical or average ecosystem which is used as a general representation of all ecosystems of a given type.

Class A models are used to get a rough estimation of where a toxic substance will be found in the environment and at approximately which concentration. This class of ecotoxicological models is useful for comparison of various alternative chemicals. It may be used to answer the following pertinent question: should we prefer chemical X or Y from an environmental point of view? This type of model has been treated very comprehensively in D. Mackay's book "Multimedia Environmental Models". Mackay distinguishes four levels of fate (fugacity, multimedia) models. This particular approach is presented in Mackay and Paterson (1982).

The first level calculates the equilibrium distribution of a chemical between phases. It assumes that each compartment is well mixed and there are no reactions or advection into or out of the considered system.

The second level considers equilibrium but includes also reaction and advection. Reaction comprises photolysis, hydrolysis, biodegradation, oxidation and so on. All the processes are assumed to be first order reactions.

The third level is devoted to a steady state non-equilibrium situation, which implies that the fugacities are different in each phase.

Level four involves a dynamic version of level three, where emissions and thus concentrations vary with time.

Levels one or two are usually sufficient, but if the ecotoxicological management problem requires prediction of the time needed for a substance to accumulate to a certain concentration in a phase after emission has started or the length of time for the system to recover after the emission has ceased, then a level four model should be applied.

Setac has in 1995 published a review of multi-media fate models in the volume named "The Multi-Media Fate Model: A vital tool for predicting the fate of chemicals". The review includes a comparison of four available software packages based on multi-media fate models: HAZCHEM, ChemCAN, SimpleBOX and CalTOX.

The review concludes that all four software packages are recommended for the development of multi-media predictions, although a proper validation in the sense applied in ecological modelling has not been carried out.

Multi-media fate models are useful in risk assessment of chemicals (compare with the procedure presented in Fig. 4.2). If the risk assessment focuses on an ecosystem or a region, comprising a few ecosystems or types of ecosystems, class B models must be recommended. In chemical ranking and scoring, where two or more chemicals are compared, multi-media fate models should be recommended as the proper tool. Application of level two or maybe level three fugacity models is recommended for comparison of the environmental fate of alternative chemicals.

Class B models are used when we have to decide on abatement of a specific pollution by a toxic substance, for instance when an organic substance is transported from a chemical plant by waste water to the environment, perhaps after the waste water has passed from a treatment plant. In this case it is necessary to include the characteristic features of the ecosystems receiving the toxic substance in the model. This type of ecotoxicological models is obviously very similar to the most widely applied ecological models, because it must include the same characteristic ecological features as ecological models in general. This class of models is generally more accurate than class A models, but is also more specific and can hardly be applied for ecosystems other than those the developed models are designed for. This class of models is presented in detail in Jørgensen (1994).

Class C models have recently found an increasing application because many uses of chemicals give impacts to specific types of ecosystems. If we want for instance to select the pesticide which is most harmless to earth worms, we have to "test" our spectrum of pesticides on a general agricultural ecosystem model with for instance a food chain which includes earth worms. We don't have a specific agricultural system in mind, but a very general agricultural ecosystem with average features characteristic for such systems. In this case we will develop a model with the characteristic features of an average agricultural system. This class of models will be less accurate than class B models but will generally give more accurate and more specifically applicable results than class A models. On the other hand the results cannot be applied with the same generality as for class A models.

Jensen, Nielsen and Halling Sørensen, (1998) recommend the use of both class A and class C models to assess the environmental risk for application of veterinary medicinal products. They emphasise the need for calculations of PEC on both a local and a regional spatial scale preferably from monitoring data.

Information on a number of processes is required to set up a class B or a class C model:

1. Evaporation
2. Adsorption
3. Adsorption to and accumulation in soil and sediment
4. Biodegradation
5. Chemical oxidation (by air, nitrate, sulphate etc.)
6. Photodegradation
7. Complex formation (several possible inorganic and organic ligands are present in all ecosystems)
8. Hydrolysis

As is seen from this list a comprehensive knowledge of physical, chemical and biological processes is required to relate the concentration to the effect on all levels in the biological hierarchy.

It is generally very difficult to find data for calibration and validation of class A models, while data for calibration and validation of class B models always should be provided to ensure acceptable accuracy of the model results. It may be possible to get data for calibration and validation of class C models, but the calibration and validation cannot be carried out with the same accuracy as for class B models, because the data will be valid only for specific ecosystems.

It is always valuable in the development of a new ecotoxicological model to build on the experience already gained in this area. A summary of the available ecotoxicological models can be found in S.E. Jørgensen, B. Halling-Sørensen and S.N. Nielsen, "Handbook of Environmental and Ecological Modelling (1995a). The handbook gives a short description of more than 400 models, including 71 ecotoxicological models: 16 models of pesticides, 28 models of other organic compounds and 27 models of inorganic compounds (heavy metals and radionuclides). The 71 models encompasses all six classes of models mentioned above.

For each model is given: model identification, model type, model purpose, a short description of state variables and forcing functions, model application, model software and hardware demands, model availability, model documentation and major references and other information relevant for other modellers. It can often save a lot of time in the development of models to learn from others' experience. It is of particular importance to get knowledge about the following modelling features:

- what is a good level of model complexity?
- which state variables are important to include in the model?
- which process descriptions are working properly?

Answers to these questions can be found in the above mentioned handbook which is published with the aim of facilitating the model development for modellers by application of experience from more than 400 different models.

4.4 ECOTOXICOLOGICAL EFFECTS ON THE LOWER LEVELS IN THE BIOLOGICAL HIERARCHY

It was mentioned in Section 4.2 that it is difficult to get sufficient information about effects of a chemical on *all* levels in the biological hierarchy from cells to ecosystems or maybe even in some few instances to the entire ecosphere. Assessment of effects is not only a question about LD- and LC-values, but will require far more comprehensive knowledge which, however, varies from case to case. This section will present some of these problems and at least for some of the problems a solution as well.

Effects, concentrations and time

Firstly, the closed interrelationship between the effect and the concentration versus time will be touched on. Exposure time is an important piece of information when results of toxicity examinations have to be interpreted, but in the context of air pollutants the relationship between concentration and time is of particularly significant importance, probably due to the following factors:

1. The effect of air pollutants on human health is very well examined for a number of combinations of concentration/exposure time.
2. The uptake through the respiratory system is proportional to the exposure time, if all other conditions are the same.
3. A number of examinations of indoor air pollution have been carried out, and here the exposure time (short exposure versus constant exposure) plays an important role.
4. The air-quality standards reflect the interrelation between concentration and exposure time. See Table 4.2, which shows one example out of many.

For water-quality standards this is usually not the case.

Fig 4.5 demonstrates this expected 2-dimensional interrelationship between concentration and time to give a certain effect.

To summarise: **The effect of toxic substances is dependent on an interrelationship between concentration and time which can be clearly demonstrated for air pollutants. Air-quality standards reflect this interrelationship between concentration and exposure time.**

The concentration can either be measured or assessed by use of the principles of mass conservation in a model. In both cases, however, we need information on the dynamics of the pollutant. What is the fate of the pollutant? In which form does it occur? This question is of utmost importance **as different forms of the same compounds do not necessarily have the same toxicity and bioavailability**. Cadmium ions are for instance more toxic to fish than the complex bound cadmium. Complexation of heavy metal ions will usually decrease their bioavailability, while forms of organic pollutants with smaller molecular weight (for instance as a result of a microbiological decomposition) will have a greater bioavailability.

pH also plays an important role. The equilibrium between the acid form and the ionic form of a compound is dependent on pH: acid = ionic form + H^+. This is demonstrated for ammonium in Chapter 2; see equations (2.21) and (2.22). Ammonia, but not ammonium is toxic to fish, consequently the toxicity is pH dependent. Similarly phenol is volatile and has a relatively low solubility in water, while phenolate, which is dominant at pH-values above the pK value for phenol = 6.2, has a high solubility in water. Phenol will bioaccumulate to a certain extent, but phenolate cannot.

This implies that many compounds have different toxicity, different uptake rates and even different water solubility (and K_{ow}) at different due to a pH-dependance. Furthermore: will the focal compound be transformed to other compounds and at what rate? Which ecotoxicological properties have these other forms, and how bioavailable are these forms? On which level in the biological hierarchy: cells, organs, organisms, population, ecosystem or ecosphere, will the effect be most dominant? The two latter levels will be treated separately in the next section, while this section will give information on the problems associated with the effects of toxic substances on cells, organs, organisms and populations.

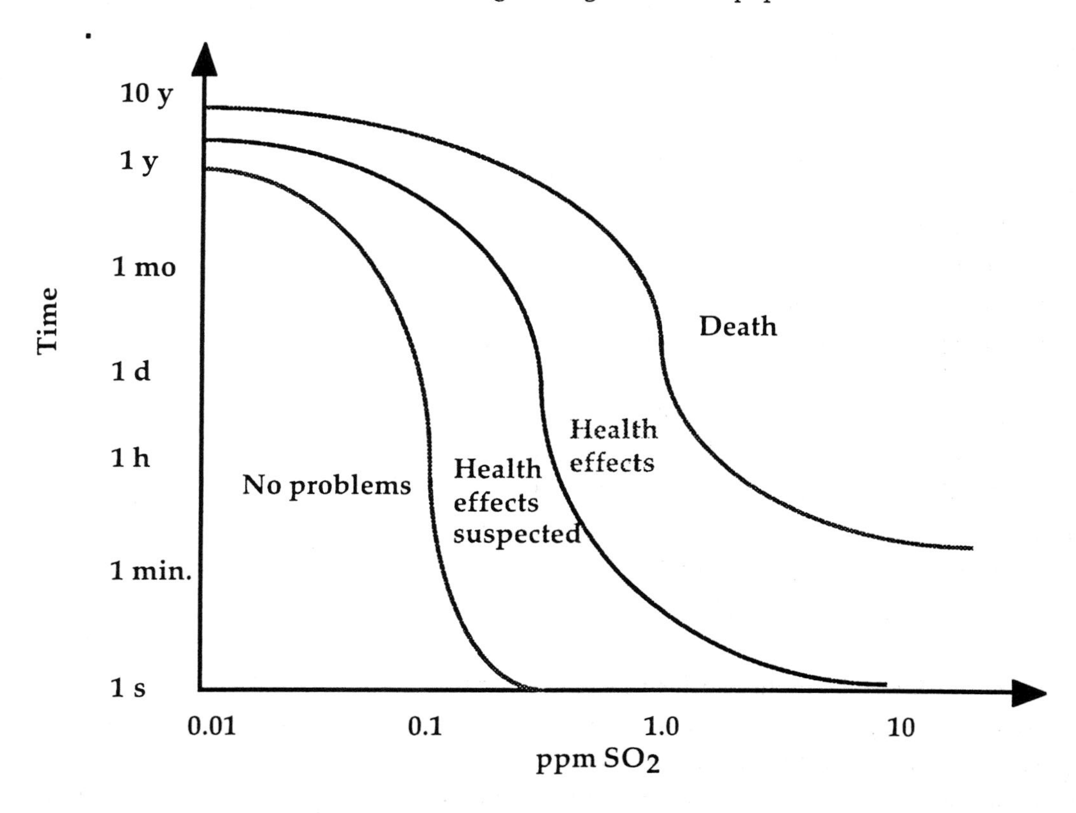

Fig. 4.5. Effects of sulphur pollution on human health .

Effects on cellular and organ level

At *the cellular level* it is necessary to know whether the toxic substances might cause mutation, which is possible by either alteration of the genes (the base sequence in DNA) or by changes in the chromosomes. The toxic substances might also cause an increased or decreased concentration of enzymes, or competitive or non-competitive inhibition.

Other biochemical effects are included in the following more comprehensive list of responses: change of enzyme activity, change of steroid hormones, hormone-like effects (these compounds are named endocrine disruptors), activation or suppression of biochemical pathways, alteration of membrane properties, changes of chromosomes, alteration of genes, alteration of chemical signals and teratogenetic effect.

TABLE 4.2

An example of air-quality standards (clean air acts amendments, U.S. 1970 (see also Council of Environmental Quality, 1982))

Pollutant	Time indications	Concentration	
		$\mu g / m^3$	ppm
Particulate matter	Annual geometric mean	75	-
Particulate matter	Maximum 24-h conc	260	-
Hydrocarbons	Maximum 3-h conc.	160	0.24
Carbon monoxide	Maximum 8-h conc.	0	9
Carbon monoxide	Maximum 1-h conc.	0	35
Sulphur oxides	Annual arithmetic mean	80	0.03
Sulphur oxides	Maximum 24-h conc.	365	0.14
Sulphur oxides	Maximum 3-h conc.	1300	0.5
Nitrogen oxides	Annual arithmetic mean	100	0.05
Photochemical oxidants	Maximum 1-h conc.	240	0.12

Our knowledge in this biochemical field is at present very limited, although intensive research is being performed all over the world to fill the gaps. Toxic substances soluble in lipids are transported in blood bound to lipoproteins and are taken up by tissue rich in lipids. Another significant binding mechanism is the formation of chemical bonds with specific groups. For instance, mercury and organic lead are bound by SH-groups in proteins. Similarly, lead is accumulated in the skeleton through replacing calcium.

Cadmium is accumulated in the kidneys due to unknown reasons. In the first instance such accumulation effects naturally cause damage to the organ, which might imply a sublethal effect at a very low level of intake.

Effects on organism level

Response to a toxic substance at *the organism level* can be categorised according to the dose rate and to the severity of damage:

1. Acute toxicity causing mortality: LD_{50}- and LC_{50}-values.
2. Chronically accumulating damage ultimately causing death.
3. Sublethal effects related to physiology: oxygen consumption, osmotic and ionic balance, feeding and digestion, assimilation, excretion, photosynthesis, N-fixation. The response can often be assessed as an increased or decreased rate.
4. Sublethal effects related to morphology: histological changes in cell and tissue, tumours, gross anatomical deformity.
5. Sublethal behavioural effects: locomotory activity, motivation and learning, equilibrium and orientation, migration, aggregation, reproductive behaviour, prey vulnerability, predator inability, chemoreception, photo-geo-taxes.

The toxicity of a compound varies with the exposure time. Time/effect relationships are of course important in understanding toxic effect. This has already been mentioned above for air pollutants where it is reflected in the air quality standards. Fig. 4.6 shows a toxicity curve of log exposure time versus log LC_{50}. Threshold LC_{50} indicates when the curve becomes asymptotic to the time axis. *The threshold* LC_{50} is usually (should be) magnitudes greater than the concentration found in nature or the threshold values reflected in environmental standards.

Toxicity experiments are carried out to determine the concentration needed to exert a particular toxic effect, for instance LC_{50} at different exposure times. 48 or 96 hours are usually applied. A typical plot (but the dose-response relationship may be more complicated; see also later in this section) would resemble Figure 4.7. NOAEL (non-observed adverse effect level) and the lowest observable adverse effect level (LOAEL) are indicated on the figure. These values may be expressed in terms of concentrations or dose. The concentration may be expressed in mg/l or mmole/l , the latter often being more useful to compare the toxicity of different chemicals. To estimate the effective concentration for a given toxicological effect (for instance LC_{50}) a horizontal line is drawn from the point of 50% effect to where it intersects the toxicity curve, as shown on the figure. Notice that the plot log concentration versus % effect is used for this purpose.

The body residue (body burden) is often a better quantity to relate to the observed effects than the concentration in water (LC-values), at least for narcotic toxicants. This is not surprising because the actual concentration in the organism determines of course to a great extent the expected effect. Carty et al., 1992 used the critical body burden (CBR) to estimate the aquatic toxicity of narcotic organic chemicals, such as halogenated hydrocarbons, ketones, alcohols, esters including phthalates, substituted benzenes and most pesticides.

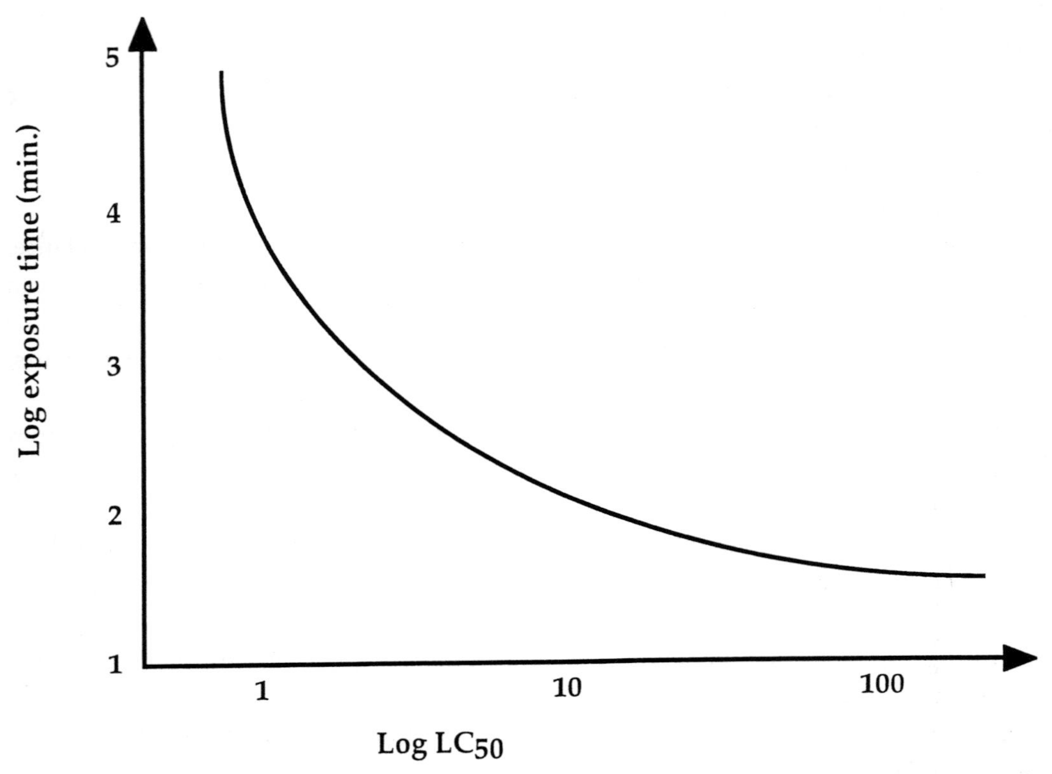

Fig. 4.6. Log exposure time versus log LC_{50} (mg copper ions/l). The plot indicates threshold 48-hr LC_{50} for green algae in hard water at pH = 7.0.

As

$$CBR = BCF^{*} \, LC_{50} \tag{4.4}$$

we can use the relationships between BCF and K_{OW} (see Section 2.5, Figure 2.13 where one possible correlation between BCF and K_{OW} is shown) and between LC_{50} and K_{OW} (a general correlation for narcotic organic compounds is $LC_{50} = 50 \; / \; K_{OW}$) to obtain an expression for the relationship between CBR and K_{OW}:

$$CBR\ (mM) = (1 + 0.05\ K_{OW}) \ (50 \; / \; K_{OW}) = 2.5\ mM + \; 50 \; / \; K_{OW} \tag{4.5}$$

This equation implies that if $K_{OW} > 1.5$ the toxicity becomes (almost) independent of the K_{OW} value. The toxicity expressed as CBR is approximately the same for all narcotic organic chemicals. Carty et al. (1992) has examined this and found that the CBR for a number of widely different narcotic organic compounds is 1-12 mmol / kg in (almost) accordance with almost the above presented hypothesis. The relationships between K_{OW} on the one hand and LC_{50}, BCF and CBR on the other are shown in Figure 4.8.

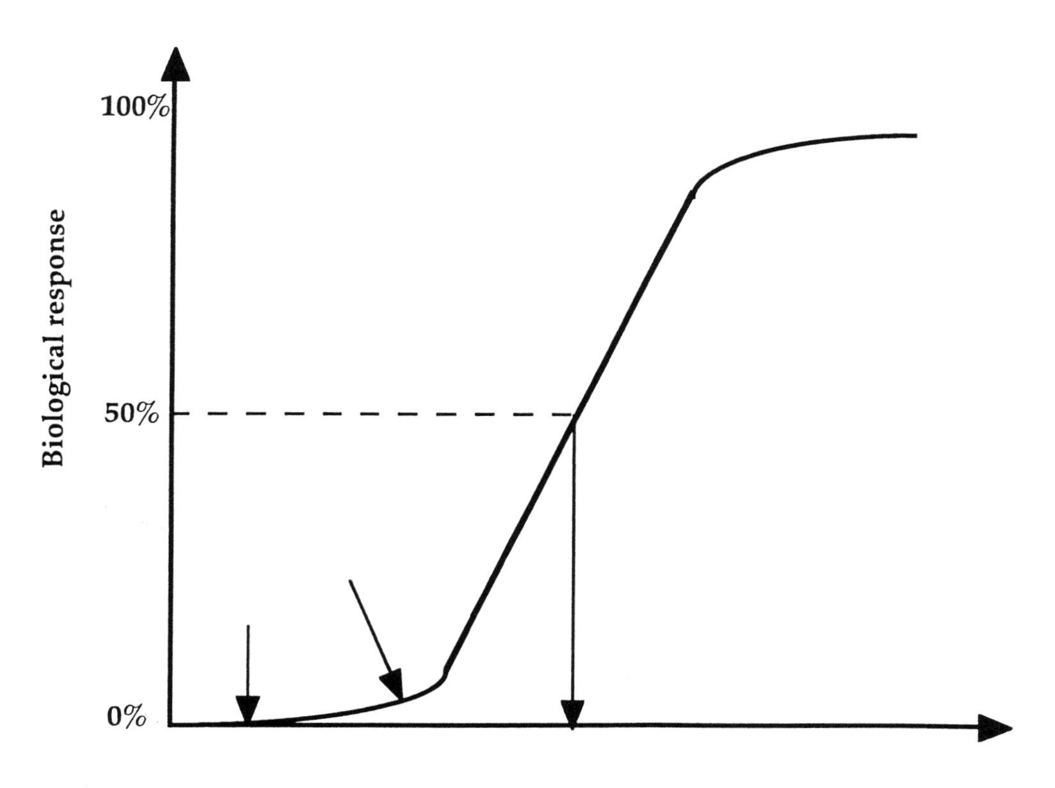

Fig. 4.7 The diagram illustrates the log/ log relationship between concentration or dose and biological response. The lowest observed adverse effect level (LOAEL) and no observed adverse effect level are indicated on the diagram.

Chemical injury to individuals resulting in premature death or reduced reproductive success and recruitment is ultimately reflected in lower abundances of exposed populations. Adaptation to toxic substances causes, furthermore, a change in the gene pool of the population. Further treatment of these important issues will be given in the next section, where the effects on the ecosystem level are discussed, but it should already be stressed here that a minor reduction in reproductivity might have a pronounced long-term effect on the population, as is observed, for instance, in lakes exposed to acid rain; see Section 4.8. The relationship between mortality rates and birth rates on the one hand and concentrations of toxicants on the other is sometimes known and can be used to set up a population dynamic model for the influence of the toxic substance on the population size.

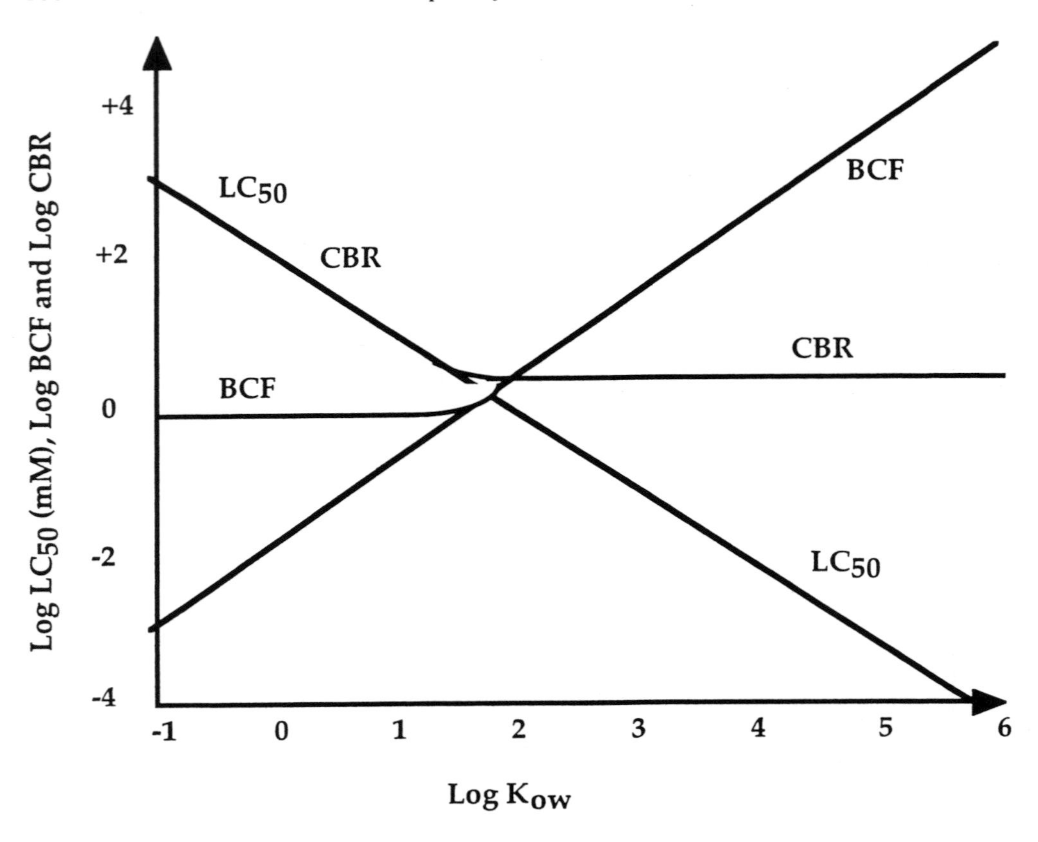

Fig. 4.8. Log LC_{50}, log BCF and log CBR are plotted versus log K_{OW}. Notice that CBR is (almost) constant when Kow is above 1.5. Log LC_{50} = 1.7 - log K_{OW} in accordance with the relationship given in the text and log BCF = log K_{OW} - 1.3.

<u>Intake, uptake and bioaccumulation</u>

It is important to distinguish between intake and uptake. Intake is the entry of a substance into the lungs, the gastrointestinal tract or subcutaneous tissues of animals. The fate of the intake in the organism will be governed by processes of absorption. Uptake is the absorption of a substance into extracellular fluid and the fate of material taken up will be governed by metabolic processes. The information about intake and uptake is different for various biota and will be presented below for four classes: mammals, fish and terrestrial and aquatic plants, but first some typical quantitative, mainly empirical descriptions of the processes involved will be presented. **The rate of uptake of a pollutant by ingestion of contaminated food is dependent on both the organism's maintenance metabolic rate and the rate of growth (Nordstrøm et al., 1975).** For freshwater fish the following expression for this relationship (at 20°C) is often used:

$$I = C_f \ (0.25\ m^{0.8} + 2\ dm\ /\ dt)\ f\ (g\ day^{-1}) \quad (4.6)$$

where
I = rate of uptake
C_f = concentration of pollutant in food (gg^{-1})
m = body mass (g)
dm / dt = growth rate ($g\ day^{-1}$)
f = fractional absorption from gastrointestinal tract

This equation expresses the food intake in brackets. It is the sum of the food required to cover respiration and growth. The uptake is found by multiplication with an efficiency coefficient, f.

The respiratory uptake of a pollutant through the gills of freshwater fish was shown experimentally to be dependent on the metabolic rate (de Freitas et al., 1975) (at 20°C):

$$I_g = 1000\ m^{0.8} * C_w * f_g\ (g\ day^{-1}) \quad (4.7)$$

where
I_g = rate of uptake through the gills ($g\ day^{-1}$)
m = body mass (g)
C_w = concentration of pollutant in water (g/g)
f_g = fractional absorption through the gills

The intake through the gills is found from the respiration = 1000 $m^{0.8}$. It is multiplied by an efficiency coefficient to find the intake.

The uptake of pollutants by terrestrial plants has been studied, because it is the first step in the transport to humans through the food chain. Burton et al. (1960) suggest the following equation:

$$C = P_d * F_d + P_r * F_r \quad (4.8)$$

where
C = yearly average uptake/ g of plant material
P_d = soil factor
F_d = concentration in soil ($g\ km^{-2}$)
P_r = rate factor
F_r = yearly fall-out rate of pollutant ($g\ km^{-2}\ y^{-1}$)

In higher aquatic plants the uptake of water borne pollutants by stems and leaves is often much more important than the absorption from the sediment or soil by the roots (Eriksson and Mortimer, 1975). **The uptake from water can often be expressed in the same simple manner for both animals and plants.** With a good approximation we have:

$$C_b \ / \ C_w = BCF \tag{4.9}$$

where

C_b = the biotic concentration (g kg^{-1})

C_w = the concentration in water (g l^{-1})

BCF = a concentration factor

There is a correlation between BCF and K_{OW} as previously presented in Section 2.5.

Equation (4.9) may be modified to account for the lipid phase in the organism. This is of importance particularly when we are using allometric principles to extrapolate the BCF value from one or a few organisms to many organisms. The allometric principles presented in Section 3.4 are strictly valid only for hydrophilic compounds (log $K_{OW} \leq 1.5$) or for organisms with the same percentage of fat tissue.

Generally we can state (see for instance Connell, 1997) that BCF = $f_{lipid} \times K_{OW}{}^{b}$, corresponding to

$$\log \ BCF = \log f_{lipid} + b \log K_{OW}, \tag{4.10}$$

where f_{lipid} is the lipid fraction in the organisms. b is usually close to 1 (often indicated to be 1.03). If C_L is the concentration of the lipophilic organic compound in the fat tissue, we have:

$$C_L = C_b \ / \ f_{lipid} \tag{4.11}$$

As $C_L \ / \ C_W = K_{OW}$, provided that we can consider the solubility in the fat tissue to be close to the solubility in octanol, we get:

$$\log BCF = \log K_{OW} + \log f_{lipid} \tag{4.12}$$

which is similar to equation (4.10) with $b \approx 1.0$.

This equation implies that the allometric principle can be used only for the same lipid fraction. However, equation (4.12) can be used to convert from one lipid fraction to another. Many fish contains about 5% lipid, or log f_{lipid} = - 1.3. If we know BCF values for fish with a lipid concentration of 5% and we want to know the BCF value for a fish of another size and with 10% lipid, we can use the allometric principles to find the BCF for the right fish size but with 5% lipid and then add 0.3 to the log BCF value to account for the higher lipid content.

The bioaccumulation factor BCF for the relationship between soil solid and biota is C_b/C_s completely parallel to equation (4.9). If the concentration in the pore water is denoted C_W, we obtain the following expression:

$$BCF = (C_b\ C_W\ /\ C_s\ C_W) = BCF_{org\text{-}water} * K_D \quad (4.13)$$

By use of equation (4.12) and equation (2.27) where the correlation of $BCF_{org\text{-}water}$ and K_D to K_{ow} are used, we get:

$$BCF = f_{lipid}\ K_{ow}{}^{b}\ /\ x{*}f{*}\ K_{ow}{}^{a} \quad (4.14)$$

where x is a proportionality constant. f is the fraction of organic carbon in the soil, as indicated in Section 2.5.

If we use the above mentioned b value of 1.03, the value corresponding to x, the proportionality constant in equation (2.27) which is antilog (- 0.006) ≈ 0.99 and the a value in (2.27) which is 0.94, we get the following expression for the BCF for the bioaccumulation factor soil - organism:

$$BCF = (1.01 * f_{lipid}\ /\ f)\ K_{ow}{}^{0.09} \quad (4.15)$$

This implies that BCF soil - organism has only a small dependence on K_{ow} and other properties of the soil. It depends more on the properties of the soil and the biota, particularly the ratio of lipid in the biota to the organic carbon content of the soil.

The retention of toxic substances is determined by the excretion rate, which can be approximated by means of the following first order equation:

$$r_e = k_e * C_b \quad (4.16)$$

where

r_e = excretion rate (g day^{-1} (body weight)$^{-1}$)

k_e = excretion rate coefficient (day^{-1})

C_b = concentration of toxic substances (g (body weight)$^{-1}$)

The excretion rate coefficient, k_e, can be approximated as:

$$k_e = a\ .\ m^b \quad \textbf{(4.17)}$$

where a and b are constants (b is close to 0.75), and m is the body weight. The retention can now be calculated as:

$$dC_b / dt = U - r_e \tag{4.18}$$

where U = uptake from food + uptake from air + uptake from water + uptake from soil.

TABLE 4.3
Excretion rates

Species	Component	Excretion rate (% of abs. amount day^{-1})
Rat (urine)	Cd	1.25
Homo sapiens (urine)	Hg	0.01
Rat (urine)	Hg	1.0
Sheep (urine)	Pb	0.5 - 1.0
Homo sapiens	Zn	8.0

This model of the concentration of toxic material in plants and animals is extremely simple and should only be used to give a first rough estimate. For a more comprehensive treatment of this problem, see Butler (1972); ICRP (1977); de Freitas et al. (1975); Mortimer and Kundo (1975); Seip (1979), Jørgensen et al., (1991) and Jørgensen (1994). Tables 4.3 and 4.4 give some characteristic excretion rates and uptake efficiencies. Note that the uptake efficiency is dependent on the chemical form of the component and on the composition of the food.

A wide variety of terms is used in an inconsistent and confusing manner to describe uptake and retention of pollutants by organisms using different paths and mechanisms. However, three terms are now widely applied and accepted for these processes:

1. *Bioaccumulation* is the uptake and retention of pollutants by organisms via *any* mechanism or pathway. It implies that both direct uptake from air and water and uptake from food are included.
2. *Bioconcentration* is uptake and retention of pollutants by organisms directly from water through gills or epithelial tissue. This process is often described by means of a concentration factor, see above, equation (4.9).
3. *Biomagnification* is the process whereby pollutants are passed from one trophic level to another and it exhibits increasing concentrations in organisms related to their trophic status. This process has been touched on in section 2.8.

TABLE 4.4
Uptake efficiencies

Component	Species	Uptake efficiency
DDT	Homo sapiens	14.4% (dairy product)
DDT	Homo sapiens	40.8% (meat product)
DDT	Homo sapiens	9.9% (fruit)
Hg	Monkey	90.0% (methyl-Hg)
Hg	Rat	90.0% (methyl-Hg)
Hg	Rat	20.0% (Hg-acetate)
Pb	Rabbit	0.8-1% (in food)
Pb	Sheep	1.3% (in food)
Zn	Pinfish	19.0% (in food)

An enormous amount of data has been published on chemical analyses of plants and animals, but much is of doubtful scientific value. The precise questions to be answered through a given examination need to be clearly formulated at the initial stage. Again the problem is very complex. It is not sufficient to set up computations for the retention of toxic substances, it is necessary to ascertain the distribution in the organism, the lethal concentration, the effect of sublethal exposure and the effects on populations over several generations (Moriarty, 1972, and Schüürmann and Markert, 1998). Our knowledge in the field of ecotoxicology is rather limited and further research in the area is urgently needed.

Dose / reponse

We distinguish between 5 classes of dose/response curves, Fig. 4.9: (Bridges, 1980 and Brown, 1978):

1. **a threshold dose** exists. It is simply indicated by dose levels at which there are no abnormal responses over controls. These threshold values are comparable with those mentioned in section 2.2.
2. **a cumulative response**, where the response increases more than linearly with the dose and any dose gives a certain response,
3. **a quasi threshold relationship**, where no effect is observed below a certain value, but the response above the threshold dose is cumulative,
4. **a linear relationship** between response and dose,
5. **a saturating effect** is observed. In this case, the curve has a steep slope at low doses, but tends to flatten out at higher levels. The following mechanisms could explain this relationship: an inducible pathway of detoxification, a progressively saturating activation pathway or differential repair processes.

Interactions of two or more pollutants

Mechanisms of interactions between two pollutants may be explained as follows:

1. The kinetic phase, by altering mechanisms of toxicant uptake, distribution, deposition, metabolism and excretion.
2. The dynamic phase, by altering toxicant receptor binding affinity and activity.
3. Chemical interactions between pollutants which produce new compounds, complexes or changes of chemical state; interactions between pollutants and substrates which alter physicochemical forms of pollutants and their toxicities (Anderson and D'Appolinia, 1978).

Some important types of pollutant interactions and related responses are illustrated in Figs. 4.10 and 4.11. These involve synergistic and antagonistic interactions for combinations of two substances. Our present knowledge about teratogenesis, mutagenesis and carcinogenesis is especially limited, and very poor when it comes to exposure of two or more chemicals simultaneously.

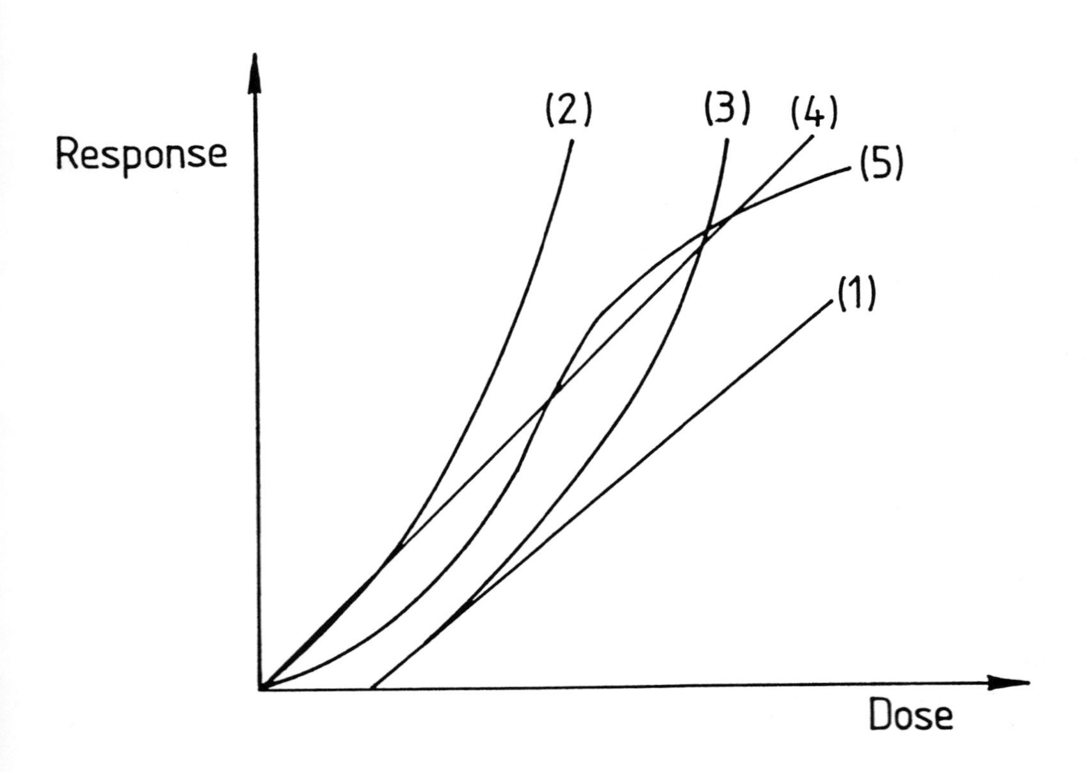

Fig. 4.9. Five types of responses are shown: (1) threshold, (2) cumulative, (3) quasi threshold, (4) linear and (5) saturating.

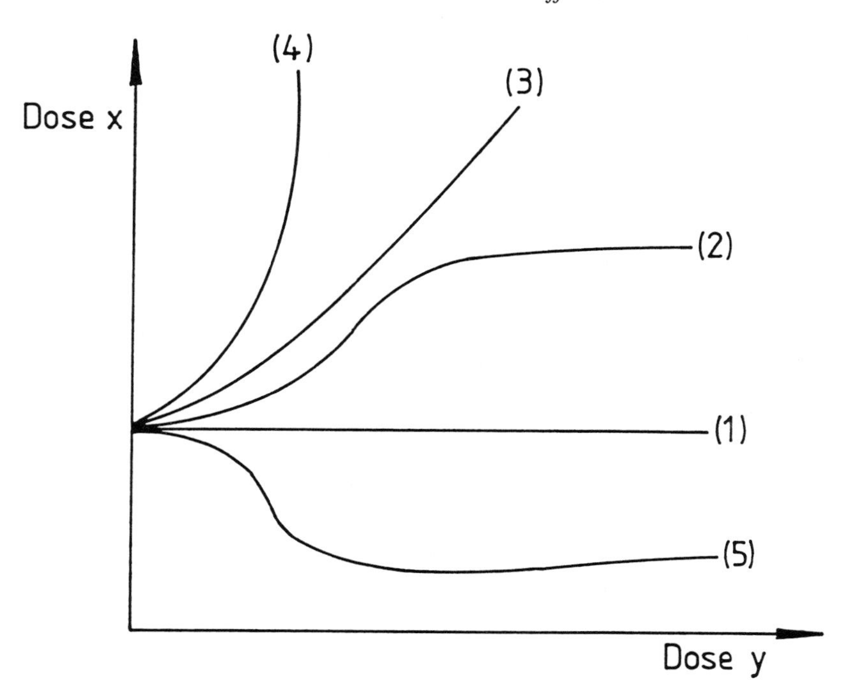

Fig. 4.10 Isoboles (curves of equal biological response) for combinations of a substance X, that is active on its own and a substance Y, that is inactive when given alone, but which influences the action of X. The following possibilities are shown: (1) Y is inert, (2) antagonistic effect by physiological function, (3) antagonistic effect by chemical competition, (4) antagonistic effect by a non-competitive irreversible process, (5) Y sensitises for X ≈ synergistic effect.

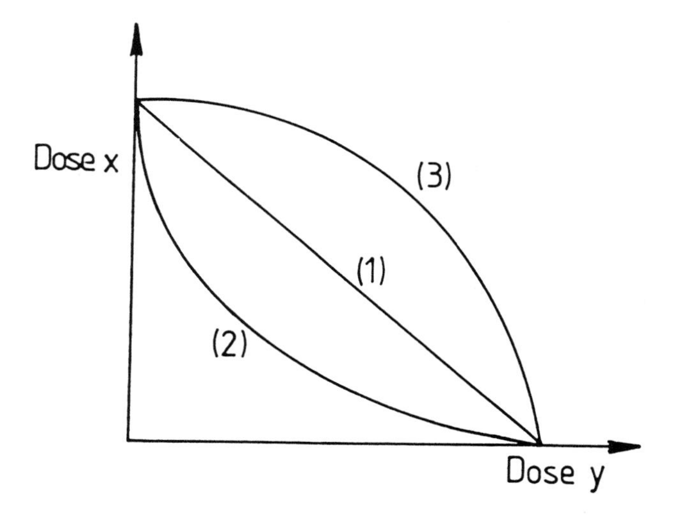

Fig. 4.11 Isoboles for combinations of two substances, X and Y, each of which is active by itself: (1) additive action, (2) synergism and (3) antagonism.

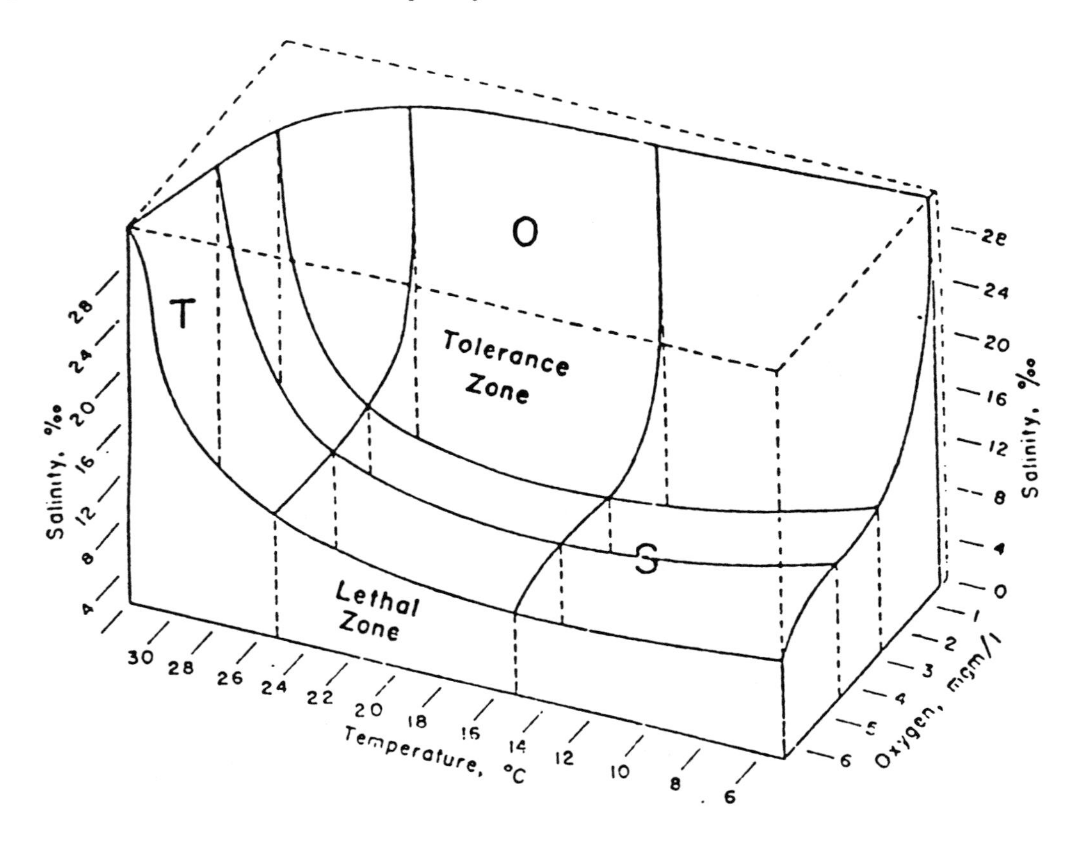

Fig. 4.12. Mortality (%) of crab larvae at various combinations of temperature and salinity (o/oo), (Sprague, 1970)

Figure 4.12 illustrates the interactions of two parameters. The uncertainty and limited knowledge in this avenue of environmental science make it necessary to treat such problems of interactions very cautiously. This does not mean, however, that we have to abandon their solution - on the contrary, we have to find one although it can only be preliminary and will be very uncertain. If the uncertainty is too great it might be necessary to avoid the use of chemicals at all, at least in certain areas. This is, however, a political decision, and one which traditionally has come down on the side of economic rather than ecological advantage.

Example 4.1

K_{OW} for pentachlorophenol is known to be approximately 5.4. Pentachlorophenol is an acid and pK is 4.75. Estimate K_D = the distribution coefficient for soil/pore water at pH = 4.45 for a soil with a organic carbon content of 2%. The adsorption of the non ionised form can be found by use of K_D, while the ionised forms are assumed to be negligibly sorbing.

Solution

k (sometimes also indicated in the literature as K_{OC}) the K_D value corresponding to 100% organic carbon, can be estimated from equation (2.27). We get:

$$\log K_D = -0.006 + 0.937 \log K_{OW} - \log 100/2 = -0.006 + 0.937x5.4 - 1.7 = 3.35$$

$$K_D = 2239$$

This value corresponds to the non-ionised form. At pH = 4.45 the ratio between the ionised and the non ionised form is 1: 2 in accordance with:

$$pH = pK + \log ([\text{ionised form}] / [\text{non ionised form}])$$

$$4.45 = 4.75 + \log (1/2)$$

This implies that it is only 2/3 of the pentachlorophenol in the pore water which is actually sorbing, while 1/3 is ionised. It means that $K_D = 2x2239/3 = 1493$.

Example 4.2

A plant with 3% oil content is cultivated in a soil with 1% organic carbon. The soil is contaminated with chlorobenzene in a concentration of 12 ppm (mg/kg). What would be the expected concentration of chlorobenzene in the plants at harvest, when the water solubility for chlorobenzene at the actual temperature is approximately 450 mg/l?

Solution

The molecular weight of chlorobenzene is 112.5. The water solubility, S = 450,000 μg/ 112.5 = 4000 μmoles/l. Log S = 3.6, which in accordance to Figure 2.12 would give a log K_{OW} = 3.0. By use of equation (4.15) we get:

$$BCF = (1.01x3/1) \times 1000^{0.09} = 5.6.$$

The concentration in the plants will therefore be $5.6x12 \approx 68$ mg/ kg.

4.5 EFFECTS ON THE POPULATION, ECOSYSTEM AND GLOBAL LEVELS

Biodegradation and Mobility

The effects on the higher levels of the biological hierarchy are to a certain extent dependent on the same properties of the chemical compounds as the effects on the lower levels. A central property of a chemical is its degradability. If the chemical is very persistent to biological degradation, chemical degradation and photolytic degradation, it will inevitably accumulate in the environment and can therefore on a long term basis cause harmful effects on all levels. If on the other hand a chemical will easily degrade, even a highly toxic effect will disappear rapidly and any harmful effect can therefore easily be eliminated by phasing out the use of the chemical. The bioavailability is also of equal importance on the ecosystem level as on the organisms level. If a chemical is (almost) not bioavailable because it is adsorbed even at small concentrations to sediment, it will hardly be able to affect organisms and as organisms are the "elements" in the ecosystems, it is most probable that also no effect will be observed at the population level and at the ecosystem level.

Similarly, high water solubility will imply high mobility and therefore the possibility of observing effects far from the discharge point - on all levels, while a high K_{OW} will increase the risk of bioaccumulation and biomagnification, giving a high probability of recording high and harmful concentrations in the higher trophic levels.

A mobility index is often used to express the mobility of organic compounds in soil. A soil mobility index, MI, is defined by the following equation:

$$MI = \log [S \cdot V / K_{OC}] \quad (4.19)$$

where S is the water solubility in mg/l, V is the vapour pressure at ambient temperature expressed in mm Hg and K_{OC} is the soil partition coefficient based on organic carbon.

The mobility index is interpreted as follows:

MI > 5.00	The compound is extremely mobile and may easily contaminate the ground water.
MI 0 to 5.00	The compound is very mobile and may contaminate the ground water, if it is not readily biodegraded.
MI 0 to -5.00	The compound is slightly mobile. It may contaminate the ground water if the compounds is very slowly biodegraded or refractory.
MI - 5.00 to -10.00	Immobile. The compound is not a threat to the ground water
MI < - 10.00	Very immobile. Will hardly move at all in soil with > 1% C.

How to find where the risk is?

An ERA on the higher levels of the biological hierarchy starts with the following pertinent question:

Where in the ecosystem is the biggest risk, when we consider the network of the system and the properties of the pollutant? An answer will require knowledge about the ecosystem and to the properties of the pollutant. Two examples are used below to illustrate these considerations in relation to lakes.

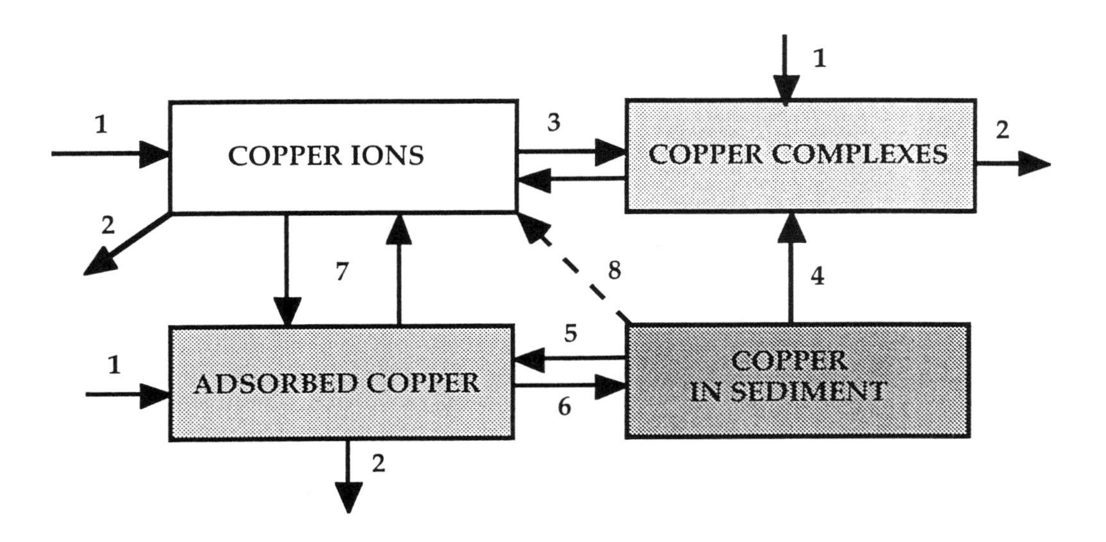

Fig. 4.13 A simple copper model applicable to an ERA for copper. The processes are: (1) and (2) inputs and outputs of copper ions, copper complexes and adsorbed copper (on suspended matter), (3) chemical equilibrium between copper ions and formation of copper complexes, (4) release of copper complexes from sediment, (5) resuspension of sediment, (6) settling of suspended matter, (7) adsorption-desorption equilibrium between free copper ions and copper adsorbed on suspended matter and (8) release of copper ions from sediment; the dotted line indicates that this release process is considered to be of minor importance.

Free copper ions are very toxic to phytoplankton, zooplankton and fish. The LC_{50} value for Daphnia magna is as low as 20 $\mu g\ l^{-1}$ and for some Salmonoid species 100-200 $\mu g\ l^{-1}$. A risk assessment for copper should consequently focus on the concentration of free copper ions in the water - copper is mainly bioavailable as copper ions and include the processes determining this concentration. In the case of copper, a lethal concentration will be reached before the concentration in fish becomes toxic to humans, and as the uptake and excretion of copper by plants and animals are insignificant for the free copper ion concentration, these processes can be omitted. This means that a model as simple as that shown in Fig. 4.13 can be applied to give at least a first estimate of how much copper, and in what form, it may be considered a risk for an aquatic ecosystem.

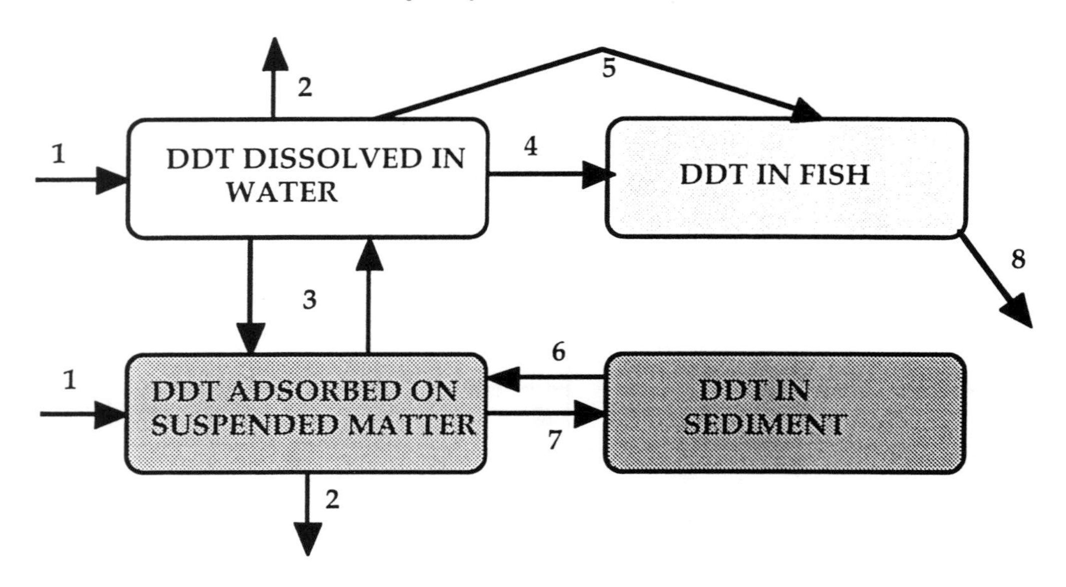

Fig. 4.14 A simple DDT model. The processes are: (1) and (2) inputs and outputs of DDT as dissolved DDT or DDT adsorbed on suspended matter, (3) adsorption- desorption equilibrium between dissolved and adsorbed DDT, (4) direct uptake of DDT from the water by respiration, (5) accumulation of DDT through the food chain, (6) resuspension of sediment, (7) settling of suspended matter containing DDT and (8) the biodegradation of DDT in the fish.

The input of copper is the forcing function in the model. The equilibrium: copper ions + ligands = copper complexes can be described to a certain extent by use of known equilibrium constants (Jørgensen et al., 1991 and Jørgensen 1979b). The adsorption process of copper ions on suspended matter requires laboratory investigation to find the adsorption capacity of the suspended matter.

The release of copper from the sediment should also be studied, although some information is available (Lu and Chen, 1977, Jørgensen et al., 1991). As can be seen from this study, a rather simple approach which, however, is complex enough to require application of a mathematical model, can be used to set up an ERA for copper, although the amount of data necessary to calibrate and validate the model is limited.

A useful, but simple model for the distribution and effect of DDT in aquatic ecosystems is illustrated in Figure 4.14. The DDT problem is different from the copper problem due to the differences in properties of the two pollutants. The main DDT problem is the concentration in fish at the highest trophic level in the lake, as DDT is accumulated mainly through the food chain.

The WHO has recommended the maximum permissible concentration of DDT in human food should be 1 mg per kg net weight, which corresponds to a daily intake of

0.005 mg/kg body weight. The major management problem is to maintain the DDT concentration in all fish species below this value, divided by a safety factor of, say, 10, which means below a concentration of 0.1 mg per kg net weight. The model shown in Fig. 4.14 is suggested for this purpose.

The direct uptake rate from water and the biodegradation rate are known for some fish (Jørgensen et al., 1991) and are not significantly different from species to species. The DDT accumulation through the food chain and the rate of photolysis and dehydrochlorination are known with acceptable accuracy. As with the copper problem, the equilibrium between DDT in the water phase and adsorbed on suspended matter must be studied in the laboratory, while the degradation rate in the sediment can be approximated with data from wet soil.

As can be seen from these two case studies, the distribution of copper and DDT can be assessed by the use of simple, workable models which might be applied to set up an ERA and to give answers to specific management problems related to discharge of these two pollutants into the aquatic ecosystem. The considerations illustrated in these two examples cover, however, only the harmful effects on the population level or on the ecosystem level, if no significant mortalities of certain species take place.

It is recommended to consider the following issues to ensure that no additional harmful effect will occur on the population and ecosystem levels:

1. Has the considered pollutant in the concentration range of importance any significant effect on the birth and mortality rates of any species? If this is the case, it is necessary to develop a population dynamic model to account for these effects.
2. What are the roles of the species which change in number as a consequence of the pollutant? Will these roles be taken over by other species? How will this change the function of the entire ecosystem network? An ecosystem can be described as a network, where all components are directly or indirectly linked to all other components. It can be shown (Patten, 1991) that the effects of the entire network (named indirect effects) often are exceeding the direct effects based on the links directly between two ecosystem components.

An illustrative, environmentally very relevant example should be mentioned in this context. Use of insecticides has sometimes an unexpected (indirect) effect, namely if the insecticide causes high mortality of insect species that are predators of the herbivorous insects which it was intended to kill. The result may be that the herbivorous insects increase in number.

3. Are the changes sufficiently slow to allow adaptation of at least some of the most important species? If this is the case, the effect will be reduced by the self-organising ability of ecosystems. If for instance pesticides are used over a long period of time, the pest may be adapted to the pesticides and the effect reduced considerably. This demonstrate that the effect of pesticides may be considerably reduced over time.
4. Which consequences will it have that the species richness is changed, in most cases decreased, as the resulting effect of the pollutant? What is the role of diversity in the

ecosystem?

A heavy modelling component is required to answer these questions as completely as possible on the basis of our present knowledge; see for instance Jørgensen (1994). Modelling will not be discussed further in this context, as it is not the topic of this volume. However, the concept of ecosystem stability in its widest sense and its relation to species diversity and other relevant population indices should be presented in this section as they are closely related to effects on the ecosystem level.

<u>Key species and species richness</u>

Out of thousands of species that might be present in an ecosystem relatively few exert a major controlling influence by virtue of their numbers, size, production or other activities. Intracommunity classification therefore goes beyond taxonomic listing and attempts to evaluate the actual importance of organisms in the community. Three types of indices are used to describe the organisation in an ecosystem:

1. **indices of dominance,**
2. **indices of similarity** and
3. **indices of diversity.**

Some of the most commonly used indices in these three groups are listed in Table 4.5. **Species diversity tends to be low in physically and man-controlled ecosystems and high in natural undisturbed ecosystems, but this rule is not always valid, and takes only the tendency into account. Two stability concepts are generally used. Stability is either understood as maintenance of constant biological structure or as resistance to changes in external factors.**

Reactions of ecosystems to perturbations have been widely discussed in relation to concepts of stability. However, this discussion has in most cases not considered the enormous complexity of regulation and feedback mechanisms.

The stability concept *resilience* is understood as the ability of the ecosystem to return "to normal" after perturbations. This concept has more interest in a mathematical discussion of whether equations may be able to return to steady state, but the shortcomings of this concept in a real ecosystem context are clear:

An ecosystem is a soft system that will *never* return to the same point again. It will be able to maintain its functions on the highest possible level, but never with exactly the same biological and chemical components in the same concentrations again. The species composition or the food web may have changed or may not have changed, but they will not be the same organisms with exactly the same properties. In addition, it is unrealistic to consider that the same conditions will occur again. We can observe that an ecosystem has the property of resilience in the sense that ecosystems have a tendency to recover after stress, but a complete recovery (understood as exactly the same situation appearing again) will never be realised.

TABLE 4.5

Some useful indices of species structure in communities

A. **Index of dominance (c) +)**

$c = \Sigma(ni/N)^2$, where ni = importance value for each species (number of individual, biomass, production, and so forth)

N = total of importance values

B. **Index of similarity (S)** between two samples o)

S = 2C / (A + B), where A = number of species in sample A

B = number of species in sample B

C = number of species common to both samples

Note: Index of dissimilarity = 1 - S.

C. **Indices of species diversity**

(1) Evenness index (e) §)

e = H / log S, where H = Shannon index (see below)

S = number of species

(2) Shannon index of general diversity (H) *)

$H = - \Sigma P_i \log P_i$ where ni = importance value for each species

N = total of importance values

P_i = importance probability for each species = ni / N

\+ See Simpson (1949)

o For a related index of "% difference" see E.P. Odum (1950)

§ See Pielou (1966); for another type of "equitability" index, see Lloyd and Ghelardi (1964).

* See Shannon and Weaver (1963); Margalef (1968).

Note: Natural logarithms ($\log_e$) are usually employed, but $\log_2$ is often used to calculate H so as to obtain "bits per individual".

The combination of external factors - the impact of the environment on the ecosystem - will never appear again and even if it did, the internal factors - the components of the ecosystem - have meanwhile changed and can therefore not react in the same way as the previous internal factors did. The concept of resilience is therefore not a realistic quantitative concept. If it is used realistically, it is not quantitative and if it is used quantitatively for instance in modelling, it is not realistic. Resilience to a certain extent covers the ecosystem property of elasticity, but in fact the ecosystem is more flexible than elastic. It will change to meet the challenge of changing external factors, not try to struggle to return to exactly the same situation.

Resistance is another widely applied stability concept. It covers the ability of the ecosystem to resist changes when the external factors are changed. This concept needs, however, a more rigorous definition and needs to be considered multidimensionally to be able to cope with real ecosystem reactions. An ecosystem will always be changed when the conditions are changed; the question is what is changed and how much?

Webster (1979) examined by use of models the ecosystem reactions to the rate of nutrient recycling. He found that an increase in the amount of recycling relative to input resulted in a decreased margin of stability, faster mean response time, greater resistance (i.e., greater buffer capacity, see the definition below) and less resilience. Increased storage and turnover rates resulted in exactly the same relationships. Increases in both recycling and turnover rates produced opposite results, however, leading to a larger stability margin, faster response time, smaller resistance and greater resilience.

Gardner and Ashby (1970) examined the influence on stability of connectance (defined as the number of food links in the food web as a fraction of the number of topologically possible links) of large dynamic systems. They suggest that all large complex dynamic systems may show the property of being stable up to a critical level of connectance and then as the connectance increase further, the system suddenly goes unstable.

O'Neill (1976) examined the role on heterotrophs on the resistance and resilience and found that only small changes in heterotroph biomass could reestablish system equilibrium and counteract perturbations. He suggests that the many regulation mechanisms and spatial heterogeneity could be accounted for when the stability concepts are applied to explain ecosystem responses. The role of variability in space and time will be discussed further below.

These observations explain why it has been very difficult to find a relationship between ecosystem stability in its broadest sense and species diversity. Compare also with Rosenzweig (1971), where almost the same conclusions are drawn.

It is observed that increased phosphorus loading gives decreased diversity (Ahl and Weiderholm, 1977 and Weiderholm, 1980), but very eutrophic lakes *are* very stable. Figure 4.15 gives the result of a statistical analysis from a number of Swedish lakes. The relationship shows a correlation between number of species and eutrophication, measured as chlorophyll-a in $\mu g/l$. A similar relationship is obtained between the diversity of the benthic fauna and the phosphorus concentration relative to the depth of the lakes.

Therefore it seems appropriate to introduce another but similar concept, named the buffer capacity, ß. It is defined as follows (Jørgensen, 1997):

$$ß = 1 / (\partial(\text{State variable}) / \partial(\text{Forcing function})) \qquad (4.20)$$

Forcing functions are the external variables that are driving the system such as discharge of waste water, precipitation, wind and so on, while state variables are the internal variables that determine the system, for instance the concentration of soluble phosphorus, the concentration of zooplankton and so on. As seen the concept of buffer capacity has a definition which allows us to quantify for instance in modelling and it is furthermore applicable to real ecosystems, as it acknowledges that *some* changes will always take place in the ecosystem in response to changed forcing functions. The question is how large these changes are relative to changes in the conditions (the external variables or forcing functions).

The concept should be considered multidimensionally, as we may consider all combinations of state variables and forcing functions. This implies that even for one type of change there are many buffer capacities corresponding to each of the state variables. Rutledge (1974) defines ecological stability as the ability of the system to resist changes in the presence of perturbations. It is a definition very close to buffer capacity, but it lacks the multidimensionality of ecological buffer capacity.

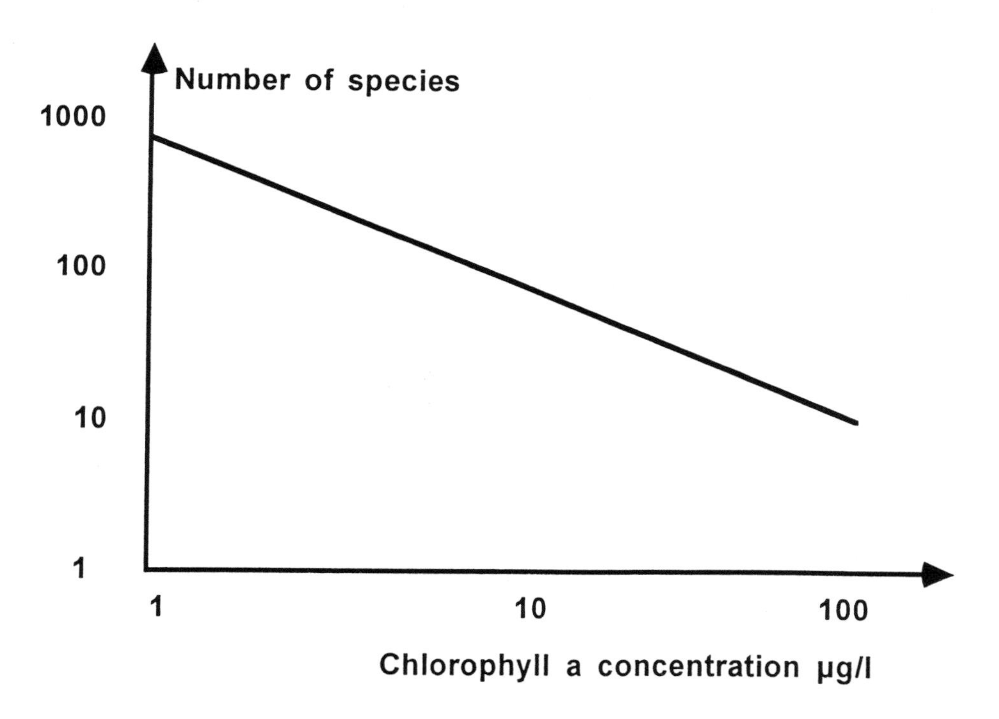

Fig 4.15 Weiderholm (1980) obtained the relationship shown for a number of Swedish lakes between the number of species and eutrophication, expressed as chlorophyll-a in $\mu g/l$.

The relation between forcing functions (impacts on the system) and state variables indicating the conditions of the system is rarely linear and buffer capacities are therefore not constant. It may therefore in environmental management be important to reveal the relationships between forcing functions and state variables to observe under which conditions buffer capacities are small or large; compare with Figure 4.16.

Model studies (Jørgensen and Mejer, 1977 and Jørgensen, 1986) have revealed that in lakes with a high eutrophication level, a high buffer capacity for nutrient inputs is obtained with a relatively small diversity. The low diversity in eutrophic lakes is consistent with the above-mentioned results by Ahl and Weiderholm (1977) and Weiderholm (1980). High nutrient concentrations = large phytoplankton species. The specific surface does not need to be large, because there are plenty of nutrients. The selection or competition is not on the uptake of nutrients but rather on escaping the grazing by zooplankton and here greater size is an advantage. The spectrum of selection becomes in other words more narrow, which means reduced diversity. This demonstrates that a high buffer capacity may be accompanied by low diversity.

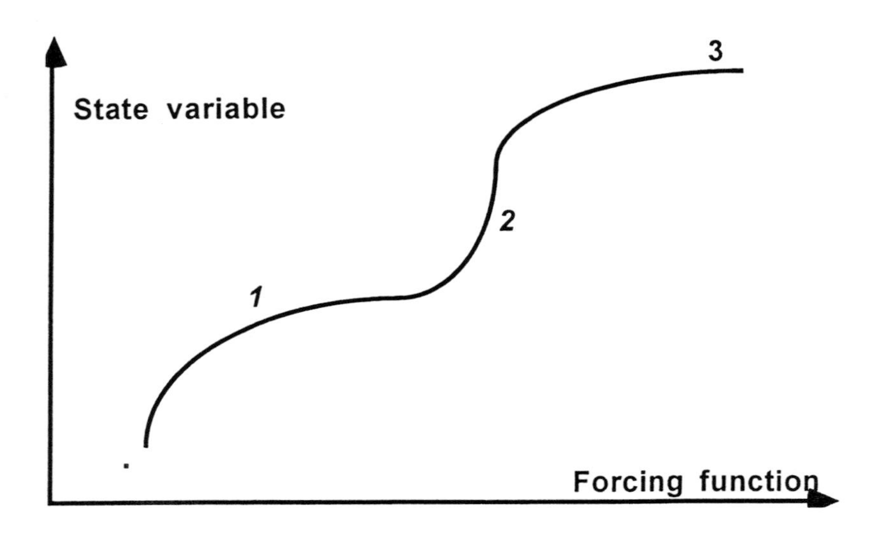

Figure 4.16. The relation between state variables and forcing functions is shown. At points 1 and 3 the buffer capacity is high; at point 2 it is low.

If a toxic substance is discharged to an ecosystem, the diversity will be reduced. The species most susceptible to the toxic substance will be extinguished, while other species, the survivors, will metabolise, transform, isolate, excrete, etc. the toxic substance and thereby decrease its concentration. We observe a reduced diversity, but simultaneously we maintain a high buffer capacity to input of toxic compounds, which means that only small changes, caused by further input of the toxic substance, will be observed. Model studies of toxic substance discharge to a lake (Jørgensen and Mejer, 1977 and Jørgensen, 1997) demonstrate the same inverse relationship between the buffer capacity to the considered toxic substance and diversity.

Ecosystem stability is therefore a very complex concept (May, 1977) and it seems impossible to find a simple relationship between ecosystem stability and ecosystem properties. Buffer capacity seems to be the most applicable stability concepts as it is based on:
1) An acceptance of the ecological complexity - it is a multidimensional concept ; and
2) Reality, i.e., that an ecosystem will never return to exactly the same situation again.

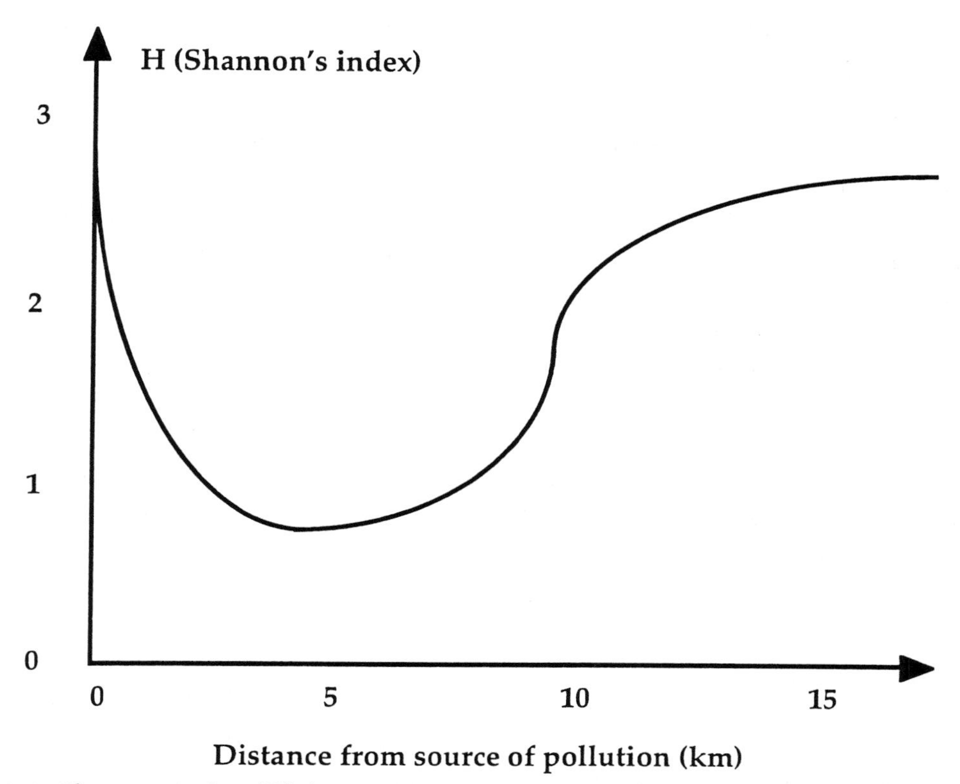

Fig. 4.17 Shannon index (H) (see Table 4.5, equation C2) plotted versus distance (0 = discharge of waste water in km). The measurements on the Danish river Suså are the result of a group work by students at the University of Copenhagen.

The most widely used index of diversity is **the Shannon index**. It is *normally distributed* (Bowman et al., 1970; Hutchinson, 1970), so *routine statistical methods* can be used to test for the magnification of differences between means. Higher diversity means longer food chains, more cases of symbioses and greater possibilities for negative feedback control, which will reduce oscillations and hence increase stability. This expresses the general interest for the ecologist in the concept diversity, as he sees in any measure of diversity an expression of the possibilities for constructing feedback systems (Margalef, 1969). **Moreover, diversity indices can be used to evaluate man-made stresses on ecosystems,** as clearly demonstrated in Figs. 4.15 and 4.17.

It is fully accepted today that there is no simple relationship between stability and species diversity as previously underlined in this section, but still we are interested as an important part of environmental management in maintaining high species diversity. Maintenance of species diversity was for instance one of the hot issues at the global environmental conference in Rio in 1992.

The interest in species diversity can easily be explained by use of the concept of ecological buffer capacity. There is no simple relationship between a high Shannon index and a high buffer capacity.

Very eutrophic lakes have a high buffer capacity towards any changes in the level of eutrophication caused by increased loading of nutrients, but eutrophic lakes have a low Shannon index, as it can be seen in Figure 4.15. A high species diversity will, however, imply that the spectrum of buffer capacities becomes wider, because the more species the higher is the probability that the number of different buffering characteristics will increase. The ecosystem will simply be represented by a wider range of properties in the number of species increases, because they are all different. A high species diversity will with other words imply that a new (unexpected) change of the forcing function (external variables) has a higher probability of being met with a adequate buffer capacity to reduce the effects of the changes.

The effects of pollutants on the population and ecosystem level with reference to all the aspects touched on here including the role of species diversity, will be illustrated in Section 4.9 of this chapter by the case of lake acidification.

The global effects of pollutants are in principle not very different from the effects at ecosystem level, except of course that the system consists of the entire ecosphere, made up of many ecosystems.

Pollutants which have an effect on the global conditions for life - for instance the global cycles, the global climate and the ultraviolet radiation penetrating the atmosphere - are not surprisingly of particular interest.

If changes in the composition of the atmosphere or the oceans take place, they will have an enormous impact on the life conditions on earth. The global effects can be illustrated by the greenhouse effect, the global dissemination of lead (or other toxic substances) and the impairment of the ozone layer. These problems have already been mentioned in Chapters 2-3.

Standards

For several pollution problems standards have been published both by governmental and by international organisation. For instance WHO has formulated water quality standards for drinking water and in addition most countries have set up their own standards to ensure that the drinking water is not creating any health problems. Standards for the acceptable maximum contamination of food have also been published by WHO, for instance 1 mg/kg for DDT and for mercury compounds. Almost all countries also have standards for the waste water treatment, for instance in Denmark:

maximum BOD_5 concentration 15 mg /l

maximum P concentration 1.5 mg/l
maximum N concentration 8 mg/l

For particular vulnerable ecosystems, it is possible to reduce the maximum concentrations to the level which is considered acceptable for the focal ecosystem.

Standards are in principle based on consideration of all levels in the biological hierarchy and should therefore also reflect an impact reduction to an acceptable level for ecosystems.

They should be based on objective scrutiny of the possible effects as a function of the loadings. This is, however, far from always the case, as standards are politically determined which implies that they also take into consideration several other factors of political interest, for instance economic aspects.

As an illustration of standards Table 4.6 shows the maximum concentrations (mg / kg dry matter) for compounds in sludge which may be applied in Denmark as soil conditioner. The standards are compared with

1) an EC-50 value for reproduction reduction for springtails

2) typical concentrations found in Danish sludge in 1997-1999, which have been a part of the material used for a final decision. Slightly less than 50% of Danish sludge can meet these standards and therefore be used as soil conditioner.

TABLE 4. 6

Maximum concentrations for compounds in sludge applied in Denmark as soil conditioner in agriculture.

Compound	Max. conc. mg/ kg d. m.	EC- 50% repr. reduct. spring-tails mg / kg soil d.m.	Range found in DK 1997-98 mg/ kg d.m.
Cd	0.8	50	0.2 - 4.0
Cr	100		11- 230
Cu	1000 (average conc. about 30)	130	8 - 89
Pb	120	2800	17 - 345
Hg	0.8		
Ni	30	450	4 - 145
Zn	4000 (average conc. about 200)	700	
LAS	1300	740	11 - 160,000
DEHP	50	>5000	4 - 170
PAHs	3		
Nonylphenol	10	44	0.3 - 67

4.6 THE NEED FOR METHODS TO ESTIMATE THE PROPERTIES OF TOXIC SUBSTANCES

Slightly more than 100,000 chemicals are produced in such an amount that they threaten or may threaten the environment. They cover a wide range of applications: household chemicals, detergents, cosmetics, medicines, dye stuffs, pesticides, intermediate chemicals, auxiliary chemicals in other industries, additives to a wide range of products, chemicals for water treatment and so on. They are (almost) indispensable in modern society and cover all some more or less essential needs in the industrialised world, which has increased the production of chemicals about 40-fold during the last 4 decades. A minor or major proportion of these chemicals is inevitably reaching the environment through their production, during their transportation from the industries to the end user or by their application. In addition, the production or use of chemicals may cause more or less unforeseen waste or by-products, for instance chloro-compounds from the use of chlorine for disinfection. As we would like to have the benefits of using the chemicals but cannot accept the harm they may cause, this conflict raises several urgent questions which we already have discussed in this volume.

We cannot answer these crucial questions without knowing the properties of the chemicals. OECD has made a review of the properties that we should know for all chemicals. We need to know the boiling point and melting point to know in which form (as solid, liquid or gas), the chemical will be found in the environment. We must know the distribution of the chemicals in the five spheres: the hydrosphere, the atmosphere, the lithosphere, the biosphere and the technosphere. This will require knowledge about their solubility in water, the partition coefficient water/lipids, Henry's constant, the vapour pressure, the rate of degradation by hydrolysis, photolysis, chemical oxidation and microbiological processes and the adsorption equilibrium between water and soil - all as function of the temperature. We need to discover the interactions between living organisms and the chemicals, which implies that we should know the biological concentration factor (BCF), the magnification through the food chain, the uptake rate and the excretion rate by the organisms and where in the organisms that the chemicals will be concentrated, not only for one organism but for a wide range of organisms. We must also know the effects on a wide range of different organisms. It means that we should be able to find the LC_{50} and LD_{50}- values, the MAC and NEC-values (for the abbreviations and the definitions used see Appendix 5), the relationship between the various possible sublethal effects and concentrations, the influence of the chemical on fecundity and the carcinogenic and teratogenic properties. We should also know the effect on the ecosystem level. How do the chemicals affect populations and their development and interactions, i.e., the entire network of the ecosystem?

Table 4.7 gives an overview of the most relevant physical-chemical properties of organic compounds and their interpretation with respect to the behaviour in the environment.

The list of properties needed to give an adequate answer to the six questions mentioned above could easily be made longer (see for instance the list recommended by OECD). To provide all the properties corresponding to the list given here is already a huge task. More than ten basic properties should be known for all 100,000 chemicals which would require 1,000,000 pieces of information. In addition we need to know at least ten properties to describe the interactions between 100,000 chemicals. Let us say modestly that we use 10,000 organisms to represent the approximately 10 million species on earth. This gives a total of 1000,000 + 100,000*10,000*10 = in the order of 10^{10} properties to be quantified! Today we have determined less than 1% of these properties by measurements and with the present rate of generating new data we can be certain that during the twenty-first century, we shall not be able to reach 10% even with an accelerated rate of ecotoxicological measurements.

TABLE 4.7

Overview of the most relevant environmental properties of organic compounds and their interpretation

Property	Interpretation
Water solubility	High water solubility corresponds to high mobility
K_{OW}	High K_{OW} means that the compound is lipophilic. It implies that it has a high tendency to bioaccumulate and be sorbed to soil sludge and sediment. BCF and K_{OC} are correlated with K_{OW}.
Biodegradability	This is a measure of how fast the compound is decomposed to simpler molecules. A high biodegradation rate implies that the compound will not accumulate in the environment, while a low biodegradation rate may create environmental problems related to the increasing concentration in the environment and the possibilities of a synergistic effect with other compounds.
Volatilisation, vapour pressure	High rate of volatilisation (high vapour pressure) implies that the compound will cause an air pollution problem
Henry's constant, H See equation (2.23)	H determines the distribution between the atmosphere and the hydrosphere.
pK	If the compound is an acid or a base, pH determines whether the acid or the corresponding base is present. As the two forms have different properties, pH becomes important for the properties of the compounds.

ERAs require, however, among other inputs also information about the properties of the chemicals and their interactions with living organisms. It is maybe not necessary to know the properties with the very high accuracy that can be provided by measurements in a laboratory, but it would be beneficial to know the properties with sufficient accuracy to make it possible to utilise the models for management and for risk assessments. Therefore estimation methods have been developed as an urgently needed alternative to measurements. They are to a great extent based on the structure of the chemical compounds, the so called QSAR and SAR methods, but it may also be possible to use allometric principles to transfer rates of interaction processes and concentration factors between a chemical and one or a few organisms to other organisms. This chapter focuses on these methods and attempts to give a brief overview on how these methods can be applied and what approximate accuracy they can offer. A more detailed overview of the methods can be found in Jørgensen et al., 1997.

It may be interesting here to discuss the obvious question: why is it sufficient to estimate a property of a chemical in an ecotoxicological context with 20%, or sometimes with 50% or higher uncertainty? Ecotoxicological assessment usually give an uncertainty of the same order of magnitude, which means that the indicated uncertainty may be sufficient from the modelling view point, but can results with such an uncertainty be used at all? The answer in most (many) cases is "yes", because we want in most cases to assure that we are (very) far from a harmful or very harmful level. We use (see also Section 4.1, on risk assessment), a safety factor of 100-1000 in many cases. When we are concerned with very harmful effects, such as for instance complete collapse of an ecosystem or a health risk for a large human population, we will inevitably select a safety factor which is very high. In addition, our lack of knowledge about synergistic effects and the presence of many compounds in the environment at the same time force us to apply a very high safety factor. In such a context we will usually go for a concentration in the environment which is magnitudes lower than corresponding to a slightly harmful effect or considerably lower than the NEC. It is analogous to civil engineers constructing bridges. They make very sophisticated calculations (develop models), that account for wind, snow, temperature changes and so on and afterwards they multiply the results by a safety factor of 2-3 to ensure that the bridge will not collapse. They use safety factors because the consequences of a bridge collapse are unacceptable.

The collapse of an ecosystem or a health risk to a large human population is also completely unacceptable. So, we should use safety factors in ecotoxicological modelling to account for the uncertainty. Due to the complexity of the system, the simultaneous presence of many compounds and our present knowledge or rather lack of knowledge, we should as indicated above use 10 - 100 or even sometimes 1000 as safety factor. If we use safety factors that are too high, the risk is only that the environment will be less contaminated at maybe a higher cost. Besides there are no alternatives to the use of safety factors. We can step by step increase our ecotoxicological knowledge, but it will take decades before it may be reflected in considerably lower safety factors. A measuring program of all processes and components is an impossibility due to the high complexity of the ecosystems. This does not of course imply that we should not use the information of measured properties available today. Measured data will almost always be more accurate than the estimated data. Furthermore the use of measured data within the network of estimation methods will improve the accuracy of estimation methods. Several handbooks on ecotoxicological parameters are fortunately available. References to the most important have already be given in Section 4.1. Estimation methods for the physical-chemical properties of chemical compounds were already applied 40-60 years ago, as they were urgently needed in chemical engineering. They are to a great extent based on contributions to a focal property by molecular groups and the molecular weight: the boiling point, the melting point and the vapour pressure as function of the temperature. In addition a number of auxiliary properties results from these estimation methods, such as the critical data and the molecular volume. These properties may not have a direct application as ecotoxicological parameters in environmental risk assessment, but are used as intermediate parameters which may be used as a basis for estimation of other parameters.

The water solubility, the partition coefficient octanol-water, K_{OW}, and Henry's constant are crucial parameters in our network of estimation methods, because many other parameters are well correlated with these two parameters. The three properties can fortunately be found for a number of compounds, or be estimated with reasonably high accuracy by use of knowledge of the chemical structure, i.e., the number of various elements, the number of rings and the number of functional groups. In addition, there is a good relationship between water solubility and K_{OW}; see Chapter 2. Particularly in the last decade many good estimation methods for these three core properties have been developed.

During the last couple of decades several correlation equations have been developed based upon a relationship between the water solubility, K_{OW} or Henry's constant on the one hand and physical, chemical, biological and ecotoxicological parameters for chemical compounds on the other. The most important of these parameters are: the adsorption isotherms soil-water, the rate of the chemical degradation processes: hydrolysis, photolysis and chemical oxidation , the biological concentration factor, BCF, the ecological magnification factor, EMF, the uptake rate, excretion rate and a number of ecotoxicological parameters. Both the ratio of concentrations in the sorbed phase and in water at equilibrium K_a and BCF may often be estimated with a relatively good accuracy from expressions likes K_a or BCF = a log K_{OW} + b. Numerous expressions with different a and b values have been published (see Jørgensen et al., 1991 and Jørgensen, 1994). Some of these relationships we mentioned in Chapter 2.

Biodegradation rates may be expressed in several ways. Microbiological biodegradation may with good approximation be described as a Monod equation:

$$dc / dt = - dB / Ydt = - \mu max^* Bc / Y (Km + c) \quad (4.21)$$

where c is the concentration of the compound considered, Y is the yield of biomass B per unit of c, B is the biomass concentration, μmax is the maximum specific growth rate and Km is the half saturation constant. If $c << Km$ the expression is reduced to a first order reaction scheme:

$$dc / dt = - K' B c \quad (4.22)$$

where $K' = \mu max / (Y Km)$. B is in nature determined by the environmental conditions. In aquatic ecosystems B is for instance highly dependent on the presence of suspended matter. B may therefore under certain conditions be considered a constant which reduces the rate expression to:

$$dc / dt = - k c \quad (4.23)$$

An indication of k in the unit 1/h, 1/24h, 1/w, 1/month or 1/y can therefore be used to describe the rate of biodegradation. If the biological half life time is denoted t we get the following relation:

$$\ln 2 = 0.7 = k t \quad (4.24)$$

This implies that also the biological half life time can be used to indicate the biodegradation rate.

The biodegradation in waste treatment plants is often of particular interest, in which case the %ThOD may be used. It is defined as the 5-day BOD as percentage of the theoretical BOD. It may also be indicated as the BOD_5-fraction. For instance, a BOD_5-fraction of 0.7 will mean that BOD_5 corresponds to 70% of the theoretical BOD. It is, however, also possible to find an indication of percentage removal in an activated sludge plant.

The biodegradation is, however, in some cases very dependent on the concentration of microorganisms as expressed in equations (7.1) and (7.2). Therefore K' indicated in the unit mg /(g dry wt 24 h) will in many cases be more informative and correct.

In the microbiological decomposition of xenobiotic compounds an acclimatisation period from a few days to 1-2 months should be foreseen before the optimum biodegradation rate can be achieved. We distinguish between primary and ultimate biodegradation. Primary, biodegradation is any biologically induced transformation that changes the molecular integrity. Ultimate biodegradation is the biologically mediated conversion of an organic compounds to inorganic compounds and products associated with complete and normal metabolic decomposition.

The biodegradation rate is expressed by application of a wide range of units:

1. As a first order rate constant (1/24h)
2. As half life time (days or hours)
3. mg per. g sludge per 24h (mg/ (g 24h))
4. mg per. g bacteria per 24 h (mg /(g 24h))
5. ml of substrate per bacterial cell per 24h (ml / (24h cells))
6. mg COD per g biomass per 24 h (mg /(g 24h))
7. ml of substrate per gram of volatile solids inclusive microorganisms (ml / (g 24h))
8. BODx / BOD∞, i.e., the biological oxygen demand in x days compared with complete degradation (-), named the BODx-coefficient.
9. BODx / COD, i.e., the biological oxygen demand in x days compared with complete degradation, expressed by means of COD (-)

The biodegradation rate in water or soil is difficult to estimate, because the number of microorganisms varies several order of magnitudes from one type of aquatic ecosystem to the next and from one type of soil to the next. Artificial intelligence has been used as a promising tool to estimate this important parameter. However, a (very) rough, first estimation can be made on the basis of the molecular structure and the biodegradability. The following rules can be used to set up these estimations:

1. Polymer compounds are generally less biodegradable than monomer compounds. 1 point for a molecular weight > 500 and ≤ 1000, 2 points for a molecular weight > 1000.
2. Aliphatic compounds are more biodegradable than aromatic compounds. 1 point for each aromatic ring.
3. Substitutions, especially with halogens and nitro groups, will decrease the biodegradability. 0.5 points for each substitution, although 1 point if it is a halogen or a nitro group.
4. Introduction of double or triple bond will generally mean an increase in the biodegradability (double bonds in aromatic rings are of course not included in this rule). - 1 point for each double or triple bond.
5. Oxygen and nitrogen bridges (- O - and - N - (or=)) in a molecule will decrease the biodegradability. 1 point for each oxygen or nitrogen bridge.
6. Branches (secondary or tertiary compounds) are generally less biodegradable than the corresponding primary compounds. 0.5 point for each branch.

Find the number of points and use the following classification:

≤ 1.5 points: the compound is readily biodegraded. More than 90% will be biodegraded in a biological treatment plant.

2.0 - 3.0 points: the compound is biodegradable. Probably about 10% - 90% will be removed in a biological treatment plant. BOD_5 is 0.1- 0.9 of the theoretical oxygen demand

3.5 - 4,5 points: the compound is slowly biodegradable. Less than 10% will be removed in a biological treatment plant. $BOD_{10} \leq 0.1$ of the theoretical oxygen demand.

5.0 - 5.5 points: the compound is very slowly biodegradable. It will hardly be removed in a biological treatment plant and a 90% biodegradation in water or soil will take ≥ 6 months.

≥ 6.0 points: the compound is refractory. The half life time in soil or water is counted in years.

Several useful methods for estimation of biological properties are based upon the similarity of chemical structures. The idea is that if we know the properties of one compound, it may be used to find the properties of similar compounds. If for instance we know the properties of phenol, which is named the parent compound, it may be used to give more accurate estimation of the properties of monochloro-phenol, dichloro- phenol, trichloro-phenol and so on and for the corresponding cresol compounds. Estimation approaches based on chemical similarity give generally more accurate estimation, but of course are also more cumbersome to apply, as they cannot be used generally in the sense that each estimation has a different starting point, namely the compound, named the parent compound, with known properties.

Allometric estimation methods presume (Peters, 1983), that there is a relationship between the value of a biological parameter and the size of a considered organism. These estimation methods were presented in Chapter 3, as they are closely related to the energy balances of organisms.The ecotoxicological parameters, LC50, LD50, MAC, EC and NEC can be estimated from a wide spectrum of physical and chemical parameters, although these estimation equations generally are more inaccurate than the estimation methods for physical, chemical and biological parameters. Both molecular connectivity and chemical similarity usually offer better accuracy for estimation of toxicological parameters.

The various estimation methods may be classified into two groups:

A. General estimation methods based on an equation of general validity for all types of compounds, although some of the constants may be dependent on the type of chemical compound or they may be calculated by adding contributions (increments) based on chemical groups and bonds.

B. Estimation methods valid for a specific type of chemical compounds for instance aromatic amines, phenols, aliphatic hydrocarbons and so on. The property of at least one key compound is known. Based upon the structural differences between the key compound and all other compounds of the considered type, - for instance two chlorine atoms have substituted hydrogen in phenol to get 2,3-dichloro-phenol - and the correlation between the structural differences and the differences in the considered property, the properties for all compounds of the considered type can be found. These methods are based on chemical similarity.

Methods of class B are generally more accurate than methods of class A, but they are more cumbersome to use as it is necessary for each type of chemical to find for each property the right correlation. Furthermore, the requested properties should be known for at least one key component which sometimes may be difficult when a series of properties are needed. If estimation of the properties for a series of compounds belonging to the same chemical class is required, it is tempting to use a suitable collection of class B methods.

Methods of class A form a network which facilitates possibilities of linking the estimation methods together in a computer software system, like for instance WINTOX (Jørgensen et al. ,1997). The software is easy to use and can rapidly provide estimations. Each relationship between two properties is based on the average result obtained from a number of different equations found in the literature. There is, however, a price for using such "easy to go" software. The accuracy of the estimations is not as good as with the more sophisticated methods based upon similarity in chemical structure, but in many, particularly modelling, contexts the results found by WINTOX can offer a sufficient accuracy. In addition, it is always useful to come up with a first intermediate guess.

The software also makes it possible to start the estimations from the properties of the chemical compound already known. The accuracy of the estimation from use of the software can be improved considerably by having knowledge about a few key parameters for instance the boiling point and Henry's constant. As it is possible to get software which is able to estimate Henry's constant and K_{OW} with generally higher accuracy than WINTOX, a combination of this software with WINTOX can be recommended. Another possibility would be to estimate a couple of key properties by use of chemical similarity methods and then use these estimation as known values in WINTOX. These methods for improving the accuracy will be discussed in the next section.The network of WINTOX as an example of these estimation networks is illustrated in Fig. 4.18. As it is a network of class A methods, it should not be expected that the accuracy of the estimations is as high as it is possible to obtain by the more specific class B methods. By WINTOX it is, however, possible to estimate the most pertinent properties directly from the structural formula.

WINTOX is based on average values of results obtained by simultaneous use of several estimation methods for most of the parameters. It implies increased accuracy of the estimation, mainly because it gives a reasonable accuracy for a wider range of compounds. If several methods are used in parallel, a simple average of the parallel results have been used in some cases, while a weighted average is used in other cases where it has been found beneficial for the overall accuracy of the program. When parallel estimation methods are giving the highest accuracy for *different* classes of compounds, use of weighting factors seems to offer a clear advantage. It is generally recommended to apply as many estimation methods as possible for a given case study to increase the overall accuracy. If the estimation by WINTOX can be supported by other recommended estimation methods, it is strongly recommended to do so.

It is also possible to achieve a higher accuracy, if some of the properties are known in addition to the structural formula. Particularly, if the boiling point, Henry's constant and / or the partition coefficient octanol-water are known, the accuracy will increase significantly, even for the toxicological properties (parameters).

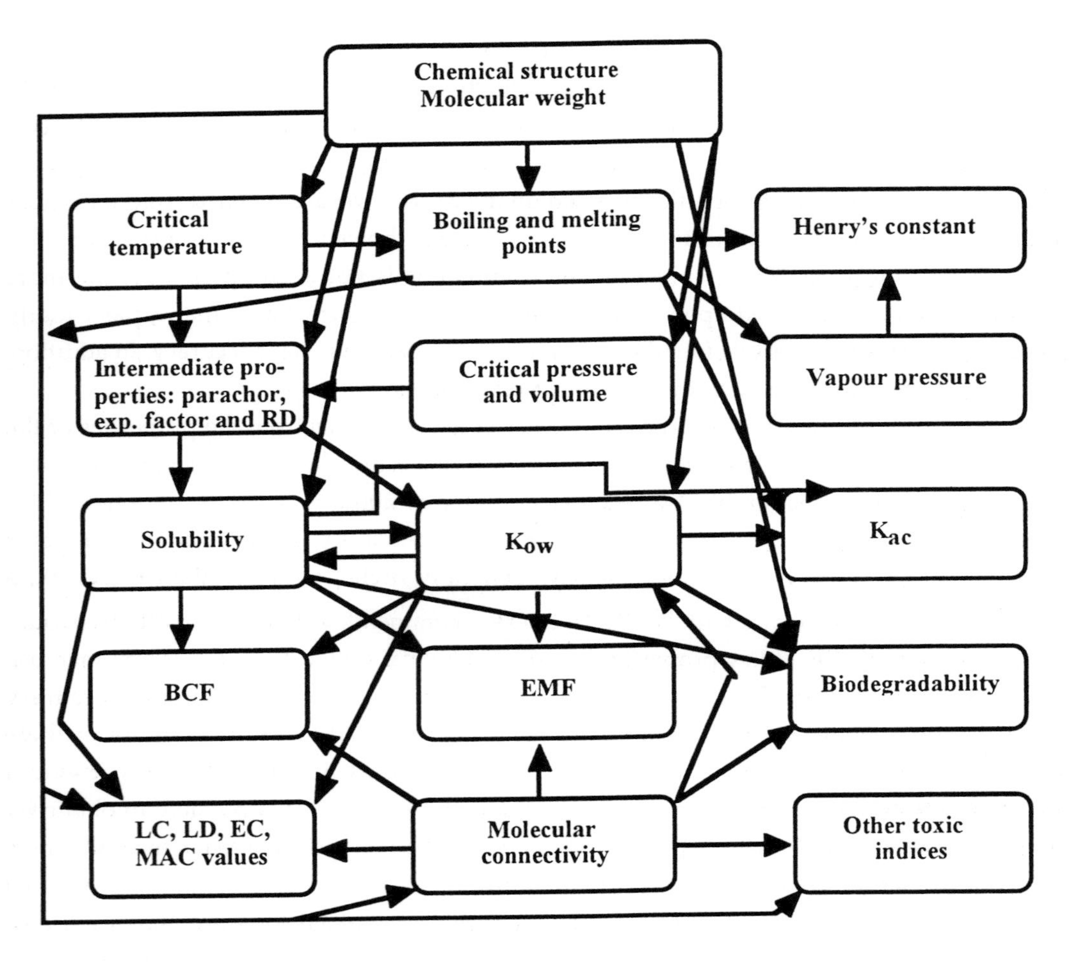

Fig. 4.18. The network of estimation methods in WINTOX is shown. An arrow represents a relationship between two or more properties.

The accuracy obtained by the software can therefore relatively easily be increased significantly by combining the use of the software with data bases and estimation methods, which are readily available, for Henry's constant and the partition octanol-water coefficient:

Henry's Law Constant Program
William M. Meylan
Philip H. Howard
Lewis Publishers, New York, U.S.
and

CLogP for Windows
BioByte Corp.
201 W. Fourth St., Suite #204
Claremont, CA 91711, U.S.

These two software packages are based on a relatively wide range of compounds and can therefore for these two properties offer higher accuracy than WINTOX. Moreover, it should be recommended to apply Beilstein to find common physical data such as boiling and melting points.

It is furthermore recommended to apply as widely as possible other available, additional estimation methods. They may for instance at least be used to confirm results obtained by WINTOX. Additional estimation methods may be found in the many references to QSAR-methods including: Jørgensen et al., 1997, Lyman et al., 1990.

This does not imply that we can reach the accuracies characterising the class B methods.

It is not possible to give general recommendations on which methods or combinations of methods to use in a given situation. It may for instance be desirable to know some properties with particularly high accuracy, while it is sufficient to know other properties within coarse ranges. It is therefore crucial in each individual situation to select the right combination of estimation methods which will meet the demands for a defined accuracy most rapidly. It is in other words necessary to find the trade off between very complex, time consuming, but accurate methods and simple, rapid, more inaccurate methods. Here the use of WINTOX, in combination with the readily available data bases and soft ware for Henry's constant and the partition coefficient octanol-water, mentioned above, may offer an attractive solution which is relatively rapid and offers medium accuracy. WINTOX does not include estimations of biodegradability, because this parameter cannot be estimated with fully acceptable accuracy. In addition, biodegradation is very dependent on a number of environmental factors, of which some can be quantified, while others cannot. The number of estimation methods for biodegradation rates is limited compared with many other properties of environmental interest due to these limitations. Some estimation methods for biodegradation rates are given in the literature, and it is of course recommended to apply them, but the results should be considered a first coarse approximation and should only be used as a preliminary value. Many methods for estimation of the properties of chemical compounds have been developed.

Estimation methods are urgently needed in environmental risk assessment and other environmental management contexts, because our knowledge about the properties and effects of toxic substances in nature is very limited. The standard deviation for estimated parameters are of course higher, sometimes much higher, than for measured values, but by a proper use of a combination of estimation methods, it is possible to make estimations with an acceptably small standard deviation.

4.7 EXAMPLES OF RISK ASSESSMENT

Environmental risk assessment follows the nine steps shown in Section 4.2, but each of the nine steps can be performed in more or less detail. They are illustrated below by presentation of specific examples.

A. Chlorinated aliphatic hydrocarbons

The first example we should use to illustrate the application of risk assessment of chemicals is associated with the discharge of a mixture of chlorinated aliphatic hydrocarbons, either straight chained or branched, with chain lengths that range between 10 and 30 C- atoms. It is based on a similar case study presented by Bartell et al., 1992. Commercial use includes additives for lubricating oils, plasticisers, traffic paints and additives for flame retardants. Approximately 15,000 metric tons enter the environment annually, mainly to the hydrosphere due to discharge into waste water. The nine ERA-steps are summarised in Figure 4.19. They were partly proposed by Bartell et al., 1992 and represent a more elaborated version of risk assessment.

Step 1. It is recommended to develop a model with the scope to predict the final concentration in surface water. It should consider biodegradation in a typical waste water treatment plant and the adsorption equilibrium between sludge and water (step 1A). The further fate of the chemicals in freshwater ecosystems should be covered by a model which includes uptake by aquatic organisms and the adsorption equilibrium between water and sediment (step 1B). It is not the aim here to give the details of the model's construction. Those who are interested in the problems associated with selection of the right model and its development for practical use in environmental management and environmental risk assessment are referred to Jørgensen, 1994.

Model results **(step 2)** and analyses **(step 3)** indicate that concentrations in surface water associated with industrial inputs may range between 0.5 and 6 ppb which is about 10 times the background concentration (see Jørgensen et al., 1991). The concentration in sediment is modelled and reported to be as much as 1000-3000 times greater than the dissolved concentrations. Due to the lipohile character bioaccumulation in aquatic organisms is considered a possible environmental problem.

Step 4. It is useful to try to obtain knowledge about all properties of the range of chemicals covered by the label "chlorinated aliphatic hydrocarbons" to get as complete an image as possible of the environmental consequences that can be expected from use of these chemicals; see also Section 4.6.

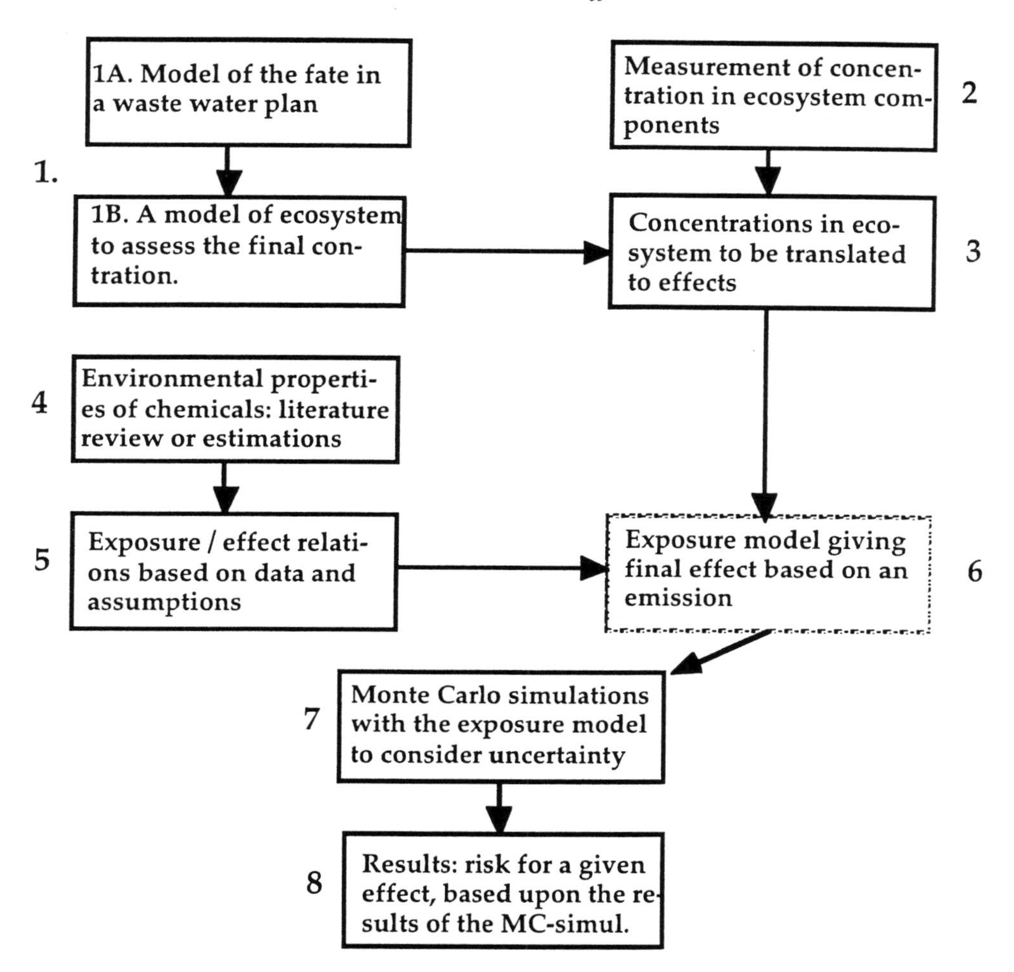

Fig. 4.19 The 8 steps in the specific example presented, chlorinated aliphatic hydrocarbons in aquatic ecosystems, of environmental risk assessment. Models and/or analyses are presumed to be applied to find the actual range of concentrations. The environmental properties are translated into an effect / exposure relation, where in most cases it is necessary to assume the relationship (linear, exponential etc.) to be able to interpolate or extrapolate to other exposures. The uncertainty of the parameters in the exposure model is used to perform Monte Carlo simulations which can give the probability of a risk for a specific effect.

The EC_{50}-concentration for phytoplankton growth is known to be 31.6 μg /l (96h) (see Jørgensen et al., 1991), LC_{50} for zooplankton is 46 μg/l and for fish of different species LC_{50} of 100-300 mg/l may be applicable (fathead minnow 100 mg/l and bluegill sunfish 300 mg/l). It is recommended that we use an exposure model to relate concentrations in surface water to their effects on the aquatic organisms. This is a useful method to fill the gap in our knowledge about the detailed relationship between concentrations and effects.

Step 5. In the case we only have one point on the response exposure curve (see

Figure 4.20, reproduced from Jørgensen, 1999), we have to use a linear approximation to find other exposure / response relations.

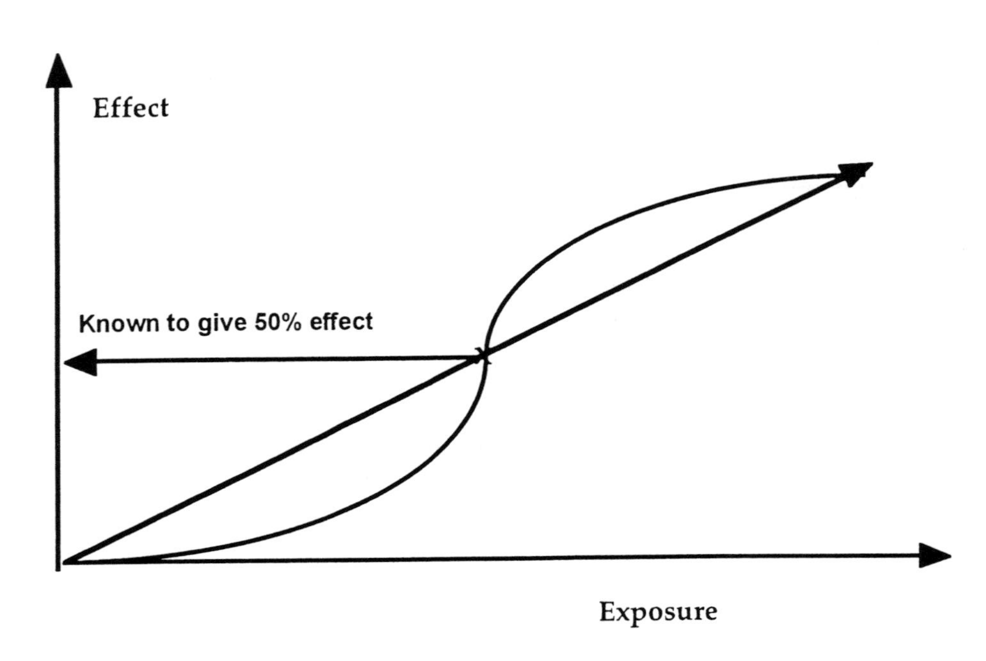

Fig. 4.20 Typical relationship between effect and exposure. If only one set of related exposure and effect is known, it is necessary to presume a linear relationship as shown. It cannot be excluded of course that the relation is different like for instance the logistic-like curve shown in the figure.

Step 6. The growth of phytoplankton in the exposure model may be expressed by an equation of the following form:

$$dB/dt = f\ (\text{light, nutrients, respiration, temperature, sinking, mortality, grazing}) \tag{4.25}$$

where B is the biomass. The usual expressions for growth as a function of light and nutrients, Michaelis-Menten's equations, are used. The sinking and mortality are first order reactions and grazing is a Michaelis- Menten's expression. The concentrations of the grazers are regulated by predation of planktivorous fish, governed by a general Michaelis-Menten's expression. The influence of the temperature is an exponential function as generally applied. The influence of the toxic substance in this method presumes that the various parameters are influenced equally by a factor x to obtain the expected effect. If we for instance want to find the parameters valid for 31.6 $\mu g/l$ chlorinated aliphatic hydrocarbons, we should find the x value which corresponds to half the increase in biomass found by integration over a period of 96 h (three days). The growth rate is x times the previous growth rate, the half saturation constant for the influence of nitrogen and phosphorus on growth in the Michaelis-Menten's expression is the previous value divided by x, the half saturation constant for light is x times the previous value, the respiration coefficient, sinking rate and mortality rate are correspondingly the previous values divided by x, and the grazing rate is also x times less than the previous value. x is found by iterations by trial and error. For a concentration of 31.6 μg/l a corresponding x value can be found giving a growth rate which is 75% of normal and so on.

TABLE 4.8

Results of the Environmental Risk Assessment for Chlorinated Aliphatic Hydrocarbons by use of an exposure model. The probability for the indicated risk is shown.

Concentration	Algal increase		Zooplankton decrease		Fish decrease	
μ g /l	200%	400%	25%	50%	25%	50%
0.1	0.33	0.14	0.0064	≈ 0	0.01	≈0
1	0.70	0.45	0.05	≈ 0	0.40	0.043
10	0.05	0.03	0.14	0.088	0.95	0.89
50	0.006	0.006	0.075	0.044	0.94	0.90
100	0.008	0.002	0.091	0.065	0.94	0.87

Step 7. All the parameters in the exposure model have, however, a standard deviation rooted in the uncertainty in our knowledge about the value of the model parameters, including the parameters translating exposure to effect. The exposure model and the translation concentration -> effect are run by Monte Carlo simulations based on the estimated uncertainties for instance 1000 times for each of the actual concentrations. If a certain effect, for instance 25% reduction of zooplankton population, is obtained by the simulations in 10% of cases, it is concluded that the risk of 25% reduction in the zooplankton concentration is 10%.

Step 8. The results of this approach for modelling the effect of chlorinated aliphatic hydrocarbons on phytoplankton, zooplankton and fish are summarised in Table 4.8 (more detailed results are presented in Bartell et al., 1992, where also different scenarios are presented). Notice that the algal increase is considered to be caused mainly by the indirect effect of the decrease in zooplankton. An exposure model is applied because it is often impossible to predict the expected effects in an ecosystem due to the complex relations. It is therefore strongly recommended to apply exposure models whenever complex relations on the ecosystem level are considered.

B. Risk assessment of benzene

This example is concerned with the risk assessment for human beings. The properties of benzene can easily be found in the handbooks listed above. The most important properties are as follows:

1) Water solubility 1.8 g/ l.
2) Log $K_{OW} = 2.3$
3) Log BCF = 1-3
4) Log $K_{OC} = 1.9$
5) Vapour pressure at room temperature: 12 700 Pa.

Based upon these properties a fate model according to McKay (1991) can be developed. It shows that 99% of benzene released to the environment will be found in the atmosphere.

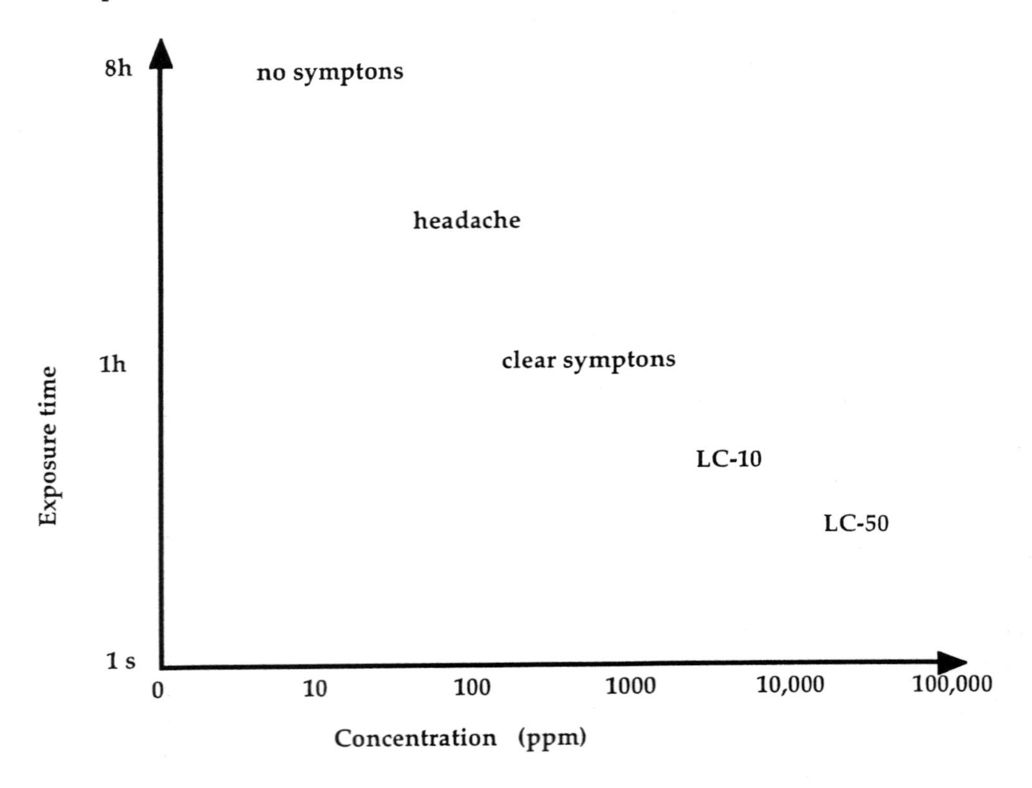

Fig. 4.21. The relation between effect and a combination of concentration and exposure time is shown.Notice that exposure is considered a combination of concentration and time.

Several effects on humans are known for benzene: reduced production of haemoglobin, narcosis, toxic effect on the central nerve system and cancer. The relation between exposure and effects is shown in Figure 4.21. Experiments on rats have shown that the no observed effect concentration in air is approximately 100 $\mu g/l$ for long term exposure 7 hours per day in 5 days per week. This is converted to constant exposure by multiplication by 7x5 and division by 24x7 and we get approximately 20μg /l. If a safety factor of 500 is used, PNEC becomes 0.04 μg /l, corresponding to 0.014 ppm.

Howard (1990) gives information about a relationship between the relative increase in mortality RIM (%) due to blood cancer and the exposure to benzene:

$$RIM = bxd, \qquad (4.26)$$

where b is a constant and d is the exposure expressed as ppm times the number of years. A b value of 1 is suggested.

Measurements of benzene concentrations show that in streets and roads with heavy traffic, concentrations in the range of 1-16 ppb are found. It yields a ratio PEC / PNEC close to or slightly greater than 1. RIM for a person exposed to 16 ppb = 0.016 ppm in 70 years is 1x70x0.016 = 1.2%, which is unacceptable. It can be concluded that a reduction of the risk is required due to an excessively high concentration in roads with heavy traffic.

C. Risk assessment of bisphenol-A (BPA).

This example is concerned with the effects on various organisms. The most important properties are:

1) Boiling point: 250°C.

2) Water solubility at room temperature: 120 mg/l

3) Henry's constant: 0.00001 Pa m^3 / mol

4) log K_{ow} = 2.2 - 3.4

5) log K_{oc} = 2.5 - 3.1

6) BCF = 5-70 (fish)

Effect / exposure data:

1) Effect on growth rate of algae from a concentration of 1-2 mg/l

2) LC_{50} for fresh water fish species $\geq$ 4.4 mg /l.

3) Oestrogen effects may be observed at 2.3 μ g/l (Andersen, 1998).

A fate model shows that the major part of BPA emissions to the environment will be present in the hydrosphere. A type C model (see Section 4.3) of a typical freshwater system is developed. It is based on industrial discharge of BPA via a waste water treatment plant in a concentration of about 0.5-2 μg /l. Measurements from freshwater systems with high concentrations are found in the range 1 -10 μg /l. A PNEC of 1 μg /l has been suggested, based upon a NOEL of 1 mg/l and a safety factor of 1000.

This value doesn't consider the oestrogen effect which would give a much lower

PNEC even by use of a safety factor of only 100. A ratio PEC / PNEC of 1-10 is found, when PNEC = 1 μ g/l is applied. As BPA is used in considerable amounts a reduction of the emission to the environment is urgently needed which can be concluded on the basis of the relatively simple environmental risk assessment described here. An effect on *the ecosystem level* is observed in several cases by characterisation of the nutrient, energy and water budgets of the ecosystem and this explains their dynamic behaviour through an understanding of the basic mechanisms governing the internal processes of the system. An ecosystem can minimise and counteract the influence of environmental stress through population shifts and interactions. Therefore, the ecosystem is characterised as persisting, in spite of perturbations, through dynamic shift in its nutrient and energy metabolism. Populations are sacrificed to preserve the integrity and function of the ecosystem, although there is a limit to the buffer capacity of the system. Another aspect of the effect on ecosystem level is the shift in the basic composition of the ecosystem and its mass-cycles. For instance, acid rain might cause a different ionic composition of the entire ecosystem by release of previously bound ions to the environment. Here, the interaction with other pollutants plays an important role as the binding capacity of soil and sediments for heavy metals is highly dependent on pH.

4.8 ACIDIFICATION AS EXAMPLE OF EFFECTS ON THE HIGHER LEVELS OF THE BIOLOGICAL HIERARCHY

The concept of buffering capacity is generally used in chemistry to express the ability of a solution to maintain its pH value. Buffering capacity in this context is defined as:

$$ß = dC / dpH \qquad (4.27)$$

where ß = buffering capacity, C = added acid or base in moles H^+ resp. OH^- per l. As pH is one of the important factors determining life conditions directly or indirectly, it is *crucial for ecosystems to have a high pH-buffering capacity,* see Figs. 4.22 and 4.23, and Tables 4.9 - 4.12. A low pH implies that hydrogen carbonate is converted into free CO_2, which is toxic to fish and other animals as respiration is controlled by the difference between CO_2- concentration in the blood and the environment. The processes involved can be described by the following chemical equations:

$$CO_3^{2-} \xrightarrow{+H^+} HCO_3^- \xrightarrow{+H^+} H_2O + CO_2 \qquad (4.28)$$

TABLE 4.9

Fish status for 1679 lakes in Southern Norway grouped according to pH

pH	No. of lakes in pH range	% of lakes with no fish	% of lakes with sparse populations	% of lakes with good populations
<4.5	111	73	25	2
4.5-4.7	245	53	41	6
4.7-5.0	375	38	41	21
5.0-5.5	353	25	40	35
5.5-6.0	164	8	36	56
>6.0	431	1	13	86

TABLE 4.10

Effects of pH values on fish (Alabaster and Lloyd, 1980)

pH-range	Effect
3.0 - 3.5	Unlikely that any fish can survive more than a few hours
3.5 - 4.0	Lethal to salmonoids. Some other fish species might survive in this range, presumably after a period of acclimation to slightly higher pH
4.0 - 4.5	Harmful to salmonoids, bream, goldfish and carp, although the resistance to this pH increases with the size and age
4.5 - 5.0	Likely to be harmful to eggs and fry of salmonoids. Harmful also to adult salmonoids and carp at low calcium, sodium and/or chloride concentration
5.0 - 6.0	Unlikely to be harmful, unless concentration of free CO_2 is greater than 20 mg l^{-1} or the water contains freshly precipitated $Fe(OH)_3$
6.0 - 6.5	Harmless unless the concentration of free $CO_2 > 100$ mg l^{-1}

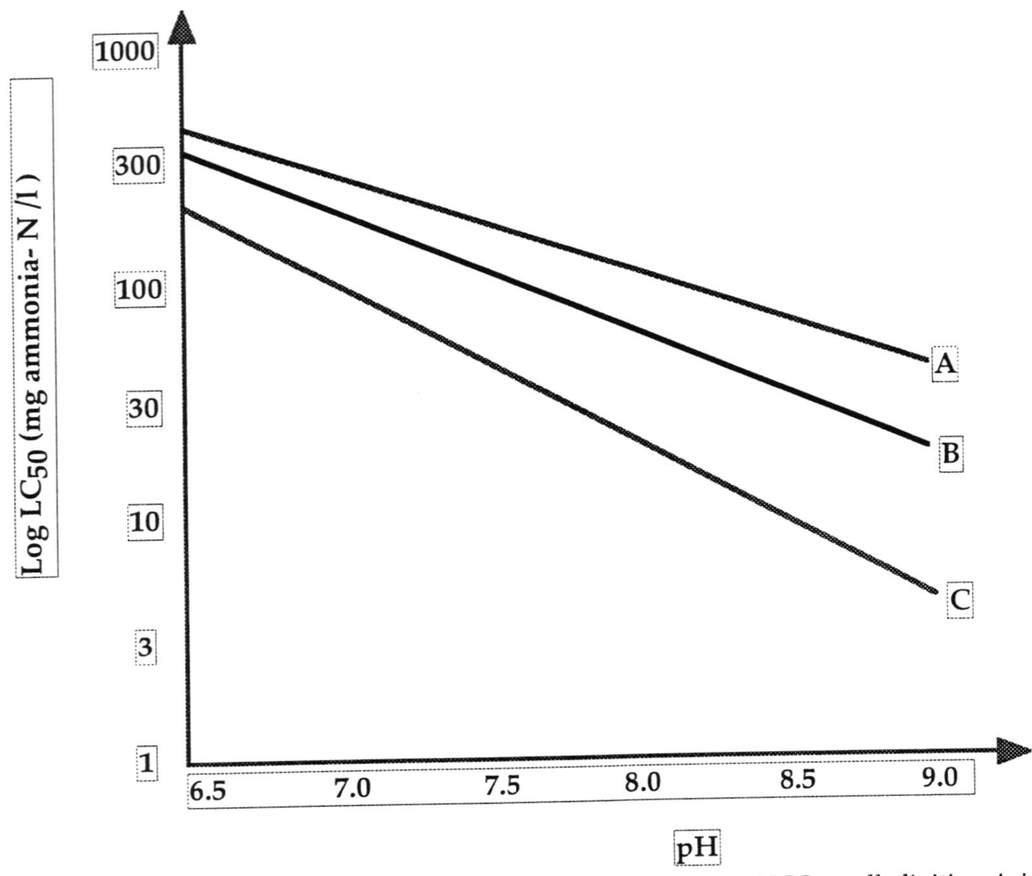

Fig. 4.22. LC_{50} of ammonia for trouts versus pH at different HCO_3^- alkalinities: A is 25 meqv. /l, B is 100 meqv. /l and C is 400 meqv. /l. Notice that the toxicity of ammonia increases with increasing pH due to a more pronounced conversion to ammonia and with decreasing alkalinity.

A lowered pH will affect the entire ecosystem and the ecological balance. Three serious environmental problems will be mentioned below to illustrate the importance of pH as an environmental factor, and consequently of a high pH-buffering capacity in ecosystems:

1. The acidification of lakes caused by SO_2- and NO_X - pollution.
2. The effect of decreased pH on soil and plants.
3. The possibility of maintaining the pH in the oceans in spite of the increased uptake of CO_2.

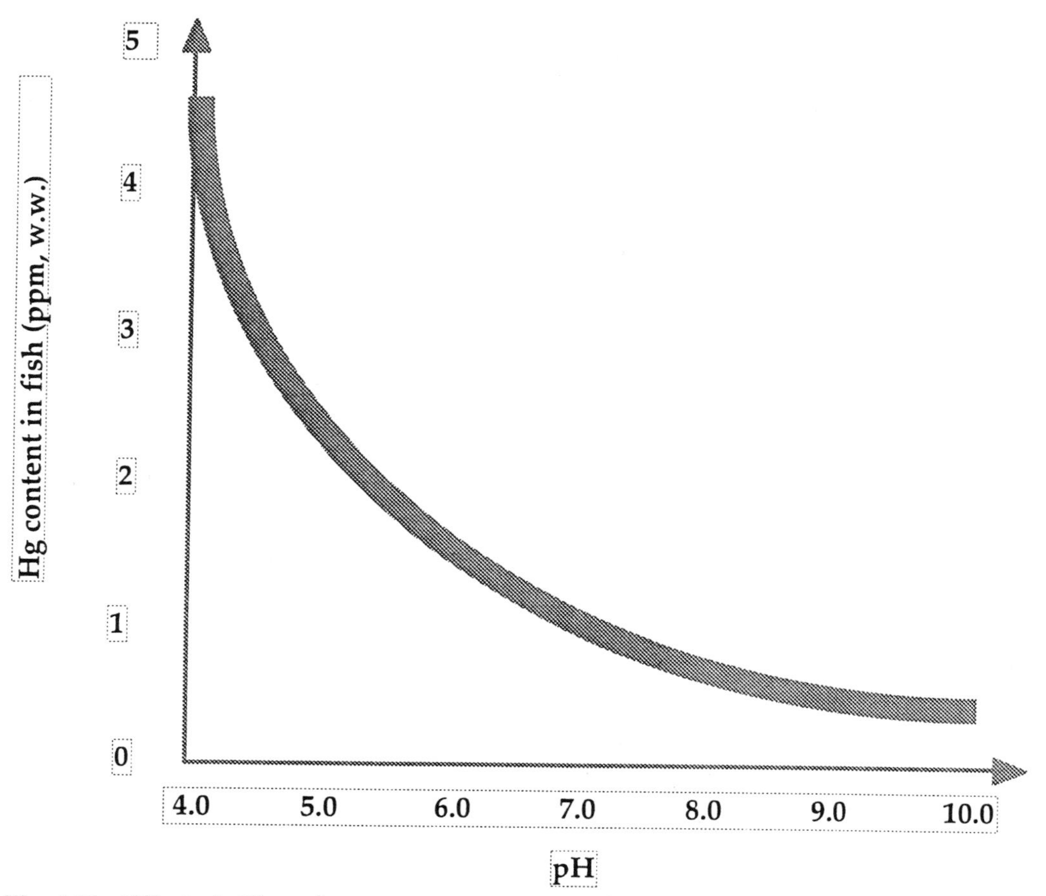

Fig. 4.23. Effect of pH on the mercury content of fish. Exposure to 1.5 ppm Hg from $HgCl_2$ solutions. The relationship is indicated as a range, based on many measurements for different fish species.

1. Acidification of lakes. Continuous acidification has been observed in many lakes in parts of North America and Scandinavia. The geographical position (dominant wind direction) and the inflow conditions of these lakes suggest that the increase in acidity is the result of deposition of an air-borne substance that lead to the formation of acids, such as SO_2 and NO_x. This deposition is, of course, not restricted to lakes, but also affects soil, forests, etc. (see point 2 below). Affected areas are characterised by a low pH-buffering capacity. Surface water in these areas is soft and has a low conductivity. The buffering capacity of surface water is mainly related to hydrogen carbonate ions, which are present in small concentrations in lakes. These ions are able to take up hydrogen ions in accordance with the following process:

$$HCO_3^- + H^+ \text{ -----> } H_2O + CO_{2(gas)} \tag{4.29}$$

During the combustion of sulphurous fuels, sulphur is primarily converted to SO_2 (97-98%), but 2-3% is oxidised to SO_3, which reacts with water to form sulphuric acid:

$$SO_3 + H_2O \text{ -------> } H_2SO_4 \tag{4.30}$$

The SO_2 on the other hand when released into the atmosphere comes into contact with very small particles covered by an aqueous film or with water droplets, and forms H_2SO_3, a medium strong acid:

$$H_2SO_3 \text{ --------> } H^+ + HSO_3^- \tag{4.31}$$

Under the influence of iron compounds acting as catalysts, HSO_3^- and SO_3^{2-} are rapidly oxidised to sulphuric acid (Brosset, 1973). Similar trends have been observed in many North American and Scandinavian lakes and even some of the great lakes of Europe have been affected by this acidification process. In smaller lakes, where precipitation forms a greater portion of lake volume, pH values as low as 4.0 have been observed. The effect of low pH on the fish population can be seen in Table 4.11 which is based on an examination carried out by Jensen and Snekvik (1972). Two other observed effects are illustrated in Table 4.10 and in Fig. 4.24, where it is shown that **the number of fish and zooplankton species is decreasing in acidified lakes.**

The decreasing pH has a striking effect on the fish population. At extremely low pH values all young fish disappear completely (Almer, 1972). Nevertheless, spawning and fertilised eggs have been observed even at low pH values, so **it seems that the development of eggs may be disturbed by the high activity of hydrogen ions in aquatic environments.** This is illustrated in Fig. 4.25, where the percentage of hatched eggs is plotted against pH . The period from fertilisation to hatching also tends to be prolonged at low pH values, as shown in Fig 4.26. Further to the primary biological effects of a continuous supply of acid substances on individuals and populations in a lake, more profound long-term changes, that force the lake into an increasingly more oligotrophic state, also take place, as suggested by Grahn et al. (1974). pH generally increases with eutrophication due to uptake (removal) of hydrogen carbonate ions and carbon dioxide by photosynthesis. In an acidified lake the phytoplankton concentration will decrease, and so will the uptake of CO_2 and HCO_3^-, and the transparency of the water will therefore increase due to an impaired primary production. (see Fig. 4.27). The biological pH-buffering capacity is reduced by this process and by means of this feedback mechanism the process of acidification is further accelerated. Finally, it should be mentioned that the concentration of free metal ions will increase with decreasing pH due to release of metal ions from sediment and their higher solubility and lower tendency to form complexes at lower pH (see also Figure 4.24). As large amounts of heavy metals are stored in sediment and soil, this problem, concerned with the bioavailability of metal ions, may be the most serious on the list of problems associated with acid rain.

TABLE 4.11

Trout population and pH in 260 lakes

No. of lakes	Population	pH 4.00-4.50 No.	%	4.51-5.00 No.	%	5.01-5.50 No.	%	≥5.51 No.	%
33	Empty	3	9.1	17	51.5	7	21.2	6	18.2
87	Sparse population	2	2.3	15	17.2	21.2	24.1	49	56.3
82	Good population	-	-	9	11.0	14	17.1	59	72.0
58	Over-populated	-	-	3	5.2	13	22.4	42	72.3
260	total	5	1.9	44	16.9	55	21.2	156	60.0

TABLE 4.12

Occurrence of fish species before acidification and species found in the late seventies

Lake	Species earlier forming permanent stocks	Species found 1978	Species re-producing 1978
Bredvatten	Pe Pi E	(E)	
Lysevatten	Pe Pi R E	Pe (E)	Pe
Gårdsjön	Pe Pi R T C E	Pe Pi E	Pe
Örvattnet	Pe St M	Pe	Pe
Stensjön	Pe Pi R St M	Pe Pi R	Pe Pi
Skitjärn	Pe Pi R L	Pe Pi R L	Pe Pi

Note to table: Pe=perch (*Perca fluviatilis*), Pi=pike (*Esox lucius*), E=eel (*Anguilla vulgaris*), R=roach (*Leuciscus rutilus*), T=tench (*Tinca tinca*), C=crucian carp (*Carassius carassius*), St=brown trout (*Salmo trutta*), L=lake whitefish (*Coregonus albula*), M=minnow (*Phoxinus phoxinus*).

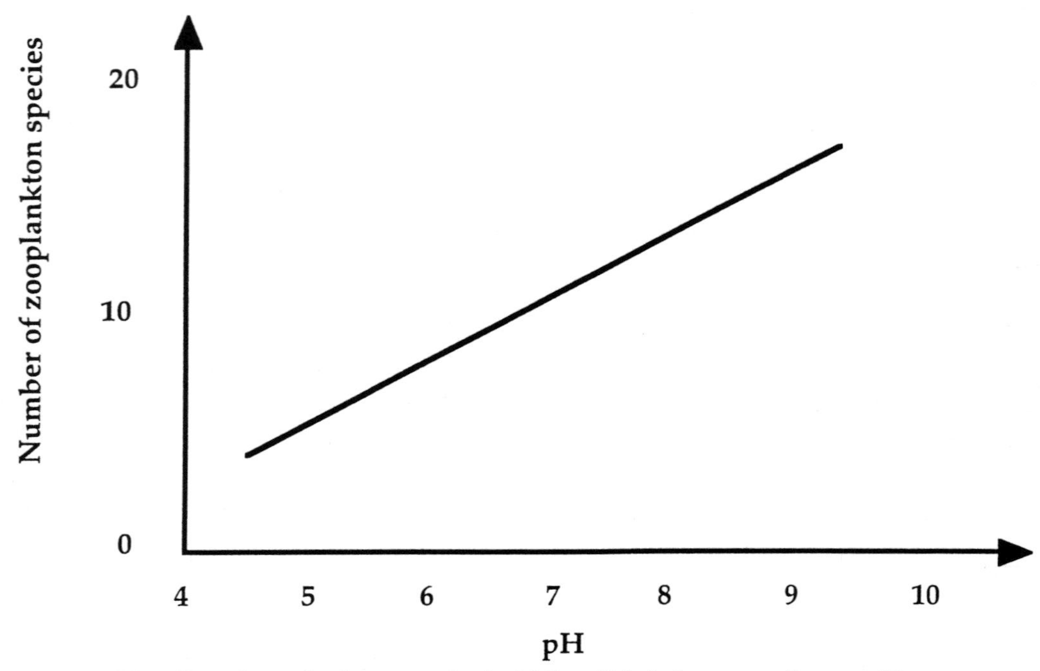

Fig. 4.24 Number of zooplankton species in 84 Swedish Lakes according to pH.

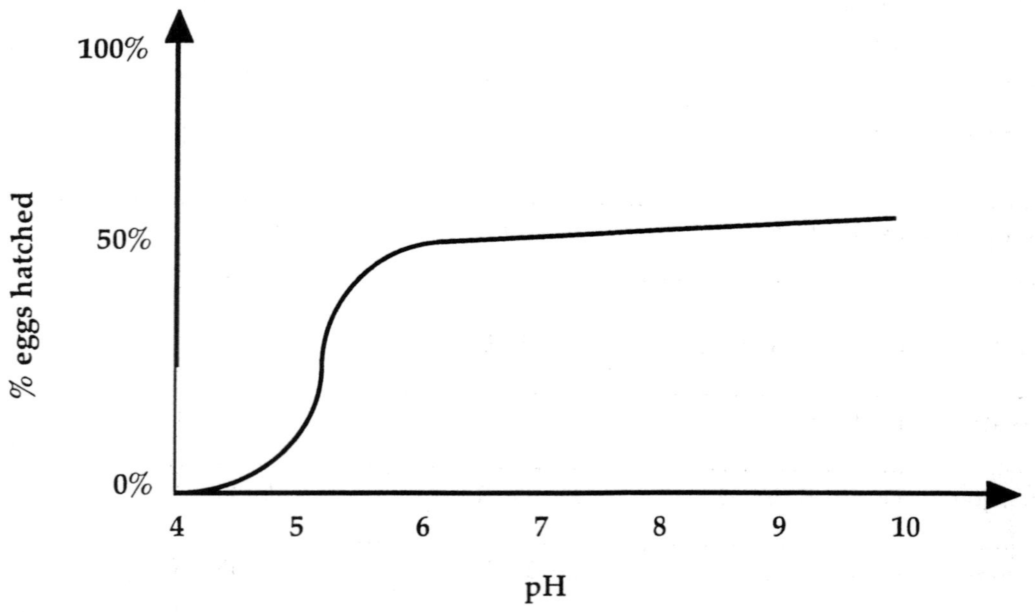

Fig. 4.25 % eggs hatched according to pH.

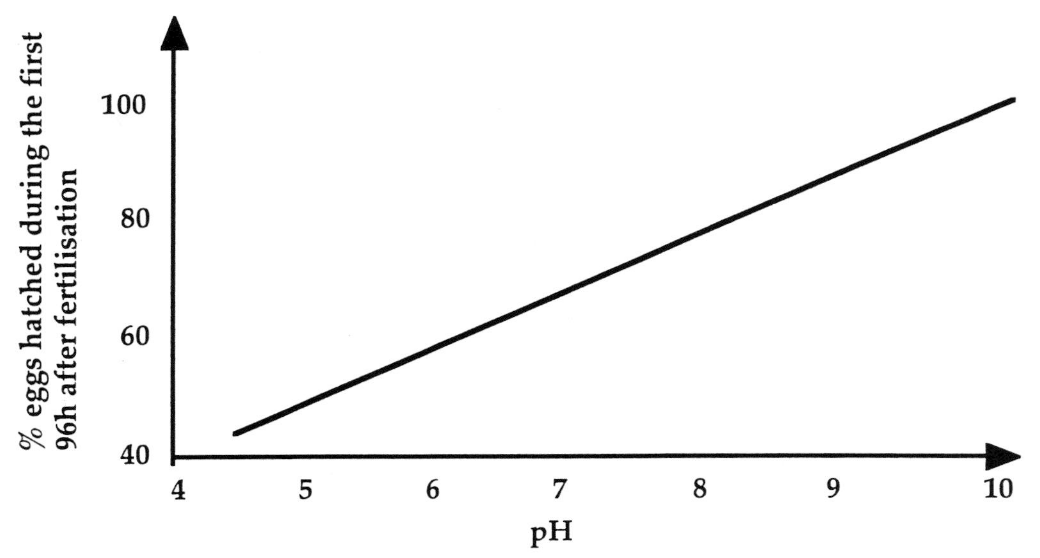

Fig. 4.26 Eggs hatched during the first 96 h after fertilisation as % of total number of hatched eggs according to pH.

The enhanced abatement of air pollution in the U.S. and Western Europe during the last 10-20 years has reduced these problems slightly, but due to still too excessive emission of sulphur dioxide and nitrogen oxides, and to previous harmful effects which have not yet been completely eliminated, acidification of lakes has not been solved in these regions. In addition, emission of acidic gases still takes place in almost unreduced amounts in Eastern Europe which implies that the total amount of acidic components in Scandinavian rain is still very significant.

However, there is hope that the observed reduction in emissions of acidic gases will continue and thereby solve the problem in the U.S. and in Europe. Meanwhile, the problem of acidification will most probably increase in other parts of the world, where increased industrialisation and use of particularly coal will take place. Recent observations indicate that the problem is rapidly under development in China.

In Sweden calcium hydroxide is widely added to rivers and lakes to reduce the damage from low pH. Hundred of millions of Swedish Crowns are spent every year on increasing the pH of natural waters. This amount should be compared with the billions of dollars it has costed to reduce the sulphur content of fossil fuel to an almost acceptable level in the U.S and Western Europe.

2. The effects of acidification on soil and plants. Decreased pH values in the lithosphere also cause an overall deterioration of the environment. **The leaching of nutrients from soil is increased at lower pH** as demonstrated for calcium ions in Fig. 4.28. **The ability of the soil to bind ions is decreased at lower pH.** Soil is a cation exchanger mainly due to its content of clay .

The leaching of ions from soil at lower pH produces therefore produces an increased concentration of metal ions. The higher concentration of aluminium ions in soil water causes the needles of conifers become brittle and easily are lost. Consequently, the coniferous forests particularly in Germany, Switzerland and Central Europe have been deteriorated.

Moreover, SO_2 has a direct effect on the degradation of plant pigments, and detritus is decomposed at a reduced rate at lower pH. The enzymatic decomposition of cellulose has its optimum at pH 5.5 -8.0 and is significantly reduced at higher or lower pH. Plant growth is furthermore dependent on pH with an optimum between 6.0 and 7.5. All in all pH has a significant impact on forestry and agriculture.

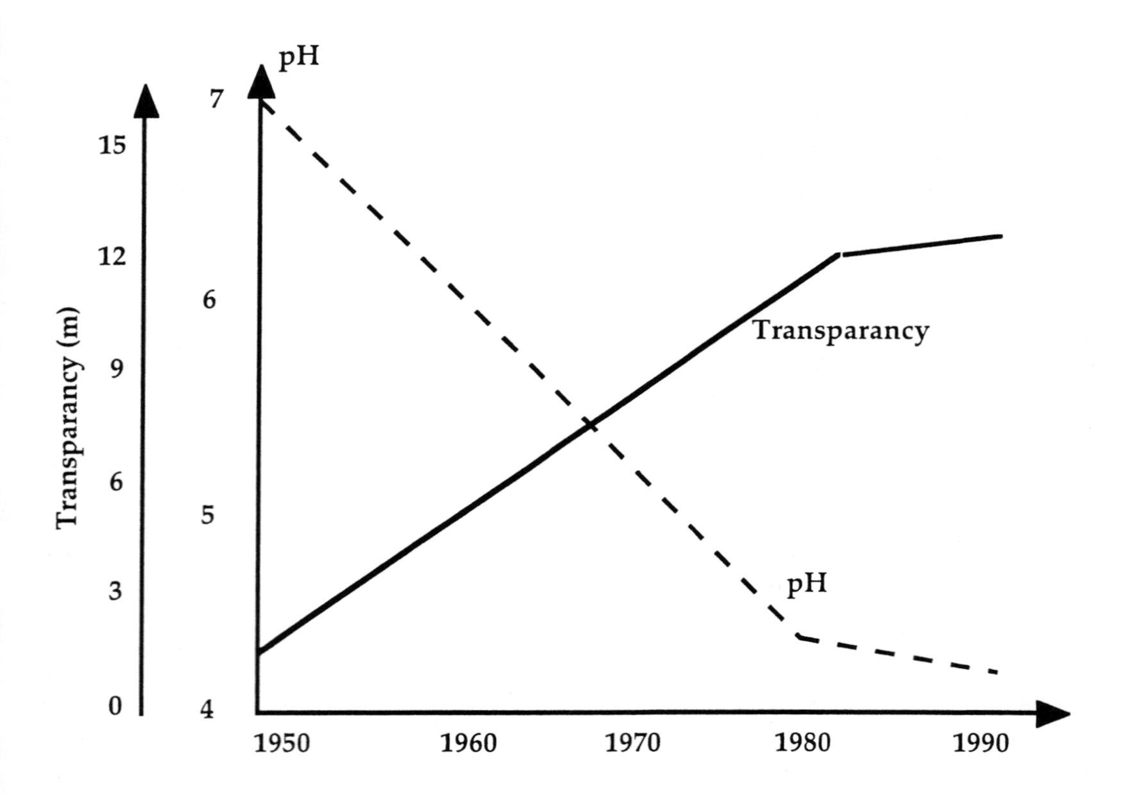

Fig. 4.27 pH and transparency of Lake Stora Skarsjön plotted against time.

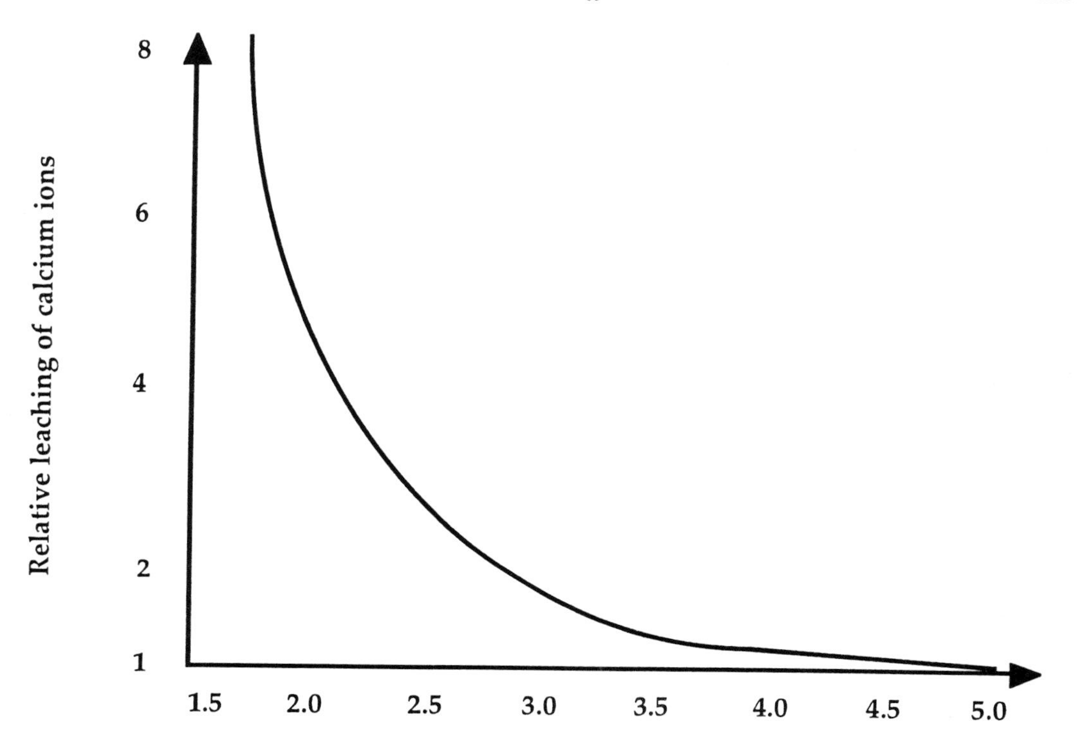

Fig. 4.28 Leaching of calcium in typical forest soil exposed to precipitation adjusted to pH values from 1.8 to 5.0 during a period of 40 days. The Ca leaching in the control (distilled water) is set at 1.0. Precipitation: 500 mm/month. Data from Jørgensen et al., 1979.

3. Will the increased uptake of carbon dioxide by the sea change pH in the oceans? Seawater contains several proteolytic species, including hydrogen carbonate (about 2.4 mM), borate (about 0.43 mM), phosphate (about 0.0023 mM), silicate and fluoride in various stages of protonation (uptake of hydrogen ions). The concentrations of proteolytic species are characterised by the total alkalinity A, and pH. The total alkalinity is determined by adding an excess of a standard acid (e.g. 0.1 M), boiling off the carbon dioxide formed and titrating back to a pH of 6. During this process all the carbonate and hydrogen carbonate are converted to carbon dioxide and expelled and all the borate is converted to boric acid. The amount of acid used (i.e. the acid added minus the base used for back titration) then corresponds to the alkalinity, Al, and the following equation is valid:

$$Al = C_{H_2BO_3^-} + 2C_{CO_3^{2-}} + C_{BO_3^-} + (C_{OH^-}\ C_{H^+}) \qquad (4.32)$$

where C = the concentration in moles per litre for the indicated species. In other words the alkalinity is the concentration of hydrogen ions that can be taken up by proteolytic species present in the sample examined. Obviously, the higher the alkalinity, the better the solution is able to maintain a given pH value if acid is added. The buffering capacity and the alkalinity are proportional (see e.g. Stumm and Morgan, 1996). The buffer capacity is dependent on the composition of the aquatic solution and the pH. The buffer capacity, ß, of seawater as a function of pH can be found by use of a double logarithmic diagram (log concentrations versus pH), showing the concentrations of the various acids and bases as functions of pH. Fig. 4.29 is a double logarithmic diagram for seawater. The proteolytic species mentioned above are represented in their appropriate concentrations. The important species are hydrogen and hydroxide ions, boric acid (HB) and carbonate ions (C^{2-}). The arrow in the diagram indicates the pH value of seawater - about 8.1. Based on such a diagram it is possible to set up another diagram, representing the buffering capacity as a function of pH, see Fig. 4.30. For those who are interested in the relationship between the two diagrams and how double logarithmic diagrams are constructed, see Hägg (1979) and Stumm and Morgan, 1996.

From Fig. 4.30 we can conclude *that the buffering capacity of seawater is very limited. About 3 mM of strong acid will change the pH from 8 to 3.* However, the pH of the entire ocean is remarkably resistant to change in pH caused by the addition of naturally occurring acids and bases, while the limited buffering capacity of an isolated litre of seawater accords with diagram 4.30. Sillen (1961) has suggested an explanation for the observed buffering capacity of the sea. A pH-dependent ion exchange equilibrium between solution and aluminosilicates (clay minerals), suspended in the sea is the main buffering system in oceans.

This buffering system may be represented by the following simplified equation:

$$3Al_2Si_2O_5(OH)_2 + 4SiO_2 + 2K^+ + 2Ca^{2+} + 12H_2O \text{ ------> } 2KCaAl_3Si_5O_{16}(H_2O)_6 + 6H^+ \tag{4.33}$$

The pH-dependence is indicated by the corresponding equilibrium expression in logarithmic form:

$$\log K = 6 \log(H^+) - 2 \log K^+ - 2 \log Ca^{2+} \tag{4.34}$$

Sillen (1961) **estimated the buffering capacity of these silicates to be about 1 mole per litre or approximately 2000 times the buffering capacity of carbonates.** However, as pointed out by Pytkowicz (1967), the buffering capacity of aluminosilicates has a much larger time scale than the buffering capacity based on the carbonate system. **In conclusion it seems that radical changes in the pH value of the oceans should not be expected as a result of increased combustion of fossil fuel, although the effect is cumulative.**

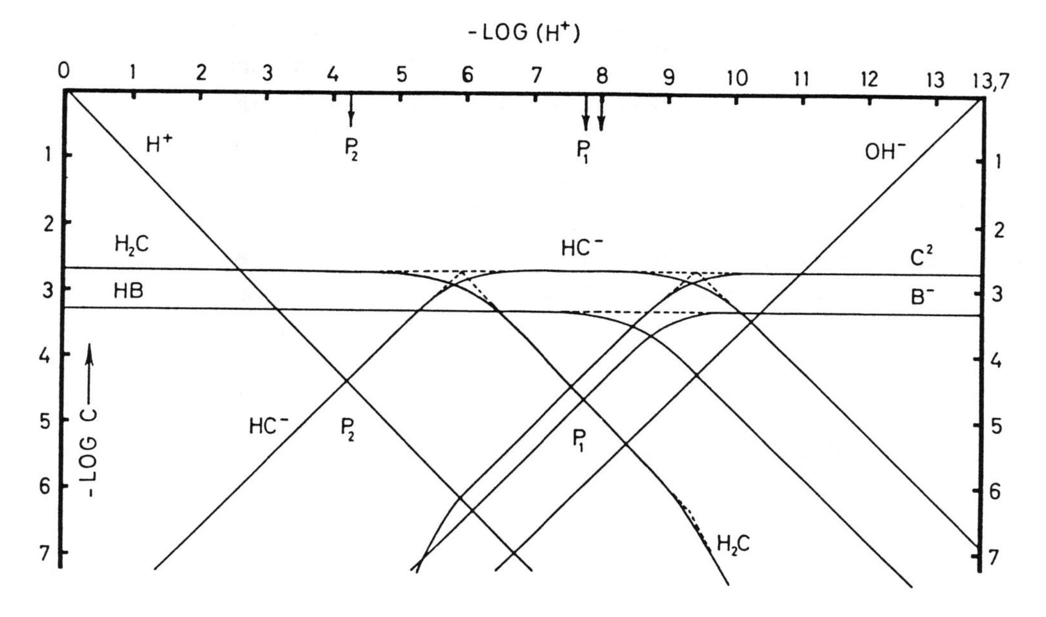

Fig. 4.29. pH-diagram. $H_2C = H_2CO_3$, $HC^- = HCO_3^-$, $C^2 = CO_3^{2-}$, B^- = borate. pH of the sea is indicated by an arrow.

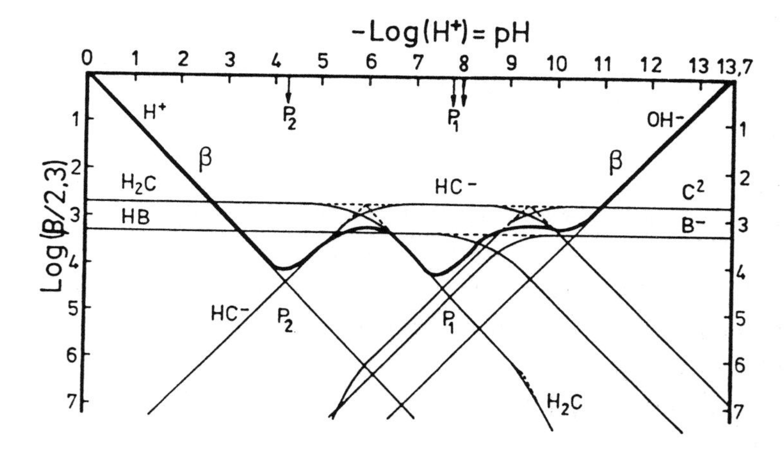

Fig. 4.30. Buffering capacity of the sea as a function of pH (ß).

However, the regional effects of combustion of fossil fuel on pH, due to deposition of sulphuric acid, is a very severe environmental problem. Its solution lies in either drastic reduction in the combustion of fossil fuel or a corresponding reduction in the sulphurous content of the fossil fuel. At present the pH values of lakes situated in districts where water has little buffering capacity are steadily decreasing and the only available remedy is the addition of calcium hydroxide to surface water. This will, however, change the chemical composition of the aquatic ecosystem in the entire region, which again will involve ecological changes. So, **the only real remedy is to reduce the emission of sulphurous compounds to an environmentally acceptable level.**

4.9 CONTAMINANTS IN THE ENVIRONMENT

This section will give a short overview of the toxic contaminants that we often detect in the environment. As we are using about 100,000 different chemicals in the modern society, we can also expect to find 100,000 different compounds plus intermediates of these compounds resulting from a wide spectrum of decomposition processes which take place in the environment. To be able to overview that many chemical compounds, their fates, effects and the associated risks, it is necessary to make a classification of the compounds. Below is briefly treated the properties, the effects, the characteristic processes and the particular risks of the following classes of contaminants:

A. Petroleum hydrocarbons
B. PCBs and dioxins
C. Pesticides
D. Polycyclic aromatic hydrocarbons (PAHs)
E. Heavy metals
F. Detergents
G. Synthetic polymers and xenobiotics applied in the plastics industry

A. Petroleum hydrocarbons include a variety of organic compounds. Hydrocarbons (compounds composed of carbon and hydrogen) constitutes only 50-90% of petroleum. They are n-alkanes, branched alkanes, cycloalkanes and aromatics. Cycloalkanes comprise usually the largest portion of hydrocarbons in petroleum, while aromatics are usually present to the extent of 20% or less. Characteristic compounds are benzene, alkyl substituted benzenes and fused ring polycyclic aromatic hydrocarbons (PAHs; see also this group, D). Some of these aromatics are carcinogenic and theseare probably the group that is of greatest environmental concern. In addition to hydrocarbons, petroleum contains sulphur and nitrogen compounds, such as, thiophene, ethanethiol and pyridine derivates.

Petroleum compounds are emitted or discharged into all spheres.

Evaporation removes the lower molecular weight, more volatile components of the petroleum mixture. Hydrocarbons with vapour pressures equal to that of n-octane (0.019 atm at room temperature) or greater will be lost quickly via evaporation.

The lower molecular weight hydrocarbons tend also to be the most water soluble, but for the same molecular weight aromatic are more soluble than cycloalkanes which are more soluble than branched alkanes with the n-alkanes being the most insoluble in water. Petroleum products are discharged directly to the sea by accidents or by violation of international regulations. In a massive discharge of petroleum products, which has been recorded in several ship accidents, most of the petroleum will initially float on the surface of marine waters as a slick. Eventually, most slicks are dispersed widely and form a 0.1 mm thick layer on the water. If drift results in landfall, coral or contact with mangrove communities, a disastrous environmental impact may occur.

Fortunately chemical transformation and degradation processes act on petroleum compounds in the environment. Microbial transformation and photo oxidation are of particular importance.

The aromatic compounds of petroleum are the most toxic substances present. They are lethal to crustaceans and fish with LC_{50} in the range of 0.1- 10 mg/l.

B. PCBs and dioxins are characterised as aromatic compounds with a high content of chlorine. As the names cover polychlorinated biphenyls and polychlorinated dibenzo (1,4) dioxins there are many different individual compounds under these labels. 209 different PCB compounds are for instance known although only about 130 are found in commercial mixtures. Figures 4.31 and 4.32 show the molecular structures and names of some of the most common PCBs and dioxins.

The applications of PCBs have been quite diverse (capacitor oil, plasticisers, printer's ink etc), but due to investigations in the 1960s and 1970s in which it was found that PCBs occurred widely in the environment and significant bioaccumulation took place, voluntary restrictions were introduced and all "open" applications banned.

Dioxins are not deliberately produced, but are byproducts of chemical processes involving chlorine for instance production of various organochlorine and bleaching of pulp and of combustion processes if chlorine containing compounds are present.

Both PCBs and dioxins are characterised by a low water solubility and a high K_{OW} (most components have log $K_{OW} > 5$). Both groups of compounds are very persistent to decomposition processes which explains why they are strong bioaccumulators, although dioxins have a UV-VIS absorption spectrum that results in significant absorption from solar radiation. Some dioxins have a half life time in thetroposphere of a few days.

C. Pesticides are used to remove pests and they have probably due to their direct use in nature been the most criticised environmental contaminants. Usage of DDT and related insecticides accelerated during the 1940s and the subsequent decades until environmental doubt occurred in the mid sixties. Since 1970 DDT has been banned in most industrialised countries, but it is still used in developing countries for instance India, where it has resulted in very high body concentrations in the Indian population. All the chlorinated hydrocarbon insecticides are banned in most industrialised countries due to their persistence and ability to bioaccumulate (K_{OW} is high).

PCB's

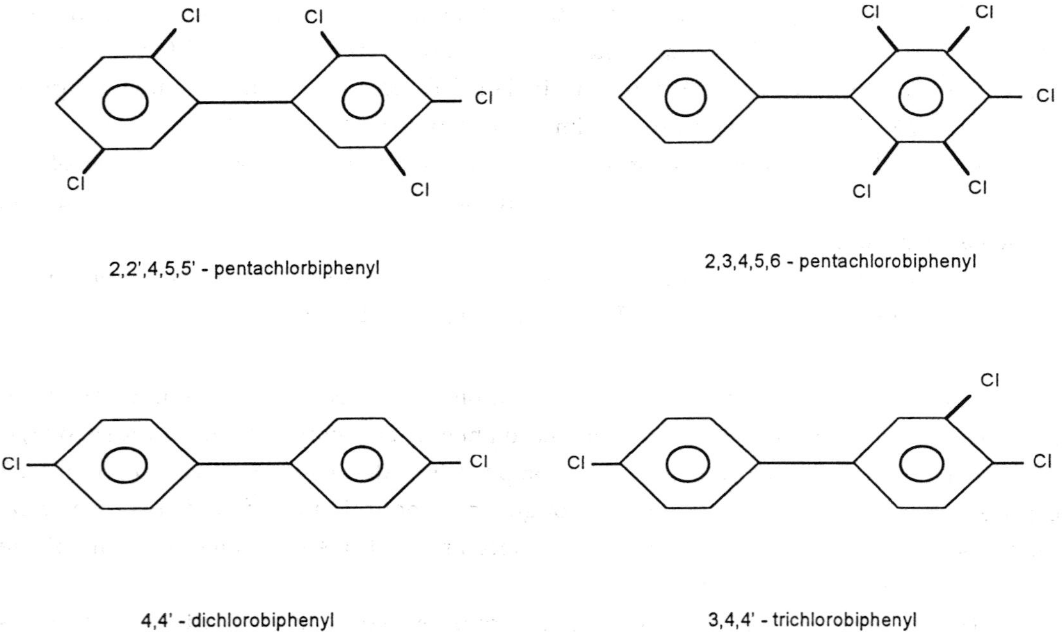

Cl Cl Cl Cl Cl

2,2',4,5,5' - pentachlorbiphenyl

Cl Cl Cl Cl Cl

2,3,4,5,6 - pentachlorobiphenyl

Cl Cl

4,4' - dichlorobiphenyl

Cl Cl Cl

3,4,4' - trichlorobiphenyl

Fig.4.31 The molecular structure and names of four common PCBs.

Pesticides can be divided into the following classes depending on their use and their chemical structure:

Herbicides comprise carbamates, phenoxyacetic acids, triazines and phenylureas.

Insecticides encompasses organophosphates, carbamates, organochlorines, pyrethrins, and pyrethroids.

Fungicides are dithiocarbamates, copper and mercury compounds.

The pesticides are chemically an extremely diverse group of substances, as they only have in common their toxicity to pests. A few of the most important molecules are shown in Figure 4.31. They are mostly produced synthetically although the natural pyrethrin pesticides have achieved commercial success.

Dioxins

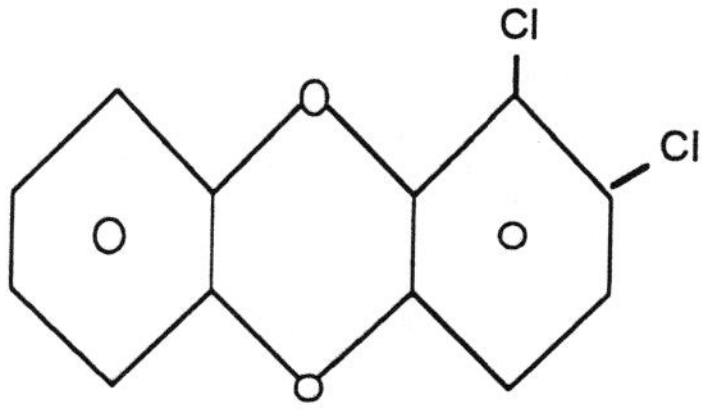

1,2 - dichloridbenzo (1,4) dioxin

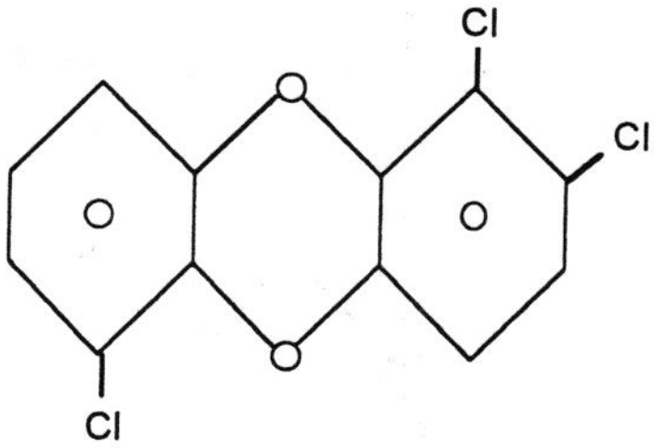

1,2,6 - trichlorodibenzo (1,4) dioxin

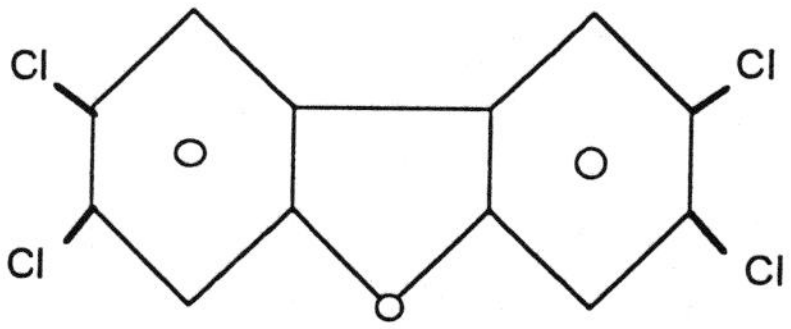

2,3,7,8 - tetrachloridbenzofuran

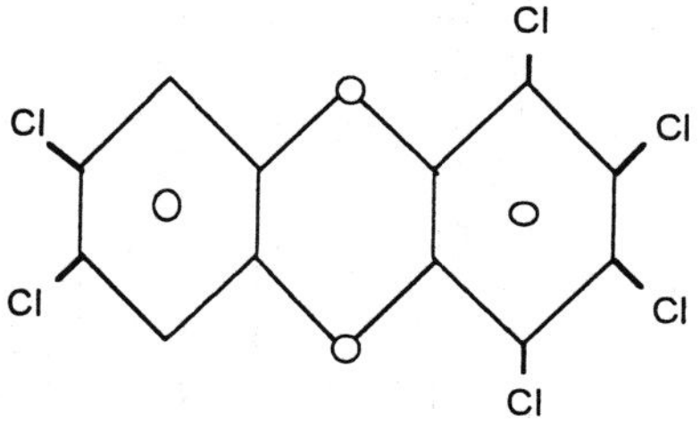

1,2,3,4,7,8 - hexachlorodibenzo (1,4) dioxin

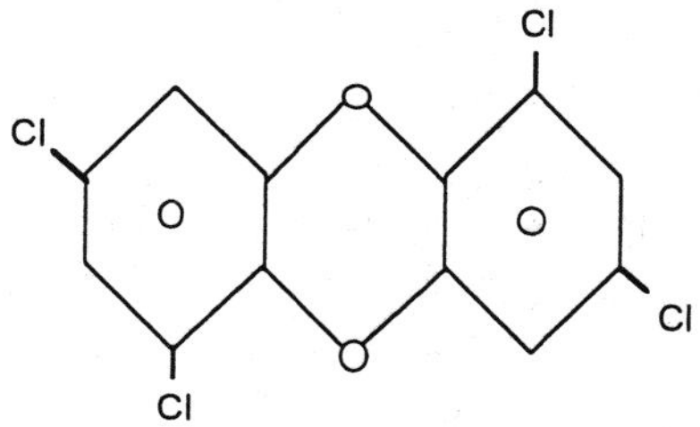

1,3,6,8 - tetrachlordibenzo (1,4) dioxin

Fig. 4.32 Molecular structure and names of five common dioxins.

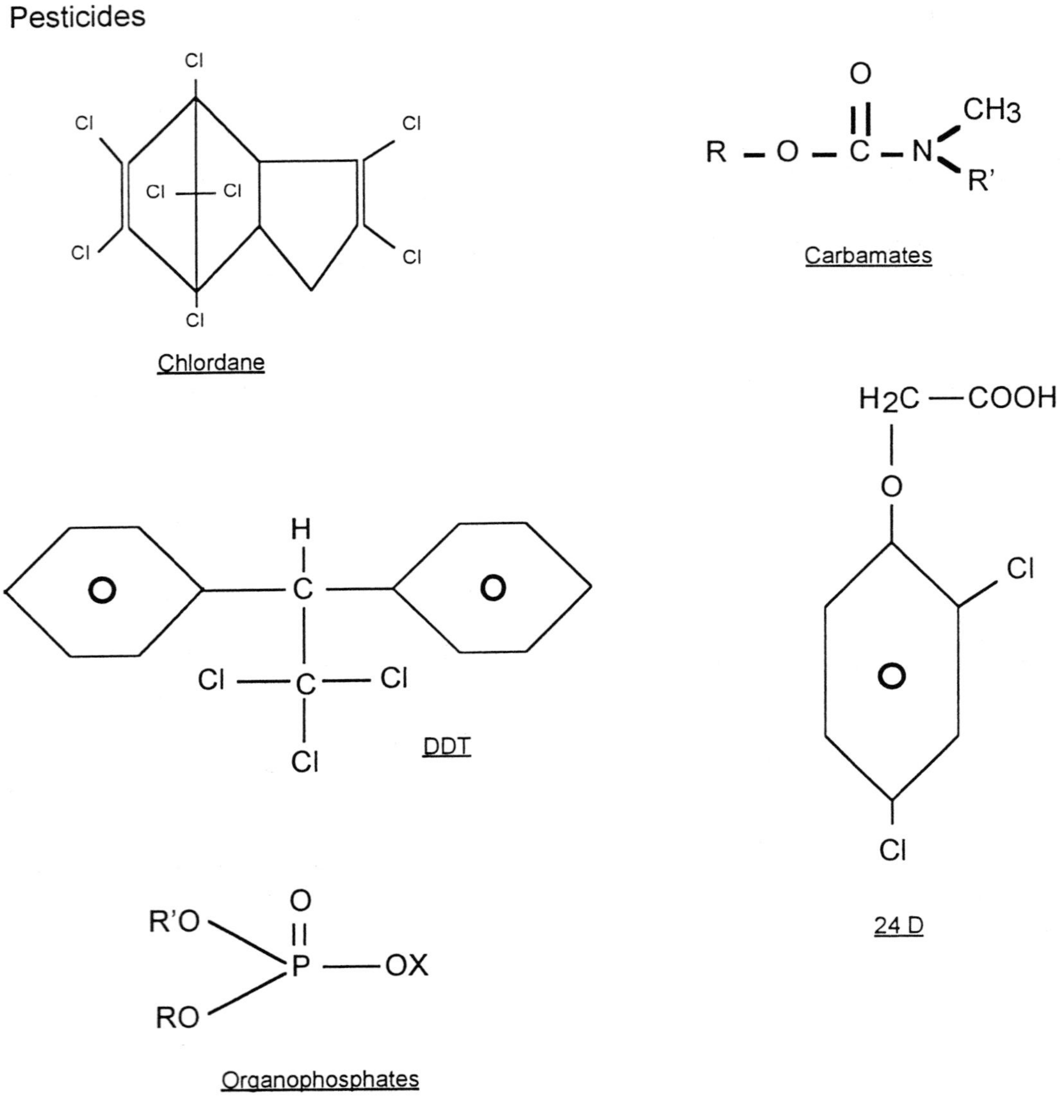

Pesticides

Chlordane

Carbamates

DDT

24 D

Organophosphates

Fig. 4.33 Molecular structure and names of five common pesticides

Chlorohydrocarbons are strongly bioconcentrated as already emphasised. In addition, they are very toxic to a wide range of biota, particularly to aquatic biota.

Organophosphates are almost equally toxic to biota, but due to these compounds, lack of persistence, higher solubility in water and bioaccumulation capacity, they are still in use.

Carbamates are relatively water soluble and have limited persistence. They are however, toxic to a wide range of biota. They act by inhibiting cholinesterase.

The pyrethins have a complex chemical structure and a high molecular weight. Thus, they are poorly soluble in water and tend to be lipophilic. They are however readily degraded by hydrolysis. They are more attractive to use than most of the other pesticides due to their very low mammalian toxicity.

Phenoxyacetic acid is a very effective herbicide but contains trace amounts of tetrachloro-dibenzo-dioxin.

Pesticides are banned in organic agriculture where they are replaced by other methods for instance mechanical and biological methods (use of predator insects).

D. PAHs are molecules containing two or more fused 6C -aromatic rings. They are ubiquitous contaminants of the natural environment, but the growing industrialisation has increased the environmental concern about these components. Two common members are naphthalene and benzo(a)pyrene; see Figure 4.32.. PAHs are usually solids with naphthalene (lowest molecular weight) having a melting point of 81°C.

The natural sources of PAHs in the environment are forest fires and volcanic activity.The anthropogenic sources are coal-fired power plants, incinerators, open burning and motor vehicle exhaust. As a result of these sources PAHs occur commonly in air, soil and biota. They are lipophilic compounds able to bioaccumulate. The low molecular weight compounds are moderately persistent; while for example benzo(a) pyren with a higher molecular weight persists in aquatic systems for up to about 300 weeks.

They are relatively toxic to aquatic organisms and have LC_{50} values for fish in the range of 0.1 - 10 mg/l. The major environmental concern of PAHs is, that that many PAHs are carcinogenic. It is has been shown (Andersen, 1998) that benzo(a)pyrene is a endocrine disrupter and it cannot be excluded that many more PAHs have the environmental adverse effect of disturbing the hormone balance of nature.

Human exposure to PAHs occurs through tobacco smoking as well as through compounds in food and the atmosphere.

E. Organometallic compounds are compounds having metal carbon bonds, where the carbon atoms are part of an organic group. The best known example is probably tetraethyl lead which is used as additive to gasoline. It has now been phased out of use in many countries due to its environmental consequences; see also Chapter 2.. Organometallic compounds can be formed in nature from metal or metal ions for instance dimethylmercury or are produced for various purposes, as catalysts e. g. organoaluminium, as pesticides, e.g. organoarsenic and organotin compounds, as stabilisers in polymers, e.g. organotin compounds and as gasoline additive e.g. organolead compounds. Organometallic compounds exhibits properties that are different from those of the metal itself and inorganic derivatives of the metal. They have for instance a relatively higher toxicity than the metals.

Most organometallic compounds are relatively unstable and undergo easily hydrolysis and photolysis. Most organometallic compounds have weakly polar carbon-metal bonds and are often hydrophobic. They therefore only dissolve in water to a small extent

PAHs

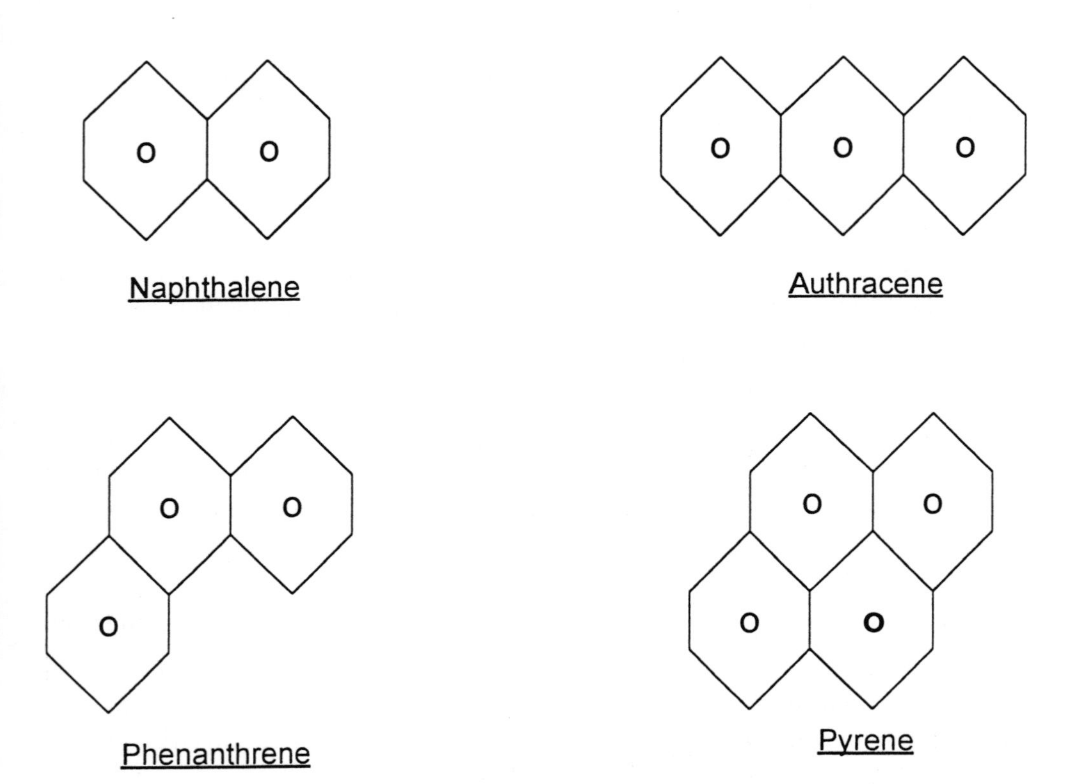

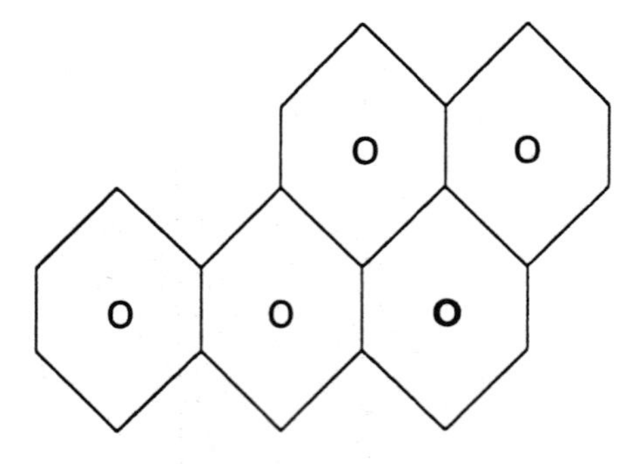

Benzo(a)pyrene

Fig. 4.34 Molecular structure and names of five common PAHs.

and are readily sorbed onto particulates and sediments.

The most important organometallic compounds from an environmental point of view are organomercury, organotin, organolead and organoarsenic which are all toxic to mammals.

F. Detergents (and soaps) contain surface active agents (surfactants) which are classified according to the charged nature of the hydrophilic part of the molecule:

Anionic: negatively charged

Cationic: positively charged

Nonionic: neutral, but polar

Amphoteric : a zwitterion containing positive and negative charges

They are produced and consumed in large quantities and are mostly discharged into the sewage system and end up in the waste water plant. The early surfactants contained highly branched alkyl hydrophobes that were resistant to biodegradation. These surfactants are largely obsolete today having been replaced by linear alkyl benzene sulphonates (LAS) and other biodegradable surfactants.

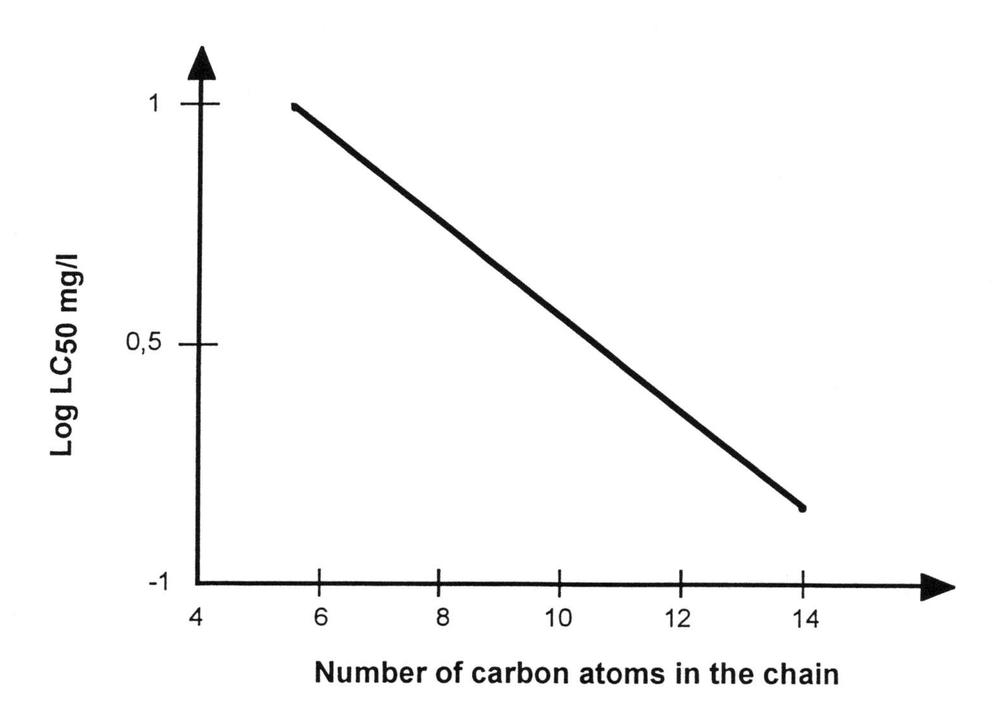

Fig. 4.35 Log LC50 is plotted versus number of carbon atoms in the chain for LASs. As seen increased chain length implies increased toxicity.

The toxicity to mammals is generally low for all surfactants, while the toxicity to aquatic organisms is relatively high (LC_{50} from about 0.1 to about 77 mg/l). The toxicity will generally increase with the carbon chain length (see Figure 4.33).

Many surfactants bind strongly to soils and sediments which implies that, to the extent that they are not biodegraded in a biological treatment plant, they will mainly be found in the sludge phase.

G. Synthetic polymers and xenobiotics applied in the plastics industry form a very diverse group of compounds from a chemical view point. Synthetic polymers are useful (plumbing, textiles, paint, floor, covering and as the basic material for a wide spectrum of products) because they are resistant to biotic and abiotic processes of transformation and degradation. These properties, however, also cause environmental management problems associated with the use of these components. In addition, several xenobiotic compounds are used as additives, softener, stabilisers and so on in synthetic polymer to improve their properties. Some of these additives are very toxic and may cause other and additional environmental problems, for instance phthalates are widely used in the plastic industry and it has been demonstrated that phthalates have effects as endocrine disrupters.

After use synthetic polymers are usually incinerated together with industrial and household garbage (solid waste). The presence of PVC will imply that hydrochloric acid is formed and also dioxins to a certain extent, but strongly dependent on the incineration conditions. As it is difficult to separate different types of plastics, it has been discussed to phase out the use of PVC, but due to PVC's unique properties this has not yet been decided.

4. 10 QUESTIONS AND PROBLEMS

1. Indicate the expected difference in diversity between a shallow and a deep lake and between a eutrophic and an oligotrophic lake.

2. What is the difference between fresh and salt water in relation to a) solubility of oxygen, b) the LC_{50} value for total ammonia (NH_3 + NH_4^+)?

3. Explain using the concept of ecological buffering capacity and energy, why anaerobic conditions give lower diversity than aerobic conditions.

4. Find the buffer-capacity of a lake with 100 mg Ca^{2+} /l and a lake with 10 mg Ca^{2+}/ l at pH = 8.0 and pH = 5.0. It is presumed that the buffer capacity is entirely related to HCO_3^- and this ion is equivalent to the calcium concentration. How much will the pH change in the 4 cases if the lakes are 10 m deep and they receive sulphuric acid by precipitation

corresponding to 0.5 g S m^{-2}?

5. A chlorophenol has a solubility in water of 1.2 mg l^{-1}. Find the approximate concentration factor for mussels (length about 5 cm) using of the methods presented in Chapters 2 and 4.

6. The water solubility of nitrobenzene is 1.27 g /l at room temperature. a) Estimate the concentration factor for fish of length 20-30 cm. b) Estimate the concentration factor for blue mussels of length 5 cm. c) Estimate the ratio between nitrobenzene adsorbed to activated sludge and dissolved in water in an activated sludge plant. d) Do you estimate nitrobenzene to be readily biodegradable, slowly biodegradable, very biodegradable or refractory? Explain your answers from the molecular formula. e) On the basis of your answers to questions c and d, where would you expect to find nitrobenzene after and activated sludge plant, in the treated water, in the sludge or would it be decomposed?

7. A considerable amount of polyaromatic hydrocarbons, PAH, has been dumped on a rubbish dump. The environmental department of the city fears that the ground water will be contaminated. a) Calculate the % of PAH adsorbed to the soil and in equilibrium with PAH in soil water. The soil is known to contain 10% carbon. Assume that log K_{OW} = 5.5. b) Henry's constant for the mixture of PAH is 0.002 atm. What is the concentration in air in equilibrium with soil water containing 80mg/l? Is it expected that PAH will be relatively rapidly removed by evaporation? c) The biodegradation can be described by a first order reaction with a rate constant of 0.005 1/ 24h. Is this biodegradation in accordance with the expected biodegradation of PAH (assume that it consists of 3 - 4 aromatic rings)? d) If the initial concentration is 80 mg/l what would the concentration be after 180 days, if we assume that biodegradation is the only removal process?

8. Explain why the concentration of most micro pollutants in an organism increases with time (and weight of the organism).

9. DDT has been banned and in many countries at least partly replaced by chlordane. Estimate the difference in physical-chemical properties between the two pesticides based on their chemical structure. What difference is expected in the bioaccumulation and persistence of these two substances?

10. What is the expected difference in environmental behaviour between phenoxyacetic acid containing the COOH group and the corresponding sodium salt containing COONa? What would be the difference between the two forms in relation to accumulation in soil and evaporation? Use the results to indicate how pH will influence these properties for

phenoxyacetic acid?

11. The following experiments on 300 mussels were carried out to assess the toxicity of dimethyl mercury for blue mussels:

Number of blue mussels surviving after 24 hours	Concentration of dimethyl mercury μg /l
300	10.6
242	14,4
185	19.2
109	22.0
33	30.3
0	59.3

Estimate LC_{50}.

12. Log K_{ow} for naphthalene is 3.36, for pyrene 5.18 and for the PAH called coronene ($C_{24}H_{12}$) 6.90. For chemicals of similar structure, a linear relationship between molecular weight and Log K_{ow} is often valid. Assume that this is correct for PAHs and use the plot (MW versus log K_{ow}) to estimate the log K_{ow} value for anthracene, phenanthrene and benzopyrene. The log K_{ow} values for these three compounds are respectively 4.55, 4.90 and 6.04. Was the assumption of a linear relationship reasonable?

13. What would be the difference in biodegradability of a branched alkylsulphonate (car- bon chain with 12 C-atoms, 7 branches) and a completely linear alkylsulphonate with the same number of carbon atoms. Use the rules presented in the text to indicate the difference semi quantitatively.

14. A spill of toluene onto soil has occurred. Analysis of the soil has revealed that the toluene concentration is 100 mg/ kg. How long time will it take to reach an acceptable concentration defined as 0.05 mg toluene / kg, when the biological half life in soil for toluene is 23 days?

15. Log K_{ow} for atrazine is 2.75. Estimate the distribution between soil with 1.8 % organic carbon and water for atrazine. What would be the estimated ratio of the concentration in carrots with 0.6 % lipid to the concentration in soil grown in atrazine contaminated soil?

16. The following contaminants have been found in the soil of an industrial site: benzene, toluene, chloropyrifos and phenol. Evaluate the potential for these four compounds to

contaminate the ground water. The following properties are available for the four compounds:

Compound	Vapour pressure (mm Hg)	Water solubility (mg /l)	Log (Soil sorption coeff.) (-)
Benzene	76	1780	3.3
Toluene	10	515	3.5
Phenol	0.2	67 000	2
Chloropyrifos	0.00002	2	4.1

17. Indicate by an x in Table 4.13, shown below, in which classes the 8 compounds in the table belong. Class 1 covers the compounds that are decomposed at least 10% in a biological treatment plant and eventually after adaptation significantly more than 10%. Class 2 comprises compounds which are 1 - 10% biodegraded in a biological treatment plant, while class 3 means that the compounds are not biodegraded (< 1%) in a biological treatment plant.

Table 4.13

Compound	**Class 1**	**Class 2**	**Class 3**
Ethylenglycol			
1,3 dichlorantracene			
2,3,4-trinitrophenol			
DDT			
PCB			
Glycerine			
Dioxin			
Pentachlorphenol			

Are the compounds that are not biodegraded (class 2 and 3) accumulated in the water or in the sludge phase?

18. Hexachlorobenzene has an octanol / water partition coefficient of $10^{6.18}$. Find the approximate concentration factor for 25-30 cm length fish.
Find also BCF for blue mussels (length approximately 5 cm) presuming the same % fat tissue. Find finally the concentration of hexachloro-benzene in soya beans with a lipid content of 8.5% cultivated in soil with 2% organic carbon and with a concentration of hexachloro-benzene of 12 mg / kg dry matter.

19 It is found that a dioxin has a BCF value for 25 cm fish oo 12, 000. The fat content of the fish used for the experiment is 7%. What would be the BCF for a fish with 1% fat content? What would be the BCF for a fish of the length 1 m with a fat content of 2%?

5. Water and Waste Water Problems

5.1. INTRODUCTION TO THE PROBLEMS OF WATER AND WASTE WATER - AN OVERVIEW

This chapter is devoted to linking the aquatic pollution problems with the various water treatment methods available in environmental technology. As shown in Figure 1.3, environmental technology is solving the pollution problems associated with point sources.

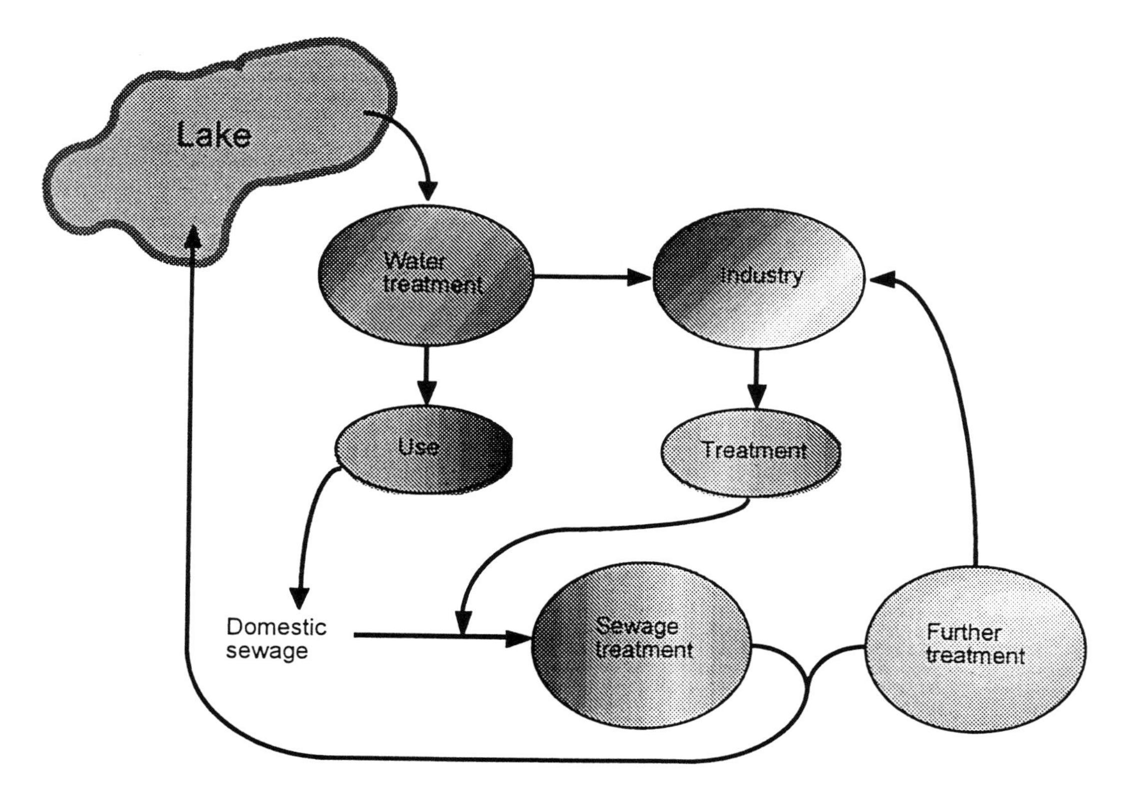

Fig. 5.1. Example of good water management practice, where 1) a part of the treated waste water is reclaimed for industries after additional treatment, 2) the industrial waste water is treated before being discharged into the public sewage system. The sewage treatment chosen pays regards to the lake as well as the water treatment.

Selection of the optimum waste water and water treatments is a very complex problem. Quantitative management requires a clear definition and quantification of the problems by application of the methods presented in Chapters 2-4, but the search for an optimum solution also requires comprehensive knowledge of the treatment methods available. Furthermore, the problem is complicated by the interdependence of water supply and waste water disposal.

Good water management practice should therefore consider not only the ecological effects in the receiving body of water, but also the effect on the quality and economy of the water supply. Fig. 5.1 illustrates these considerations.

Effluents fall into seven groups, but many have, to a greater or lesser degree the polluting properties of at least two of these categories:

1. **Organic residues,** including domestic sewage, effluent from food-processing-industries, ensilage, manure heaps and cattle yards, laundries, paper mills, etc. These effluents vary a great deal, but they have much in common. They all contain complex organic compounds in solution and/or suspension, sometimes with toxic substances and various salts. Their basic property, however, is that they contain unstable compounds, which are readily oxidised and so use up the dissolved oxygen in the water. Some of these compounds are more readily decomposed than others: for example, slaughterhouse wastes oxidise rapidly while wood pulp is comparatively stable. Section 5.2 focuses on the technical solutions to this problem.

2. **Nutrients,** including ammonia, nitrates, other nitrogenous compounds, orthophosphates, other phosphorous compounds, silica and sulphates. The main sources are domestic sewage and effluents from fertiliser manufacture. Discharge of nutrients may cause undesirable eutrophication as described in section 2.8. Section 5.3 discusses the methods available for nutrient removal.

3. **Toxicants** in solution occur in the waste waters from many industries. They include acids, alkalis, oil, heavy metals and toxic organic compounds, mainly from chemical industries, gas works and use of insecticides; see the overview presented in Section 4.9. Their effects on ecosystems are described in Sections 4.4 and 4.5. Section 5.5 reviews the methods available for the removal of heavy metals and Section 5.4 deals with removal of toxic organic compounds.

4. **Inert suspensions** of finely divided matter result from many types of mining and quarrying and from washing processes, such as those of coal and root crops. The effect of these pollutants on the ecosystem can be evaluated from the principles mentioned in chapter 2. Removal of inert finely divided matter can be carried out by the mechanical treatment methods mentioned in Section 5.2.

5. **Other inorganic agents,** such as salts or reducing agents (e.g. sulphides, sulphites and ferrous salts) occur as constituents of the effluent of several types of industry. Minor discharges of salts are generally harmless to the environment, but reducing compounds use up the oxygen in the receiving body of water, and have the same effect as organic residues (see Section 2.7). This effect can, however, easily be eliminated by aeration, a process which

will only be discussed in relation to biological treatment methods (see Section 5.2). For a more comprehensive account, see Jørgensen (1979d).

6. **Hot water** is produced by many industries that use water for cooling purposes. They often use river water, which is pumped through the cooling system and sometimes raised to very high temperatures during part of its journey. The effects of this process are described in section 2.4. The methods available to meet this problem are a) use of cooling towers, b) use of heat exchangers, c) use of alternative technology, d) use of an alternative receiving water body which is less susceptible to damage. These solutions will not be discussed in this context, as they either must be considered as purely industrial engineering problems, or have already been mentioned in chapter 2.

7. **Bacteriological contamination** of waters originates mainly from domestic sewage, but food-processing industries, manure heaps and cattle yards are also sources of this type of pollution. Methods to meet this problem are used in the production of potable water, for process water in industry and in the treatment of waste water. The available methods are mentioned in Section 5.6.

Waste waters emanate from four primary sources: 1) municipal sewage, 2) industrial waste waters, 3) agricultural runoff and 4) storm water and urban runoff. The problem of the first group is related mainly to organic residues and nutrients, while the second group encompasses the entire spectrum of pollution problems, although discharge of heavy metals and organic toxic compounds is the most serious one. It is not possible to give a comprehensive review of industrial waste water problems here; for detailed discussion see Jørgensen (1979d).

TABLE 5.1
Pollution from urban and agricultural runoff

Constituent	Urban runoff (Storm water)	Agricultural runoff
Suspended solids (mg/l)	5-1200	-
Chemical oxygen demand COD (mg/l)	20-610	-
Biological oxygen demand BOD (mg/l)	1-173	-
Total phosphorus (mg/l)	0.02- 7.3	1.1-0.65
Nitrate nitrogen (mg/l)	-	0.03 -5.0
Total nitrogen (mg/l)	0.3 -7.5	0.5-6.5
Chlorides (mg/l)	3-35	-

TABLE 5.2

Generally applied waste water treatment methods

Method	Pollution problem	Efficiency %	Costs ($\$/100m^3$)
Mechanical treatment	Suspended matter	75-90	3-8
	BOD_5 reduction	20-35	
Biological treatment	BOD_5 reduction	70-95	25-40
Flocculation	P-removal	30-60	6-10
	BOD5 reduction	40-60	
Chemical precipitation	P-removal	65-95	10-18
$Al2(SO4)_3$ or $FeCl_3$	Reduction of		
	heavy metals	40-80	
	BOD_5 reduction	50-65	
Chemical precipitation	P-removal	85-95	12-18
$Ca(OH)_2$	Reduction of		
	heavy metals	80-95	
	BOD5 reduction	50-70	
Chemical precipitation	P-removal	90-98	12-18
and flocculation	BOD_5 reduction	60-75	
Ammonia stripping	Ammonia removal	70-95	25-40
Nitrification	Ammonium		
	-> nitrate	75-95	20-30
Active carbon	COD	40-95	60-90
adsorption	BOD_5 reduction	40-70	
Denitrification	Nitrogen removal	75-95	15-25
Ion exchange	BOD_5 reduction		
	(e.g.proteins)	20-40	40-60
	P-removal	80-95	70-100
	Nitrogen removal	80-95	45-60
	Heavy metals	80-95	10-25
Chemical oxidation	Oxidation of toxic		
(e.g. with Cl_2)	compounds	0.90-0.98	60-100
Extraction	Heavy metals and other		
	toxic compounds	50-95	80-120
Reverse osmosis	Removes pollutants with high efficiency, but is expensive		100-200

TABLE 5.2 (continued)

Disinfection methods	Microorganisms	High, cannot be indicated	6-30
Waste stabilisation ponds	Microorganism	High	2-8
	Reduction of BOD_5	70-85	
	Nitrogen removal	50-70	
Constructed wet-land	Reduction of BOD_5	20-60*)	5-15
	Nitrogen removal	70-90	
	P-removal	0-80**)	

*) presumes a pretreatment ($BOD_5 \leq$ about 75 mg/l)
**) The removal is dependent on the adsorption capacity of the soil applied and whether harvest of the plants is foreseen

As municipal and industrial waste waters receive treatment, increasing emphasis is being placed on the pollution effects of urban and agricultural runoff. The range of pertinent characteristics of these waste waters is given in Table 5.1. In many places sewage continues to be discharged into systems of drains intended also for the removal of surface runoff from rainstorms and melting snow or ice. This is called combined sewerage. However, in most modern developments, sewage and runoff are each collected into a separate system of sanitary sewers and storm drains in order to avoid pollution of water course by the occasional spillage of sewage and storm water mixtures. This is called separate sewerage. Often the receiving body of water also serves as an important source of supply for many purposes. It is this multiple use of natural waters that creates the most impelling reasons for sound water-quality management, as already mentioned above. This part of the problem is discussed in section 5.6. Agricultural pollution problems are related to:

1. the extensive use of natural and industrially produced fertilisers to increase yield,
2. use of pesticides to eliminate damage by pests
3. waste from domestic animals.

It seems only possible to solve the non-point pollution problems 1) and 2) by use of sound ecological engineering, which has been treated in Chapter 8. Almost all unit operations applied for water and waste water treatment are mentioned in this chapter. Some of the unit operations can, however, be used to solve more than one pollution problem. An overview of the general applied waste water treatment methods is given in Table 5.2 with indication of efficiency and approximate cost to treat 100 m^3 waste water including interest and depreciation of the capital costs. The two last methods waste stabilisation ponds and constructed wetlands, are at least partially ecological engineering methods.

They will therefore not be presented in this chapter but in Chapter 8, although

TABLE 5.3

Efficiency ratio (0.0-1.0) matrix relating pollution parameters and waste water treatment methods (reproduced from Hansen and Jørgensen, 1988)

	Susp. matter	BOD_5	COD	P-total	NH4+
Mechanical treatment	0.75-0.90	0.20-0.35	0.20-0.35	0.05-0.10	~0
Biological treatment*	0.75-0.95	0.65-0.90	0.10-0.20	0.05-0.10	~0
Chemical precipitation	0.80-0.95	0.50-0.75	0.50-0.75	0.80-0.95	~0
Ammonia stripping	~0	~0	~0	~0	0.70-0.96
Nitrification	~0	~0	~0	~0	0.80-0.95
Active carbon adsorption*	-	0.40-0.70	0.40-0.95	~0.1	high**
Denitrification after nitrification	~0	-	-	~0	-
Ion exchange	-	0.20-0.40	0.20-0.50	0.80-0.95	0.80-0.95
Chemical oxidation	-	corresp. to oxidation	~0	~0	~0
Extraction	-	corresp. to extraction of toxic substance		~0	~0
Reverse osmosis*		see Table 5.2			
Disinfection methods	-	corresp. to appl. of chlorine, ozone, etc.			

those in Table 5.2 are foreseen to be applied for point pollution sources. Table 5.3. gives an efficiency matrix relating pollution parameters and waste water treatment methods.

A selection of methods is already possible on the basis of Tables 5.2 and 5.3, for instance two or more methods can be combined and the efficiency for the total treatment, e, can be calculated approximately by the following equation:

$$1 - e = (1 - e_1)(1 - e_2)(1 - e_3) \ldots 1 - e_n) \qquad (5.1)$$

TABLE 5.3 (continued)

	N-total	Heavy metals	E. coli	Colour	Turbidity
Mechanical treatment	0.10-0.25	0.20-0.40	-	~0	0.80-0.98
Biological treatment*	0.10-0.25	0.30-0.65	fair	~0	-
Chemical precipitation	0.10-0.60	0.80-0.98	good	0.30-0.70	0.80-0.98
Ammonia stripping	0.60-0.90	~0	~0	~0	~0
Nitrification	0.80-0.95	~0	~0	~0	~0
Active carbon ads.*	high**	0.10-0.70	good	0.70-0.90	0.60-0.90
Denitrification after nitrification	0.70-0.90	~0	good	~0	-
Ion exchange	0.80-0.95	0.80-0.95	very good	0.60-0.90	0.70-0.90
Chemical oxidation	~0	~0	~0	0.60-0.90	0.50-0.80
Extraction	~0	0.50-0.95	~0	~0	~0
Reverse osmosis*					
Disinfection methods			very high	0.50-0.90	0.30-0.60

*) depends on the composition, **) as chloramines.

where e_i ($i = 1, 2, 3, \ldots n$) are the efficiencies of the individual steps.

The function of the processes will be treated in some details, and how various sets of conditions determine the efficiency and the costs necessary to make the right selection. These issues are therefore treated in the following sections. Design of treatment plant will not be covered. This volume is written for non-engineers who want to have an overview of the possibilities to abate pollution with all available tools and to select good solutions for environmental management problems. Those interested in the design of treatment plants are referred to numerous books on environmental engineering.

Costs and efficiencies of alternative combinations of methods can easily be compared. The cost indications in Table 5.2 are, however, very preliminary and cannot be used for final decisions but only for the very first estimations.

5.2 REDUCTION OF THE BIOLOGICAL OXYGEN DEMAND

When organic matter is added to an aquatic ecosystem it is immediately attacked by bacteria, which break it down to simpler substances, using up oxygen in the process, see section 2.7. The rate at which a particular type of effluent is able, in the presence of ample oxygen, to satisfy its oxygen demand depends on what it contains. Industrial effluents which contain only chemical reducing agents, such as ferrous salts and sulphides, take up oxygen purely by chemical reactions. They do this very rapidly, exerting what is called immediate oxygen demand. Organic substances, such as carbohydrates, proteins, etc., become oxidised by the activities of bacteria. The rate at which they are broken down therefore depends first on the presence of suitable bacteria and second on how satisfactory and balanced a food they are for microorganisms. Compounds which are more or less refractory are decomposed at a low or very low rate. They will have a ratio of BOD_5 to theoretical oxygen demand which is $<<$ 1.0 (see also the discussion in Section 4.6). They will therefore not make a significant contribution to the BOD_5. These compounds are more or less toxic to the aquatic flora and fauna, and Section 5.5 deals with methods of removing toxic organics, which obviously cannot be removed by the same methods as biodegradable organic matter. Readily biodegradable matter has a biological oxygen demand which can be calculated theoretically by using the reaction scheme. Waste water containing 100 mg l^{-1} of glucose will have a BOD_5 of 110 mg l^{-1}, which is in accordance with the following reaction:

$$C_6H_{12}O_6 + 6O_2 \text{ --------> } 6CO_2 + 6H_2O \qquad (5.2)$$

Glucose is easily broken down to CO_2 and H_2O, oxidation being complete in less than 5 days. However, other components readily oxidised but at a lower rate will not give a BOD_5 in accordance with the reaction of decomposition because a part of their mass is synthesised into new bacterial substances, which will not be broken down during the 5-day period. For a more complex mixture of organic compounds the decomposition reaction might be written as follows:

$$COHN + O_2 \xrightarrow{\text{(cells)}} CO_2 + H_2O + NH_3 + \text{cells} \qquad (5.3)$$

The theoretical oxygen demand may be found by determination of COD for most chemicals, although there are exceptions (see Baker et al., 1999). As mentioned in Section 5.1 the major sources of organic residues are domestic sewage and the food industry. Table 5.4 gives the range of concentration of pertinent characteristics of domestic waste water in the case of separate sewerage.

TABLE 5.4

Concentration of pertinent characteristic of domestic waste water (mg l^{-1})

BOD_5	150- 300
N_{total}	25- 453
P_{total}	6- 12

The consumption of water in the food industry is is very high, although its has been reduced during the last two decades considerably due to higher water costs. In addition, the the concentrations of biodegradable organic matter (for many food industries in the range 1000 - 2000 mg/l) and nutrients are several times the corresponding concentrations in untreated municipal waste water. The loading of waste discharge from the industries is often expressed in person equivalents. 1 person equivalent corresponded previously to 60 g BOD_5/24h per inhabitant, but the many industrialised countries where the cost of water has increased during the last decade, the water consumption has been reduced without a corresponding increase in the BOD-concentration. Therefore, it is probably more correct today to consider 45- 50 g BOD_5/24h per inhabitant as 1 person equivalent. When the actual cost of waste water treatment is charged to the food industry, the cost of discharging untreated waste water to a municipal waste water treatment plant may be as high as several $ / m^3. In Denmark where the polluter must pay principle is applied, untreated abattoir waste water will be charged 3-5 $/$m^3$ or 3-5 times the normal price for domestic waste water. Consequently, it is profitable for many food industries to make a preliminary treatment for instance a chemical precipitation or a biological treatment which will reduce the discharge costs considerably.

The composition of domestic sewage varies surprisingly little from place to place, although, to a certain extent, it reflects the economic status of the society. A typical organic composition of domestic waste water is given in Table 5.5. A more detailed analysis reveals that domestic sewage is a very well balanced food for microorganisms. It contains sufficient amounts of essential amino acids, nutrients and vitamins. Consequently its biological decomposition causes few problems; most difficulties with biological treatment plants are related to the discharge of more or less uncontrolled amounts of industrial waste water.

Reduction of BOD_5 in domestic waste water and other types of waste with a similar composition is carried out by a combination of mechanical and biological treatment methods. **Mechanical methods remove suspended matter in one or more steps, with the aim of not only removing the coarser inorganic particles, but also reducing the BOD_5 in accordance with the amount of suspended organic matter.** In principle, the entire BOD_5-reduction could be carried out using biological treatment methods alone, but this will often prove more expensive than a combined mechanical and biological treatment. The results of the combined treatment are outlined later in this section.

TABLE 5.5
Organic composition of domestic waste water

Organic constituent	Concentration		
	Soluble (mg/l)	Particulate (mg/l)	Total (mg/l)
Total Carbohydrate	30.5	13.5	44.0
Free Amino Acids	3.5	0	3.5
Bound Amino Acids	7	21.25	28.25
Higher Fatty Acids	0	72.5	72.5
Soluble Acids	22.75	5	27.75
Esters	0	32.7	32.7
Anionic Surfactants	11.5	4	15.5
Amino Sugar	0	0.7	0.7
Amide	0	1.3	1.35
Creatinin	3.	0	3.1
Fraction sum	78.35	151	229.35
Present in Waste Water	94	211.5	305.5

<u>Mechanical treatment methods.</u>

Mechanical treatment methods comprise screening, sand trap, sedimentation, filtration and flotation.

Screening. Screens are used for removing larger particles, such as branches, paper, etc. They are made from iron bars or gratings and can be classified according to the distance between the bars as coarse (40-100 mm) and fine (10-40 mm). Screens need to be cleaned frequently, either manually or mechanically by rakes. The purpose of the screens is to protect pumps and other mechanical equipment. The removed material is either broken down and added to the waste water or disposed of by controlled composting or by incineration.

Sand traps. The sand trap or grit chamber removes mainly inorganic particles 0.1-3 mm in size. It operates by sedimentation, but due to a short retention time (10-20 minutes) and air stirring, the finer organic particles are prevented from settling. Therefore only inorganic sand particles are removed which is beneficial as they don't need to be treated as sludge, but can be used directly for construction.

Sedimentation. Sedimentation is used to remove suspended solids from waste water. In principle, it is the same process as is used in nature, see section 2.7.

A settling tank has three main functions:

1. It must provide for effective removal of suspended solids so that its effluent is clear.
2. It must collect and discharge the subnatant stream of sludge.
3. It must thicken the sludge to a certain concentration of solid.

Three distinct types of sedimentation may be considered:

1. Discrete settling. This is the settling of a dilute suspension of particles which have little or no tendency to flocculate.
2. Flocculent settling, which occurs when the settling velocity of the particles increases as they fall to the bottom of the tank, due to coalescence with other particles.
3. Zone settling, which happens when inter particle forces are able to hold the particles in a fixed position relative to each other. In this case the particles sink as a large mass rather than as discrete particles.

Plug flow is never achieved in practice. Some of the particles will be short circuited and will therefore be held in the tank for a time less than V/Q, where V = the tank volume (m^3), and Q = the flow rate (m^3/h). **Wind effects, hydraulic disturbances and density and temperature effects** will all result in deviation from the ideal plug flow. Short circuiting in a tank can be characterised by tracer techniques. Dye, salt or radioactive materials are introduced into the inlet and the concentration distribution in the effluent stream indicates the flow patterns.

It is frequently possible to improve the performance in an existing settling tank by making modifications based on the results of a dispersion test. The addition of stream-deflecting baffles, an inflow dividing mechanism and velocity dispersion feed wells may decrease short circuiting and increase efficiency.

The design of a tube settler incorporates the use of very small diameter tubes in an attempt to apply the shallow depth principle as suggested by Camp (1946). Flow through tubes with a diameter of 5-10 cm offers optimum hydraulic conditions and maximum hydraulic stability. Culp et al. (1968) have reported excellent results using tube settlers with a retention time of less than 10-15 minutes. In practice the retention time is usually slightly longer (25-25 minutes). A tube settler costs more than a common settling tank of conrete, but the shorter settling time means that the volume and the required area are much less. It has therefore found the widest application in industries where space may be of importance.

The usual clarifier may be designed as a rectangular or circular tank, and may utilise either centre or peripheral feed. The tank can be designed for centre sludge withdrawal or for withdrawal over the entire tank bottom. The different types of tank are shown in Figs. 5.2 and 5.3 . The first one is designed for small flow rates, where the height of the tank is only moderate in spite of the angle of the cone. The second clarifier is made of concrete and is able to deal with a considerably larger flow rate. An inlet device is designed to distribute the flow across the width and the depth of the settling tank, and correspondingly an outlet device is designed to collect the effluent uniformly at the outlet end of the tank.

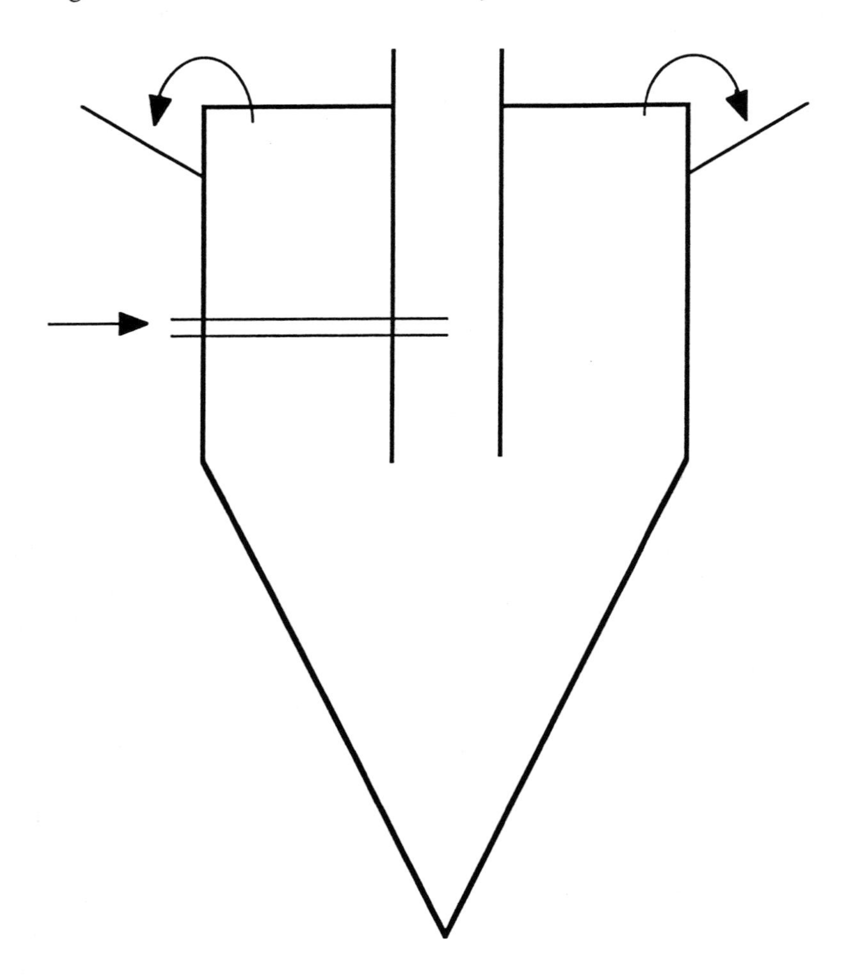

Fig. 5.2. Settling tank for small flow rates.

It is very difficult to design a full-scale sedimentation tank based on settling experiments, as several important factors influencing particle behaviour in a full-scale operation are neglected in such experiments. Tanks are subject to eddies, currents, wind action, resuspension of sludge, etc. A full-scale clarifier will therefore show a slightly reduced efficiency compared to settling experiments, but this can be taken into consideration by incorporating a safety factor. The choice of an acceptable safety factor requires experience. The practical factor might vary from 1.5 when the tank is very small, baffled and protected from wind, to 3.0 in the case of a large tank, unbaffled and unprotected from wind. Even with the use of the safety factor, however, perfect performance should not be expected.

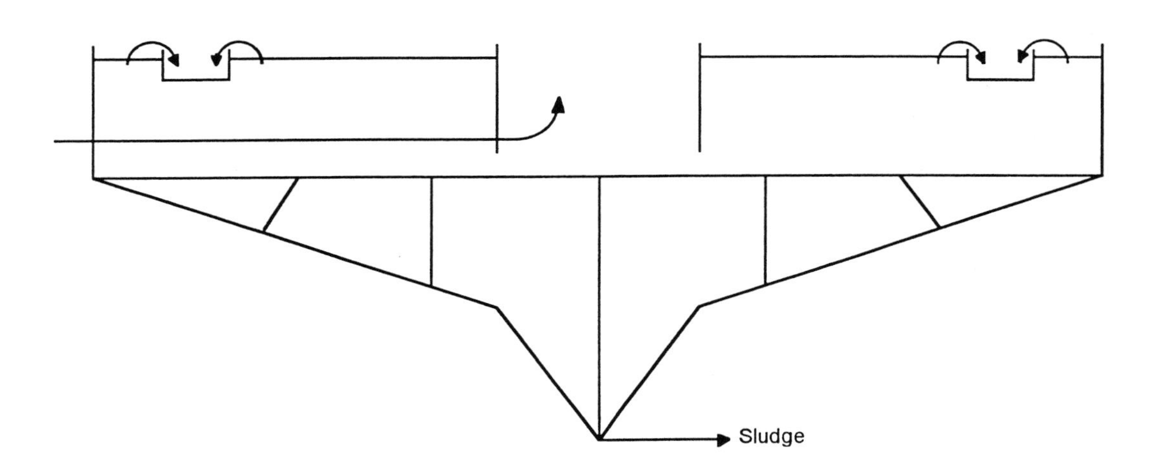

Fig. 5.3. Settling tank for large flow rates.

Filtration. Water treatment by filtration uses principally either deep granular filters or precoat filters. Deep granular filters are either silica sand or a dual medium or multi-media filters. A dual medium filter of coal over sand is widely used, and multi-media filters consisting, for example, of coal over silica sand over garnet sand are finding increasing application. Stevenson (1997) has recently developed an easy-to-use model to optimise the filtration rate though granular media. The model includes the effect of particle size which makes it useful for an optimisation of the granular filter design.

Precoat filters use diatomaceous earth, perlite or powdered activated carbon. Sand filters were developed in England in the middle of the 19th century. These filters operated at a relatively low rate, between 0.1 and 0.3 m/h. Today the same filters are used at rates of up to 0.6 m/h, and are known as slow sand filters in contrast with the rapid sand filters developed later in the 19th century in the U.S.A., which operate with a filtration velocity of 3 to 10 m/h.

The present filters, which consist of a number of porous septa in a filter housing, have found wide application since the Second World War. The septa support is a thin-layer filter medium, which is deposited on the outside of the septa at the beginning of the filtration cycle. As mentioned above, sand filters can be divided into two classes - slow filters and rapid filters. There are two main differences between them:

1. As shown in Table 5.6, the properties of the filter media are different. The effective grain size is the diameter of the largest grain of sand in that 10% of the sample by weight which contains the smallest grains. The uniformity coefficient is the ratio of the largest grain in the 60% of the sample by weight which contains the smallest grain, to the effective size.

As can be seen rapid filters operate with a higher effective size and a smaller uniformity coefficient. The finer the sand used, the smaller will be the turbidity of the treated water and the flow rate.

TABLE 5.6
Typical properties of filter media

	Slow sand filter	Rapid sand filter
Effective size (mm)	0.45-0.60	0.6-1.0
Uniformity coefficient	1.50-1.80	1.2-1.8
Material	sand and/or cru-shed anthracite	multi media

2. Slow filters operate for 10 to 30 days. By then the head loss will be 1 m of water or more. The filtration is interrupted and 1.5 to 4 cm of the filter sand is removed. When the sand layer reaches a height of about 40 cm, new or washed sand is added to replace up to 30 cm of the sand layer removed. In rapid filtration, impurities are removed by back-washing, usually by reversing the flow of water through the filter at a rate adequate to lift the grains of the filter medium into suspension. The deposited material thus flushed up through the expanded bed is washed out of the filter.

The rapid filter can be either an open filter or a pressure filter. Open filters are mainly built of concrete, whereas pressure filters are water-tight steel tanks which are usually cylindrical and may stand either horizontally or vertically. The most common use of pressure filters is in small cities treating ground water supplies for iron and manganese removal, in swimming pool filtration or for polishing industrial water.

The filtration cycle by use of precoat filtration consists of three steps:

1. Precoating
2. Filtration
3. Removal of the spent filter cake

A precoat thickness of 1.3-3 mm is generally used. During filtration the suspended solids are removed on the precoat surface resulting in an increasing pressure drop across the filter. Due to the hydraulic compression of the solid, the filtration cycle may be very short unless additional filter aid is used during filtration. The amount required varies with the type and concentration of suspended solids in the treated water. A typical pressure filter flow is shown in Fig. 5.4.

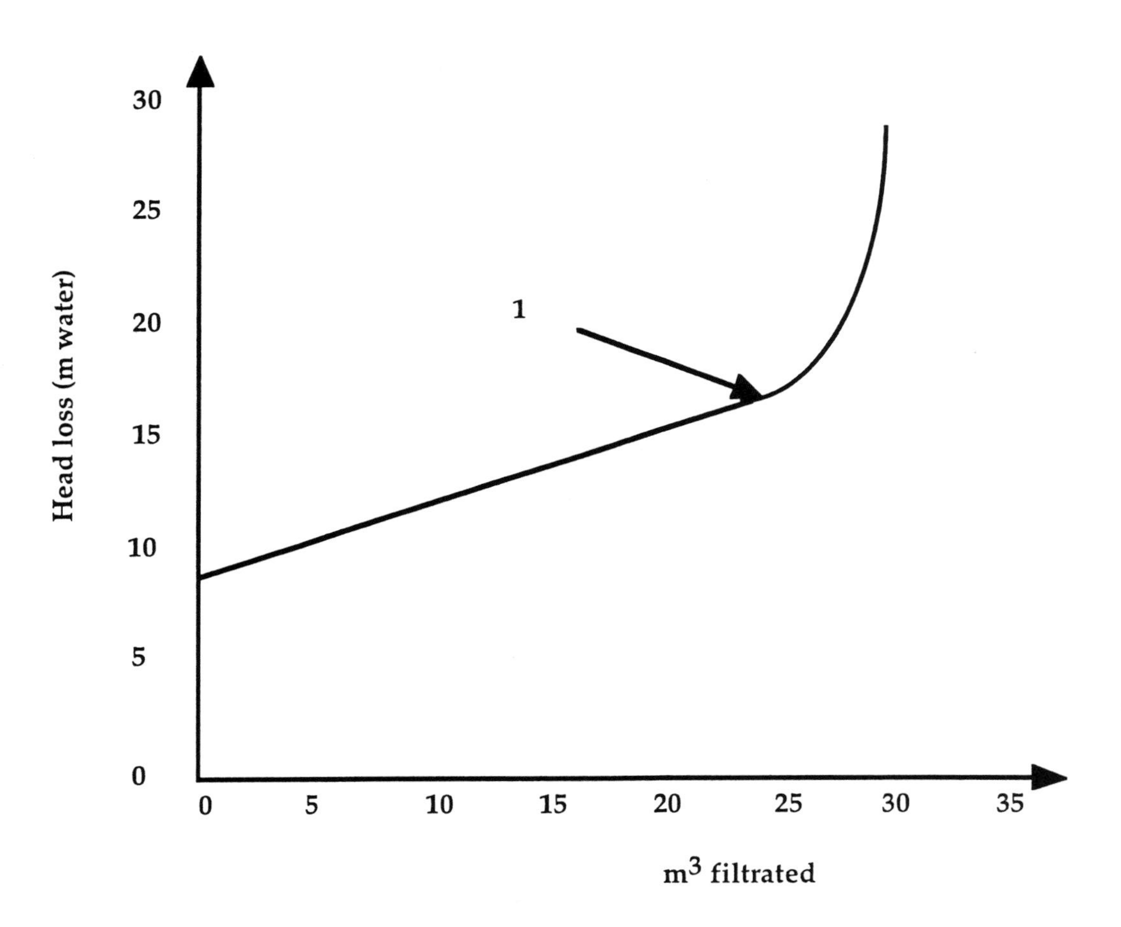

Fig. 5.4. Head loss plotted against volume for a precoat filter. Filtration should be interrupted at (1). The head loss at time = 0 which is $0 m^3$ filtered corresponds to head loss of the filter + precoat.

Flotation. Flotation is used to remove suspended solids from waste water and to concentrate sludge. Thus flotation offers an alternative to sedimentation, especially when the waste water contains fat and oils.

Either a portion of the waste water or the clarified effluent is pressurised at 3-10 atm. When the pressurised water is returned to normal atmospheric pressure in a flotation unit air bubbles are created. The air bubbles attach themselves to particles and the air-solute mixture rises to the surface, where it can be skimmed off, while the clarified liquid is removed from the bottom of the flotation tank.

Fig. 5.5 shows a flotation system with partial recirculation of the effluent. Generally it is necessary to estimate the flotation characteristics of the waste water by use of a laboratory flotation cell:

1. The rise of the sludge interface must be measured as a function of time.
2. The retention time must be varied and the corresponding saturation of pressurised water determined.
3. The effluent quality must be determined as a function of the air/solids ratio.

Based on such results it is possible to scale up.

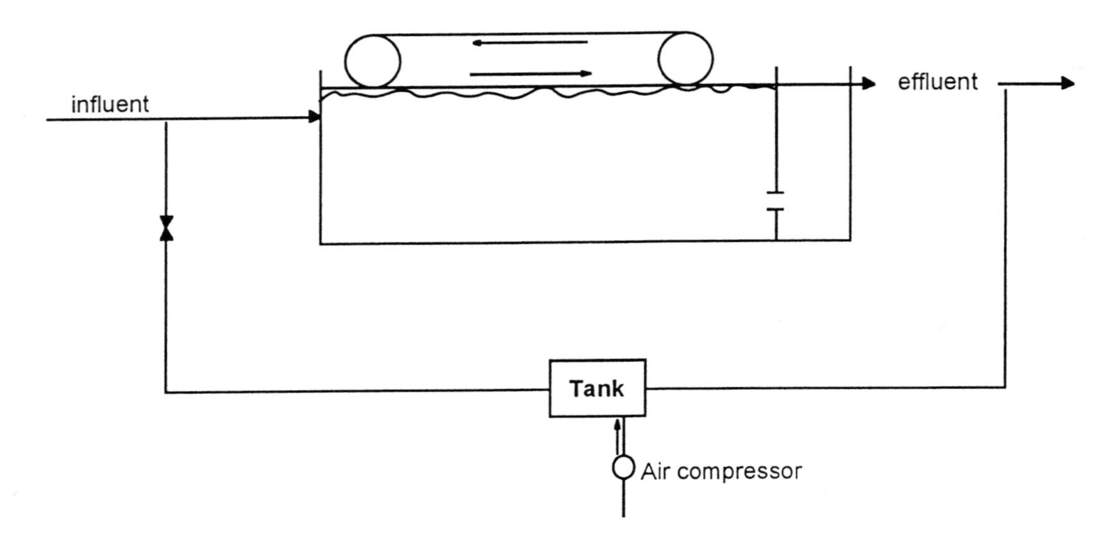

Fig. 5.5. Flotation unit.

Biological treatment processes.

Many types of biological processes are active in the breakdown of organic matter. A nutritional classification of organisms is given in Table 5.7.

Most biological systems used to treat organic waste depend upon heterotrophic organisms, which use organic carbon as their energy source. As seen in the table, the organisms can be either strictly aerobic, strictly anaerobic or facultative anaerobic. Anaerobic breakdown is used in the treatment of sludge, or denitrification where nitrate is the oxygen source. Chemolithotrophic organisms are also used in biological treatment processes. These comprise specialised groups of bacteria which are able to oxidise inorganic compounds such as those of hydrogen, sulphur or ammonia. **Of the various types of metabolism in which the redox reaction provides the ultimate source of energy, there are three major classes of energy-yielding processes:**

TABLE 5.7

Nutritional classification of organisms

Class	Nutritional requirements
Autotrophic	The organisms depend entirely on inorganic compounds
Heterotrophic	Organic compounds are required as nutrient
Phototrophic	Use radiant energy for growth
Chemotrophic	Use dark redox reaction as energy source
Lithotrophic	Use inorganic electron donors (e.g. hydrogen gas, ammonium ions, hydrogen sulphate and sulphur)
Organotrophic	Require organic compounds as electron donors
Strictly aerobic	Cannot grow without molecular oxygen, which is used as oxidant
Strictly anaerobic	Use compounds other than oxygen for chemical oxidation. Sensitive to the presence of minor traces of molecular oxygen
Facultative anaerobic	Can grow either in the presence or absence of air.

Fermentation, in which organic compounds serve as the final electron acceptors;
Respiration (aerobic), in which molecular oxygen is the ultimate electron acceptor;
Respiration (anaerobic), in which inorganic compounds - not oxygen - are the ultimate electron acceptors.
These reactions can be described by the following overall process:

$$\text{Organic matter} + O_2 + \text{cells} \text{ -----> } CO_2 + NH_3 + H_2O + \text{new cells} \quad (5.4)$$

Nitrification results from a two-step oxidation process. First, ammonia is oxidised to nitrite by Nitrosomonas. Second nitrite is oxidised to nitrate by Nitrobacter:

$$2NH_4^+ + 3O_2 \text{ -----> } 2NO_2^- + 2H_2O + 4H^+ \quad (5.5)$$

$$2NO_2^- + O_2 \text{ ------> } 2NO_3^- \quad (5.6)$$

Respiration and nitrification are the same processes as those used in nature to oxidise organic matter and ammonia. The processes are of importance for the oxygen balance of streams, as presented in section 2.7. Nitrate can be used as an oxygen source for the biological decomposition of organic matter. The reaction - called denitrification - is:

$$2NO_3^- + H_2O \longrightarrow N_2 + 2OH^- + 5O \qquad (5.7)$$

In comparison with molecular oxygen supplied by the aeration method, the use of nitrate as an oxygen source is undoubtedly easier because of its extremely high solubility. Further, it can be expected that satisfactory biodegradation of organic matter may be carried out with microorganisms and waste water containing nitrate. Industrial waste water, especially from petrochemical plants, sometimes contains a large amount of nitrate as well as highly concentrated organic matter. The application of a biological treatment method for treating such waste water using nitrate as the oxygen source is therefore attractive. Studies by Miyaji and Cato (1975) have shown that the amount of BOD_5 removed by biological treatment with nitrate as an oxygen source is linearly related to the amount of nitrate removed in the reaction tank. Studies have shown that one of the microorganism involved is Pseudomonas denitrificans. **Cellular growth can often be described as a first-order reaction:**

$$\frac{dX}{dt} = \mu_m * X \qquad (5.8)$$

where

X = concentration of volatile biological solid matter

μ_m = the maximum growth rate

t = time

Integration of this equation where $X = X_0$ and $t = 0$, gives:

$$\ln X / X_o = \mu_m * t \qquad (5.9)$$

This equation is only valid during the so-called logarithmic growth phase in which the substrate (the organic matter) is unlimited. μ_m is the maximum growth rate which anticipates that the substrate is not limiting. The relationship between the growth rate and the substrate concentration can be expressed by the Michaelis-Menten equation (see equation (2.17)):

$$\mu = \mu_m \frac{S}{S + K_S} \qquad (5.10)$$

where

μ = growth rate ($= \frac{dX * 1}{dt * X}$)

μ_m = maximum growth rate
S = substrate concentration
K_S = Michaelis-Menten constant

As can be seen, when S ----> ∞ the equation becomes (5.8) - the growth rate is independent of the substrate concentration and when $K_S >> S$ the equation is transformed into $dX/dt = k' S$ - the growth rate increases linearly with S. Fig. 2.9 shows the Michaelis-Menten relationship; the growth rate is plotted against the substrate concentration. Three levels of an additional important substrate are shown on this figure. It is often convenient to illustrate the Michaelis-Menten equation by means of a Lineweaver-Burk plot. The reciprocals of the growth rate and of the substrate concentration are plotted against each other (see Fig. 2.11). The relationship is linear as can be seen from equation (5.10) which can be transformed to:

$$\frac{1}{\mu} = \frac{K_S}{\mu_m} * \frac{1}{S} + \frac{1}{\mu_m} \quad (5.11)$$

Equation 6.8 is incomplete without an expression to account for depletion of biomass through endogenous decay (respiration). A first-order expression could be used:

$$\frac{dX}{dt}(\text{end}) = -k_d * X \quad (5.12)$$

Incorporation of endogenous decay results in:

$$\frac{dX}{dt} = \mu_m \frac{S}{K_S + S} * X - k_d * X \quad (5.13)$$

By use of the yield constant, a (a = mg biomass produced per mg of substrate used), dX can be expressed in terms of substrate removal:

$$dX = -a * dS \quad (5.14)$$

Combining equations (5.10) and (5.14) gives:

$$\frac{-dS}{dt} = \frac{\mu_m}{a} * \frac{S}{S + K_S} * X \quad (5.15)$$

The set of equations is not valid for complex substrate mixtures, but in many cases the equations can be used as good approximations using BOD as a measure of the substrate concentration.

Temperature influences these processes significantly. The effect of temperature on the reaction rate is listed in Table 2.7. The biological processes in a treatment plant has as seen here a k value of 1.02 - 1.09. A more specific k value for various forms of biological treatments which are presented below, can be obtained from Table 5.8

TABLE 5.8

Temperature effects on biological processes

Process	k
Activated sludge (low loading)	1.00 -1.01
Activated sludge (high loading)	1.02 -1.03
Trickling filter	1.035
Lagoons	1.05 -1.07
Nitrification	1.08- 1.12

Example 5.1

The BOD_5 of a waste water sample is determined to be 150 mg l^{-1} at 20°C. Find BOD_7 (which is used in Sweden to indicate biological oxygen demand) at 15°C. As natural conditions are simulated, the constants from section 2.4 are used not those valid for biological treatment methods.

Solution

Since it is waste water, $K_1 = 0.35$ day^{-1} and $K_T = 1.05$ (range 1.02- 1.09) are used.
The following equation is used :

$$L_t = L_0 * e^{-k_1 * t}$$

150 mg l^{-1} corresponds to $L_0 - L_5 = L_0 (1-e^{-0.35*5})$, which gives:

$$L_0 = \frac{150}{1 - e^{-0.35*5}} = 181 \text{ mg } l^{-1}$$

$K_{15} = (1.05^{15-20}) * 0.35 = 0.27$
At 15°C:
$BOD_7 = L_0 - L_7 = L_0(1 - e^{-0.27*7}) = 154 \text{ mg l}^{-1}$

The various biological treatment processes can be summarised as follows:

1. The conventional **activated-sludge process** is defined as a system in which flocculated biological growth is continuously circulated and put in contact with organic waste water in the presence of oxygen, which is usually supplied in the form of air bubbles injected into the liquid sludge mixture. The process involves an aeration step followed by sedimentation. The separated sludge is partly recycled back to be removed with the waste water. The following processes occur:

a. Rapid adsorption and flocculation of suspended organics,
b. Oxidation and decomposition of adsorbed organics, and
c. Oxidation and dispersion of sludge particles.

Sometimes, depending on the retention time and amount of oxygen introduced, ammonium ions are oxidised to nitrate by nitrifying organisms. This is seen particularly during the summer, and is due to the influence of the temperature on the rate constant for the nitrification process; see above and Table 2.7. An activated sludge which is not overloaded usually provides an effluent with a soluble BOD_5 of 10-20 mg/l. The process necessitates the treatment of excess sludge before disposal. A high rate activated sludge treatment may in some cases be an attractive solution to pretreat waste water with high concentrations of biodegradable organics, for instance brewery waste water; see Bloor et al., 1995.

2. **The extended aeration** process works on the basis of providing sufficient aeration time for oxidising the biodegradable portion of the sludge produced form the organics removed from the process. Fig. 5.6 shows the process schematically. The excess sludge in the process contains only non-biodegradable residue remaining after total oxidation. The total BOD_5 provided by this process is 20 mg/l or less.

Oxidation ditches have been developed as self-sufficient structures for the extended aeration of waste waters from small communities. Fig. 5.7 demonstrates the principles of an oxidation ditch. The retention time may be from 1 to 3 days with the design of the aerotor being significant for the over-all efficiency of the treatment (Dudley, 1995). No effluent is withdrawn until the water level in the channel has built up to the highest operation level. The influent is then cut off and the rotor stopped. Solids are allowed to settle for an hour or two, then the clarified supernatant is withdrawn through an effluent launder, and, if desired, excess sludge is lifted from a section of the ditch to drying beds. The effluent is then cut off and the operation routine is repeated. Because the solids are well stabilised during the long aeration time, they are no longer putrescent and water is readily removed from them. The sludge can, if not contaminated by toxic substances (compare with the standards presented in Table 4.6) be used as soil conditioner.

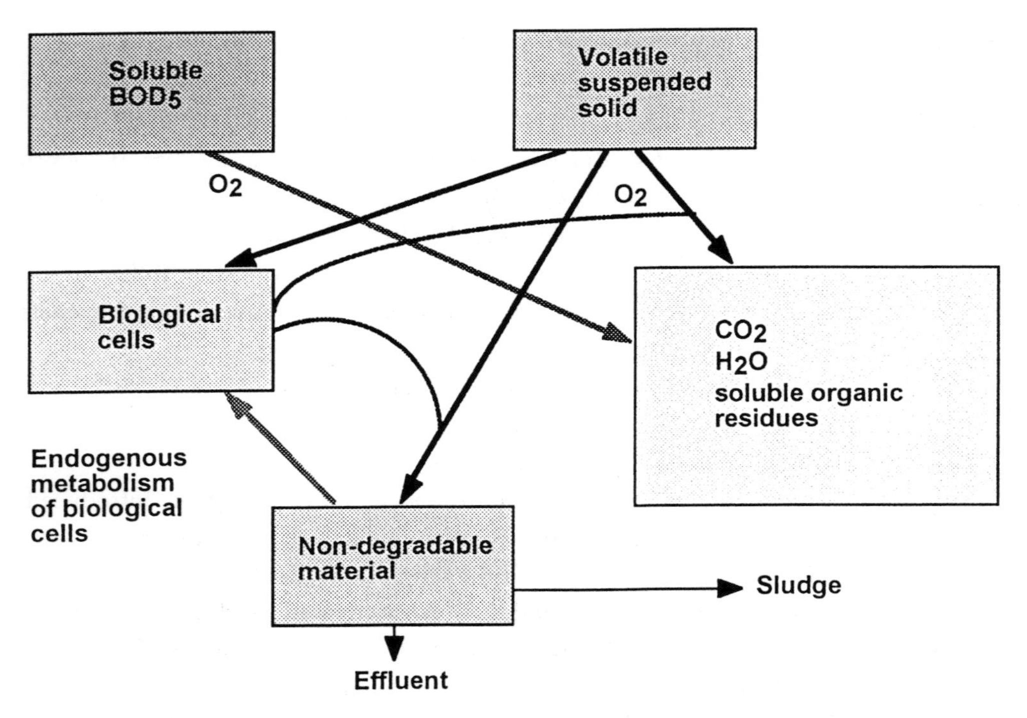

Fig. 5.6. Extended aeration process.

3. In **the contact stabilisation process** the waste water is aerated with stabilised sludge for a short period of 1/2 - 1 hour. The mixed liquid is then separated by sedimentation and when settled the sludge is transferred to a sludge stabilisation tank where aeration is continued to complete the oxidation. This process is used to advantage when a high percentage of BOD is removed rapidly by bioadsorption after contact with the stabilised sludge. The extent of removal depends on the characteristics of the sludge and of the waste water. As general rule the process should give an efficiency of 85% BOD_5 removal.

4. A **trickling filter** is a bed packed with rocks, although, more recently, plastic media celite pellets (Sorial et al., 1998) or bio-blocks (blocks with a large surface area due to the high porosity) have been used. The medium is covered with a slimy microbiological film. The waste water is passed through the bed, and oxygen and organic matter diffuse into the film where oxidation occurs. In many cases recirculation of the effluent improves the BOD removal, especially when the BOD of the effluent is relatively high. A high-rate trickling filter provides an 85% reduction of BOD for domestic sewage, but 50-60% is the general figure for BOD_5 reduction in the treatment of organic industrial waste water. Good results have, however, been reported for treatment of dairy waste water (Jørgensen, 1981) and styrene containing waste water (Sorial et al., 1998)

A plastic-packed trickling filter will require substantially less space than a stone-packed one due to its bigger depth and specific surface area. According to Wing et al. (1970), plastic media packed to a depth of 6.5 m in a trickling filter will require less than one-fifth of the land required by those packed with stones to the usual depth of 2-4 m. The specific surface area of rock-trickling filters is 40-70 m^2/m^3 and the void space is 40-60%, while plastic filters or bio-blocks have a specific surface of 80-120 m^2/m^3 or even more, with a void space of 94-97%.

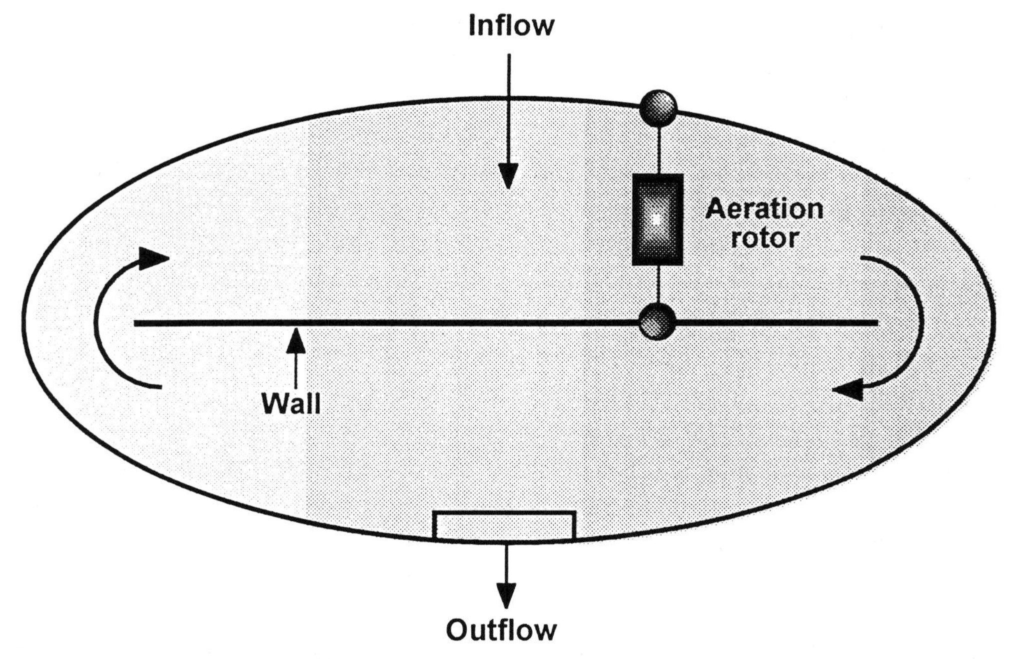

Fig. 5.7 Oxidation ditch.

5. **Lagoons** are commonly applied methods of organic waste treatment, when sufficient area is available. They can be divided into four classes:

a. *Aerobic algal ponds,* which depend upon algae to provide sufficient oxygen.

b. *Facultative ponds,* which have an aerobic surface and an anaerobic bottom.

c. *Anaerobic ponds,* which are loaded to such an extent that anaerobic conditions exist throughout the liquid volume.

d. *Aerated lagoons,* which are basins where oxidation is accomplished by mechanical or diffused aeration units and induced surface aeration. The turbulence is usually insufficient to maintain solids in suspension, thus most inert solids settle to the bottom where they undergo anaerobic decomposition. The basin (2-4 m deep) may include a sedimentation compartment to yield a more clarified effluent. If the turbulence level in the basin is increased to maintain solids in suspension, the system becomes analogous to an activated-sludge system.

An aerated lagoon can provide an effluent with less than 50 mg/l BOD_5, depending on the temperature and the characteristics of the waste water. Post-treatment is necessary when a highly clarified effluent is desired, and large areas are required compared with the activated sludge process.

Lagoons without aeration are also named waste stabilisation ponds (WSP). They are mainly used in developing countries today due to their moderate cost. They are soft technological solutions to waste water problems and may therefore be considered as ecotechnology. They may be combined with wetlands to provide a proper solution to waste water problems in developing countries. This combination of WSP and a wetland will be discussed in Chapter 8.

A pond can be designed in accordance with the mass-balance equations given in section 2.3. If several ponds are arranged in series, equation (2.11) is valid. An example will be used to illustrate the application of mass balances to design of lagoons.

6. **Fluidised bed bioreactor (FBBR)** is a novel waste water treatment process which was introduced in the late seventies. The basis is that most organic matter adsorbed by activated carbon can be degraded by biological treatment processes. A bed containing activated carbon is fluidised by application of a suitable waste water flow. The large surface of the porous medium has an active biofilm which yields a high efficiency per unit volume. This method demonstrates by use of a proper design a high efficiency measured as the ratio BOD_5 removed / BOD_5 of untreated waste water, also for relatively slowly biodegraded organic compounds. It shows a good ability to absorb fluctuations in the in-flowing water quality and to offer a stable water quality.

Example 5.2

A waste water flow of 500 m^3/d from a small community is treated by use of 4 ponds in series. How large must the ponds be to remove 90% BOD_5 when the average BOD_5 is 200 mg l^{-1} and the temperature is 18°C?

Solution

K_1 (20°C) = 0.35 day^{-1} (see Table 2.10)

O = 1.06 (see aerated lagoons, Table 5.8)

$$K_T = K_{18°} = 0.35 * 1.06^{-2} = 0.31$$

$$\frac{C_{im}}{C_{io}} = \frac{1}{10} = \left(\frac{1}{1 + 0.31 * tr}\right)^4$$

$$(1 + 0.31 * tr)^4 = 10$$

1 + 0.31 * tr = 1.78

tr = 2.52

Each of the 4 ponds should have a volume of 2.52 * 500 = 1258 m^3.

If a depth of 1.5 m is used, the area will be 839 m^2 * 4 = 3356 m^2.

The oxygen must most probably be supplied by aeration, i.e. a minimum of 180 * 5000 = 900 kg/day or, if an efficiency of only 50% is assumed, 1800 kg/day.

7. **Anaerobic digestion.** The anaerobic breakdown of organic matter to harmless end-products is very complicated. Fig. 5.8 summarises some of the more general processes.

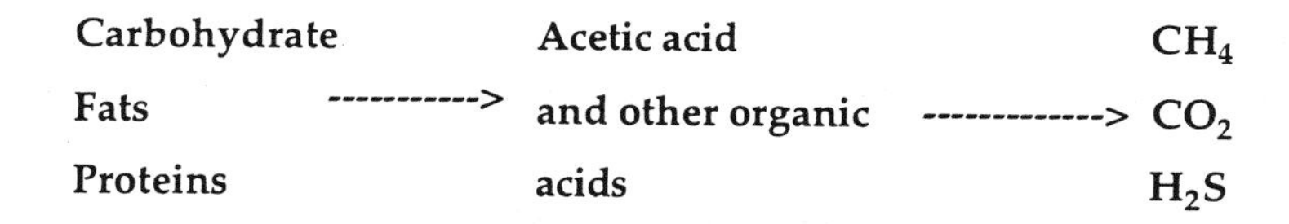

Fig. 5. 8 Anaerobic degradation of organics.

Methane-producing organisms convert long-chain volatile acids to methane, carbon dioxide and other volatile acids with a short carbon chain, which are then fermented in a similar fashion. Acetic acid is directly converted into carbon dioxide and methane. The rate of methane fermentation controls the overall reaction rate. Sufficient time must be available in the reactor to permit growth of the organisms or they will be washed out of the system. This means that the retention time must be greater than that corresponding to the growth rate of the methane-producing organisms. It is possible, by the use of extracellular enzymes, to cut down the resistance time considerably, but the use of such enzymatic processes is still only in its infancy.

Many factors, such as the composition of the sludge and the waste water. pH and temperature, influence the reaction rate, but it is generally shown that the overall rate is controlled by the rate of conversion of volatile acids to methane and carbon dioxide. Digestion fails to occur when there is an imbalance in the rate of the successive processes, which might result in a build-up of volatile acids. The optimum conditions can be summarised as follows: pH 6.8-7.4; redox potential -510 to -540 mV; concentration of volatile acids 50-500 mg/l; alkalinity (as calcium carbonate) 1500-5000 mg/l; temperature 35-40°C. It should be possible to obtain effective digestion with a retention period as low as 5 days, but increasing the retention time to 10 days should assure 90% degradation of organic matter. Anaerobic digestion is used for the treatment of sludge from biological processes as well as for the treatment of industrial waste water with an extremely high BOD_5, e.g. industrial waste water from the manufacture of yeast or sheep tallow.

The major part of the gas produced by anaerobic treatment processes comes from the breakdown of volatile acids. The gas is composed of methane, carbon dioxide, hydrogen sulphide and hydrogen. The higher the resistance time, the lower the percentage of carbon dioxide and the higher the percentage of methane in the gas produced.

Lawrence and McCarty (1967) have shown that methane gas production, at a good approximation, is 0.4 m^3 gas per kg BOD removed. This value must be considered as the maximum obtained by complete conversion of the solid into methane.

8. **Effective nitrification** occurs when the age of the sludge is greater than the growth rate of the nitrifying microorganisms. FFor further details on this process, see Section 5.3

9. **Nitrate can be reduced to nitrogen** and dinitrogen oxide by many of the heterotrophic bacteria present in activated sludge. For further details, see Section 6.3

The mechanical- biological treatment systems.

Many variations in flow pattern for mechanical-biological systems, are used in the treatment of municipal waste water.

Fig. 5.9 shows a flow diagram of a conventional activated-sludge plant.

In the process sewage is mixed with a portion of returned activated-sludge to facilitate the primary settling (improved flocculation). For typical municipal sewage an aeration time of 4-6 hours must be applied if a BOD_5-removal of 90% or more is required. Table 5.9 gives some typical results.

Trickling filters are generally used at smaller sewage treatment plants (1000 - 10,000 inhabitants). They are classified according to the hydraulic and organic loading used, see Table 5.10.

TABLE 5.9

Typical results obtained from activated-sludge plant (mg/l). Numbers refer to Fig. 5.9

	1	2	3
BOD_5	150 - 300	100 - 180	10 - 15
Suspended matter	50	5 - 10	2 - 8
N_{total}	25 - 45	22 - 40	18 - 32
P_{total}	6 - 12	6 - 12	5 - 10

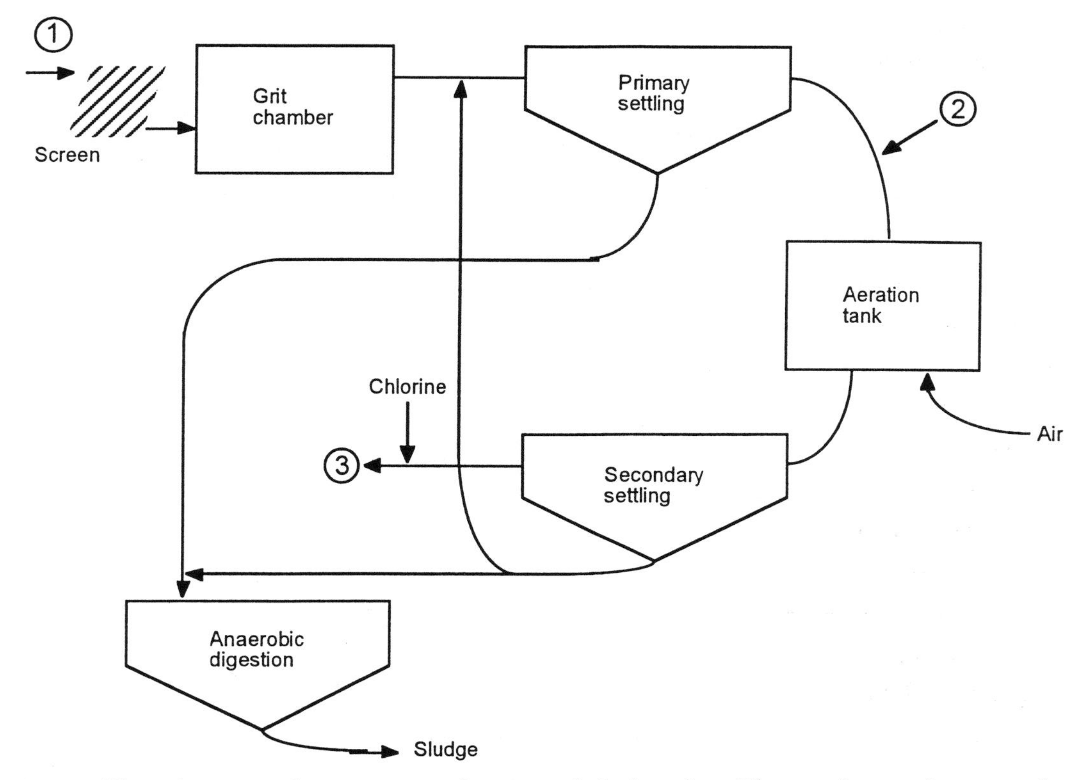

Fig. 5.9 Flow diagram of a conventional activated-sludge plant. The numbers refer to results in Table 5.9

An activated sludge plant is able to meet the standards in most countries for instance in Denmark, where a maximum BOD_5 of 15 mg/l is required, while only a trickling filter at (very) low-rate operation is able to provide an acceptable effluent. Consequently, biological filters (trickling filters) are mostly used today to reduce the BOD_5 from high to lower levels for instance as a preliminary treatment of industrial waste water.

A combination of an activated-sludge plant and a trickling filter is often applied where high quality effluent is required. In addition to 95% or more BOD_5-removal, 80-90% nitrification of ammonia will take place, except perhaps during the winter (the nitrification process is highly dependent on temperature, see Table 5.8). Fig. 5.10 illustrates an example of a sewage plant using such a two-step biological treatment.

Lagoons are only used where considerable space is available, and as, in most cases, they do not give an effluent of sufficient quality by today's standards, this method is now used mainly in developing countries. Table 5.11 summarises design factors and results for lagoons. The final selection of biological treatment method is highly dependent on the receiving water, which determines the acceptable BOD_5 of the effluent, see section 2.7.

TABLE 5.10
Trickling filters

	Low-rate operation	High-rate operation
Hydraulic loading ($m^3/m^2/24h$)	1 - 4	8 - 40
Process loading ($kg/m^3/24h$)	0.2 - 1.5	2 - 20
Depth (m)		
Single stage	1.5 - 3	1 - 3
Multi stage	0.7 - 1.5	0.5 - 1.5
Relative recirculation in term of inflow		0.5 - 10
BOD_5-removal	85 - 90%	75 - 85%
Suspended solid removal	90 - 95%	80 - 90%

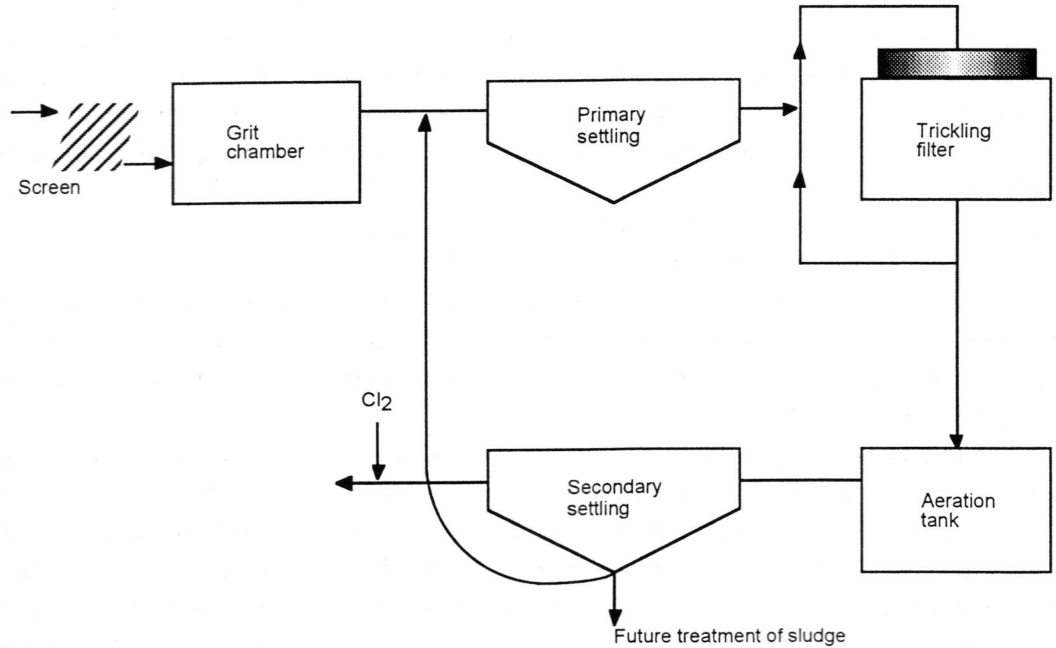

Fig. 5.10 Sewage plant combining trickling filter and activated sludge treatment.

TABLE 5.11
Design factors for lagoons

	Aerobic	Facultative	Anaerobic	Aerated
Depth (m)	0.2 - 0.4	0.75 - 2	2.5 - 4	2 - 4
Retention time (days)	2 - 6	7 - 30	30 - 50	2 - 10
BOD loading (kg/ha/day)	100 - 200	20 - 50	300 - 500	dependent upon waste characteristics and aeration
BOD removal(%)	80 - 90	75 - 85	50 - 70	50 - 90

Other methods used for BOD-removal.

Mechanical-biological systems are widely used for BOD removal, but other methods are also available for the treatment of municipal and industrial waste waters - combinations of physical and chemical methods and irrigation. They are relatively more attractive for the treatment of waste water with high BOD_5, such as waste water from the food industry, but can also be used for municipal waste water. It is difficult to provide guidelines on when physical-chemical methods are more advantageous than mechanical-biological ones, but the following issues should be considered:

1. Physical-chemical methods are not *susceptible to shock loadings* and the presence of toxic compounds.
2. *The space* required for physical-chemical methods is, in most cases, less than for mechanico-biological treatment plants.
3. *Recovery of fat, grease and proteins can be achieved* using physical- chemical methods in the treatment of waste water from the food-processing industry.
4. Although *operation costs are slightly higher* for physical-chemical methods, *investment cost will generally be slightly lower.*

The following physical-chemical methods are used for BOD-reduction: chemical precipitation, ion exchange, adsorption and reverse osmosis.

Apart from obtaining a substantial reduction of the phosphate concentration organic matter is also precipitated by using lime, aluminium sulphate or iron(III) chloride, due to a reduction of the zeta potential of organic flocs (Balmer et al., 1968). Normally, a direct precipitation from municipal waste water will reduce the potassium permanganate number and the BOD_5 by 50-65%, which has to be compared with the effect obtained by plain settling without the addition of chemicals (Davidson and Ullman, 1971). Furthermore, precipitants (lignosulphonic acid, activated bentonite and glucose sulphate) have been developed for precipitation of proteins, permitting their recovery from food-processing industries (see Tønseth and Berridge, 1968 and Jørgensen, 1971). Use of polymer flucculants enhances the settling rate. Lately, it has been proposed to use a mixture of bentonite and a bioflocculant produced by *Anabaenopsis circularis* strains by the presence of calcium chloride (Levy et al, 1992). Generally, it is recommended that one considers all relevant combinations of precipitants and polyflocculants to ensure that a close to optimum solution is selected.

Ion exchangers are able to take up ionic organic compounds, such as polypeptides and amino acids. Macroporous ion exchangers are even able to remove protein molecules from waste water. Cellulose ion exchangers designed for protein removal have been developed (Jørgensen, 1969, 1970 and 1973a). They are inexpensive compared with other ion exchangers and are highly specific in uptake of high molecular weight ions.

Adsorption on activated carbon can remove not only refractory organic compounds (see Section 5.4), but also biodegradable material. This has been used for the treatment of municipal waste water (see below).

Reverse osmosis is an expensive waste water treatment process, which explains why its application has been limited to cases in which other processes are inappropriate or where recovery is possible.

The applications of purely physical-chemical methods for BOD-removal are best illustrated by some examples. *The so-called Guggenheim process uses chemical precipitation with lime and iron sulphate followed by a treatment on zeolite for ammonia removal* (Gleason and Loonam, 1933). 90% removal of BOD_5, phosphate and ammonia is achieved (Culp, 1967a and 1967b).

A combination of chemical precipitation by aluminium sulphate and ion exchange on cellulose ion exchangers and clinoptilolite also seems promising for the treatment of municipal waste water. 90% removal of BOD_5 and ammonia is obtained in addition to 98% removal of phosphate and suspended matter (see Jørgensen, 1973a and 1976a).

The so-called AWT system is *a combination of precipitation with lime and treatment on activated carbon.* Zuckermann and Molof (1970) claim that by using lime the larger organic material is hydrolysed, giving better adsorption because activated carbon prefers molecules with a molecular weight of less than 400. A flowchart of this process is shown in Fig. 5.11.

Fig. 5.12 is a flow diagram of *the combination of chemical precipitation and ion exchange* used in the treatment of waste water from the food industry (Jørgensen, 1968, 1969, 1973a and 1978). This process allows recovery of fat, grease and proteins. Table 5.12 gives the analytical data obtained when this process was used on waste water from herring filetting after centrifugation of the raw waste water to recover fish oil. Table 5.13 gives a before and after analysis of this process for waste water from an abattoir. For comparison the table also includes the results obtained from using a biological plastic filter. The combination methods have generally not found a wide application, as the discharge of industrial waste water to municipal treatment plants seems to be more attractive. Many industries have, however, during the last 10-15 years found it advantageous to use chemical precipitation to reduce the waste water discharge costs. If the polluter must pay principle is used it will be beneficial for industries to reduce the BOD_5 concentration to what is (almost) normal for municipal waste water which is possible (see Table 5.13) by chemical precipitation. Abattoirs in Denmark for example may be able to reduce their discharge fee from about 4 $ to less than 1 $ by introduction of chemical precipitation.

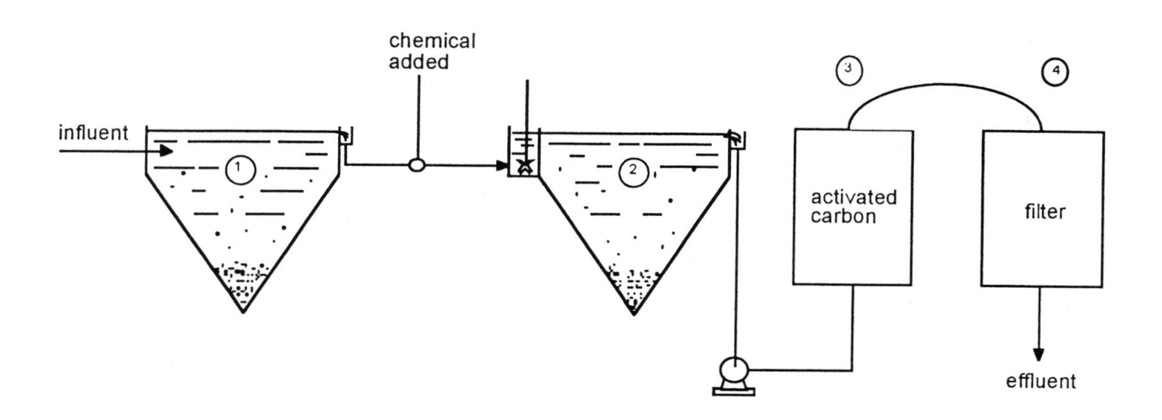

Fig. 5.11. The AWT system consists of 1) mechanical treatment, 2) settling after chemical precipitation, 3) treatment on activated carbon, 4) filtration.

TABLE 5.12

Analytical data of waste water from herring filleting

	Raw waste water 1. step	After centrifugation 2. step	After chem. precipitation 3. step	After cellulose ion-exchanger
BOD_5 (mg/l)	11000	5800	2000	1100
N (mg/l)	180	162	60	23
Suspended matter (mg/l)	400	170	40	2
$KMnO_4$ (mg/l)	8000	4000	1200	600

TABLE 5.13
Analysis of waste water from an abattoir (mg/l)

	Raw water	After biological plastic filter	After chem. precipitation (glucose sulphate is used)	After chem. precipitation and ion exchange
BOD_5	1500	400	600	50
$KMnO_4$	950	350	460	60
Total N	140	42	85	15
NH_3-N	20	15	18	2
NO_3^--N	4	5	4	1
P	45	38	39	1.5

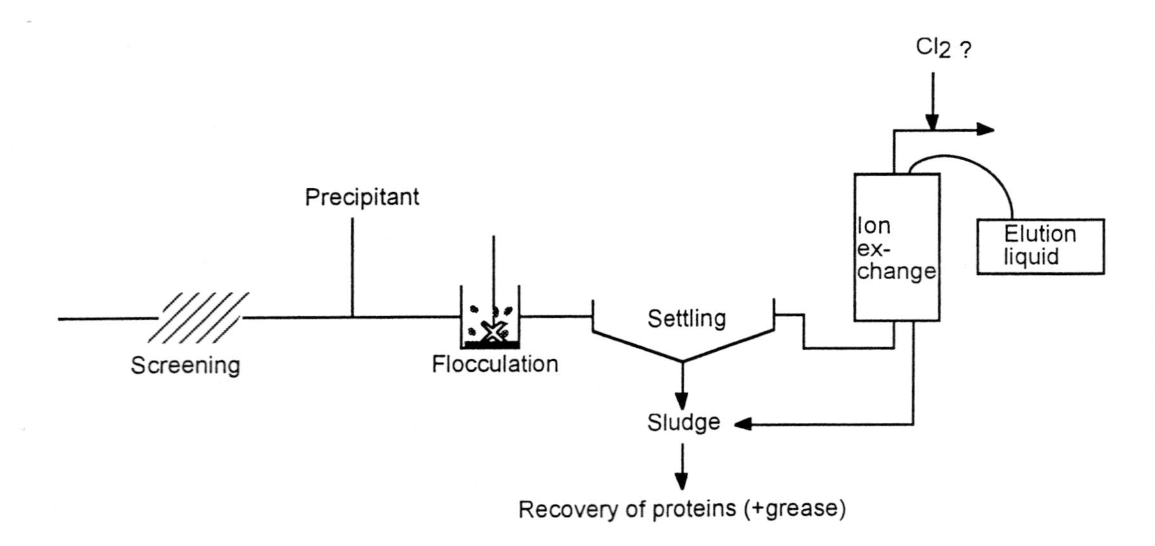

Fig. 5.12. Recovery of proteins (+ grease).

In addition to mechanico-biological treatment or a combination of physical-chemical methods the use of spray irrigation can also be considered as a solution to the problem of the biological oxygen demand of waste water. Spraying and irrigation of waste water onto fields and meadows is widely considered to be the best and cheapest way of treating waste water from the food-processing industry. The industry favours this solution as it ensures maximum utilisation of the fertilising properties of these waste waters (Sanborn, 1953). This treatment method may be considered ecotechnology even it is used on point sources. It is however included in this chapter covering almost entirely environmental technological methods.

As early as 1942 an outline of the agricultural use of waste water from dairies was published. Raw waste water was stored for 5-6 hours in a tank situated on an irrigated area. Irrigation was recommended at a rate of 150-400 mm/year or 75-200 m^3/day of effluent per ha of irrigated fields. The storage tank should be cleaned continuously, by means of compressed air for example. After irrigation the tanks and pipes should be rinsed with fresh water in an amount equal to 1/3 of the daily discharge of waste water.

The method is not usually recommended in areas subject to persistent frosts, or where the soil has a natural high moisture content or a high ground water level (see further below).

Spraying the waste water on pastures carries the risk of affecting cattle with tuberculosis. Therefore a period of 14 days is recommended between irrigation and putting cows out to pasture. The limitations of this process can be summarised as follows (Eckenfelder et al., 1958).

1. **Depth of the ground water.**

The quantity of waste water that can be sprayed onto a given area will be proportional to the depth of the soil through which the waste water must travel to the ground water. A certain soil depth must be available if contamination of the ground water is to be avoided.

2. **Initial moisture content.**

The capacity of the soil to absorb is dependent on its initial moisture content.

Sloping sides will increase the runoff and decrease the quantity of water which can be absorbed by a given area.

3. **Nature of the soil.**

Sandy soil will give a high filtration rate, while clay will pass very little water. A high filtration rate will give insufficient biological degradation of the organic material and too low a filtration rate will reduce the amount of water which can be absorbed by a given area.

The capacity is proportional to the coefficient of permeability of the soil and can be calculated by the following equation:

$$Q = K * N \tag{5.16}$$

where Q = the quantity of water (m^3) which can be absorbed per m^2 per 24h; K = the permeability coefficient, expressed as m/24h; N = the saturation of the soil for which a value of 0.8-1.0 can be used in most cases.

If the soil has different characteristics at different depths an overall coefficient can be calculated:

$$K = \frac{H}{H_1/K_1 + H_2/K_2 + ... H_n/K_n} \quad (5.17)$$

where H = the total depth and H_1, H_2 ...H_n = the depth of the different types of soil with permeability coefficients K_1, K_2, etc.

Table 5.14 gives the permeability coefficients for different soil types (Eckenfelder et al., 1958). In addition to soil properties several characteristics require consideration in a spray irrigation system. Suspended matter must be removed by screening or by sedimentation before the water is sprayed. Otherwise the solids will clog the spray nozzles and may mat the soil surface rendering it impermeable to further percolation (Canham, 1955). The pH must be adjusted as excess acid or alkali may be harmful to crops. High salinity will also impair crop growth. A maximum salinity of 0.15% has been suggested to eliminate this problem.

TABLE 5.14
Permeability coefficients for different soil types

Description	Permeability (m per 24h)
Fine sand	100 - 500
Trace silt	15 - 300
Light agricultural soil	1 - 5
50% clay and 50% organic soils	0.15 - 0.40
Predominating clay soil	< 0.1

Spray irrigation has been successfully used for waste water from dairies, pulp and paper industries (Gellman and Blasser, 1959; Wisneiwski et al., 1956), cannery waste (Luley, 1963; Williamson, 1959) and fruit and vegetable processing plants. The data from various spray irrigation plants are given in Table 5.15

Disposal of industrial waste water by irrigation can be carried out in several ways:

1. Distribution of the water over sloping land with runoff to a natural water source.
2. Distribution of the water through spray nozzles over relatively flat terrain.
3. Disposal to ridge and furrow irrigation channels.

TABLE 5.15

Data from spray irrigation plants

Waste water from the manufacture of:	Loading kg BOD_5 per year	Application rate (m/year)
Asparagus and beans	30	5
Tomatoes	600	3
Starch	750	3
Cherries	700	1.6
Paper board and hard board	500	1.6

5.3 NUTRIENT REMOVAL

Environmental Management, Eutrophication and Nutrient Removal

Eutrophication is generally considered to be undesirable because:

1. the green colour of eutrophic lakes makes swimming and boating more unsafe,
2. an aesthetic point: clear water is preferred to turbid water,
3. the oxygen content of the hypolimnion is reduced due to decomposition of algae, especially in the autumn, but also in the summer, when the stratification of deeper lakes is most pronounced.

The eutrophication is, however, not controlled merely by determining the limiting element or elements and then reducing their concentration in the discharged waste water - the problem is far more complex (see Section 2.8 and Section 2.11). It is important to consider the sources of the nutrients and for a balanced overview of the situation a mass balance must be set up for each case. However, in some situations it is not possible to obtain the necessary data for setting up a mass balance. In these cases information about the water discharge, the waste water treatment, the number of persons living in houses connected to the sewage system, the drainage area, the soil characteristic of the drainage area and how the area is used (agricultures, forestry, towns, industries and other applications) should be provided. Based upon this information it is possible to calculate approximately the inputs of phosphorus and nitrogen to the lake from the different sources. Generally each person in industrialised countries discharges to the sewage system 200 l of waste water with the concentrations indicated in Table 5.4. When the treatment is known it is relatively easy to calculate the discharge from the treatment plant to the lake. Table 5.16 can be used to estimate the input of phosphorus and nitrogen to the lake from the drainage area. The figures are approximate, based on an interpretation of the following references: Dillon and

Kirchner (1975), Lønholt (1973) and (1976), Vollenweider (1968) and Loehr (1974), Schindler and Nighswander (1970), Dillon and Rigler (1974), Lee and Kluesener (1971), Jørgensen et al. (1973).

The concentrations of nitrogen and phosphorus in rain water are given in Table 5.17.

TABLE 5.16

Sources of nutrients. Export scheme of phosphorus E_P and nitrogen E_N ($mg\,m^{-2}y^{-1}$)

	E_P		E_N	
	Geological classification		Geological classification	
Land use	Igneous	Sedimentary	Igneous	Sedimentary
<u>Forest runoff</u>				
Range	1-10	5-20	130 - 320	180 - 560
<u>Forest + pasture</u>				
Range	5-20	10-40	200 - 600	300 - 900
<u>Agricultural areas</u>				
Citrus		10-25		1000-3000
Pasture		15 - 75		100 - 1000
Crop land		20 - 100		500 - 1500

TABLE 5.17

Nutrient concentration in rain water ($mg\,l^{-1}$)

	C_{PP}	C_{NP}
Range	0.015 - 0.1	0.3 - 1.6
Mean	0.075	1.1

Phosphorus is mainly derived from sewage, although the input from agriculture in some cases may also be significant for instance when hilly land close to a lake is used for agriculture. It is often much more difficult to control the input of nitrogen to an ecosystem, because a significant part of the nitrogen will usually come from non-point sources (agriculture mainly). Such processes as nitrogen fixation by algae may also play a major role. Control of the eutrophication of lakes requires removal of one or more nutrients from waste water, and the final decision as to the right management strategy - what nutrient should be removed and to what extent - requires the use of ecological models, see Jørgensen (1994)). A simple model (for instance the very simple Vollenweider plot shown in Figure 2.26) might, however, be the right model in a data-poor situation.

Example 5.3

A lake has a surface area of 5 km^2 and a catchment area of 120 km^2, mainly agricultural. The average depth of the lake is 5 m. It receives an annual precipitation of 0.8 m and mechanical-biologically treated waste water from 20,000 inhabitants. Set up an approximate mass balance for the lake and suggest what should be done to reduce its eutrophication.

Solution

E_P and E_N are chosen to be 60 mg $m^{-2}y^{-1}$ and 1000 mg $m^{-2}y^{-1}$, respectively (see Table 5.16).

This gives an annual input of:

A. Runoff — P-input $60 * 120 * 10^6$ mg = 7200 kg
N-input $1000 * 120 * 10^6$ mg = 120,000 kg

B. Precipitation — P-input $0.075 * 0.8 * 5 * 10^6$ g ≈ 300 kg
N-input $1.1 * 0.8 * 5 * 10^6$ =4400 kg

C. Waste water — 80 m^3/inh. 10 mg l^{-1} P and 30 mg l^{-1} N
corresp. to 800 g P/inh. and 2400 g N
P-input: 16000 kg
N-input: 48000 kg

More than 2/3 of the P-input comes from waste water, while only about 40% of the N-input comes from this source.

The total P-loading at present corresponds to:

$$\frac{(7200 + 300 + 16000) * 10^3}{5 * 10^6} \text{ g P/m}^2\text{ y} \approx 4.7 \text{ g P/m}^2\text{ y}$$

: A very high value (see Fig. 2.26).

A considerable improvement could be expected if the P-input from waste water were to be reduced by 95-99% (see Fig. 2.26). The removal of nitrogen would not reduce the eutrophication correspondingly. A **simultaneous** removal of nitrogen would not improve the conditions, as:

1. 120,000 kg would still be the input and this is more than 10 times as much as the P-input after P-removal. The use of P and N by algae is estimated to be in the ratio of about 1:8.5 (see Table 2.14).
2. Nitrogen-fixing algae might be present in the lake which would reduce the effect of nitrogen removal.

In addition to the eutrophication problem discharge of nitrogen compounds into receiving waters can cause two environmental problems:

1. *Ammonium compounds may be nitrified, influencing the oxygen balance in accordance with the following equation:*

$$NH_4^+ + 2O_2 \text{ -------> } NO_3^- + H_2O + 2H^+ \qquad (5.18)$$

Municipal waste water, which is treated by a mechanico-biological method, but not nitrified, will normally contain approximately 28 mg ammonium-nitrogen per litre. In accordance with equation 5.18, this concentration might cause an oxygen consumption of 128 mg oxygen (see also section 2.7).

2. *Ammonia is toxic to fish* as mentioned in Section 2.3

We can conclude that under all circumstances methods for the removal of both nitrogen and phosphorus must be available, and these methods are discussed in this section.
Phosphorus can be removed by chemical precipitation or ion exchange, while nitrogen can be removed by nitrification + denitrification, ammonia-stripping, ion exchange or oxidation by chlorine followed by treatment on activated carbon. Both nutrients can be removed, however, at low efficiencies using algal ponds.

<u>Chemical precipitation of phosphorous compounds.</u>

Chemical precipitation is widely used as a unit process for the treatment of waste water.

Municipal waste water is treated by precipitation with:
aluminium sulphate ($Al_2(SO_4)_3$, $18H_2O$) supplemented during recent years with many other aluminium compounds, for instance polyaluminates,
calcium hydroxide ($Ca(OH)_2$) or
iron chloride ($FeCl_3$) or ($FeSO_4$)
for removal of phosphorus.

When chemical precipitation is combined with a mechanico-biological treatment, the precipitation can be applied after the sand-trap using the primary settling for sedimentation, simultaneously with the biological treatment or after complete mechanico-biological treatment. In these cases the precipitation is described respectively as *direct precipitation, simultaneous precipitation and a post-treatment,* see Fig 5.13.

Chemical precipitation is today widely used in many countries. However, the process was already in use for this purpose in Paris in 1740 and in England more than 100 years ago. Just before the Second World War, Guggenheim introduced the process as a treatment preceding an activated-sludge plant (Rosendahl, 1970). However, the method did not find favour until the late 1960s due to the cost of chemicals and to the low reduction in BOD_5, which was supposed to be the major concern.

The main objective of applying chemical precipitation is the removal of phosphorus for the control of eutrophication. The efficiency of phosphorus removal is generally between 75 and 95% (Statens Naturvårdsverk, 1969). However, *in addition to this effect, chemical precipitation can also achieve the following:*

1. Substantial *reduction in the number of microorganisms,* particularly when calcium hydroxide is used (Buzzell and Sawyer, 1967).
2. *Reduction in* BOD_5 of 50-65% (Buzell and Sawyer, 1967).
3. *Reduction of non-biodegradable material* (so-called refractory material) in the same ratio as biodegradable material (Jørgensen, 1976b).
4. *Reduction in the concentration of nitrogenous compounds* due mainly to the removal of organic compounds.
5. *A reduction in heavy metal concentration,* which is essential (see also Section 5.5).

The removal of phosphorus entails the use of three chemicals: aluminium sulphate, other derivates of aluminium (see below), iron(III) chloride or calcium hydroxide.

As the composition of waste water and the cost of the chemicals vary from place to place, it is difficult to give a general recommendation on the choice of chemicals. It seems necessary to examine each case separately, although it is possible to draw some conclusions and set up some general guidelines.

TABLE 5.18

Approximate concentrations in typical raw domestic sewage

Form	mg P l^{-1}
Total	8-10
Orthophosphate	5-6
Tripolyphosphate	3
Pyrophosphate	1
Organic phosphate	≤ 1

Source: Jenkins et al. (1971)

Municipal waste water contains orthophosphate, and other phosphorus species. Table 5.18 shows the approximate concentration of the major phosphorous compounds.

They have a different bioavailability for different species of phytoplankton. Dissolved reactive phosphorus appears to be entirely bioavailable, whereas only an average of 22% of dissolved unreactive phosphorus and 25% of particulate phosphorus are utilised by algae (Ekholm and Krogerus, 1998).

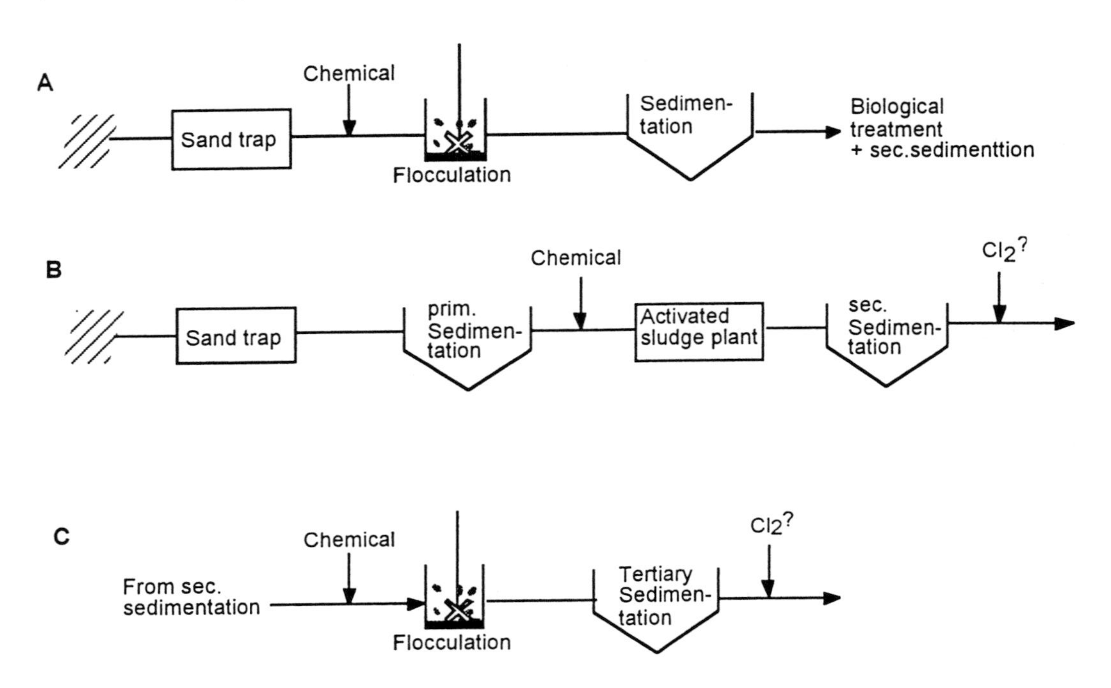

Fig. 5.13. A Direct precipitation. B Simultaneous precipitation. Recirculation of sludge is not shown. C. Post-treatment.

The chemical precipitations form insoluble compounds with phosphorus species, but the solubility is naturally dependent on pH. For aluminium and iron phosphates the pH dependence is probably explained to a certain extent by the following process:

$$(MePO_4)_n + 3OH^- \text{------>} Me_n(OH)_3(PO_4)_{n-1} + PO_4^{3-} \qquad (5.19)$$

Increasing OH^- concentration will move this process to the right causing dissolution of the precipitated phosphate. The precipitation of phosphates using calcium hydroxide is based on the formation of $Ca_{10}(PO_4)_6(OH)_2$, which will involve a high pH value: 10.5 - 12.0 . However, many of the processes interact, and the phosphate species are able to form complexes with some metal ions, thereby significantly increasing phosphate stability. Table 5.19 shows some of the more important processes which may influence precipitation. The precipitation chemistry is as can be seen rather complex. It is therefore hardly possible to base the optimum precipitation conditions on theoretical calculations. It is recommended that we find for any specific waste water the relationships between dose, pH and efficiency by laboratory experiments

TABLE 5.19

Phosphate equilibria (Source: Sillen and Martell (1964))

Equilibrium	Log equilibrium constant 298° K
$H_3PO_4 = H^+ + H_2PO_4^-$	-2.1
$H_2PO_4^- = H^+ + HPO_4^{2-}$	-7.2
$HPO_4^{2-} = H^+ + PO_4^{3-}$	-12.3
$H_3P_2O_7^- = H^+ + H_2P_2O_7^{2-}$	-2.5
$H_2P_2O_7^{2-} = H^+ + HP_2O_7^{3-}$	-6.7
$HP_2O_7^{3-} = H^+ + P_2O_7^{4-}$	-9.4
$H_3P_3O_{10}^{2-} = H^+ + H_2P_3O_{10}^{3-}$	-2.3
$H_2P_3O_{10}^{3-} = H^+ + HP_3O_{10}^{4-}$	-6.5
$HP_3O_{10}^{4-} = H^+ + P_3O_{10}^{5-}$	-9.2
$Ca^{2+} + PO_4^{3-} = CaPO_4^-$	6.5
$Ca^{2+} + HPO_4^{2-} = CaHPO_4$	2.7
$Ca^{2+} + H_2PO_4^- = Ca\,H_2PO_4^+$	1.4
$Ca^{2+} + P_2O_7^{4-} = CaP_2O_7^{2-}$	5.6
$Ca^{2+} + HP_2O_7^{3-} = Ca\,HP_2O_7^-$	3.6
$Ca^{2+} + P_3O_{10}^{5-} = CaP_3O_{10}^{3-}$	8.1
$Ca^{2+} + HP_3O_{10}^{4-} = CaHP_3O_{10}^{2-}$	3.9
$Ca^{2+} + H_2P_3O_{10}^{3-} = Ca\,H_2P_3O_{10}^-$	3.9

Aluminium sulphate is the precipitant most widely used in Scandinavia, where a technical grade named AVR is used. In Table 5.20 the composition of AVR is compared with an iron-free quality (Boliden, 1967 and 1969). Many other grades are available today and several other aluminium compounds including polyaluminates compounds may be used to obtain higher efficiency of the precipitation process. They are, however, more expensive to use.

The relationship between the addition of AVR-quality aluminium sulphate and the phosphorus concentration in municipal waste water is shown in Fig. 5.14 for different concentrations of phosphorus in the treated water. The dose of the chemical depends on the phosphorus concentration in the waste water and on the required concentration in the treated water.

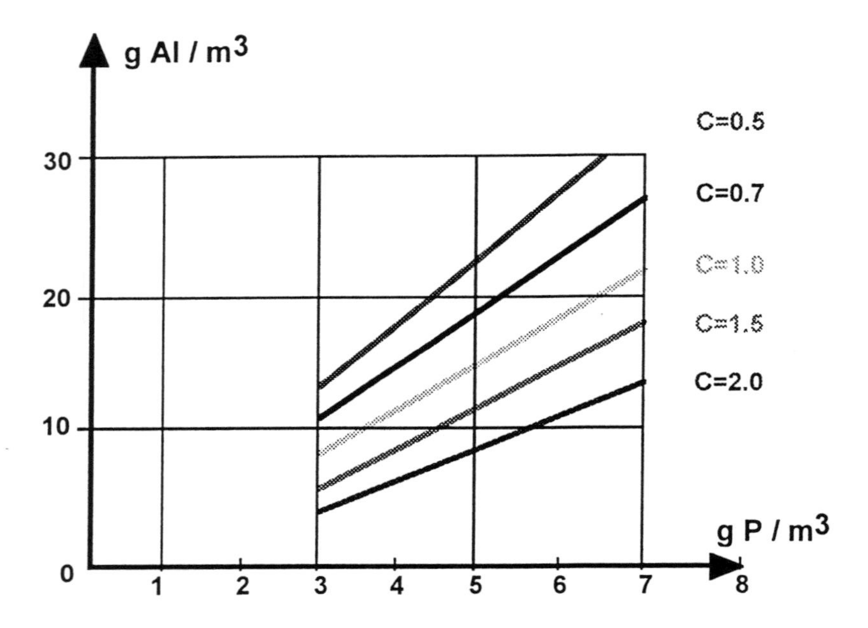

Fig. 5.14 The dependence of AVR addition on the P concentration of the waste water and the desired P concentration, indicated as C. Source: Gustafson and Westberg (1968).

The relationship, that describes the effect of chemical precipitation, can be interpreted by the Freundlich adsorption isotherm (see Fig. 5.14):

$$\frac{C_o - C}{n} = a * C^b \qquad (5.20)$$

where C_o = initial concentration of P (mg l^{-1})
C = final concentration of P (mg l^{-1})
n = dose of chemical mg l^{-1} Fe or Al
a and b = characteristic constants. The constants a and b can be found in Table 5.21.

TABLE 5.20

Composition of AVR compared with an iron-free quality

	Composition percentage	
	Iron-free	AVR
Al_2O_3 total	17.1	15.1
Fe_2O_3	0.0	2.1
Fe	0	0.2
Water insoluble	0.05	2.5
H_2O, chemical	43.7	43.9
pH 1% water soluble	3.6	3.6
Size (mm)	0.25 - 2.0	0.25 - 2.0
	granular	granular
Colour	white	brownish
Weight kg/l	0.90	0.95

TABLE 5.21

a and b in Freundlich's adsorption isotherm for aluminium sulphate and iron(III) chloride

Precipitation with	a	b
Aluminium sulphate	0.63	0.2
Iron(III) chloride	0.26	0.4

Example 5.4

Find the dose of Al required to remove 90% P from waste water with a P-concentration of 10 mg l^{-1} and hypolimnic water with a P-concentration of 1 mg l^{-1}.

Solution.

$$\frac{C_o - C}{n} = a * C_b$$

or

$$n = \frac{C_o - C}{a * C^b} = \frac{C_o - C}{0.63 * C^{0.2}}$$

A. Waste water $C_o = 10 \quad C = 1$

$$n = \frac{10 - 1}{0.63 * 1^{0.2}} = \frac{9}{0.63} = 14.3$$

B. Hypolimnic water $C_o = 1 \quad C = 0.1$

$$n = \frac{1 - 0.1}{0.63 * 0.1^{0.2}} = \frac{0.9}{0.63 * 0.63} = 2.3$$

As seen, about 6 times as much precipitant is needed in case A, but 10 times more phosphorous is removed. In practice, even relatively more precipitant will be used in case B than the theoretical calculations show here.

The values given in Fig. 5.14 and Table 5.21 are based on the assumption that the pH is close to the optimum (for aluminium sulphate 5.5 - 6.5, for iron(III) chloride 6.5 - 7.5).

Iron(III) chloride is more expensive to use than aluminium sulphate, but iron(II) sulphate is a waste product of the chemical industry and can easily be oxidised to iron(III) by the use of chlorine or by aeration, for instance in the activated-sludge process (Särkka, 1970). Therefore iron(II) sulphate has been widely used as a precipitant, especially for simultaneous precipitation.

The relationship between dose, phosphorus concentration and the final concentration in the treated water is similar to that for aluminium sulphate (see Fig. 5.15). The precipitation in this case also follows the Freundlich adsorption isotherm, but the constants a and b are different from those for aluminium sulphate.

Hydrated lime or calcium hydroxide is the cheapest of the three precipitants used for removal of phosphorus. However, greater amounts of calcium hydroxide are required to obtain a given efficiency. While 120 - 180 mg per litre of aluminium sulphate or iron(III) chloride are sufficient in most cases to achieve 85% removal or more, 2 to 5 times more calcium hydroxide will be needed. For the precipitation of phosphorus compounds in waste water with aluminium sulphate and iron(III) chloride more than the stoichiometric amount is used, as pure orthophosphate is not precipitated, but adsorption is also involved. However, for calcium hydroxide precipitation a pH value of at least 10 is required to achieve a sufficiently high OH^- concentration. Thus the amount of calcium hydroxide is determined by the alkalinity of the water. (Alkalinity is a measure of the concentration of alkaline ions, mainly hydrogen carbonate and carbonate; it can be expressed as mg l^{-1} of calcium carbonate). The practical relationship between the alkalinity and the mg of calcium hydroxide required to reach pH 11, should be applied to assure effective precipitation. Most waste water has an alkalinity in the range of 150 - 300 mg calcium carbonate per litre, which, as can be seen from the figure, corresponds to about 300 to 480 mg calcium hydroxide per litre. Generally higher efficiencies are obtained by precipitation with hydrated lime than by the other two precipitants - 90 - 95% efficiency being common in practice. The orthophosphate is precipitated with an efficiency of almost 100%, while polyphosphates are precipitated with more modest efficiency giving an overall efficiency close to 95%.

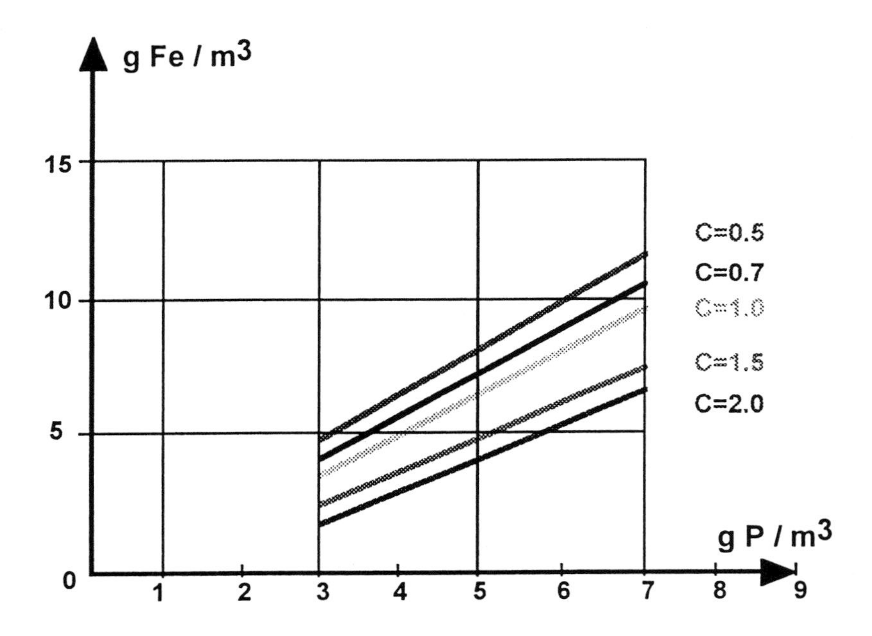

Fig. 5.15 The dependence of iron(III) salt addition on the P concentration of the waste water and the desired P concentration, indicated as C. Source: Gustafson and Westberg (1968).

The disadvantage of precipitating with lime is the high pH, which makes an adjustment after precipitation necessary. Carbon dioxide produced by incineration of the sludge can be used for this purpose (see Fig. 5.16), where the calcium oxide produced by the incineration is also reused to a certain extent at the same time as the carbon dioxide is applied to adjust the pH. In this process the calcium oxide can be recycled 3 - 5 times corresponding to a more stoichiometric ratio between calcium and phosphate. The calcium hydroxide produced in this manner can be used as fertiliser (Dryden and Stern, 1968).

The amount of sludge produced by chemical precipitation is dependent on several factors, of which the major factor is the dose of chemicals (see Table 5.22). The sludge formed from the precipitation of phosphorus with aluminium sulphate can be treated together with anaerobic biological sludge, without difficulty, while that formed from precipitation with iron compounds may be dissolve due to reduction of iron(III) to iron(II) phosphate, which can react in accordance with the following process:

$$Fe_3(PO4)_2 + 3H_2S \text{ -------> } 3FeS + 2H_2PO_4^- + 2H^+ \qquad (5.21)$$

The same process occurs when iron rich sediment releases phosphorus due to anaerobic conditions.

TABLE 5.22
Additional sludge by precipitation
D = dose expressed as g/m³ Al, Fe or hydrated lime

Chemical	Additional sludge in g per m³ waste water
Aluminium sulphate	4 D
Iron(III) salt	2.5 D
Hydrated lime	1 - 1.5 D

Without a pH-adjustment the sludge produced by the calcium hydroxide precipitation can hardly be treated by anaerobic digestion. If the chemical sludge is mixed with a large volume of biological sludge the pH might be sufficiently low to allow anaerobic treatment. The same considerations are valid for an aerobic sludge treatment with the exception that the above-mentioned reduction of iron, of course, cannot take place. Filtration, centrifugation or drying of chemical sludge does not cause any additional problems. Recirculation of sludge is practised in many plants and often means a reduced consumption of chemicals. Recovery of aluminium sulphate has even been suggested. With treatment of filtered sludge with sulphuric acid after aluminium sulphate precipitation the aluminium ions go into solution as aluminium sulphate, which can be reused. However, the economical advantage is in most cases too modest to pay for this extra process.

Rapid flocculation, precipitation and settling are possible with a process which uses a combination of micro sand, polymers and aluminium or iron salts as precipitant. A process using this combination of precipitants, named Aciflo, is able to operate with a surface loading of 50- 120 m/ t which is 50 - 240 times the traditional settling. The process can be used for waste water, rain water (storm water) and surface water. It offers a particularly interesting solution to the problem of phosphorus removal from lake water which is needed when ecotechnological methods are applied to restore lakes.

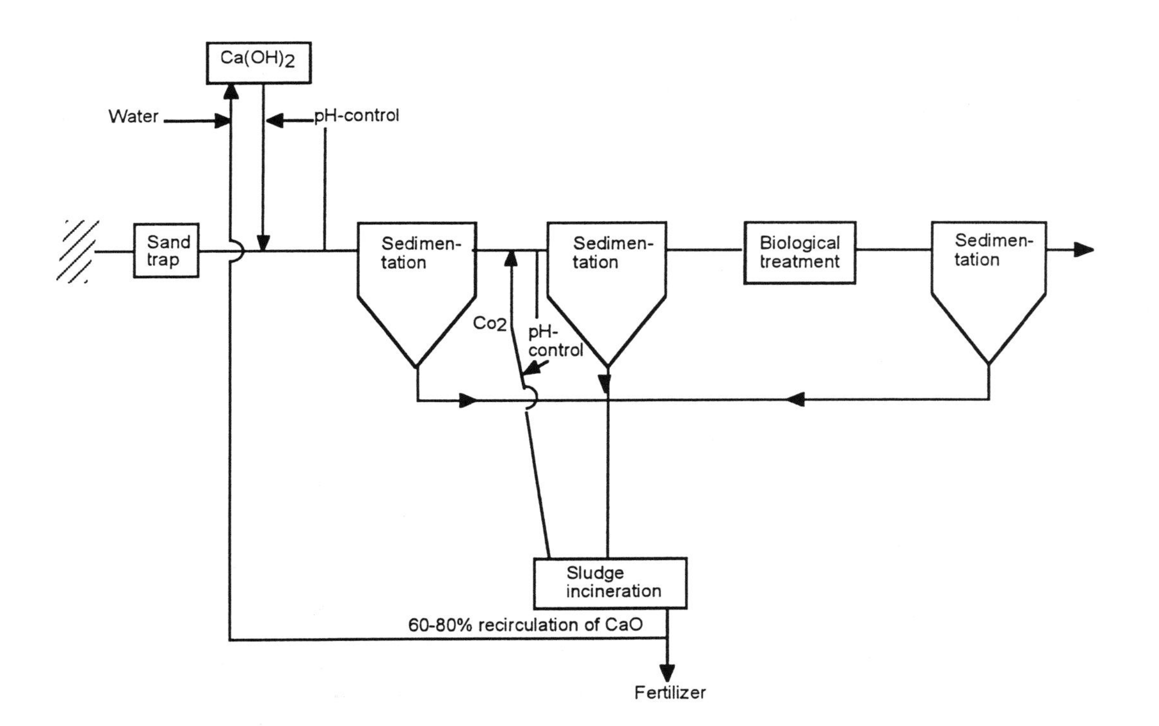

Fig. 5.16. Chemical precipitation with partial recirculation of calcium oxide and use of CO_2 from incineration for adjustment of pH.

Several heavy metals are removed quite effectively by precipitation. Aluminium sulphate is able to precipitate lead, copper and chromium with high efficiency, while cadmium and zinc are only precipitated partially. Iron-based chemicals produce the same effects at the same pH, but as the pH is often lower with this precipitation, in practice iron salts often give a lower efficiency than aluminium sulphate. Calcium hydroxide removes almost all heavy metals very efficiently, as most metal hydroxides have very low solubility at pH 10-11. Calcium hydroxide is used (see Section 6.4) to remove heavy metals from industrial waste water. Municipal waste water will, however, often contain compounds that are able, through formation of complexes, to increase the solubility of most metal ions. This is the case for instance, when water contains NTA (Nitrilo Three Acetate) chloride which is used in some washing powders.

These additional effects of the precipitation process must not be overlooked. For treatment of municipal waste, it might also be important to precipitate toxic organics and heavy metals in addition to achieving a reduction of BOD_5 of approximately 60% and a 80-95% removal of phosphorus compounds.

TABLE 5.23
Synthetic organic polymeric flocculants

A. Cationic polyelectrolytes

CH2
----- CH CH — CH2 -----
H2C CH2
N+
H3C CH3

Polydiallyldimethylammonium

B. Anionic polyelectrolytes

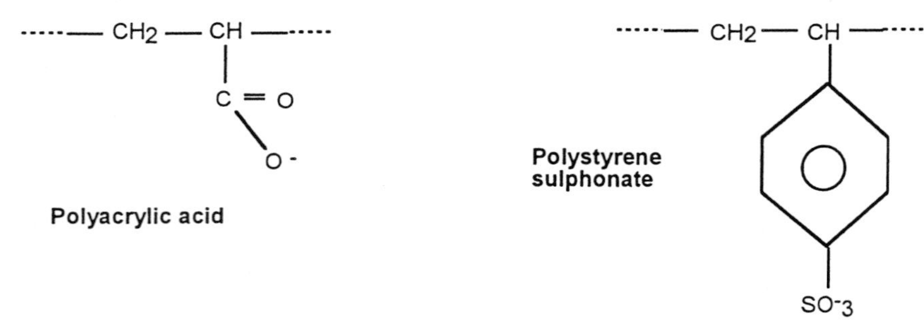

C. Nonionic polymers

----- CH2 — CH -----
C = O
NH2

Polyacrylamide

----- CH2 — CH2 — O -----------

Polyethylene oxide

If the municipal waste water contains more or less treated industrial waste water, in some cases it might be a better solution to use chemical precipitation rather than mechanical-biological treatment, depending on the receiving water. It seems worthwhile, at least as a primary step, to consider the chemical precipitation treatment of waste water for discharges to lakes or the sea. The best solution is of course a combination of biological and chemical treatment.

The effects mentioned, however, can only be obtained with current process control. An unacceptable level of efficiency results from:

1. **Too low a dose.** Recirculation of the chemical sludge might be an attractive remedy.
2. **Wrong pH.** In some cases it might be necessary to adjust the pH by the addition of calcium hydroxide or sulphuric acid.
3. **The flocculation is insufficient.** Flocs that are too small and will not settle are formed. It might be necessary to change the design of the flocculator (Jørgensen, 1979a). The optimum design of a flocculator may be based on a mathematical model, see Dharmappa et al., 1993 or Thomas et al., 1999.

The flocculation can be further improved by the application of synthetic polyflocculants, as already mentioned above in discussion of the Actiflo process.

Clay, starch and gelatin can be used to accelerate and stimulate the flocculation, but the synthetic polyelectrolytes that have appeared on the market during the last decade or two seem to give better efficiency (Black, 1960). They work mainly on flocs in the range of 1-50 μm. Table 5.23 shows the molecular structure of the polyelectrolytes most applied.

The long-chain polymers are able to collect many small flocs and agglomerate them into larger groups. The polymers are able to form a bridge between the small particles. For flocculation of municipal waste water anionic polyelectrolytes are chiefly used, while the cationic polyelectrolytes are used for flocculation of different types of industrial waste water or even for improving the dewatering of municipal sludge. Several synthetic polymers are on the market, and non-ionic, anionic or cationic ones are available, see Table 6.23. For waste water treatment polyacrylamide is recommended as polyflocculant (see Table 5.23).

It is possible to produce polyacrylamides with a molecular weight of 4-10 million. By a copolymerisation of acrylamide and acrylic acid it is possible to prepare an anionic polymer of this type. The main effects of using polymers for flocculation are (Boeghlin, 1972 and Ericsson and Westberg, 1968):

1. Gaining a higher settling rate,
2. Reduction in turbidity, caused by small flocs.

Phosphorus removal by biological treatment

In recent decades activated sludge systems with anaerobic and aerobic zones in sequence have been developed to achieve phosphorus removal. Such an activated sludge system is generally referred to as an enhanced biological phosphorus removal (EBPR) system. The shift between aerobic and anaerobic conditions activates the microorganisms to take up considerably more phosphorus than under aerobic conditions, particularly if the waste water contains relatively high concentrations of easily biodegraded organic matter. With a BOD_5 / P ratio of more than 20, a phosphorus removal of between 80 and 90% can be obtained which often may be sufficient. The sludge production (dry matter) is in the order of 4-5 times the phosphorus concentration. The process may be controlled by use of a dynamic model (see Chang et al, 1998)

Nitrification and denitrification

The nitrification process eliminates oxygen consumption related to the oxidation of ammonium (also shown as equation (2.33):

$$NH_4^+ + 2O_2 \longrightarrow NO_3^- + H_2O + 2H^+ \qquad (5.22)$$

As can be seen it is the same process as otherwise would occur in the ecosystem (see Section 2.7). Effective nitrification occurs when the age of the sludge produced by biological treatment is greater than the reciprocal rate constant of the nitrifying micro-organisms. Bernhart (1975) has demonstrated that it is possible to oxidise ammonia in a complex organic effluent by biological nitrification.

The sludge age, *f*, is defined as:

$$\frac{X}{\Delta X} = f \qquad (5.23)$$

where X = the mass of biological solid in the system and ΔX = the sludge yield.

The relationship between nitrification and sludge age is shown in Fig. 5.17.

Nitrification results from a two-step oxidation process, see Section 5.2, equations (5.5) and (5.6). The optimum pH range for Nitrosomonas is 7.5-8.5 and for Nitrobacter 7.7-7.9. The rate seems to be dependent on the ammonium ion concentration at concentrations in excess of 0.5 mg/l, which is considerably lower than those generally found in waste water containing ammonium ions. The oxygen concentration also influences the nitrification rate significantly and a concentration of at least 2 mg O_2 /l is required (Beccari et al., 1992).

Heavy metals are toxic at rather low concentrations. Toxic levels of about 0.2 mg/l are reported for chromium, nickel and zinc.

Temperature and pH exert a profound effect on nitrification (see also Sections 2.4 and 2.7). pH optimum is 6.7- 7.0 (Groeneweg et al.,, 1994) for the overall nitrification process. Downing (1966) has reported that the influence of temperature on the rate coefficient can be expressed as follows:

$$K_N = 0.18 * 1.128^{T-15} \quad (24h^{-1}) \qquad (5.24)$$

As can be seen, K_N = 0.18 at 15°C. Compare with Section 2.4, Table 2.7. However, the amount of nitrogen which can be removed per m^3 is limited. It is not possible to remove more than about 21 mg/l per 24h. Even the ammonium concentration in the inflow is increased to about 1000 mg/l.

Nitrate can be reduced to nitrogen and dinitrogen oxide by many of the heterotrophic bacteria present in activated sludge, but the process requires anaerobic conditions.

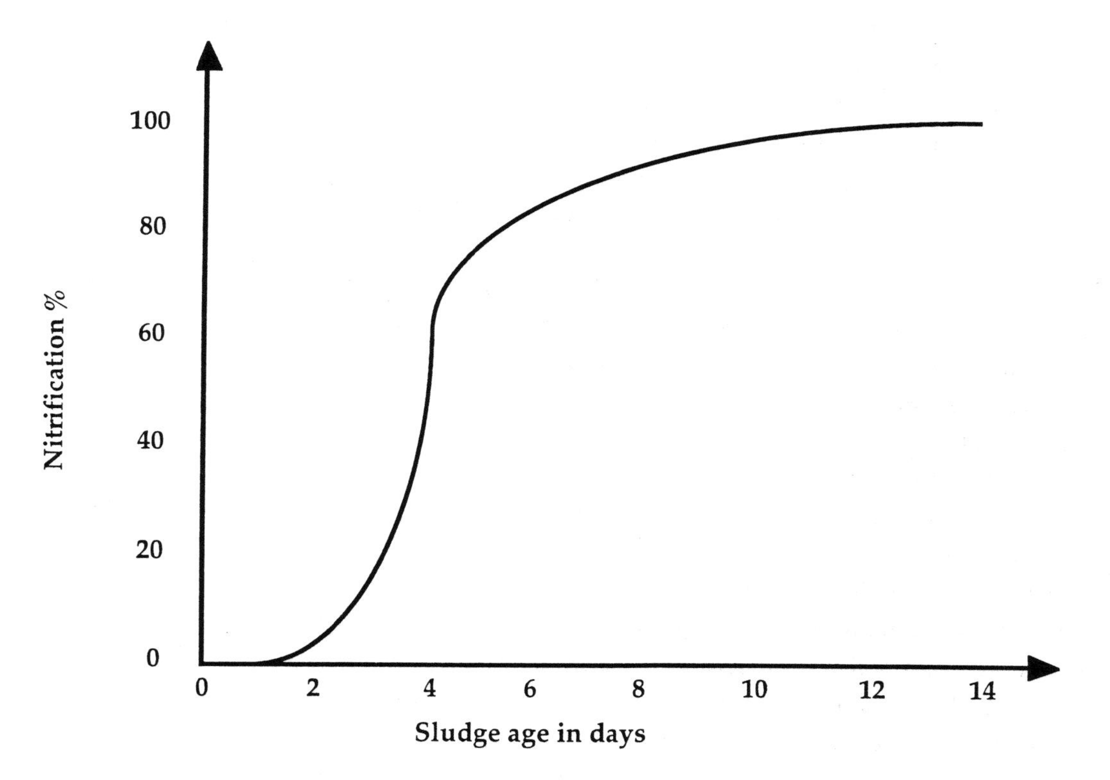

Fig. 5.17. Relationship between nitrification and sludge age.

The pH affects the process rate, the reported optimum being above 7.0. The same process is observed in anaerobic sediment. As the denitrifying organisms are heterotrophic, they require an organic carbon source. It is possible either to add the carbon source, e.g. by using methanol or molasses, or to use the endogenous by-product as the food supply.

If acetate is used as a carbon source the process is:

$$5CH_3COO^- + 8NO_3^- \text{------>} 4N_2 + 7HCO_3^- + 3CO_3^{2-} + 4H_2O \quad (5.25)$$

The influence of the COD/NO_3^--N ratio on the denitrification efficiency corresponds to the stoichiometric relationship in equation (5.25). The amount of acetate to nitrate-N can be calculated from equation (5.25) to be about 3. As the amount of oxygen to oxidise acetate to carbon dioxide and water is 64/ 60 in accordance with the stoichiometry, about 3 times as much BOD_5 should be used as mg nitrate-N to ensure an adequate denitrification. In practice the amount is slightly higher, and if the usual ratio between BOD_5 and organic matter (dry matter) is used, it is 1.3 for normal municipal waste water, the BOD_5 /nitrate-N ratio should be approximately 4.0. The rate of denitrification increases with increasing concentrations of available carbon and of nitrate. Francis et al. (1975) report a successful denitrification of waste water that contained more than 1000 ppm nitrate-N, which should be compared with the concentration of nitrate-nitrogen in municipal waste water of 20-40 ppm. Denitrification can be carried out even in water with a high salinity and high nitrate concentration (Glass and Silverstein, 1999). It is obviously not attractive to add organic matter to treated waste water to ensure denitrification. It is, however, possible to apply organic carbon to the waste water by imposing a denitrification process on the waste water before the treatment is complete - when there is still some organic matter left. This is generally possible by two methods:

1. To switch between aerobic and anaerobic conditions in the activated sludge plant. This is called alternating operation (see for instance Diab et al., 1993)
2. To recycle a part of the waste water containing nitrate after the treatment is complete to the inlet of the activated sludge plant.

Both methods are working properly (Halling Sørensen and Jørgensen, 1993). It is also possible to accomplish simultaneous nitrification and denitrification by use of zeolite ion exchange material. This possibility will be mentioned together with the corresponding ion exchange process.Denitrification by electrolysis (Isalam and Suidan, 1998) or by use of metallic iron (Huang et al., 1998) have been proposed as alternatives to biological denitrification. An effective denitrification with simultaneous high removal of phosphorus is achievable in a packed bed biofilm reactor (Æsøy et al., 1998). Enhanced denitrification can be obtained on activated carbon (Sison et al., 1996).

Simultaneous nitrification and denitrification is possible by playing on the pertinent parameters according to model results (Yoo et al., 1999) or by use of clinoptilolite as packing material in a trickling filter (Sørensen and Jørgensen, 1993 and Hjuler, 1997).

Stripping

The stripping process is used to remove volatile gases, such as hydrogen sulphide, hydrogen cyanide and ammonia by blowing air through the waste water. The removal of ammonia by stripping is used in the treatment of municipal waste water, but it has also been suggested for the treatment of industrial waste water or for regeneration of the liquid used for elution of ion exchangers (Jørgensen, 1976b). **The rate at which carbon dioxide, hydrogen sulphide, ammonia and hydrogen cyanide can be removed by air stripping is highly dependent on pH, since all four of these volatile gases are acids or bases.**

Ammonia stripping is based on the following process: $NH_4^+ \longrightarrow NH_3 + H^+$ which is shown in Section 2.4 as equation (2.21). If we introduce pK = 9.3 in equation (2.22), we get:

$$pH = 9.3 + \log [NH_3] / [NH_4^+] \quad (5.26)$$

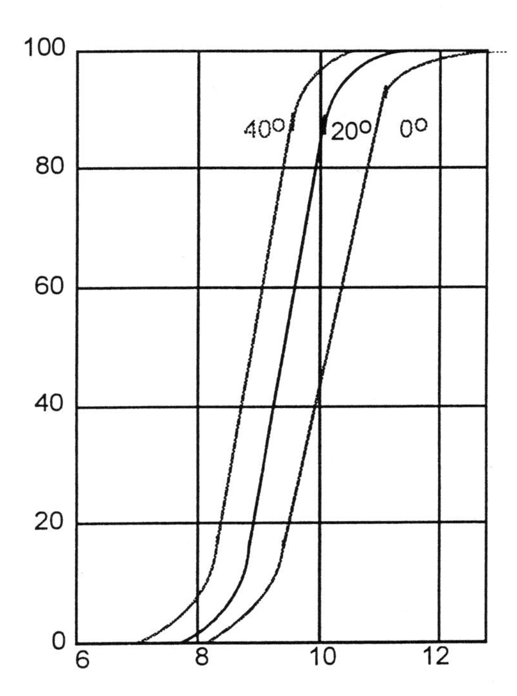

Fig. 5.18. Stripping efficiency as a function of pH at three different temperatures.

The same considerations are used when the ammonium concentration is found in an aquatic ecosystem to estimate the toxicity level of the water, see section 2.4 and Table 2.13. From equation (5.26) we can see that at pH 9.3 50% of the total ammonia-nitrogen is in the form of ammonia and 50% in the form of ammonium. Correspondingly the ratio between ammonia and ammonium is 10 at pH 10.3 and 100 at pH 11.3. Consequently it is necessary to adjust the pH to 10 or more before the stripping process is used. Due to the very high solubility of ammonia in water a large quantity of air is required to transfer ammonia effectively from the water to the air.

The efficiency of the process depends on:

1. **pH**, in accordance with the considerations mentioned above.
2. **The temperature.** The solubility of ammonia decreases with increasing temperature. The efficiency at three temperatures - 0°C, 20°C and 40°C - is plotted against the pH in Fig 5.18.

3. **The quantity of air per m^3 of water treated.** At least 3000 m^3 of air per m^3 of water is required (see Figure 5.19).

4. **The depth of the stripping tower.** The relationship between the efficiency and the depth shows that the stripping tower should be 7-8 m deep (see Figure 5.19).

5. **The specific surface of the packing** (m^2/m^3). The greater the specific surface the greater the efficiency.

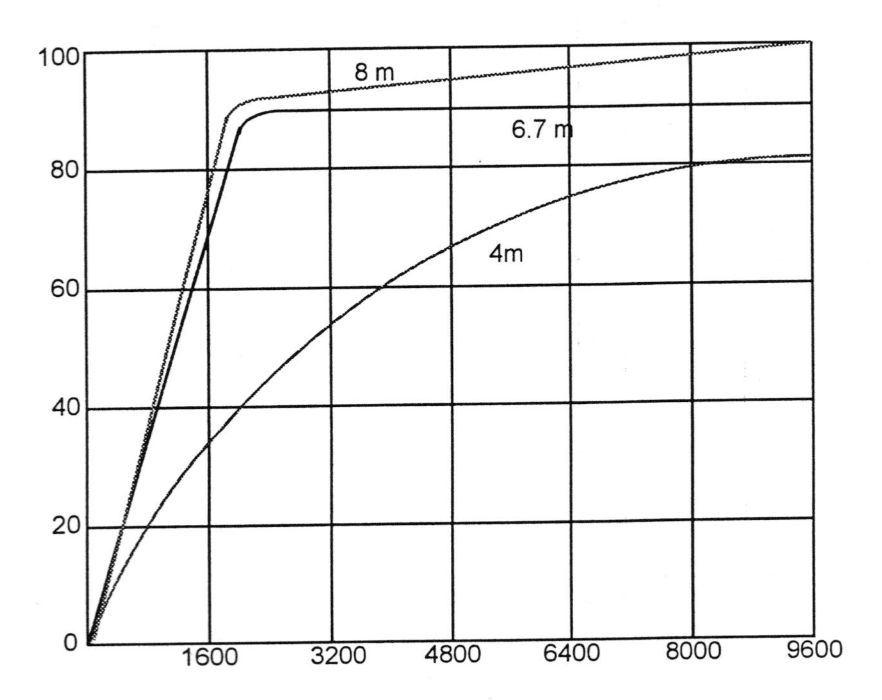

Efficiency versus m^3 air/ m ^{3} water for three different tower depths

The three tower depths 4 m (insufficient) , 6.7 m and 8 m.

Tower height of 7-8 m is recommended.

Fig 5.19 Efficiency of the stripping process versus air/water ratio and tower height.

The cost of stripping is relatively small, but the process has two crucial limitations:

1. It is practically impossible to work at temperatures below 5-7°C. The large quantity of air will cause considerable evaporation, which means that *the water in the tower is freezing.*

2. *Deposition of calcium carbonate* can reduce the efficiency or even block the tower.

Due to limitation 1) it will be necessary to use warm air for the stripping during winter in temperate climates, or to install the tower indoors. This makes the process too costly for plants in areas with more than 10,000 inhabitants and limits the application for treatment of bigger volumes to tropical or maybe subtropical latitudes.

Stripping of solutions with higher ammonium concentrations than we find in waste water is more beneficial, because the air/water ratio is fairly independent of the ammonium concentration. Consequently, stripping has only found application in colder climates for the stripping of filtrate resulting from filtration of sludge (containing often 50-100 mg ammonium-N /l) and for industrial waste water with high ammonium concentrations.

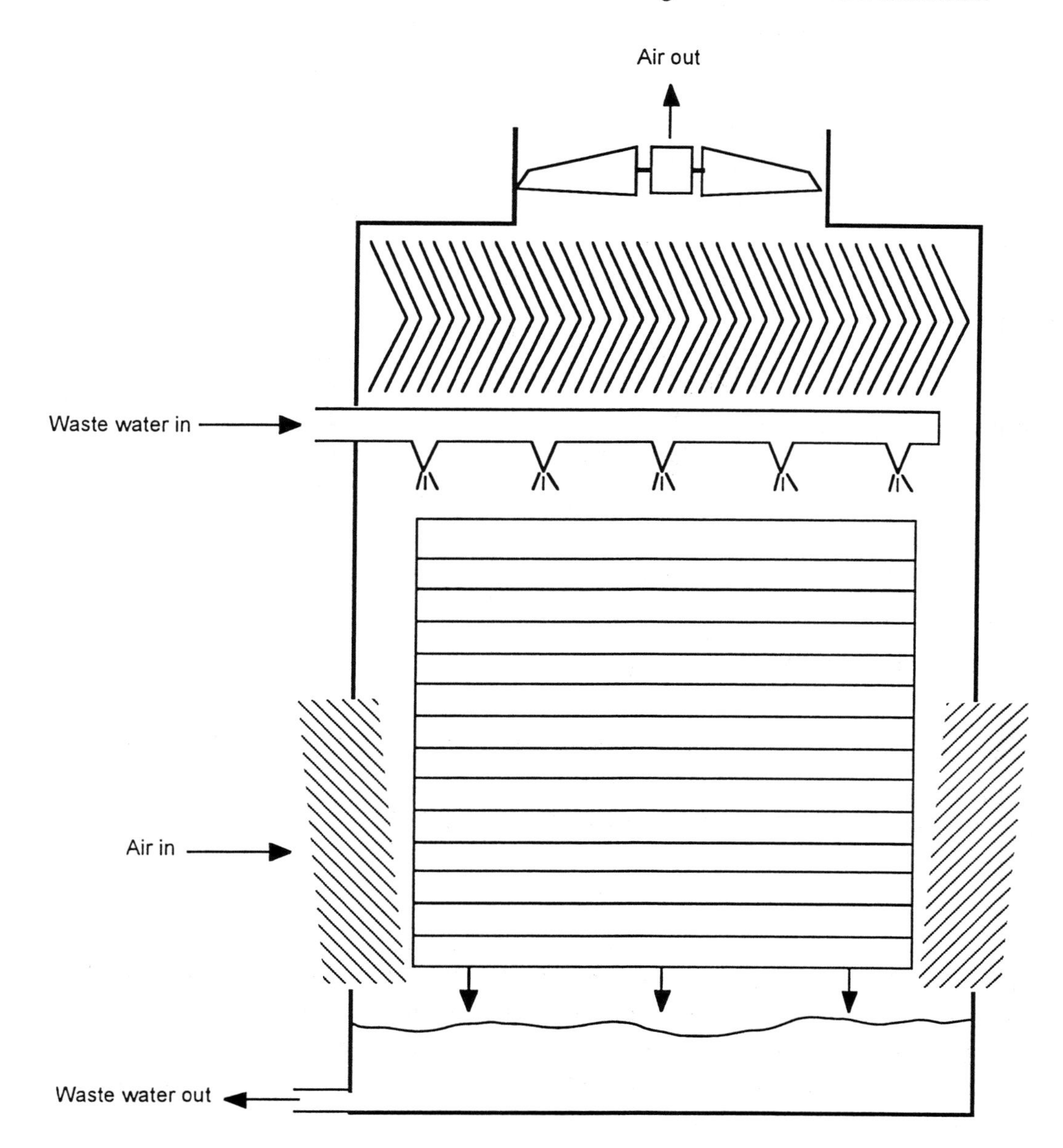

Fig. 5.20 The principle of a stripping tower.

However, it seems advantageous to combine the process with ion exchange. By ion exchange it is possible to transfer ammonium from 100 m^3 of waste water to 1 m^3 of elution liquid. As elution liquid is used as a base (sodium hydroxide), a pH adjustment is not required. Furthermore, the ratio of air to water is increased slightly although the elution liquid has 100 times greater ammonia concentration.

Consequently a stripping tower 100 times smaller with preconcentration by ion exchange can be used. Recovery of ammonia, for use as fertiliser, also seems easier after the preconcentration. The stripped ammonia could, for example, easily be absorbed into a sulphuric acid solution, yielding production of ammonia sulphate.

As mentioned in the introduction, the shortcoming of some of the technological solutions is, that *they do not consider a total environmental solution, as they solve one problem but sometimes create somewhere else a new one.*

The stripping process is a characteristic example, since *the ammonia is removed from* the waste water *but transferred to the atmosphere, unless recovery of* ammonia is carried out. In each specific case it is necessary to assess whether the air pollution problem created is greater that the water pollution problem solved.

Chlorination and adsorption on activated carbon

Chlorine can oxidise ammonia in accordance with the following reaction scheme:

$$\begin{aligned} &Cl_2 + H_2O \text{ ------> } HOCl + HCl \\ &NH_3 + HOCl \text{ ------> } NH_2Cl + H_2O \\ &NH_2Cl + HOCl \text{ -------> } NHCl_2 + H_2O \\ &NHCl_2 + HOCl \text{ -------> } NCl_3 + H_2O \end{aligned} \tag{5.27}$$

Activated carbon is able to adsorb chloramines, and a combination of chlorination and adsorption on activated carbon can be applied for removal of ammonia.

The most likely reaction for chloramine on activated carbon is:

$$C + 2NHCl_2 + H_2O - N_2 + 4H^+ + 4Cl^- + CO \tag{5.28}$$

Further study is needed, however, to show conclusively that surface oxidation results from this reaction. Furthermore, it is important to know that the Cl_2/NH_3-N oxidised mole ratio is 2:1, which is required for ammonium oxidation by this pathway.

The monochloramine reaction with carbon appears to bemore complex. On fresh carbon the reaction is most probably:

$$NH_2Cl + H_2O + C \text{ -------> } NH_3 + H^+ + Cl^- + CO \tag{5.29}$$

After this reaction has proceeded to a certain extent, partial oxidation of monochloramine is observed, possibly according to the reaction:

$$2NH_2Cl + CO - N_2 + H_2O \quad 2H^+ + 2Cl^- + C \tag{5.30}$$

It has been observed that fresh carbon is necessary before monochloramine can be oxidised.

In the removal of ammonia with a dose of chlorine followed by contact with activated carbon, pH control can be used to determine the major chlorine species. The studies reported herein indicate that a pH value near 4.5 should be avoided, because $NHCl_2$ predominates and thus 10 parts by weight of chlorine are required for each part of NH_3-N oxidised to N_2.

At a slightly higher pH and carbon concentration, the proportion of monochloramine increases and the chlorine required per unit weight of NH_3-N oxidised should approach 7.6 parts, ignoring the chlorine demand resulting from other substances. However, further testing should be used to verify this conclusion in each individual case.

When accidental overdosing of chlorine has occurred or after an intentional addition of large quantities of chlorine to accelerate disinfection, it isdesirable to remove the excess chlorine. This is possible using a reducing agent, such as sulphur dioxide, sodium hydrogen sulphite or sodium thiosulphate.

Complete removal of the 25-40 mg per litre ammonium-N is far too costly by this method. Chlorine costs about 30-35 US cents per kg, which means that the chlorine consumption alone will cost about 7-8 US cents per m^3 waste. When the capital cost and other operational costs are added the total treatment cost will be as high as 25 US cents per m^3, which is considerably more expensive than other removal methods. It is possible to use chlorine to oxidise ammonium compounds to free nitrogen, but this process involves even higher chlorine consumption and, therefore, is even more expensive.

The method has, however, one advantage: by using sufficient chlorine it is possible to obtain a very high efficiency. This has meant that the method has found application *after other ammonium removal methods,* where high efficiencies are required. This is the case when the waste water is reclaimed, for example, in the two plants shown in Figs. 5.21 and 5.22. As can beseen it is necessary to use several treatment processes to achieve sufficient water quality after the treatment.

Chlorination and treatment on activated carbon are used as the last treatment to ensure good ammonium removal and sufficient disinfection of the water. Additional chlorination is even used after the treatment on activated carbon to ensure there is a chlorine residue in the water supply system.

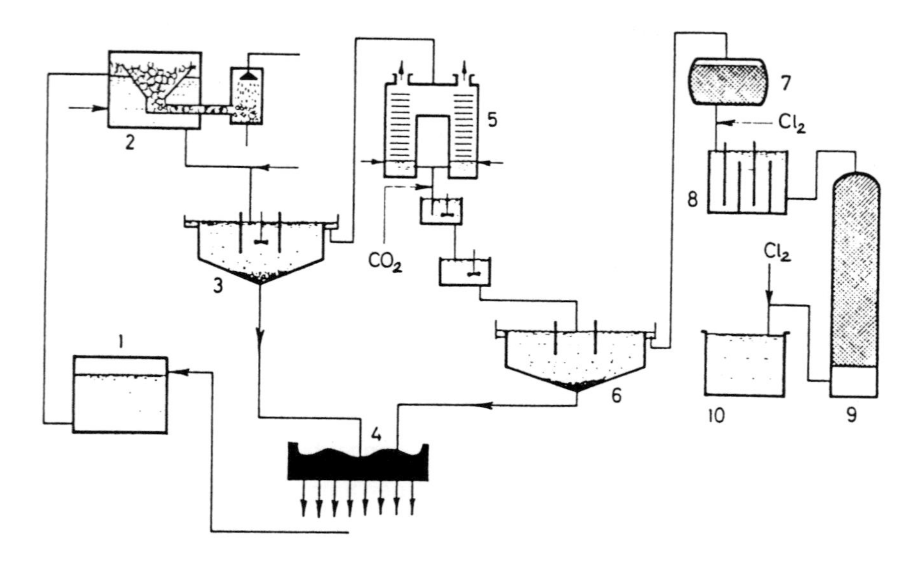

Fig. 5.21 Waste water treatment plant, Pretoria. After mechanical- biological treatment: 2) aeration, 3) lime precipitation, 4) sludge drying, 5) ammonia stripping, 6) recarbonisation, 7) sand filtration, 8) chlorination, 9) adsorption on activated carbon, 10) chlorination..

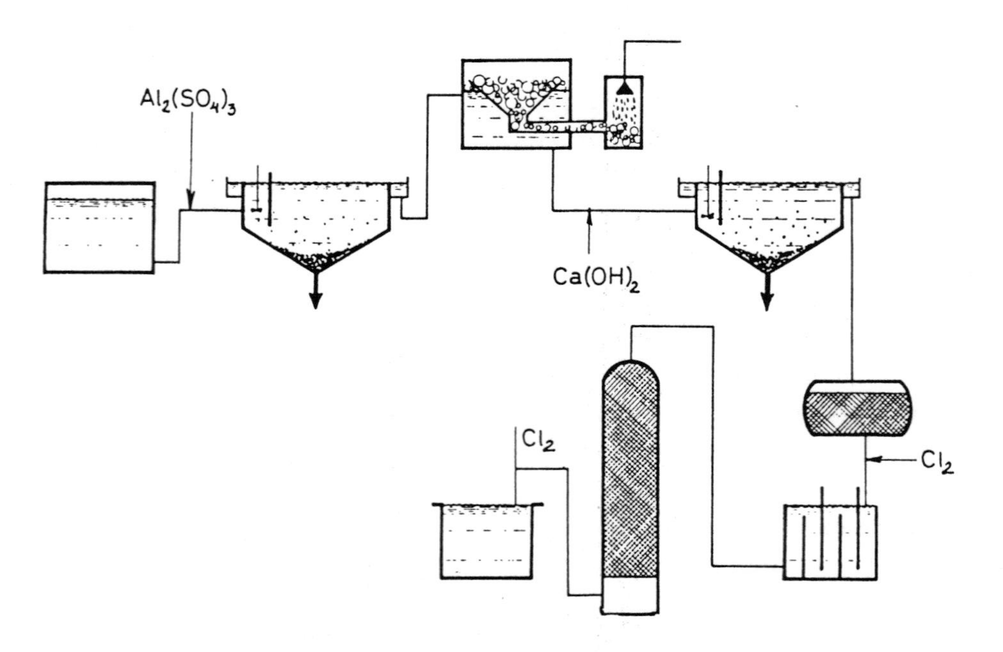

Fig. 5.22. Production of potable water from waste water (Windhoek).

Application of ion exchange for removal of nutrients

Ion exchange is a process in which ions on the surface of the solid are exchanged for ions of a similar charge in a solution with which the solid is in contact. Ion exchange can be used to remove undesirable ions from waste water. Cations (positive ions) are exchanged for hydrogen or sodium and anions (negative ions) for hydroxide or chloride ions. The cation exchange on a hydrogen cycle can be illustrated by the following reaction, using, in this example, the removal of calcium ions, which are one of the ions (Ca^{2+} and Mg^{2+}) that cause water hardness:

$$H_2R + Ca^{2+} \text{ -----> } CaR + 2H^+ \qquad (5.31)$$

where R represents the resin.

The anion exchange can be similar illustrated by the following reaction:

$$SO_4^{2-} + R(OH)_2 \text{ --------> } SO_4R + 2OH^- \qquad (5.32)$$

When all the exchange sites have been replaced with calcium or sulphate ions, the resin must be regenerated. The cation exchanger *can* be regenerated by passing a concentrated solution of sodium chloride through the bed, while the anion exchanger, which in this case is of hydroxide form, must be treated with a solution of hydroxide ions, e.g. sodium hydroxide. Ion exchange is known to occur with a number of natural solids, such as soil, humus, metallic minerals and clay. *Clay,* and in some instances other natural materials, can be used for demineralisation of drinking water. In the context of adsorption, the ability of aluminium oxide to achieve surface ion exchange should be mentioned, but the natural clay mineral, *clinoptilolite,* can also be used for waste water treatment as it has a high selectivity for removal of ammonium ions.

Synthetic ion exchange resins consist of a network of compounds of high molecular weight to which ionic functional groups are attached. The molecules are cross-linked in a three-dimensional matrix and degree of the cross-linking determines the internal pore structure of the resin. Since ions must diffuse into and out of the resin, ions larger than a given size may be excluded from the interaction through a selection dependent upon the degree of cross-linking. However, the nature of the groups attached to the matrix also determines the ion selectivity and thereby the equilibrium constant for the ion exchange process.

The cation exchangers contain functional groups such as sulphonic $R\text{-}SO_3\text{-}H$ -, carboxylic R-COOH -, phenolic R-OH, and phosphonic $R\text{-}PO_3H_2$ (R represents the matrix). It is possible to distinguish between strongly acidic cation exchangers derived from a strong acid, such as H_2SO_4, and weakly acidic ones derived from a weak acid, such as H_2CO_3.

It is also possible to determine a pK-value for the cation exchangers in the same way as it is for acids generally.

This means: $R\text{-}SO_3H - R\text{-}SO_3^- + H^+$

$$[[H^+] * [R\text{-}SO_3^-] / R\text{-}SO_3H] = K; \; pK = -\log K \qquad (5.33)$$

Anion exchange resins contain such functional groups as *primary amine,* $R\text{-}NH_2$, *secondary amine,* $R\text{-}R_1NH$, and *tertiary amine* $R\text{-}R_1\text{-}R_2N$ groups and the *quarternary ammonium group* $R\text{-}R_1R_2R_3N^+OH^-$.

It can be seen that the anion exchanger can be divided into weakly basic and strongly basic ion exchangers derived from quarternary ammonium compounds. It is also possible to introduce ionic groups onto natural material. This is done using cellulose as a matrix, and due to the high porosity of this material it is possible to remove even high molecular weight ions. Preparation of cation exchange resin, using hydrocarbon molecules as a matrix, is carried out by polymerisation of such organic molecules as styrene and metacrylic acid.

The degree of cross-linking is determined by the amount of divinylbenzene added to the polymerisation. This can be illustrated by the example shown below.

It is characteristic that the exchange occurs on an equivalent basis. The capacity of the ion exchanger is usually expressed as equivalent per litre of bed volume. When the ion exchange process is used for reduction of hardness, the capacity can also be expressed as kg of calcium carbonate per m^3 of bed volume. Since the exchange occurs on an equivalent basis, the capacity can be found based either on the number of ions removed or the number of ions released. Also, the quantity of regenerant required can be calculated from the capacity. However, neither the resin nor the regeneration process can be utilised with 100% efficiency.

The Fig. 5.23 plot is often used as an illustration of the preference of an ion exchange resin for a particular ion. As can be seen, the percentage in the resin is plotted against the percentage in solution.

The selectivity coefficient, SK, is not actually constant, but is dependent upon experimental conditions. A selectivity coefficient of 50% in solution is often used = $SK_{50\%}$. If we use concentration and not activity, it will involve, for monocharged ions A and B:

$$C_B = C_A$$

$$SK_{50\%} = C_{RA} / C_{RB} \qquad (5.34)$$

The plot in Fig. 5.23 can be used to read C_{RA} = A expressed as % and C_{RB} = B expressed as %. The selectivity of the resin for the exchange of ions is dependent upon the ionic charge and the ionic size. An ion exchange resin generally prefers counter ions of high valence. Thus, for a series of typical anions of interest in waste water treatment one would expect the following order of selectivity:

$PO_4^{3-} > SO_4^{2-} > Cl^-$.

Similarly for a series of cations:

$AL^{3+} > Ca^{2+} > Na^+$.

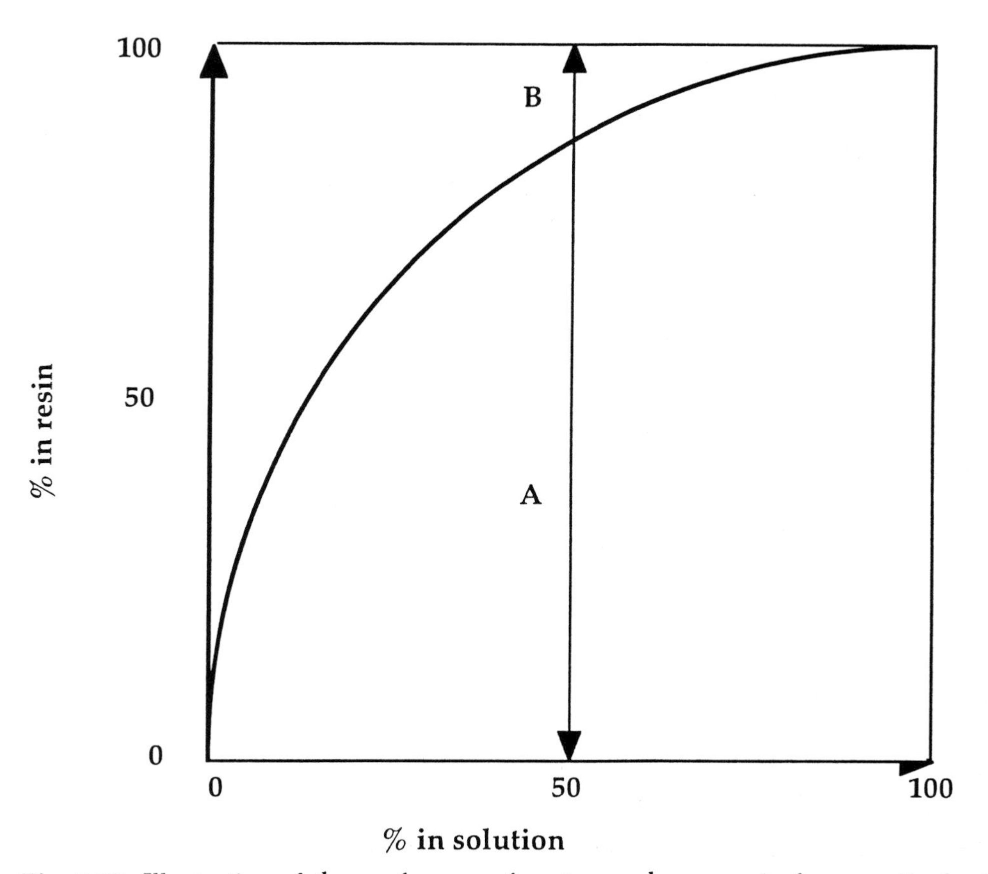

Fig. 5.23 Illustration of the preference of an ion exchange resin for a particular ion. It is shows how the selectivity at 50% is found: $SK_{50\%}$ = A / B. 100% corresponds to 100% for ion A and 0% for ion B.

But this is under circumstances where the internal pore structure of the resin is sufficiently large to allow the reactions of the considered ions. Organic ions are often too large to penetrate the matrix of an ion exchange, which is, of course, more pronounced when the resins considered have a high degree of cross-linking. As most kinds of water and waste water contain several types of ions besides those which must be removed it is naturally a great advantage to have a resin with a high selectivity for the ions to be removed during the ion exchange process.

The resin utilisation is defined as the ratio of the quantity of ions removed during the actual treatment to the total quantity of ions that could be removed at 100% efficiency; this is the theoretical capacity. The regeneration efficiency is the quantity of ions removed from the resins compared to the quantity of ions present in the volume of the regenerant used. Weakly base resin has significant potential for removing certain organic compounds from water, but the efficiency is highly dependent upon the pH.

It seems reasonable to hypothesise that an adsorption is taking place in the formation of a hydrogen bond between the free amino groups of the resin and the hydroxyl groups of the organic substance taken up. As pH decreases, so that the amino groups are converted to their acidic form, the adsorption capacity significantly decreases.

The exchange reaction between ions in solution and ions attached to the resin matrix is generally reversible. The exchange can be treated as a simple stoichiometric reaction. For cation exchange the equation is:

$$A^{n+} + n(R^-)B^+ - nB^+ + (R^-)_nA^{n+} \tag{5.35}$$

The ion exchange reaction is selective, so that the ions attached to the fixed resin matrix will have preference for one counter ion over another. Therefore the concentration of different counter ions in the resin will be different from the ratio of concentrations in the solution. According to the law of mass action, the equilibrium relationship for reaction (5.36) will give:

$$K_{AB} = \frac{a_B{}^n * a_{RA}}{a_A * a_{RB}{}^n} \tag{5.36}$$

where a_B and a_A are the activities of the ions B^+ and A^{n+} in the solution and correspondingly a_{RB} and a_{RA} are the activities of the resin in B- and A-form, respectively.

The clay mineral, clinoptilolite, can take up ammonium ions with a high selectivity. This process is used for the removal of ammonium from municipal waste water in the U.S.A., where good quality clinoptilolite occurs. Clinoptilolite has less capacity than the synthetic ion exchanger, but its high selectivity for ammonium justifies its use for ammonium removal. The best quality clinoptilolite has a capacity of 1 eqv. or slightly more per litre. This means that 1 litre of ion exchange material can remove 14 g ammonium-N from waste water, provided all the capacity is occupied by ammonium ions. Municipal waste water contains approximately 28 g (2 eqv.) per m^3, which means that 1 m^3 of ion exchange material can treat 500 m^3 waste water (which represents a capacity of 500 bed volumes). The practical capacity is, however, considerably less - 150-250 bed volumes - due to the presence of other ions that are taken up by the ion exchange material, although the selectivity is higher for ammonium than for the other ions present in the waste water. The concentration of sodium, potassium and calcium ions might be several eqv. per litre, compared with only 2 eqv. per litre of ammonium ions.

Clinoptilolite is less resistant to acids or bases than are synthetic ion exchangers. Good elution is obtained by the use of sodium hydroxide, but as the material dissolves in sodium hydroxide a very diluted solution should be used for elution to minimise the loss of material. A mixture of sodium chloride and lime is also suggested as an alternative elution solution. The flow rate through the ion exchange column is generally smaller for clinoptilolite than for synthetic material resin - 10 m as against 20-25 m. The elution liquid can be recovered by air stripping as previously mentioned. The preconcentration on the ion exchanger makes this process attractive - the sludge problem is diminished and the cost of chemicals reduced considerably. For further details about this method of recovery, see Jørgensen (1975). A biological regeneration which implies that there will be no elution liquid is also possible as reported by Lahav and Green (1998). This regeneration method is more cost effective than a chemical regeneration but it may also be more time consuming and more difficult to control at optimum conditions.

Phosphate can also be removed by an ion exchange process. Synthetic ion exchange material can be used, but due to the presence of several eqv. of anions, it seems a better solution to use activated aluminium, which is a selective adsorbent (ion exchanger) for phosphate. In accordance with Neufeld and Thodos (1969) the phosphorus adsorption on aluminium oxide is a dual process of ion exchange and chemical reaction, although many authors refer to the process as an adsorption. The selectivity for phosphorus uptake is very high (Yee, 1960), and only a minor disturbance from other ions is observed. Municipal waste water contains about 10 mg phosphorus or 1/3 eqv. P per 1 m^3. If the aluminium oxide is activated by nitric acid before use, the theoretical capacity will be as high as 0.561 eqv. per litre or 1500-3000 bed volumes. The practical capacity is lower - about 1000 bed volumes (Ames, 1969). Also activated aluminium oxide can be regenerated by sodium hydroxide, but again the use of a low concentration is recommended to reduce loss of material (Ames and Dean, 1970). Aluminium oxide has a high removal efficiency for all inorganic phosphorus species in waste water, but the removal efficiency for organic phosphorus compounds is somewhat lower. Jørgensen (1978) has suggested the use of an anionic cellulose exchanger together with activated aluminium oxide as a mixed bed ion exchange column to improve the overall efficiency for removal of all phosphorus species present in waste water, including the organic ones. A combination of chemical precipitation and ion exchange has developed as an alternative to the mechanical-biological-chemical treatment method, but it is only used at a few places.

5.4 REMOVAL OF TOXIC ORGANIC COMPOUNDS

The problem and source of toxic organic compounds

The presence of toxic organic compounds in waste water is causing severe problems:

1. Since toxic organics are scarcely decomposed in the biological plant, life in the receiving water will suffer from their toxic effects. If the compound is lipophilic it will mainly be accumulated in the sludge which will contaminate terrestrial ecosystems if it is applied as soil conditioner. The fate of toxic substances in aquatic ecosystems has already been covered in Chapter 4
2. Inhibition of the biological treatment might reduce the efficiency of this process considerably.

Here the inhibition of biological treatment will be considered in more detail, in relation to heavy metals as well as toxic organic compounds. **The biological reactions are influenced by the presence of inhibitors. In the case of competitive inhibition, the Michaelis-Menten equation (see Sections 2.3, 2.4 and 6.2) becomes:**

$$\mu = \mu_m \frac{S}{S + K_S * (K_{S,I} + I)/K_{S,I}} \tag{5.37}$$

where $K_{S,I}$ = inhibition constant, and I = concentration of the inhibitor(s). μ symbolises the growth rate and μ_m the maximum growth rate. S represents as usually the substrate concentration. Competitive inhibition occurs when the inhibitor molecule has almost the same structure as the substrate molecule, which means that the micro-organism is able to break down the inhibitor and the substrate by the same, or almost the same, biochemical pathways. The resulting influence on the Lineweaver-Burk plot is shown in Fig. 5.24. If the plot can be drawn sufficiently accurately, it will be possible to read the inhibition constant from the plot.

If a toxic compound (non-competitive inhibitor) is present, only the maximum growth rate will be influenced, and it is reduced according to the relationship:

$$\mu_m^{+} = \mu_m * \frac{K_{S,I}}{K_{S,I} + I} \tag{5.38}$$

In this case the Lineweaver-Burk plot is also changed (see Fig. 5.24). $K_{S,I}$ can be found from the plot, provided this is sufficiently accurate to allow a proper reading of the difference between the presence and non-presence of the toxic substance. Heavy metals and cyanide are examples of toxic materials that inhibit non-competitively. The approximate values of $K_{S,I}$

for some non-competitive inhibitors are given in Table 5.24.

Uptake of toxic substances (organic compounds as well as heavy metals) by the biomass including sorption to the formed bioflocs depends on several factors, including pH and the concentration of organic matter and metals present in the system. For organic compounds with a high K_{OW} the sorption may be very significant (see also equation (2.28), Section 2.5) A higher initial concentration of the toxic substance or sludge increases the overall uptake. In general, the uptake capacity increases with increasing pH. Although the affinity of the biomass for the sorption of toxic substances is relatively small, it is generally higher than that of competing organics in a supernatant (Jacobsen et al., 1993). The large-scale accumulation of some toxic substances by activated sludge with its subsequent removal in a secondary clarifier explains the significant reduction observed in many treatment plants.

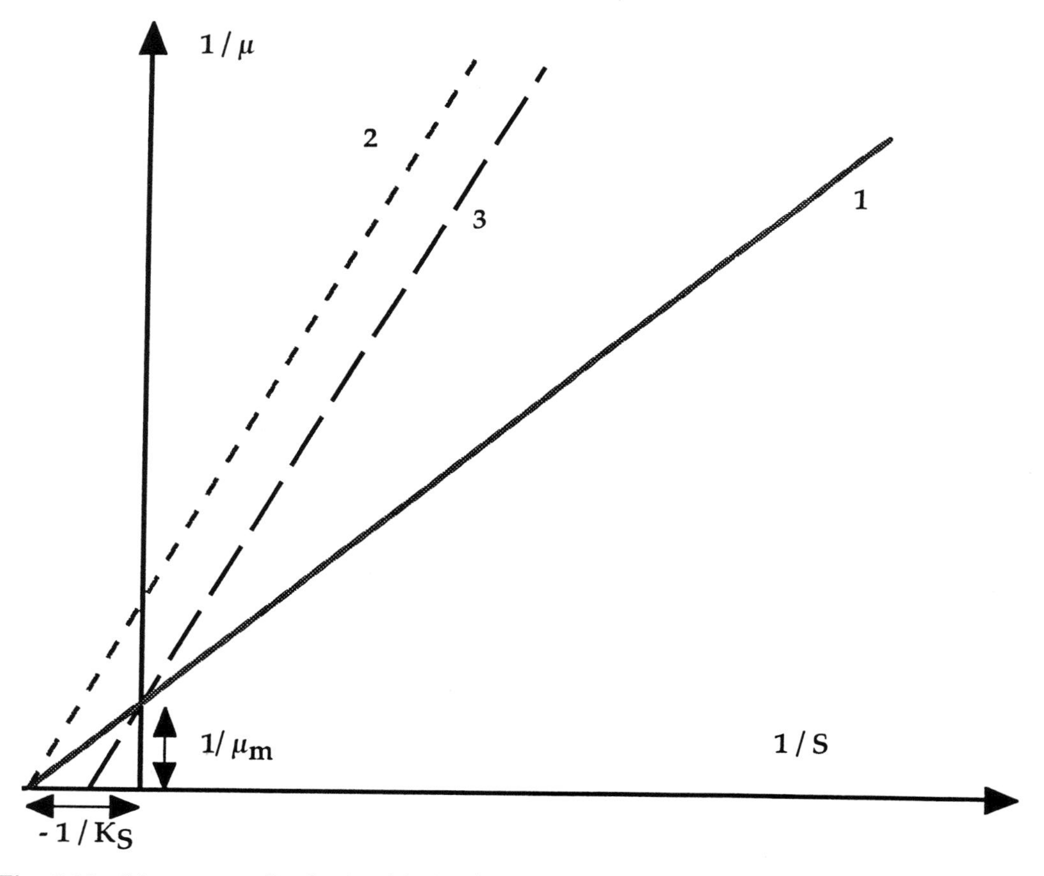

Fig. 5.24. Lineweaver-Burk plot (1); (2) the plot when a toxic compound is present; (3) the plot when competitive inhibition takes place.

In general industrial waste water contains toxic organics, which consequently **must be removed** before this waste water is discharged into the municipal sewage system. Only in this way can the two problems mentioned above be solved simultaneously.

Compounds of minor toxicity and with a biodegradation rate slightly slower than average for municipal waste water can be tolerated in municipal waste water at low concentrations. However, industrial waste water containing medium or high levels of toxic compounds should not be discharged into the municipal sewage system without effective treatment. A first estimation of the relationship between the composition of an organic compound and its biodegradability, has already been presented in Section 4.6.

TABLE 5.24
Effect of inhibitors

Non-competitive inhibitor	$K_{S,I}$ (mg/l)
Hg	2
Ag	5
Co	10
Cu	20
Ni	40
Cr^{6+}	200
CN^-	200-2000
2-nitrophenol	20
PCB mixture	3

Apart from biological treatment with acclimatised sludge, which is applicable in some cases, the following methods might come into consideration for treatment of toxic organics:

1. **Separators** for removal of oil.
2. **Flocculation.** With flocculation it is possible to remove a wide range of organic colloids.
3. **Extraction.** This method is generally rather costly, but recovery of chemicals will often justify the high cost. The method is mostly used when high concentrations are present in the waste water.
4. **Flotation** is of special interest when impurities with a specific gravity of less than 1 are present in the waste water.
5. **Adsorption.** A wide range of organic compounds, such as insecticides and dye stuffs, can be adsorbed by activated carbon.
6. **Sedimentation** is only used in conjunction with removal of suspended matter or in

combination with chemical precipitation or flocculation and biological treatment.

7. **Chemical oxidation and reduction** are mainly used for the treatment of cyanides, some toxic dye stuffs and chromates. The process might be rather costly. Wet oxidation with oxygen at a high temperature has been proposed (Mishara et al., 1994). Oxidation in combination with use of ultrasonic sound (sonolysis) seems to enhance the decomposition processes (Olson and Barbier, 1994)

8. **Distillation** of waste water is used when the recovery of solvents is possible or when other methods are not available; for the treatment of radioactive waste water, for example. The method is very costly.

TABLE 5.25

Use of chemical precipitation for treatment of industrial waste water

Type of waste water	Chemical used	Reference
Metal plating and finishing industry	Lime	Schjødtz-Hansen, 1968
Iron industry and mining	Lime, aluminium sulphate	S.E. Jørgensen, 1979a
Electrolytic industry	Hydrogen sulphide	S.E. Jørgensen, 1979d
Coke and tar industry	Lime or sodium hydroxide	S.E. Jørgensen, 1979d
Cadmium mining	Xanthates	Hasebe & Yamamoto, 1970
Manufacturing of glass- and stone wool	Sodium hydroxide	Schjødtz-Hansen & Krogh, 1968
Oil refineries	Aluminium sulphate, iron(III) chloride	S.E. Jørgensen, 1979d
Manufacture of organic chemicals	Aluminium sulphate, iron(III) chloride	S.E. Jørgensen, 1979d
Photo chemicals	Aluminium sulphate	S.E. Jørgensen, 1979d
Dye industry	Iron(II) salts, aluminium sulphate, lime	S.E. Jørgensen, 1979d
Fertiliser industry	PO_4^{3-}: iron(II) salts, aluminium sulphate, lime, NH_4^+: magnesium sulphate + phosphate	S.E. Jørgensen, 1979d
Plastics industry	Lime	S. E. Jørgensen, 1979d
Food industry	Lignin sulphonic acid, dodecylbenzen-sulphonic acid, glucose trisulphate, iron(III) chloride, aluminium sulphate	S.E. Jørgensen, 1979d
Paper industry	Bentonite, kaolin, polyacrylamide	S.E. Jørgensen, 1979d
Textile industry	Bentonite, aluminium sulphate	S.E. Jørgensen, 1979d

9. Organic acids and bases can be removed by **ion exchange.**

10. **Filtration** can be used for the removal of suspended matter from small quantities of waste water.

11. **Neutralisation**. In all circumstances it is necessary to discharge the waste water with a pH between 6 and 8. Calcium hydroxide, sulphuric acid and carbon dioxide are used for the neutralisation process.

12. **Biodegradation (aerobic or anaerobic)** by use of adapted microorganisms (for examples see Arcangeli et al., 1996 and Ettala et al., 1992.)

The processes numbered 2, 5 and 7 are discussed in more detail below, as they are the most important processes for treatment of industrial waste water containing toxic organics.

Application of chemical precipitation for treatment of industrial waste water

Chemical precipitation for the removal of phosphorus compounds has been mentioned in Section 5.3. At the same time as it precipitates phosphorus compounds, the process will reduce the BOD_5 by approximately 50-70%. This effect is mainly a result of adsorption on the flocs formed by the chemical precipitation. As this is a chemical-physical effect it is understandable that the COD is also reduced by 50-70%, while the biological treatment will always show a smaller effect on COD removal than on BOD_5 reduction. The chemical precipitation used for municipal waste water - precipitation with aluminium sulphate, iron(III)chloride and calcium hydroxide - is therefore also able to remove many organic compounds, toxic as well as non-toxic, which means that the process is applicable for the treatment of some types of industrial waste water. In addition to these three precipitants a wide range of chemicals is used for the precipitation of industrial waste water: hydrogen sulphide, xanthates, sodium hydroxide, bentonite, kaolin, starch, polyacrylamide, lignin, sulphonic acid, dodecylbenzensulphonic acid, glucose trisulphate. Table 5.25 gives a survey, with references, of the use of chemical precipitation for different types of industrial waste water.

Application of adsorption for treatment of industrial waste water

Adsorption involves accumulation of substances at an interface, which can either be liquid-liquid, gas-liquid, gas-solid or liquid-solid. The material being adsorbed is termed the adsorbate and the adsorbing phase the adsorbent. The word sorption, which includes both adsorption and absorption, is generally used for a process where the components move from one phase to another, but particularly in this context when the second phase is solid. A solid surface in contact with a solution has the tendency to accumulate a surface layer of solute molecules, because of the imbalance of surface forces, and so an adsorption takes place. The adsorption results in the formation of a molecular layer of the adsorbate on the surface. Often *an equilibrium concentration is rapidly formed at the surface* and is generally followed by slow diffusion onto the particles of the adsorbent.

The rate of adsorption is generally controlled by the rate of diffusion of the solute molecules. The rate varies reciprocally with the square of the diameter of the particles and increases with increasing temperature (Weber and Morris, 1963). For practical application either Freundlich's isotherm or Langmuir's isotherm provides a satisfactory relationship between the concentration of the solute and the amount of adsorbed material (these adsorption isotherms are mentioned in Section 2.5).

The values of k and b (see equation (2.25)) are given for several organic compounds in Table 5.26 (Rizzo and Shepherd, 1977).

Langmuir's adsorption isotherm is based on equation (2.26). The Langmuir constant for several organic compounds that can be adsorbed on activated carbon was found by Weber and Morris (1964). Most types of waste water contain several substances that will be adsorbed, and in this case a direct application of Langmuir's adsorption isotherm is not possible. Weber and Morris (1965) have developed equations (5.38) and (5.39) for competitive adsorption of two substances (A and B). In other words, competitive adsorption can be described in almost the same way as a competitive enzymatic reaction, although also the pore size relative to the size of the molecules may also be of importance (Pelekani and Snoeyink, 1999):

$$a_A = \frac{A_{Ao} * C_A}{1 + b_A * C_A + b_B * C_B} \qquad (5.39)$$

$$a_B = \frac{A_{Bo} * C_B}{1 + b_A * C_A + b_B * C_B} \qquad (5.40)$$

When it is necessary to find whether Freundlich's adsorption isotherm or Langmuir's adsorption isotherm gives the best fit to a set of data, the two plots in Figs. 2.10 and 2.11 should be applied.

It is possible, to a certain extent, to predict the adsorption ability of a given component. The solubility of the dissolved substance is by far the most significant factor in determining the intensity of the driving forces.The greater the affinity of a substance for the solvent, the less likely it is to move towards an interface to be adsorbed. For an aqueous solution this means that the more hydrophilic the substance is, the less likely it is to be adsorbed. Conversely, hydrophobic substances will be readily adsorbed from aqueous solutions. This is in complete accordance with equation (2.27).

Many organic components, e.g. sulphonated alicylic benzenes, have a molecular structure consisting of both hydrophilic and hydrophobic groups. The hydrophobic parts will be adsorbed at the surface and the hydrophilic parts will tend to stay in the water phase.

TABLE 5.26
Freundlich's constant for adsorption of some organic compounds on activated carbon*

Compound	k	b
Aniline	25	0.322
Benzene sulphonic acid	7	0.169
Benzoic acid	7	0.237
Butanol	4.4	0.445
Butyraldehyde	3.3	0.570
Butyric acid	3.1	0.533
Chlorobenzene	40	0.406
Ethyl acetate	0.6	0.833
Methyl ethyl ketone	24	0.183
Nitrobenzene	82	0.237
Phenol	24	0.271
Phenol	24	0.271
TNT	270	0.111
Toluene	30	0.729
Vinyl chloride	0.37	1.088

* a in mg/g when C in mg/l

The sequential operation is frequently called "contact filtration", because the typical application includes treatment in a mixing tank followed by filtration, but more frequently settling is used for removal of the used adsorbents in industrial waste water engineering. The sequential adsorption operation is limited to treatment of solutions where the solute to be removed is adsorbed relatively strongly when compared with the remainder of the solution. This is often the case when colloidal substances are removed from aqueous solutions using carbon, as in the production of process water.

The method for dealing with the spent adsorbent depends upon the system under consideration. If the adsorbate is valuable material, it might be desorbed by contact with a solvent other than water. If the adsorbate is volatile, it may be desorbed by reduction of the partial pressure of the adsorbate over the solid by passing steam or air over the solid. In the case of most sequential adsorption operations in the context of waste water treatment, the adsorbate is of no value and it is not easily desorbed. The adsorbent may then be regenerated by burning off the adsorbate, followed by reactivation.

In the continuous operation, the water and the adsorbent are in contact throughout the entire process without a periodic separation of the two phases. The operation can either be carried out either in strictly continuous steady-state fashion by movement of the solid as well as the fluid, or in a semi continuous fashion characterised by moving fluid but stationary solid so-called fixed adsorption. Due to the inconvenience and relatively high cost of continuously transporting solid particles, it is generally found more economical to use a stationary bed of adsorbent for waste water treatment.

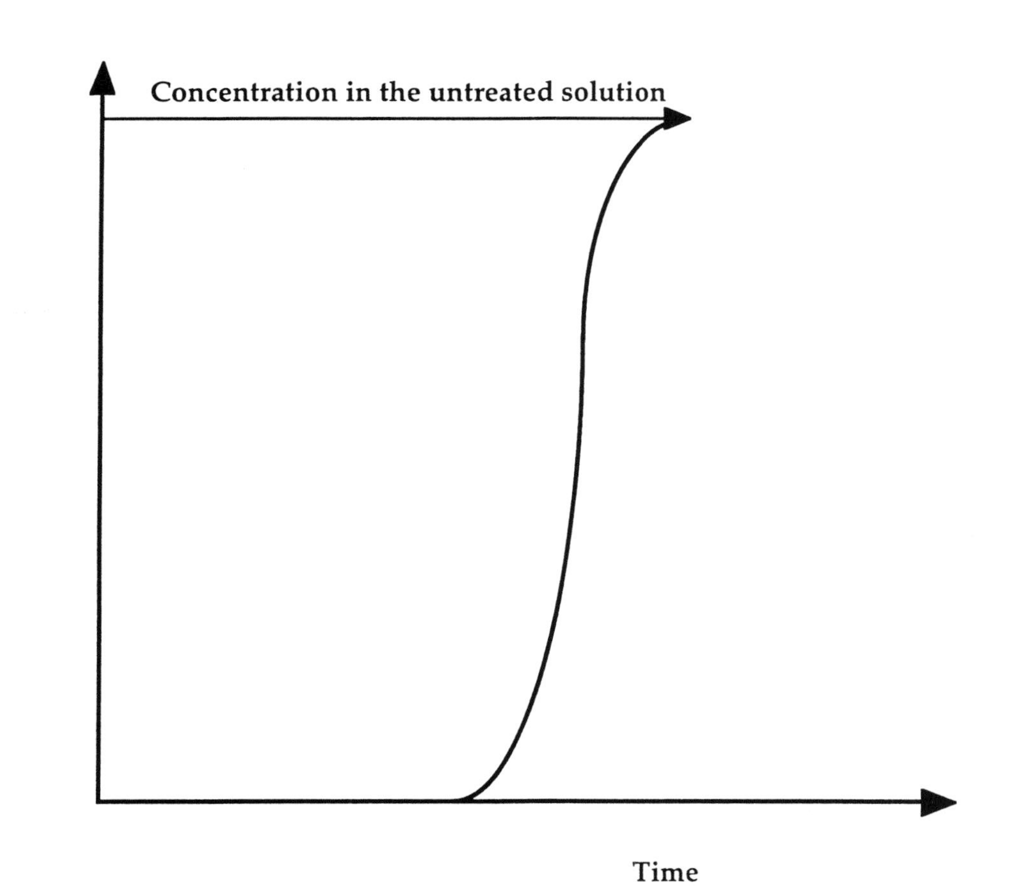

Fig. 5.25. Idealised breakthrough curve. Concentration versus time or treated volume.

The design of a fixed bed adsorber and the prediction of the length of adsorption cycle require knowledge of the percentage approach to saturation at the break point. Fig. 5.25 shows an idealised breakthrough curve. The extent of adsorption is proportional to the surface area. In order to compare different adsorbents a specific surface area, defined as that portion of the total surface area that is available for adsorption per unit of adsorbent, is used. This means that the adsorption capacity of a non-porous adsorbent should vary inversely with the particle diameter, while in the case of highly porous adsorbent, the capacity should be almost independent of the particle diameter. However, for some porous material, such as activated carbon, the breaking up of large particles to form smaller ones opens some tiny sealed channels in the column, which might then become available for adsorption (Weber and Morris, 1964).

The nature of the adsorbate also influences the adsorption. **In general an inverse relationship can be anticipated between the extent of adsorption of a solute and its solubility in the solvent (water) from which adsorption occurs. This is the so-called Lundilius' rule,** which may be used for semi quantitative prediction of the effect of the chemical character of the solute on its uptake from solution (water) (Lundilius, 1920).

Ordinarily, the solubility of any organic compound in water decreases with increasing chain length because the compound becomes more hydrophobic as the number of carbon atoms increases. This is Traube's rule.

Adsorption from aqueous solution increases as homologous series are ascended, largely because the expulsion of increasingly large hydrophobic molecules from water permits an increasing number of water-water bonds to form which illustrates very well the above-mentioned Traube's rule. The molecular weight is also related to the rate of uptake of solutes by activated carbon, if the rate is controlled by intraparticle transport. **The molar rate of uptake decreases with increasing molecular weight.** pH strongly influences the adsorption as hydrogen and hydroxide ions are adsorbed, and the charges of the other ions are influenced by the pH of the water. This is the same role of pH as mentioned in Section 4.4., namely the change from acid to ionic form of organic acids. The acid form is more readily adsorbed to activated carbon or soil as it is less soluble in water than the ionic form. For typical organic pollutants from industrial waste water, **the adsorption decreases therefore with increasing pH.** Normally, the adsorption reactions are exothermic, which means that the adsorption will increase with decreasing temperature, although small variations in temperature do not tend to alter the adsorption process to a significant extent.

Adsorption can be used to remove several organic compounds, such as phenol, alkylbenzene-sulphonic-acid, dye stuffs and aromatic compounds from waste water by the use of activated carbon (Edwards, 1995).

Scaramelli and Dibiano (1973) have examined the effect (on effluent quality) of adding powdered activated carbon to an activated sludge system.

They found that 100 to 200 mg/l were able to reduce TOC from about 20 mg/l to 7 mg/l.

Activated carbon has also been suggested as an adsorbent for the removal of refractory dye stuffs (Eberle et al., 1976).

In very small plants it may be feasible to use granular carbon on a **use and throw away basis**, although economics probably favour the use of powdered carbon in a sequential operation. The use of granular activated carbon involves the regeneration and reuse of carbon, with some exceptions. This **regeneration can be carried out with sodium hydroxide** provided that high molecular weight colloids are removed before the treatment of the activated carbon. It is possible with this **chemical regeneration** to *recover phenols,* for example (Jørgensen, 1976b), and *to remove coloured bodies* (Chamberlin et al., 1975; Mulligan and Fox, 1976). Also **solvent** can be used for regeneration of activated carbon, as indicated by Rovel (1972). When sodium hydroxide or solvent is used, the adsorbate is passed through the carbon bed in the opposite direction to that of the service cycle, until all is removed. The bed is then drained and the regenerated carbon is ready to go back into the stream.

Juhola and Tupper (1969) have studied the **thermal regeneration** of granular activated carbon, which consists of three basic steps, (1) **drying**, (2) **baking of adsorbate** and (3) **activation by oxidation** of the carbon residues from decomposed adsorbates. Drying requires between 100 and 700°C and activation a temperature above 750°C. All three steps can be carried out in a direct fired, multiple hearth furnace. This is the best commercial equipment available for regeneration of carbon for use in combination with waste water treatment. The capacity of the activated carbon will generally decrease by approximately 10% during the first thermal regeneration and another 10% during the next 5-10 regenerations (Mazet a et al., 1994).

Bioregeneration of powdered activated carbon (Jonge et al., 1996) or use of bacterial inoculation for removal of adsorbed components (Jones et al., 1998) have lately been proposed as alternatives to chemical and thermal regeneration.

Electrochemical regeneration is claimed to give a lower reduction of the adsorption capacity than chemical or thermal regeneration (Narbaitz and Cen, 1998).

<u>Application of chemical oxidation and reduction for Treatment of industrial waste water</u>

By chemical oxidation and reduction the oxidation stage of the substance is changed. Oxidation is a process in which the oxidation stage is increased, while chemical reduction is a process in which the oxidation stage is decreased. To illustrate the use of chemical oxidation a chlorination can be considered: e.g. $Cl_2 + C_2H_6$ ------> $C_2H_5Cl + HCl$.

Several aspects are considered in the selection of a suitable oxidising agent for industrial waste water treatment. These are:

1. Ideally, no residue of oxygen should remain after the treatment and there should be no residual toxic or other effects.
2. The effectiveness of the treatment must be high.
3. The cost must be as low as possible.
4. The agent should be easy o handle.

Only a few oxidants are capable of meeting these requirements. The following oxidants are in use today for treatment of water and waste water:

1. **Oxygen or air, including wet oxidation at elevated temperature.**
2. **Ozone.**
3. **Potassium permanganate.**
4. **Hydrogen peroxide.**

5. **Chlorine.**
6. **Chlorine dioxide.**

Oxygen is significant in biological oxidation, but also plays an important role in chemical oxidation. The primary attraction of oxygen is that it can be applied in the form of air. Organic material, including phenol, can be catalytically oxidised by the use of suitable catalysts, such as oxides of copper, nickel, cobalt, zinc, chromium, iron, magnesium, platinum and palladium, but, as yet, this process has not been developed for use on a technical scale. Wet oxidation at elevated temperature and pressure has lately been proposed as a possible methods to remove very toxic substances (Thomsen, 1998).

Ozone is a more powerful oxidant than oxygen and is able to react rapidly with a wide spectrum of organic compounds and microorganisms present in waste water. It is produced from oxygen by means of electrical energy, which is a highly attractive process since air is used. Ozonators are today available in relatively small units which makes it attractive to use ozone as oxidator also for small water volumes. One of the advantages of using ozone is that it does not impart taste and odour to the treated water. Ozone is used in the following areas of water treatment:

1. **Removal of colour, taste and odour** (Holluta, 1963 and Hwang et al., 1994).
2. **Disinfection.**
3. **For the oxidation of organic substances, e.g. phenol, surfactants** (Eisenhauer, 1968; Wynn et al., 1972), **quinoline** (Andreozzi et al., 1992), **diazinon** (Ku et al., 1998), **atrazine** with or without the use of UV radiation (Beltran et al., 1994a and b), **p-nitrophenol** (Beltran et al., 1992), **pyrene** (Yao et al., 1998) **and cyanides** (Anon, 1958; Khandelwal et al., 1959).

The ozonation may be catalysed by various metals (for instance manganese, Andreozzi et al., 1998). Sonolysis is able to enhance rates of decomposition of organic matter by ozone (Olson and Barbier, 1994).

The solubility of ozone is dependent upon the temperature. Henry's coefficient as a function of temperature for ozone and oxygen is shown in Table 5.27. Henry's coefficient, H, is used in Henry's law (see equation (2.23) Section 2.5).

Ozone has low thermodynamic stability at normal temperature and pressure (Kirk-Othmer, 1967). It decomposes both in the gas phase and in solution. Decomposition is more likely in aqueous solutions, where it is strongly catalysed by hydroxide ions; see (Stumm and Morgan 1996). The oxidation of phenol by gaseous ozone has been studied by Gould and Weber. (1976) under a number of conditions. Virtually complete removal of phenol and its aromatic degradation products is realised when 4-6 moles of ozone have been consumed for each mole of phenol originally present. At this point approximately 1/3 of the initial organic carbon will remain and 70-80% reduction of the COD-number will have been achieved. Concentrations of the non-aromatic degradation products will be less than 0.5 mg/l. Subsequent dilution of the discharge of this effluent should reduce the concentration of the various components in the receiving body of water to tolerable levels. Ozone is extremely toxic, having a maximum tolerable concentration of continuous exposure of 0.1 ppm. However, the half-life of ozone is reduced by high pH as demonstrated in Table 5.28.

TABLE 5.27
Henry's coefficient for oxygen and ozone in water

	H * 10^{-4}	
Temperature (°C)	Oxygen	Ozone
0	2.5	0.25
5	2.9	0.29
10	3.3	0.33
15	3.6	0.38
20	4.0	0.45
25	4.4	0.52
30	4.8	0.60
35	5.1	0.73
40	5.4	0.89
45	5.6	1.0
50	5.9	1.2

TABLE 5.28
Influence of pH on half-life of ozone in water (Stumm, 1958)

pH	Half-life (min)
7.6	41
8.5	11
8.9	7
9.2	4
9.7	2
10.4	0.5

Permanganate is a powerful oxidising agent and is widely used by many municipal water plants for taste and odour control and for the removal of iron and manganese. Furthermore, it can be used as an oxidant for the removal of impurities such as Fe^{2+}, Mn^{2+}, S^{2-}, CN^{-} and phenols present in industrial waste water.

In strongly acidic solutions permanganate is able to take up five electrons:

$$MnO_4^- + 8H^+ + 5e^- \text{------>} 4H_2O + Mn^{2+} \quad (5.41)$$

Whereas in the pH range from approximately 3 to 12 only three electrons are transferred and the insoluble manganese dioxide is formed:

$$MnO_4^- + 4H^+ + 3e^- \text{----->} 2H_2O + MnO_2 \quad (5.42)$$

or

$$MnO_4^- + 2H_2O + 3e^- \text{------->} 4OH^- + MnO_2 \quad (5.43)$$

The stoichiometry for the oxidation of cyanide in a hydroxide solution of pH 12-14 is:

$$2MnO_4^- + CN^- + 2OH^- \text{------>} 2MnO_4^{2-} + CNO^- + H_2O \quad (5.44)$$

In a saturated solution of calcium hydroxide (Posselt, 1966) the reaction takes the form:

$$2MnO_4^- + 3CN^- + H_2O \text{------>} 3CNO^- + 2MnO_2 + 2OH^- \quad (5.45)$$

The presence of calcium ions affects the rate of manganate(IV) production disproportionately:

$$3MnO_4^{2-} + 2H_2O \text{------->} 2MnO_4^- + 4OH^- + MnO_2 \quad (5.46)$$

Chlorine is known to be a successful **disinfectant** in waste water treatment, but it is also able to oxidise effectively such compounds **as hydrogen sulphide, nitrite, divalent manganese and iron and cyanide.** The oxidation effectiveness usually increases with increasing pH. Cyanide, which is present in a number of different industrial waste waters, is typically oxidised with chlorine at a high pH. The oxidation to the much less toxic cyanate (CNO^-) is generally satisfactory, but in other cases complete degradation of cyanide to carbon dioxide and nitrogen is required. The disadvantage of chlorine is that it forms aromatic chlorocompounds, which are highly toxic, e.g. chlorophenols, when phenol-bearing water is treated with chlorine (Aston, 1947). This fact has resulted in the more widespread use of chlorine dioxide. Chlorine dioxide is as unstable as ozone and must therefore be generated in situ.

The industrial generation of chlorine dioxide is carried out by means of a reaction between chlorine and sodium chlorite in acid solution (Granstrom and Lee, 1958):

$$Cl_2 + 2NaClO_2 - 2ClO_2 + 2NaCl \quad (5.47)$$

As chlorine dioxide is a mixed anhydride of chlorous and chloric acid, disproportionation to the corresponding anions occurs in basic solutions:

$$2ClO_2 + 2OH^- \text{------->} ClO_2^- + ClO_3^- + H_2O \qquad (5.48)$$

This process becomes negligible under acidic conditions. The equilibrium:

$$2ClO_2 + H_2O \text{------->} HClO_2 + HClO_3 \qquad (5.49)$$

shifts to the left at lower pH.

Chlorine dioxide is used for **taste and odour control**. It has been reported to be a selective oxidant for industrial waste water containing **cyanide, phenol, sulphides and mercaptans** (Whaler, 1976).

Hydrogen peroxide can be used as an oxidant for **sulphide** in water (Cole et al., 1976). Recently this oxidising ability has been applied to control odour, toxic substances and corrosion in domestic and industrial waste water. The oxidation rate may be enhanced by use of UV (Crittenden et al., 1999) or by use of ultrasound in combination with copper oxide as catalyst (Drijvers et al., 1999)

5.5 REMOVAL OF (HEAVY) METALS

The problem of heavy metals.

Heavy metals are the most harmful metals, but in the title the word heavy is shown in brackets to indicate that toxic metals other than heavy metals (sometimes defined as metals with a specific gravity > 5.0) are dealt with in this chapter.

1. **Toxic metals are harmful to aquatic ecosystems**. (For further details see Section 2.1 and 4.4). Treatment by a mechanical-biological process is, to a certain extent, able to remove metals, but the efficiency is only in the order of 30-70%, depending on the metal. The heavy metals removed are concentrated in sludge, which, even for plants treating solely municipal waste water, shows measurable concentrations. This is reflected in standards for sludge used as soil conditioner; see Section 4.5. As mentioned in Section 6.2, chemical precipitation offers high efficiency in the removal of heavy metals.

2. **Heavy metals are harmful to biological treatment and are example of non-competitive inhibitors**. The presence of significant concentrations of heavy metal ions has a particular pronounced effect on anaerobic digestion processes, for instance chromium(III) ions at concentrations of 1 g/l (Alkan et al., 1996). No adverse effect was, however, observed by the combined presence of Ni(II) and Cr (VI) (Dilek et al., 1998), while an effect on both the maximum growth rate and on the half saturation constant is observed for denitrifiation

process rates (Mazierski, 1994).

The effect on the results of activated sludge treatment is presented in Section 5.4.

Minor amounts of the least toxic metals (Zn for instance) may be discharged provided their contribution to the concentration in the municipal waste water is of little importance, but as demonstrated in Section 6.4 a small concentration of toxic metals can be tolerated.

Chiefly, four processes are applied for removal of heavy metals from industrial waste water:

1. **Chemical precipitation**
2. **Ion exchange**
3. **Extraction**
4. **Reverse osmosis**
5. **Adsorption** has also been suggested for the removal of mercury by Logsdon and Symons, (1973). 1 mg per litre of powdered carbon is needed for each 0.1 μg per litre of mercury to be removed.

It is often economically as well as feasibly preferable from the resource and ecological point of view, to recover heavy metals. The increasing cost of metals, foreseen for the coming decades, will probably provoke more industries to select such solutions to their waste water problems. Partial or complete recirculation of waste water should also be considered, since a decrease in water consumption and in loss of material is achieved simultaneously; in many cases there is even a considerable saving in water to be obtained by reorganisation of processes.

<u>The application of chemical precipitation for removal of heavy metals.</u>

One of the processes most used for the removal of metal ions from water is precipitation, as metal hydroxide. Table 5.29 lists the solubility products for a number of metal hydroxides.

From the solubility product it is possible to find the pH value at which precipitation will start for a given concentration of the metal ions. Fig. 5.26 shows the solubilities of metal ions at various pH values. By means of this diagram it is possible to find the concentration in solution at any given pH. For example, at pH 6.0 the concentration of Cr^{3+} is 10^{-6} M. The same concentration for Zn^{2+} is obtained at pH 8.0. The slopes of the lines in Fig. 65.26 correspond to the valency of the metal ions. The slope of the Cr^{3+} and Al^{3+} lines, for example, is +3, while the other ions in the figure have lines with a slope of +2. This is obtained from the solubility product:

$$[Me^{n+}]\,[OH^-]^n = K_s \qquad (5.50)$$

$$\log\,[Me^{n+}] = -pK_s + n * pOH = -pK_s + 14 - n * pH \qquad (p = \text{"-log"})$$

From the basis of the solubility product it is possible to find one point though which the line passes. For example, for the solubility product of iron(II) hydroxide, which is 10^{-15}, we have $[Fe^{2+}]\ [OH^-]^2 = 10^{-15}$. The line will therefore go through the point -log c = 5 and pOH = 5 (c = concentration of Fe^{2+}), since $(10^{-5})^3 = 10^{-15}$. Control of pH makes it possible to achieve easy automatic control of the dosage of chemical (either sodium hydroxide or calcium hydroxide is applied in most cases).

The presence of ligands which may form complexes with the heavy metal ions may prevent the precipitation or inhibit the growth of metal hydroxide particles and thereby reduce the settling rate significantly (Luo et al, 1992)

TABLE 5.29

pK_s values at room temperature for metal hydroxides

$pK_s = -\log K_s$, where $K_s = [Me^{z+}]\ [OH^-]^z$

Hydroxide	z = charge of metal ions	pK_s
$AgOH(1/2\ Ag_2O)$	1	7.7
$Cu(OH)_2$	2	20
$Zn(OH)_2$	2	17
$Ni(OH)_2$	2	15
$Co(OH)_2$	2	15
$Fe(OH)_2$	2	15
$Mn(OH)_2$	2	13
$Cd(OH)_2$	2	14
$Mg(OH)_2$	2	11
$Ca(OH)_2$	2	5.4
$Al(OH)_3$	3	32
$Cr(OH)_3$	3	32

Table 5.30 gives the pH value at which the solubility is 10 mg or less of the metal ion per litre and 1 mg or less per litre, respectively.

From these considerations it is presumed that other ions present do not influence the precipitation, but in many cases it is necessary to consider the ionic strength:

$$I = \Sigma 1/2\ C Z^2 \qquad (5.51)$$

where C = the molar concentration of the considered ions and Z = the charge.

On the basis of the ionic strength, it is possible to find the activity coefficient, f, from

$$-\log f = \frac{0.5 * Z^2 * \sqrt{I}}{\sqrt{I} + 1} \qquad (5.52)$$

where I = ionic strength, Z = charge and f = activity coefficient

Table 5.31 gives the activity coefficients for different charges of the ions considered, calculated from the equation (5.52).

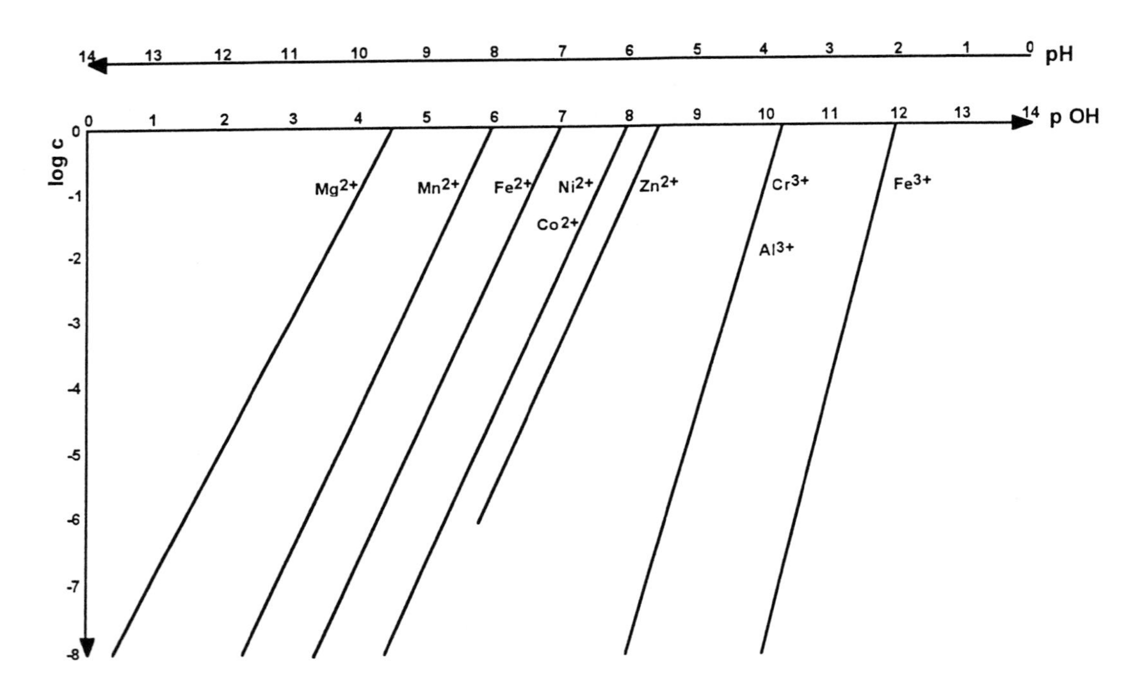

Fig. 5.26. log(solubility) against pH and pOH.

Since calcium hydroxide is the cheapest source of hydroxide ions, it is used most often for the precipitation of metals as hydroxides, but as sodium hydroxide is available in 28% solutions, while a calcium hydroxide slurry has to be produced, sodium hydroxide is the preferred precipitant for relatively small volumes. In most cases it is necessary to determine the amount of calcium or sodium hydroxide required in the laboratory. Flocculation is carried out after the addition of the chemical and before settling has occurred. If the amount of waste water that must be treated per 24 hr is 100 m^3 or less, it is preferable to use discontinuous treatment. The system in this case consists of two tanks with a stirrer.

TABLE 5.30
The pH value indicates where the solubility is ≤ 10 mg/l and ≤ 1 mg/l

Metal ion	Solubility ≤ 10 mg/l	≤ 1 mg/l
Mg^{2+}	11.5	12.0
Mn^{2+}	10.1	0.6
Fe^{2+}	8.9	9.4
Ni^{2+}	7.8	8.3
Co^{2+}	7.8	8.3
Zn^{2+}	7.2	7.7
Cr^{3+}	5.1	5.4
Al^{3+}	5.0	5.3
Fe^{3+}	3.2	3.5

TABLE 5.31
Activity coefficient f at different ionic strengths

I	$\frac{\sqrt{I}}{1+\sqrt{I}}$	f for Z = 1	f for Z = 2	f for Z = 3
0	0	1.00	1.00	1.00
0.001	0.03	0.97	0.87	0.73
0.002	0.04	0.95	0.82	0.64
0.005	0.07	0.93	0.74	0.51
0.01	0.09	0.90	0.66	0.40
0.02	0.12	0.87	0.57	0.28
0.05	0.18	0.81	0.43	0.15
0.1	0.24	0.76	0.33	0.10
0.2	0.31	0.70	-	-
0.5	0.41	0.62	-	-

I = ionic strength, Z = charge, f = activity coefficient

The waste water is discharged into one tank while the other is used for the treatment process. If chromate is present the treatment processes will follow the scheme below:

1. The concentration of chromate and the amount of acid required to bring the pH down to 2.0 are determined.
2. On the basis of this determination the required amounts of acid and reducing compound (sulphite or iron(II) are used) are added (about the reduction processes, see below).
3. Stirring for 10-30 minutes.
4. The concentration of chromate remaining is determined and the pH checked. If chromate is present further acid or reducing agent is added. In this case the stirring (10-30 minutes) must be repeated.
5. The amount of calcium hydroxide necessary to obtain the right pH for precipitation is measured. If Cr^{3+} is present a pH value of 8-9.5 is normally required.
6. Flocculation for 10-30 minutes.
7. Settling for 3-8 hours.
8. The clear phase is discharged into the sewer system.

The sludge can be concentrated further by filtration or centrifugation.

Stages 1-5 can, of course, be left out, if chromate is not present and the metal ions can be readily precipated in their present oxidation stage. Generally, all 8 steps are used in the treatment of plating waste water containing chromate. This water, from chromate acid baths used in electroplating and anodising processes, contains chromate in the form of CrO_3 or $Na_2Cr_2O_7$, $2H_2O$. The pH of such waste water is low and the Cr(IV) concentration is often very high - up to 20,000 ppm or more. The most commonly used reducing agents are iron(II) sulphate, sodium meta-hydrogen sulphite or sulphur dioxide. Since the reduction of chromate is most effective at low pH, it is, of course, an advantage if the waste water itself contains acid, which is often the case. Iron(II) ions react with chromate by reducing the chromium to a trivalent state and the iron(II) ions are oxidised to iron(III) ions. The reaction occurs rapidly at a pH below 3.0, but since the acidic properties of iron(II) sulphate are low at high dilution, acid must often be added for pH adjustment. The reactions are:

$$CrO_3 + H_2O \text{ ------> } H_2CrO_4 \quad (5.53)$$

$$2H_2CrO_4 + 6FeSO_4 + 6H_2SO_4 \text{ ------> } Cr_2(SO_4)_3 + 3Fe_2(SO_4)_3 + 8H_2O \quad (5.54)$$

$$Cr_2O_7^{2-} + 6FeSO_4 + 7H_2SO_4 \text{ ------> } Cr_2(SO_4)_3 + 3Fe_2(SO_2)_3 + 7H_2O + SO_4^{2-} \quad (5.55)$$

It is possible to show that 1 mg of Cr will require 16 mg $FeSO_4$, $7H_2O$ and 6 mg of H_2SO_4, based on stoichiometry. Reduction of chromium can also be accomplished by the use of meta-hydrogen sulphite or sulphur dioxide. When meta-hydrogen sulphite is used, the salt hydrolyses to hydrogen sulphite:

$$S_2O_5^{2-} + H_2O \text{ ------> } 2HSO_3^- \quad (5.56)$$

The hydrogen sulphite reacts to form sulphurous acid:

$$HSO_3^- + H_2O \text{ -------> } H_2SO_3 + OH^- \quad (5.57)$$

Sulphurous acid is also formed when sulphur dioxide is used, since:

$$SO_2 + H_2O \text{ ------> } H_2SO_3 \quad (5.58)$$

The reaction is strongly dependent upon pH and temperature.
The redox process is:

$$2H_2CrO_4 + 3H_2SO_3 \text{ ------> } Cr_2(SO_4)_3 + 5H_2O \quad (5.59)$$

Based on stoichiometry, the following amounts of chemicals are required for 1 mg of chromium to be precipitated: 2.8 mg $Na_2S_2O_5$ or 1.85 mg SO_2. Since dissolved oxygen reacts with sulphur dioxide, excess SO_2 must be added to account for this oxidation:

$$H_2SO_3 + 1/2\,O_2 - H_2SO_4 \quad (5.60)$$

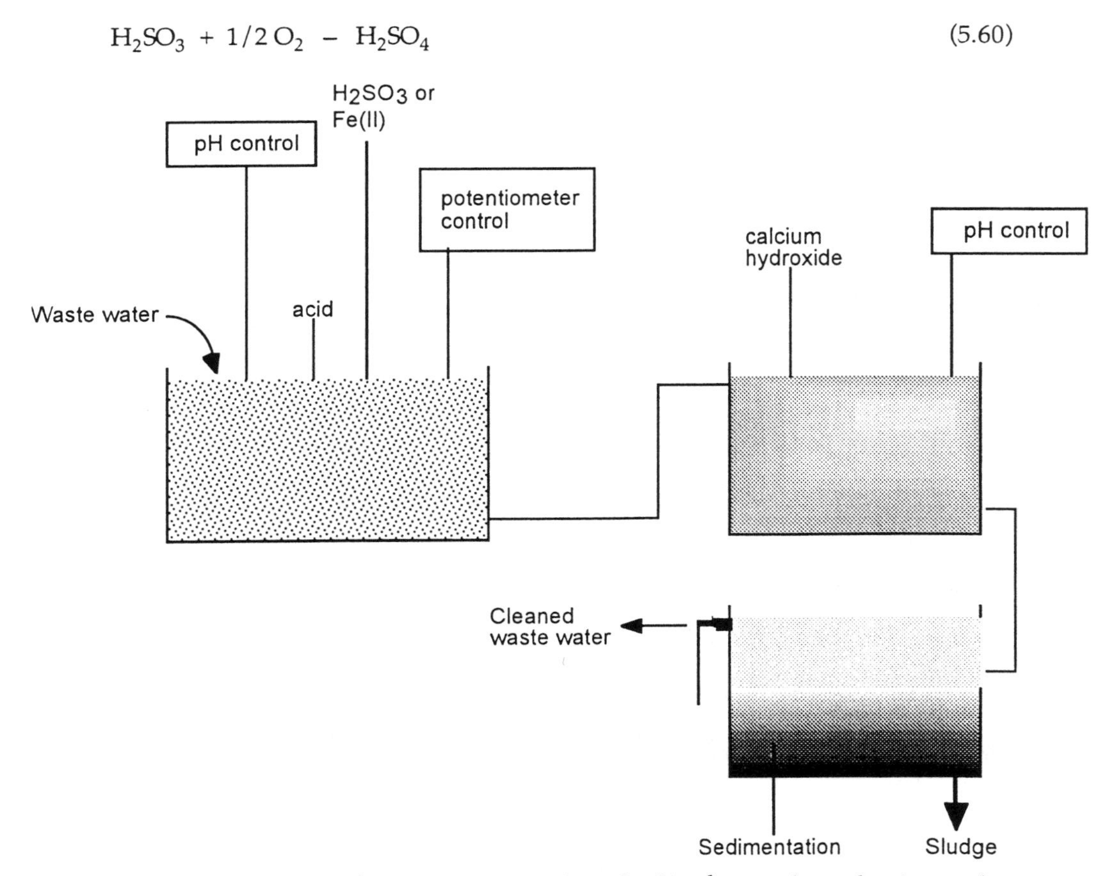

Figure 5.27 A flowchart for the 8 step process described in the text for reduction and precipitation of chromate.

If the amount of waste water exceeds 100 m^3 in 24 h, the plant shown in Fig. 5.27 can be used. A potentiometer and pH meter are used to control the addition of acid and reducing agent. These instruments can be coupled to an automatic dosing control. As seen from the figure, the plant consists of a reaction tank, a flocculation tank and a sedimentation tank. Usually a settling time of 24 h or more is used. A concentration of 1 1/3 - 3% dry matter for the sludge is usually obtained, which is slightly less than that obtained by discontinuous treatment. The flow chart of the process is shown in Figure 5.27.

The treatment might not operate effectively because of the complexing or stabilising effects of high carbonate or pyrophosphate concentrations in the water. It is possible to eliminate the problem caused by carbonate and phosphate by modifying the treatment scheme for the combined waste. Prime consideration should be given to the modification consisting simply of using lime instead of caustic soda to neutralise the waste after the chemical reduction of chromate to chromium(III) has taken place. The use of lime will cause precipitation and removal of most of the carbonate and pyrophosphate species from the solution while also providing doubly charged counter-ions to aid in coagulating the negatively charged chromium(III) hydroxide colloids that exist at pH values of about 8. Other treatment modifications, such as the use of polyelectrolytes to flocculate the stabilised chromium(III) hydroxide system, may or may not be effective alternatives, depending on the amount of complexation present in a particular waste system.

Data presented indicate that a hydrogen carbonate alkalinity of 250 mg/l as $CaCO_3$ and a pyrophosphate concentration of 30 mg/l as P together cause appreciable complexation and may make alternatives other than lime neutralisation impractical. The formation of calcium carbonate flocs during lime neutralisation could also aid in the removal of chromium(III) hydroxide colloids by the enmeshment mechanism and, at the same time, make the resulting sludge easier to dewater because of the presence of the large volume of calcium carbonate. The increased treatment efficiency and better sludge handling characteristic may make the use of lime a more favourable solution even though the quantity of sludge may be somewhat increased. The last treatment modification to be mentioned is the use of Fe^{2+} ions as the reducing agent in the chromate-reduction step. The end product of this reaction is Fe^{3+} ions. In the neutral pH range this is very effective in removing chromium(III) hydroxide colloids.

The application of Ion Exchange for Removal of Heavy Metals

A cation exchanger can be used for the removal of metal ions from waste water, such as Fe^{2+}, Fe^{3+}, Cr^{3+}, Al^{3+}, Zn^{2+}, Cu^{2+}, etc. (Spanier, 1969) The most common type of cation exchanger consists of a polystyrene matrix, which is a strong acidic ion exchanger containing the sulphonic acid group. The practical capacity of the cation exchanger is 1-1.5 eqv/l of ion exchange material (Rüb, 1969). To ensure high efficiency, a relatively low flow rate is recommended, often below 5 bed volumes/hour.

A recent development has brought a **starch xanthate and a cellulose**

polyethyleneimine ion exchanger onto the market (WRL, 1977). These have a lower capacity than the conventional cation exchangers, but since they are specific for heavy metal ions they have a higher capacity when measured as the volume of water treated between two regenerations per volume of ion exchanger.

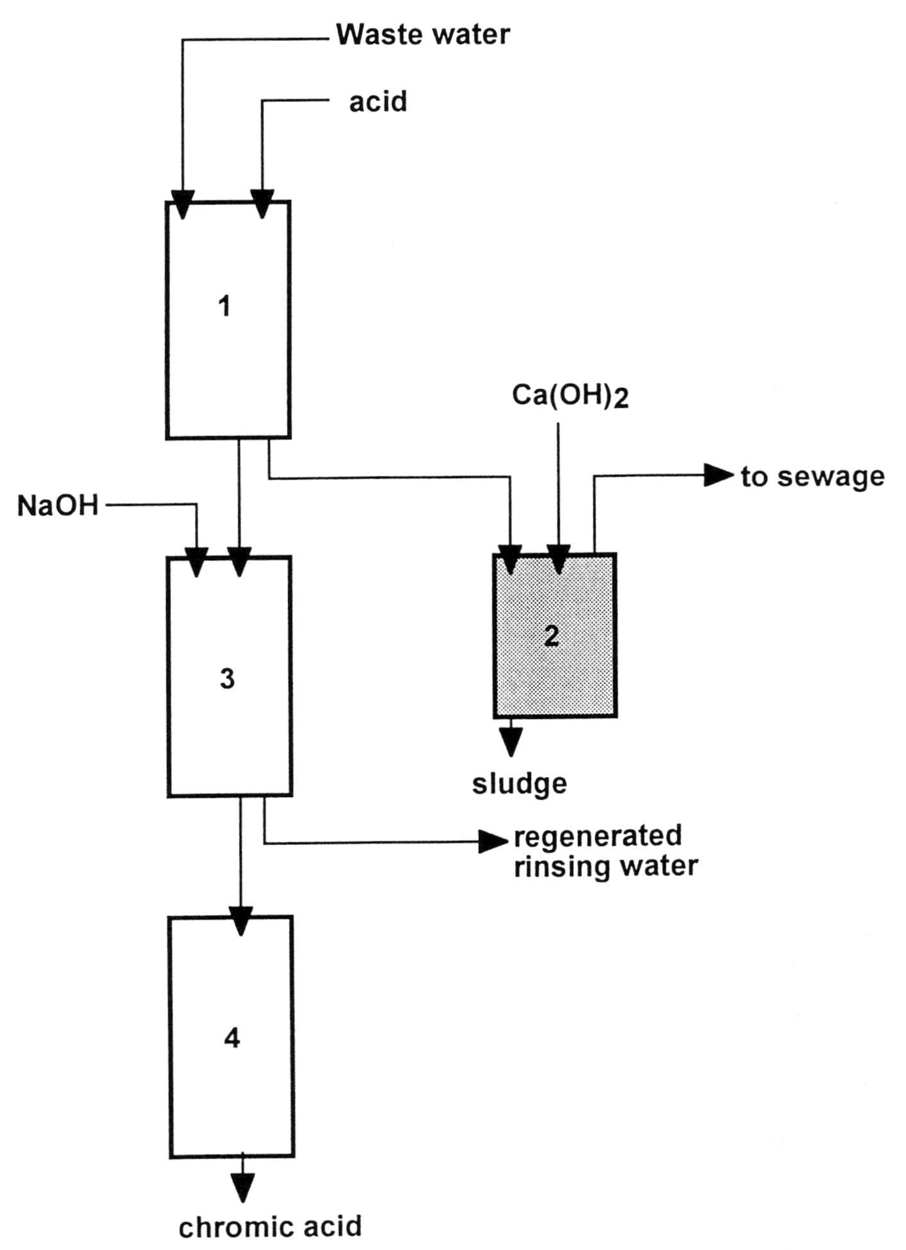

Fig. 5.28. Ion exchange of waste water containing chromate. (1) cation exchanger; (2) precipitation tank; (3) anion exchanger; (4) exchange of Na^+ with H^+.

It is often preferable to separate waste water from the chromate baths and the rinsing water, since this opens up the possibility of reusing the water. *Chromate can be recovered* and the waste water problem can be solved in an effective way (Mohr, 1969). Fig 5.28 is a flow diagram of an ion exchange system in accordance with these principles. Reuse of the elution liquid from the anion exchanger and of the treated waste water might be considered. The pH of the elution liquid from the cation exchanger must be adjusted and the metal ions precipitated for removal by the addition of calcium hydroxide (see Fig. 5.28).

In some instances it is too costly to separate the different types of waste water and rinsing water. In such cases a simpler ion exchange system is used (Schaufler, 1969), although it is, of course, possible to recover the chromate from the elution liquid. Since, as well as inorganic impurities, the waste water contains organic material, such as oil, fat, dust, etc., the ion exchanger often becomes clogged if the waste water is not pretreated before being passed into the ion exchange system. *Macro porous ion exchangers* have only partly solved this problem since suspended matter, emulsions and high molecular weight organic compounds will also affect this type of ion exchanger. Consequently, it is an advantage to pretreat the waste water on sand filters or activated carbon. The sand filter will remove the suspended particles and the activated carbon most of the organic impurities. Since most of the ions are eluted in the first 60-75% of the eluting volume, it is often of advantage to use the last portion for the subsequent regeneration.

Recently, a very cost moderate ion exchanger has been used to polish the waste water after a chemical precipitation. The ion exchanger (Dancraft Management, 1997) is based on bark - which is a waste material - as the matrix for sulphone and sulphate groups. The ion exchanger has a high selectivity for heavy metals and as it is used to remove the last few mg /l it has a high capacity measured as the volume of water which can be treated per volume of ion exchanger before saturation. The ion exchanger has found application in the electroplating industry, in the photo chemical industry (to remove silver) and for removal of mercury from waste water discharged from dentist clinics. The ion exchanger is mostly used on a throw-away basis as it is too expensive to install regeneration equipment, giventhat regeneration is needed only a few times per year. This means that the saturated ion exchanger is incinerated and dust and ash which contains the heavy metal must be properly deposited; see also the discussion in Chapter 6.

<u>The application of extraction for removal of heavy metals</u>

Most heavy metal ions can be recovered from aqueous solutions by extraction. Complexes of the metal ions are formed, for example by reaction with Cl^-, and these metal complexes are then extracted by means of organic solvents. By using different concentrations of the ligand it is possible to separate different metals by extraction, again opening up the possibility of recovering the metals. The calculation of the equilibrium is based on the following two reactions:

Metal + ligand ------> metal complexes (5.61)

Metal complex)$_{water}$ -------> (Metal complex)$_{organic\ solvent}$ (5.62)

On increasing the concentration of the ligand and increasing the volume of the organic phase, $[Me]_{org.}$ will increase and [Me] decrease.

Recovery of copper from a solution containing Cu^{2+}, SO_4^{2-}, Na^+ and H^+ (pH = approx. 2.0) is possible by extraction with acetone. The efficiency should be 99.9% according to patent DDR 67541. On recovering the acetone by distillation a loss of 0.3 kg acetone/100 kg copper is recorded. An amine extraction process has also been developed for the recovery of cyanide and metal cyanide from waste streams of plating processes (Chemical Week, 1976).

Application of membrane process for removal of heavy metals

Membrane separation, electrodialysis, reverse osmosis, ultrafiltration and other such processes are playing an increasingly important role in waste water treatment. *A membrane is defined as a phase which acts as a barrier* between other phases. It can be a solid, a solvent-swollen gel or even a liquid. The applicability of a membrane for separation depends on differences in its permeability to different compounds. Table 5.32 gives a survey of membrane separation processes and their principal driving forces, applications and their useful ranges.

Osmosis is defined as spontaneous transport of a solvent from a dilute solution to a *concentrated solution across a semi-permeable membrane.* At a certain pressure - the so-called osmotic pressure - equilibrium is reached. **The osmotic pressure** can vary with concentration and temperature, and depends on the properties of the solution.

For water, the osmotic pressure is given by:

$$\pi = \frac{n}{V} RT \qquad (5.63)$$

where

n = the number of moles of solute

V = the volume of water

R = the gas constant

T = the absolute temperature

This equation describes an ideal state and is valid only for dilute solutions. For more concentrated solutions the equation must be modified by **the van Hoff factor** by using an osmotic pressure coefficient:

$$\pi = \varnothing * \frac{n}{V} RT \qquad (5.64)$$

For most electrolytes the osmotic pressure coefficient is less than unity and will usually decrease with increasing concentrations. This means that equation (5.64) is usually conservative and predicts a higher pressure than is observed. **If the pressure is increased above the osmotic pressure on the solution side of the membrane, as shown in Fig. 5.29, the flow is reversed. The solvent will then pass from the solution into the solvent. This is the basic concept of reverse osmosis.** Reverse osmosis can be compared with filtration, as it also involves the moving of liquid from a mixture by passing it through a filter.

TABLE 5.32
Membrane separation processes

Process	Driving force	Range (μm) particle size	Function of membrane
Electrodialysis	Electrical potential gradient	< 0.1	Selective to certain ions
Dialysis	Concentration	< 0.1	Selective to solute
Reverse osmosis	Pressure	< 0.05	Selective transport of water
Ultrafiltration	Pressure	$5 * 10^{-3}$ - 10	Selective to molecular size and shape

However, one important difference is that *the osmotic pressure,* which is very small in ordinary filtration, plays an important role in reverse osmosis.

Second, a filter cake *with low moisture content cannot* be obtained in reverse osmosis, because the osmotic pressure of the solution increases with the removal of solvents.

Third, the filter separates a mixture on the basis of size, whereas reverse osmosis membranes work on the basis of other factors. Reverse osmosis has sometimes also been termed hyperfiltration. The permeate flux, F, through a semipermeable membrane of thickness, d, is given by:

$$F = \frac{D_W * C_W * V}{R\,T\,d} (\Delta P - \pi) \qquad (5.65)$$

where

D_W = the diffusion coefficient
D_W = the concentration of water
V = the molar volume of water
ΔP = the driving pressure (see Fig. 5.29)

Equation (5.65) indicates that the water flux is inversely proportional to the thickness of the membrane. These terms can be combined with the coefficient of water permeation, W_p, and equation (5.65) reduces to:

$$F = W_p * (\Delta P - \pi) \qquad (5.66)$$

where

$$W_p = \frac{D_W * C_W * V}{R T d} \qquad (5.67)$$

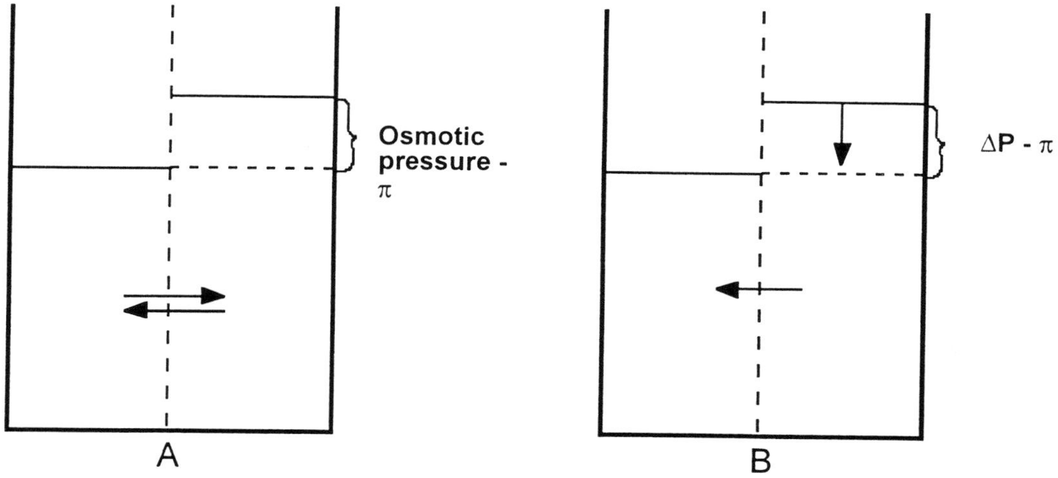

Fig. 5.29 A - illustrates equilibrium. An osmotic pressure appears. B - illustrates the principle of reverse osmosis.

For the solute flux, F_s, the driving force is almost entirely due to the concentration gradient across the membrane, which leads to the following equation (Clark, 1962):

$$F_s = D_s \frac{dC'_i}{dx} = D_s \frac{\Delta C'_i}{d} \qquad (5.68)$$

where

C'_i = the concentration of species, i, within the membrane
$\Delta C'_i$ = measured across the membrane.

In constructing a system for reverse osmosis many problems have to be solved:

1. The system must be designed to give a **high liquid flux** reducing the concentration potential.
2. **The packaging density must be high** to reduce pressure vessel cost.
3. **Membrane replacement costs must be minimised.**
4. The usually fragile **membranes must be supported** as they have to sustain a pressure of 20-100 atmospheres. Four different system designs have been developed to meet the solution to the latter problem: *The plate and frame technique, large tube technique, spiral wound technique and the hollow fine fibre technique.* The most widely used membrane is the *cellulose acetate membrane.* This membrane is asymmetrical and consists of a thin dense skin of approximately 0.2 μm on an approximately 100 μm thick porous support. Cellulose acetate membranes are not resistant to high or low pH, and a temperature range of 0-50°C must be recommended. Several other membranes (made by polyamide, poly-acetate-butyrate, or polyacrylacid for instance) are offered on the market today which makes it possible to get a membrane suitable for high and low pH values and for extreme temperature Several natural materials could also be of use as membranes and extensive laboratory investigations may hold promise for the application of such natural membranes in the near future (Kraus et al., 1967). Reverse osmosis is particularly well suited for the treatment of *nickel-plating rinsing water.* Cellulose acetate membranes are recommended (Hauck and Sourirajan, 1972). The treatment can contribute significantly to both *water pollution control and recovery of nickel.*

Push and Walcha (1975) have demonstrated that it is possible to fractionate metal salts by reverse osmosis. By tailoring membranes for specific applications and by optimising the process parameters, such as pressure applied, solute concentrations, and added polyelectrolyte, it is possible to fractionate metals from waste water and recover valuable materials. By using polyamide membranes for reverse osmosis experiments at 100 atmospheres with an equal molar solution of $AgNO_3$ and $Al(NO_3)_3$, the membrane was able to reject 99.98% of aluminium salt, while the silver salt was enriched in the permeate.

5.6 WATER RESOURCES

Introduction

Life in the form we know on earth cannot exist without water, so there has always been a demand for water and all the nearest and more obvious sources have already been exploited. Any future demand must inevitably be met from more remote and increasingly less attractive sites, and for this reason alone development costs will continue to rise in the future.

The relationship between the hydraulic cycle and the water demand now and in the near future was touched on in Section 2.10. Most waters have to be purified before they can be used for human consumption. Raw water is so infinitely variable in quality that there is no fixed starting point in the treatment process. Many countries have their own standards of acceptable purity for potable water and these vary. **WHO lays down two standards**, which are widely applied in developing countries.

Virtually no water is impossible to purify to potable standards, but some raw waters are so bad as to merit rejection, because of the expense involved.

Water from underground sources is generally of better quality than surface water, but it may be excessively hard and/or contain iron and manganese. Full treatment of water may comprise pretreatment, mixing, coagulation, flocculation, settling, filtration and sterilisation. Not all waters require full treatment, however.

The differences between the two sources are summarised in Table 5.33.

TABLE 5.33
Characteristics of ground water and surface water

Properties	Ground water	Surface water
Salt concentration	high	low
Iron concentration	high	low
Manganese concentration	high	low
$KMnO_4$ number	low	high
Hardness	high	low
pH	6-8	7-9
Turbidity	low	high
Temperature	low	high
Number of E. coli	0	>0
Colour	none	yellowish

Pretreatment includes screening, raw water storage, prechlorination, aeration, algal control and straining.

Aluminium sulphate, sodium aluminate and iron salts are used as coagulants. They are usually applied in combination with coagulant acids, which include lime, sodium carbonate, activated silica and polyelectrolytes. After coagulation, mixing, flocculation, and sedimentation, take place, as described in Sections 5.2 and 5.3. For filtration sand filters are widely used either after coagulation, mixing, flocculation and sedimentation of surface water or after aeration of ground water for removal of iron and manganese compounds.

Two typical flow diagrams for treatment of ground and surface water are shown in Figs. 5.30 and 5.31. Most of the unit processes used in water treatment have already been mentioned.

Water management is closely linked to waste water management (see the discussion early in this chapter). Insufficient treatment of waste water can have several consequences, of which two will be mentioned:

1. Low quality surface water that **requires more advanced treatment of the raw water** for production of potable water,
2. The treated waste water is discharged to the sea to avoid a deterioration of the quality of the surface water. Consequent- ly, it is necessary **to use supplementary sources of raw water,** which may be difficult close to cities or in areas suffering from water shortage.

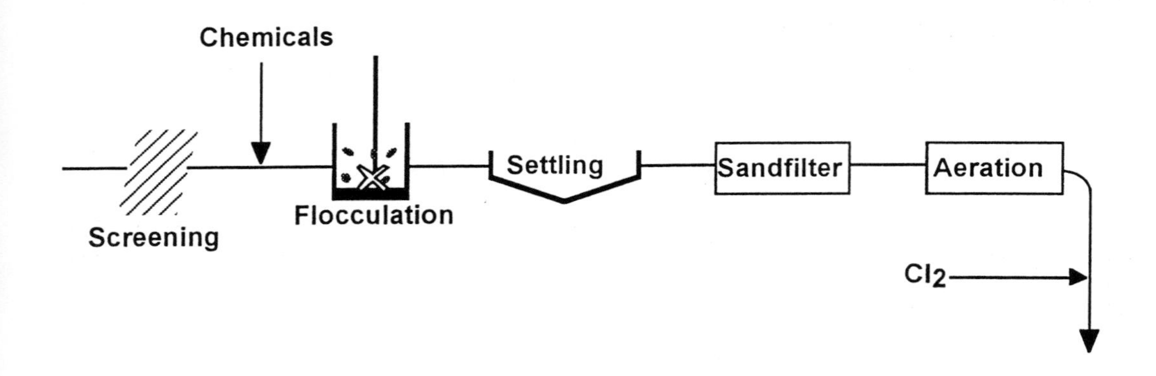

Fig. 5.30. Typical treatment of surface water.

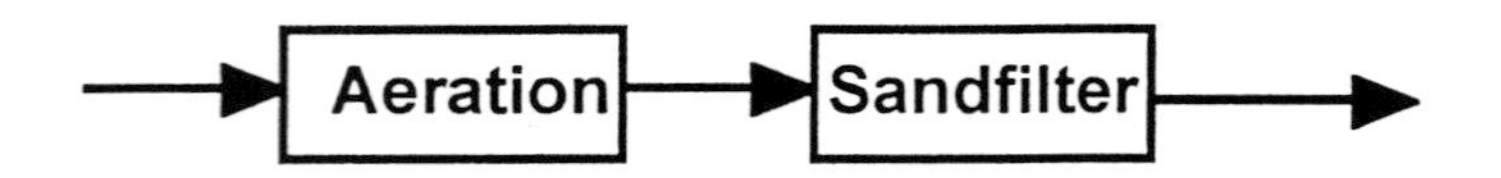

Fig. 5.31 Typical treatment of ground water.

Softening

The presence of polyvalent cations in water causes hard water, so called because with this water it is hard to form a lather with soap. Polyvalent salts of the long-chain fatty acids present in soap are insoluble. In the days when all washing was done with soaps, the problem of hard water was more of a nuisance than it is today, when soaps have been replaced largely with synthetic detergents, whose polyvalent metal salts are relatively soluble. The predominant metal ions in waters used for potable supplies are calcium and magnesium.

Hardness can be either calcium hardness (including magnesium hardness), **temporary hardness, which is carbonate and hydrogen carbonate hardness, or permanent hardness, which is the difference between the calcium hardness and the temporary hardness,** i.e. calcium hardness that has no carbonate or hydrogen carbonate as counter ions. **Temporary hardness can be removed by boiling the water,** since:

$$Ca(HCO_3)_2 \text{ -------> } CaCO_3 + CO_2 + H_2O \quad (5.69)$$

Hardness is often quantified as calcium carbonate equivalents in mg per litre or as **hardness degrees, which uses 10 mg l^{-1} CaO as its base unit.** The content of magnesium ions is taken into consideration by calculating the equivalent amount of calcium carbonate and calcium oxide, respectively. It turns out that 83 mg of magnesium carbonate correspond to 100 mg calcium carbonate and 39 mg of magnesium oxide give the same number of hardness degrees as 56 mg of calcium oxide = 5.6 degrees of hardness.

There are two principal methods of softening water for municipal purposes: by lime and lime-soda and by ion exchange.

The first method is based upon precipitation of calcium as calcium carbonate and magnesium as magnesium hydroxide.

A classification of water in terms of hardness is given in Table 5.34. When lime is used, only carbonate hardness is reduced. The additional use of soda can reduce the permanent hardness as well.

The process can be described in five steps:

1. The reaction between free carbon dioxide and added lime:

$$CO_2 + Ca(OH)_2 \text{ -------> } CaCO_3 + H_2O \quad (5.70)$$

2. The reaction of calcium carbonate hardness with lime:

$$Ca(HCO_3)_2 + Ca(OH)_2 \text{ -------> } 2CaCO_3 + 2H_2O \quad (5.71)$$

3. The reaction of magnesium carbonate hardness with lime:

$$Mg(HCO_3)_2 + Ca(OH)_2 \text{ --------> } CaCO_3 + MgCO_3 + 2H_2O \quad (5.72)$$

$$MgCO_3 + Ca(OH)_2 \text{ -------> } CaCO_3 + Mg(OH)_2 \quad (5.73)$$

Combining the above two equations gives:

$$Mg(HCO_3)_2 + Ca(OH)_2 \longrightarrow 2CaCO_3 + Mg(OH)_2 + 2H_2O \quad (5.74)$$

From these equations it can be concluded that 2 moles of lime are required to remove 1 g atom of magnesium, or twice as much as is required for calcium removal.

4. Non-carbonate calcium hardness is removed by soda. The reaction is:

$$CaSO_4 + Na_2CO_3 \longrightarrow CaCO_3 + Na_2SO_4 \quad (5.75)$$

The non-carbonate hardness is represented here as sulphate.

5. The reaction of non-carbonate magnesium hardness with lime and soda is:

$$MgSO_4 + Ca(OH)_2 \longrightarrow Mg(OH)_2 + CaSO_4 \quad (5.76)$$

$$CaSO_4 + Na_2CO_3 \longrightarrow CaCO_3 + Na_2SO_4 \quad (5.77)$$

TABLE 5.34
Classification of water in terms of hardness

mg/l calcium carbonate	Hardness degrees	Specification
0 - 80	0 - 4	Very soft
80 - 160	4 - 8	Soft
160 - 240	8 - 12	Medium hard
240 - 600	12 - 30	Hard
>600	>30	Very hard

From these reactions it can be seen that the addition of lime always serves three purposes and may serve a fourth. It removes, in order, carbon dioxide, calcium carbonate hardness and magnesium carbonate hardness. Furthermore, when magnesium non-carbonate hardness must be reduced, lime converts the magnesium hardness to calcium hardness. Soda then removes the non-carbonate hardness in accordance with equations (5.76) and (5.77).

Since softening is usually accomplished at high pH values and the reactions do not go to completion, the effluent from the treating unit is usually supersaturated with calcium carbonate. This would cement the filter medium and coat the distribution system. In order to avoid these problems, the pH must be reduced, so that insoluble calcium carbonate can be converted to the soluble hydrogen carbonate. This is accomplished in practice by recarbonation, i.e. the addition of carbon dioxide. The equations may be as follows:

$$CaCO_3 + CO_2 + H_2O \text{ -------> } Ca(HCO_3)_2 \qquad (5.78)$$

$$CO_3^{2-} + CO_2 + H_2O \text{ ------> } 2HCO_3^- \qquad (5.79)$$

The large amount of sludge containing calcium carbonate and magnesium hydroxide, produced by softening plants, presents a disposal problem, since it can no longer be discharged into the nearest stream or sewer. Some of this sludge can be recycled to improve the completeness of reaction, but a large quantity must be disposed of. The methods used are lagooning, drying for landfill, agricultural liming and lime recovery by recalcination.

In ion exchange softening, which is widely used in the industry, the calcium and magnesium ions are exchanged for monovalent ions, usually sodium and hydrogen. This is termed "cation exchange softening", and current practice uses ion exchange resins based on highly cross-linked synthetic polymers with a high capacity for exchangeable cations. In cases where complete demineralisation is required anion exchangers are also employed. In this case, chloride, hydrogen carbonate and other anions are replaced by hydroxide ions, which together with the hydrogen ions released by the cation exchanger form water:

$$OH^- + H^+ \text{ ------> } H_2O \qquad (5.80)$$

<u>Disinfection processes.</u>

Micro-organisms are destroyed or removed by a number of physical- chemical waste water treatment operations, such as coagulation, sedimentation, filtration and adsorption. However, inclusion of a disinfection step has become common practice in water and waste treatment to ensure against transmission of waterborne diseases. **The disinfection process must be distinguished from sterilisation. Sterilisation involves complete destruction of all micro-organisms including bacteria, algae, spores and viruses, while disinfection does not provide for the destruction of all micro-organisms,** e.g. the hepatitis virus and polio virus are generally not inactivated by most disinfection processes.

The mechanism of disinfection involves at least two steps:

1. **Penetration of the disinfectant through the cell wall.**
2. **Reaction with enzymes within the cell** (Fair et al., 1968).

Chemical agents, such as ozone, chlorine dioxide and chlorine, probably cause disinfection by direct chemical degradation of the cell matter, including the enzymes, while application of thermal methods or degradation accomplishes essentially physical

destruction of the micro-organisms.

The large number of organic and inorganic chemicals exert a poisoning effect on the micro-organisms by interaction with enzymatic proteins or by disruptive structural changes within the cells.

The rate of destruction of micro-organisms has been expressed as a first order reaction referred to as Chick's law:

$$dN / dt = -k * N \quad (5.81)$$

where N is the number of organisms per volume and k is a rate constant. This differential equation has the following solution:

$$N = N_0 * e^{-kt} \quad (5.82)$$

As seen, Chick's law states that the rate of bacterial destruction is directly proportional to the number of organisms remaining at any time. This relationship indicates a uniform susceptibility of all species at a constant concentration of disinfectant, pH, temperature and ionic strength. Many deviations from Chick's law have been described in the literature. In accordance with Fair et al. (1968) chlorination of pure water shows typical deviations from Chick's law. Often deviation from the first order rate expression is due to autocatalytic reaction. In this case the expression can be transformed to:

As pointed out the disinfection rate expression (Chick's law) does not include the effect of disinfectant concentration. The relationship between disinfectant concentration and the time required to kill a given percentage of organisms is commonly given by the following expression:

$$C^n * t = \text{constant} \quad (5.83)$$

Berg (1964) has shown that the concentration/time relationship for HOCl at 0-6°C to give a 99% kill of E. coli is expressed as:

$$C^{0.86} * t = 0.24 \ (\text{mg l}^{-1} \text{min}) \quad (5.84)$$

Temperature influences the disinfection rate first by its direct effect on the bacterial action and secondly by its effect on the reaction rate. Often an empirical temperature expression is used, such as:

$$k_t = k_{20} * d^{(t-20)} \quad (5.85)$$

where

k_t = the rate constant at t°C

k_{20} = the rate constant at 20°C

d = an empirical constant

Most micro-organisms are effectively killed by extreme pH conditions, i.e. at pH below 3.0 and above 11.0.

The effect of disinfection is also strongly dependent on the coexistence of other matter in the waste water, e.g. organic matter. The disinfectant may (1) react with other species to form compounds which are less effective than the parent compounds, or (2) chemically oxidise other impurities present in the water, reducing the concentration of the disinfectants.

The application of heat is one of the oldest, and at the same time most certain, methods of water disinfection. In addition, *freezing and freeze- drying are effective methods for the preservation of bacteria.* However, these techniques are of little practical significance for waste water treatment, as they are too costly. Disinfecting large volumes of water by heating is clearly not suitable for economic reasons.

The wavelength region 250-265 nm, beyond the visual spectrum, has bactericidal effects. Mercury vapour lamps emit a narrow band at 254 nm and can be used for small-scale disinfection. It is assumed that the nucleic acids in bacterial cells absorb the ultraviolet energy and are consequently destroyed. These nucleic acids include deoxyribonucleic acid (DNA) and ribonucleic acid (RNA). The main problem in the application of ultraviolet irradiation for disinfection is to ensure that the energy is delivered to the entire volume of the water. Even distilled water will absorb only 8% of the applied energy to a depth of 3 cm, and turbidity, dyes and other impurities constitute barriers to the penetration of ultraviolet radiation.

This means that only a thin layer of clear water without impurities able to absorb ultraviolet light can be treated. For a 99% level of kill, the use of a 30 W lamp would allow flows of from 2 to 20 m^3/h to be disinfected. The use of ultraviolet lamps for disinfection has some important advantages. As nothing is added to the water no desirable qualities will be changed. No tastes or odours result from the treatment. The disadvantage of ultraviolet irradiation is that it provides no residual protection against recontamination as the application of chlorine does.

Gamma and X rays are electromagnetic radiation of very short wavelength and have an excellent capacity for destroying microorganisms. However, their use is relatively expensive. The application of this method requires care, and this will restrict its use considerably.

Chlorine is produced exclusively by electrolytic oxidation of sodium chloride in aqueous solutions:

$$2Cl^- \text{------>} Cl_2 + 2e^- \quad (5.86)$$

After generation, the chlorine gas is purified by washing in sulphuric acid and the product usually has a purity of more than 99%. The gas is liquefied by compression to 1.7 atmospheres between -30°C and -5°C, and stored in steel cylinders or tanks. Chlorine should be handled with caution, as the gas is toxic and has a high chemical activity, with danger of fire and explosion. In the presence of water chlorine is highly corrosive. When chlorine is added to an aqueous solution it hydrolyses to yield Cl^- and OCl^-:

$$Cl_2 + 2H_2O \text{ ------> } H_3O^+ + Cl^- + HOCl \tag{5.87}$$

As can be seen, the process is a disproportionation, since chlorine in zero oxidation state turns into oxidation states -1 and +1. Hypochlorous acid is a weak acid:

$$HOCl + H_2O \text{ ------> } H_3O^+ + OCl^- \tag{5.88}$$

The acidity constant K_a:

$$K_a = \frac{[H_3O^+]\,[OCl^-]}{[HOCl]} \tag{5.89}$$

is dependent on the temperature as illustrated in Table 5.35. HOCl is a stronger disinfectant than OCl^- ions, which explains why the disinfection is strongly dependent on pH.

TABLE 5.35
The acidity constant for HOCl

Temperature	K_a
0	$1.5 * 10^{-8}$
5	$1.7 * 10^{-8}$
10	$2.0 * 10^{-8}$
15	$2..2 * 10^{-8}$
20	$2.5 * 10^{-8}$
25	$2.7 * 10^{-8}$

The ammonium present in the water is able to react with the chlorine or hypochlorous acid (see Section 3 where this process is applied to remove ammonium/ ammonia).

The rates of chlorine formation from chloramines, which are also used as disinfectants, depend mainly on pH and the ratio of the reactants employed. Moore (1951) infers that the distribution of chlorine is based on the equation:

$$2NH_2Cl + H_3O^+ - NH_4^+ + NHCl_2 + H_2O \qquad (5.90)$$

for which:

$$K = \frac{[NH_4^+][NHCl_2]}{[H_3O^+][NH_2Cl]^2} = 6.7 * 10^5 \ (25°C) \qquad (5.91)$$

The disinfection power of chloramines measured in terms of contact time for a given percentage kill, is less than that of chlorine.

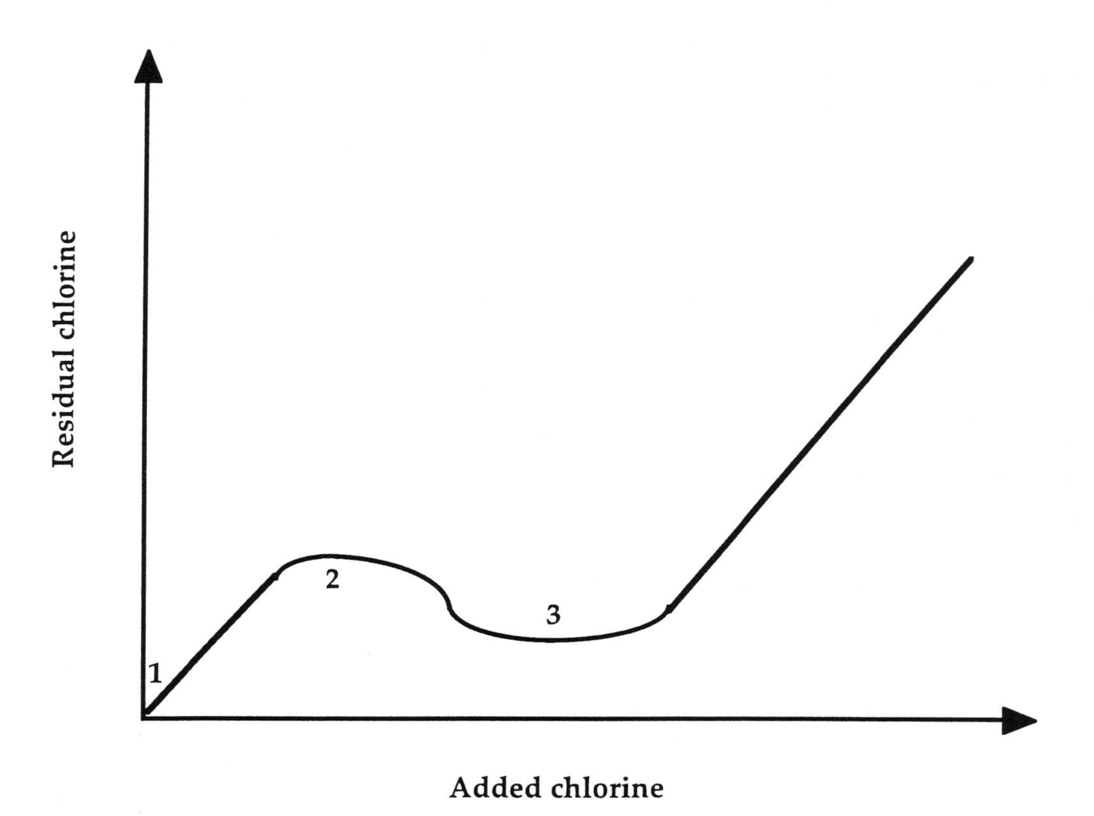

Fig. 5.32. Residual (free) chlorine plotted versus total added chlorine. "Break chlorination" takes place beyond point 3. The difference at that point between added chlorine and residual chlorine corresponds to the chlorine used for oxidation processes, including the formation of chloramines.

The bacterial properties of chlorine are probably based on the formation of free hypochlorous acid:

$$NH_2Cl + H_2O \text{ ------> } HOCl + NH_3 \quad (5.92)$$

$$NH_2Cl + H_3O^+ \text{ ------> } HOCl + NH_4^- \quad (5.93)$$

However, in practice the reaction of chlorine with ammonia or amino compounds presents a problem in the chlorination of waste water containing such nitrogen compounds. Fig. 5.32 shows the residual chlorine as a function of the chlorine applied. Between points 1 and 2 in the figure, mono- and di-chloramine are formed, if ammonia is present. The oxidation processes of organic and nitrogen compounds present in the water occur between points 2 and 3 give a decline in residual chlorine. Point 3 is called the breakpoint. Addition of chlorine in this interval probably produces free nitrogen gas as the predominant product of oxidation. Further addition of chlorine beyond the breakpoint gives an increasing residue of free chlorine. Only beyond the breakpoint, we can be certain that the water contains sufficient chlorine to ensure a disinfection, because before point 3 there is still a wide range of possible processes which may react with the chlorine and thereby reduce the concentration of free chlorine.

Table 5.36
Actual and available chlorine in pure chlorine-containing compounds

Compound	Mol. weight	Chlorine equiv. (moles of Cl_2)	Actual chlorine (%)	Available chlorine (%)
Cl_2	71	1	100	100
Cl_2O	87	2	81.7	163.4
ClO_2	67.5	2.5	52.5	260
NaOCl	74.5	1	47.7	95.4
CaClOCl	127	1	56	56
$Ca(OCl)_2$	143	2	49.6	99.2
HOCl	52.5	1	67.7	135.4
$NHCl_2$	86	2	82.5	165
NH_2Cl	51.5	1	69	138

Oxidative degradation by chlorine is limited to a small number of compounds. Nevertheless, oxidation of these compounds contributes to overall reduction of BOD_5 in wastes treated with chlorine. A disadvantage is that chlorinated organic compounds may be formed in large quantities. A variety of chlorine compounds is applied in waste water treatments. For these compounds the available chlorine can be calculated. Generally this is expressed as the percentage of chlorine having the same oxidation ability. Data for the different chlorine-containing compounds are given in Table 5.36.

It can be seen that the actual chlorine percentage in chlorine dioxide is 52.5, but the available chlorine is about 262.5 %. This is, of course, due to the fact that the oxidation state of chlorine in chlorine dioxide is +4 which means that five electrons are transferred per chlorine atom, while Cl_2 only transfers one electron per chlorine atom.

Hypochlorite can be obtained by the reaction of chlorine with hydroxide in aqueous solution:

$$Cl_2 + 2NaOH \text{ ------> } NaCl + NaOCl + H_2O \quad (5.94)$$

Chlorinated lime, also called bleaching powder, is formed by reaction of chlorine with lime:

$$Ca(OH)_2 + Cl_2 \text{ ------> } CaCl(OCl) + H_2O \quad (5.95)$$

A higher content of available chlorine is present in calcium hypochlorite, $Ca(OCl)_2$. Chlorine dioxide is generated in situ by the reaction of chlorine with sodium chlorite:

$$2NaClO_2 + Cl_2 \text{ ------> } 2ClO_2 + 2NaCl \quad (5.96)$$

Toxic trihalomethan gases are formed in swimming pools (Judd and Jeffrey, 1995). The formation is enhanced by the presence of urine (Judd and Jeffrey, 1995).

Theoretically, fluorine could be used for disinfection, but almost nothing is known about the bactericidal effectiveness of this element at low concentrations. However, bromine is used mainly for disinfection of swimming pools, the reason being **that monobromamine, unlike chloramine, is a strong bactericide.** There is therefore no need to proceed to breakpoint bromination. Bromine has a tendency to form compounds with organic matter, resulting in a high bromine demand. This and the higher cost are the major factors limiting the use of bromine for treatment of waste water.

Iodine can also be used as a disinfectant (Ellis et al., 1993). It dissolves sparingly in water unless iodide is present:

$$I^- + I_2 \text{ ------> } I_3^- \quad (5.97)$$

It reacts similarly with water in accordance with the scheme for chlorine and bromine:

$$I_2 + H_2O \text{ ------> } HOI + HI \quad (5.98)$$

It has a number of advantages over chlorination. Iodine does not combine with the ammonium to form iodomines, but rather oxidises the ammonia. Also it does not combine with organic matter very easily, e.g. it oxidises phenol rather than forming iodo-phenols. However, iodine is costly and it has, up till now, found a use only for swimming pool disinfection.

Ozone is produced by passing compressed air through a commercial electric discharge ozone generator. From the generator the ozone travels through a gas washer and a coarse centred filter. A dispersion apparatus produces small bubbles with a large surface area exposed to the solution. Ozone is used extensively in water treatment for disinfection and for removal of taste, odour, colour, iron and manganese. Ingols and Fetner (1957) have shown that the destruction of Escherichia coli cells with ozone is considerably more rapid than with chlorine when the initial ozone demand of water has been satisfied. The activity of ozone is a problem in the disinfection of water containing high concentrations of organic matter or other oxidisable compounds. A further problem arises from the fact that the decomposition of ozone in water does not permit long-term protection against pathogenic regrowth. *However, ozone has the advantage of being effective against some chlorine-resistant pathogens, like certain virus forms* (Stumm, 1958 and Stumm and Morgan, 1996).

The simultaneous removal of other compounds makes ozonation an advantageous water treatment process. Ozone can be used to alleviate the toxic and oxygen demanding characteristics of waste water containing ammonia by converting the ammonia to nitrate. The oxidation is a first order reaction with respect to the concentration of ammonia and is catalysed by OH^- over the pH range 7-9. The average value of the reaction rate constant at pH 9.0 is $5.2 \pm 0.3 * 10^{-2}$ min^{-1}. Ammonia competes for ozone with the dissolved organic constituents comprising the BOD and is oxidised preferentially relative to the refractory organic compounds, provided alkaline pH values can be maintained. Due to the elevated pH required, ammonia oxidation by ozone is attractive for the process of lime clarification and precipitation of phosphate. The reaction of ozone with simple organic molecules has been extensively studied in recent years. The reactions are usually complex, subject to general and specific catalysts and yield a multitude of partially degraded products.

5.7 QUESTIONS AND PROBLEMS

1. Find the approximate BOD_5 for waste water containing 120 mg l^{-1} carbohydrates, 80 mg l^{-1} proteins and 200 mg l^{-1} fats.

2. Calculate the total oxygen demand (BOD_5 + nitrification) for domestic waste water with $BOD_5 = 180$ mg l^{-1} and 38 mg l^{-1} ammonium.

3. Write the chemical reaction for denitrification of nitrate using acetic acid as a carbon

source.

4. Calculate BOD_1, BOD_2, BOD_{10}, and BOD_{20} when BOD_5 = 200 mg l^{-1} and the biological decomposition is considered to be a first order reaction. Room temperature is presumed.

5. Nitrification (99%) in a biological plant requires 6 h retention time at 20°C. What retention time is necessary at 0°C?

6. Calculate the area necessary for spray irrigation of 100,000 m^3 of waste water per year with a BOD_5 of 1000 mg l^{-1}. The maximum is 20,000 kg BOD_5 per ha per year and maximum application rate is 200-300 cm y^{-1}.

7. What disadvantages has the use of iron(III) chloride as a precipitant compared with calcium hydroxide and aluminium sulphate?

8. A waste water has an alkalinity corresponding to its hardness. It contains 120 mg l^{-1} Ca^{2+} and 28 mg l^{-1} Mg^{2+}. How much calcium hydroxide should be used to adjust the pH to 11.0?

9. A waste water contains 6.5 mg P l^{-1}. How much 1) $FeCl_3$, $6H_2O$ and 2) $Al_2(SO_4)_3$, $18H_2O$, must be used for chemical precipitation when 95% P-removal is required?

10. A biological plant is designed to give 90% nitrification. The retention time and the aeration are sufficient, but the observed nitrification is only 80%. What should be done?

11. Calculate the inhibition effect (as %) on the biological treatment of municipal waste water containing 2 mg l^{-1} Hg^{2+} and 10 mg l^{-1} Cu^{2+}. Use an equation similar to equation (6.1) to answer this question.

12. Calculate the BOD_5 of a waste water that contains 25 mg l^{-1} acetone, 20 mg l^{-1} acetic acid, 10 mg l^{-1} citric acid and 100 mg l^{-1} glucose.
What is the number of person eqv. of discharging 150 m^3 24 h^{-1} ?

13. Which of the following components can be treated at a mechanical-biological treatment plant (no adaptation is foreseen):

1) Butyric acid, 2) 2,4-dichlorophenol, 3) polyvinyl-chloride, 4) pentanone, 5) stearic acid, 6) butadiene. Suggest a treatment method for each of the six components in the concentration range 10-100 mg l^{-1}.

14. Suggest a treatment method for waste water containing diethyl-disulphide.

15. How much Ni^{2+} remains in industrial waste water after precipitation with calcium hydroxide at pH = 9.5? What methods are available if a concentration of 0.005 mg l^{-1} or less is required?

16. Calculate the stoichiometrical consumption of chemicals for treatment of 100 m^3 24 hr^{-1} of waste water containing 120 mg l^{-1} $Cr_2O_7^{2-}$. Reduction to Cr^{3+} using Na_2SO_3 + HCl precipitation with calcium hydroxide is suggested.

17 Design an aerated pond (5 steps should be considered) for treatment of 250 m^3 d^{-1} waste water at room temperature from an aqua culture plant. The ponds should bring the BOD_5 from 80 mg l^{-1} to 10 mg l^{-1}. How much oxygen must be supplied to the ponds?

18. A lake has a catchment area of 12 km^2. It covers an area of 500 ha and has an average depth of 15 m. Annual precipitation is 600 mm.

Set up an approximate N- and P-balance for the lake, when it is known that a waste-water plant with mechanical-biological treatment discharges 2000 m^3 d^{-1} to the lake.

Characterize the eutrophication of the lake and consider what improvement one should expect after 90% removal of P or N from the waste water.

19. Consider the following three systems:

System 1: ozone, an organic contaminant and air

System 2: ozone, the same organic contaminant and water

System 3: as 2 but with a stirrer.

Rank the three systems in accordance with how fast the organic contaminant will be decomposed.

20. A stream has an excessive high concentration of the following components: phosphate, ammonium, nitrite, PCBs, PAHs, carbohydrates and cadmium ions. All the mentioned contaminants are discharged from an overloaded municipal waste water treatment plant. List all the initiatives to be taken to ensure good water quality in the stream. In addition, mention the treatment methods to be applied.

6. THE PROBLEMS OF SOLID WASTE

6.1. SOURCES, MANAGEMENT AND METHODS

Classification of solid waste.

The disposal of solid waste has become a galloping problem in all highly developed countries due to a number of factors:

1. Increased urbanisation has increased the concentration of solid waste.
2. Increased use of toxic and refractory material.
3. Increased use of "throw away" mass produced items.

Solid wastes include an incredible miscellany of items and materials, which makes it impossible to indicate one simple solution to the problem. It is necessary to apply a wide spectrum of solutions to the problems according to the source and nature of the waste. Table 6.1 shows a classification of solid waste. Typical quantities per inhabitant in a technological society are included. The numbers represent the situation in EU 1995 and are taken from various sources.

TABLE 6.1
Classification of solid waste

Type of waste	kg/inhabitant/day (approx.)
Domestic garbage	4.5
Agricultural waste (1)	12
Mining waste (2)	18
Wastes from construction (3)	0.12
Industrial waste	1.35
Junked automobiles	0.15

(1) Not mentioned further in this context. Recycling is recommended and also widely in use.

(2) A substantial part is used for land filling.

(3) Not mentioned further in this context. Recycling is possible and recommended, but a part is also used for land filling.

Examination of mass flows.

A universal method for treatment of solid waste does not exist, and it is necessary to analyse each individual case to find a relevant solution to the problem. The analysis begins with an examination of mass flows (see Chapter 2) to ascertain whether there is an economical basis for **reuse** (for example, bottles), **recovery of valuable raw materials** (paper and metals) or **utilisation of organic matter**, as a soil conditioner or for the production of energy. These considerations are illustrated in Fig. 6. 1. Mass balances for important materials, as demonstrated in Chapter 2, must therefore be set up.

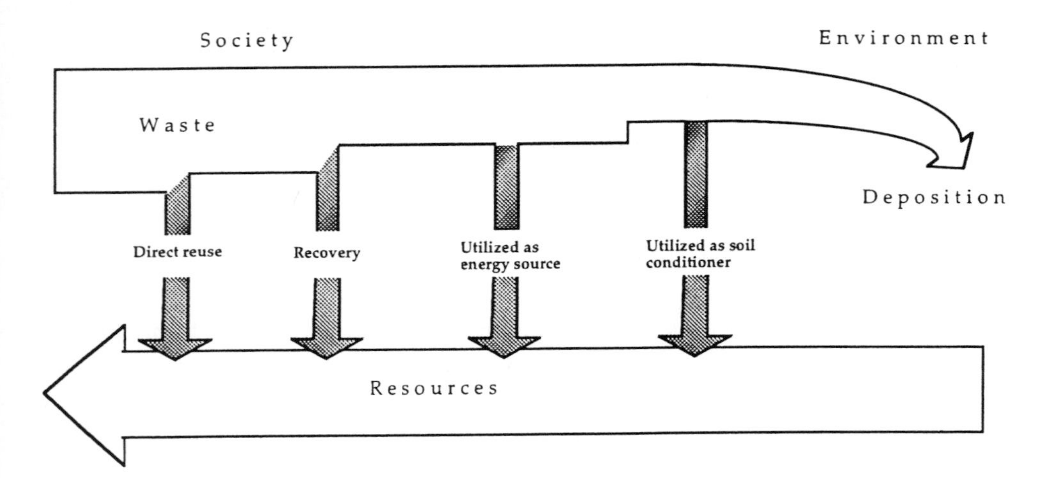

Fig. 6.1 Principles of recycling illustrated by mass flows.

The analysis of mass flows is the framework for a feasible solution, which takes economic as well as environmental issues into consideration. However, legislation and economical means are required to achieve the management goals. For instance, the use of returnable bottles can be realised **either by banning the use of throw-away bottles or by placing a purchase tax on throw-away bottles and not on returnable bottles.** The management problems are highly dependent on the technological methods used for solid waste treatment. The classification mentioned in the introduction to this book (Chapter 1, Section 1.6, principal methods 1-6) might be used in this context. Methods based on cleaner technology (see also Chapter 9) and recycling, to reduce the amount of waste produced, often require environmental legislation or economical means (green tax) to guarantee success.

Methods based on deposition will require a comprehensive knowledge of environmental effects on the ecosystems involved. Complete decomposition to harmless components is not possible for most solid waste. For instance incineration will produce a slag, which might cause a deposition problem, and smoke, which involves air pollution.

Methods for treatment of solid waste: an overview.

Table 6.2 gives an overview of the methods applicable to solid waste. The table indicates on which principle the method is based - recycling, deposition or decomposition to harmless components. It also shows the type of solid waste the method is able to treat.

TABLE 6.2

Methods for treatment of solid waste

Method	Principle	Applicable to
Conditioning and composting	Decomposition before deposition	Sludge, domestic garbage, agricultural waste
Anaerobic treatment	Decomposition, deposition, utilisation of biogas	Sludge, agricultural waste
Thickening, filtration, centrifugation and drying	Dewatering before further treatment and deposition	Sludge
Combustion and incineration	Decomposition and deposition, utilisation of energy	All types of waste
Separation	Recycling	Domestic garbage
Dumping ground	Decomposition, deposition	Domestic garbage
Pyrolysis	Recycling, utilisation of energy, decomposition	Domestic garbage
Precipitation and filtration	Deposition or recycling	Industrial waste
Land filling	Deposition	Industrial-, mining- and agricultural waste
Aerobic treatment	Decomposition and deposition	Sludge, agricultural waste

In the following sections the problems of solid waste will be discussed in four classes: A. Sludge, B. Domestic garbage, C. Industrial, Mining and Hospital waste, D. Agricultural waste. The methods used for the four classes are, to a certain extent, different, although there is some overlap, between them.

6.2 TREATMENT OF SLUDGE

Sludge handling.

In most waste water treatment the impurities are not actually removed, but rather concentrated in the form of solutions or a sludge. Only when a chemical reaction takes place does real removal of the impurities occur, e.g. by chemical or biochemical oxidation of organics to CO_2 and H_2O, or denitrification of nitrate to nitrogen gas. Sludge from industrial waste water treatment units in most cases requires further concentration before its ultimate disposal. In many cases two- or even three-step processes are used to concentrate the sludge. It is often an advantage to use further thickening by gravity, followed by such treatments as filtration or centrifugation. There are a number of ways of reducing the water content of the sludge that might be used to provide the most suitable solution of how to handle the sludge in any particular case. The final arrangement must be selected not only from consideration of the cost, but also by taking into account that the method used must not cause pollution of air, water or soil.

Characteristics of sludge.

The characteristics of a sludge are among the factors that influence the selection of the best sludge-treatment method. The sludge characteristics vary with the waste water and the waste water treatment methods used. One of the important factors is **the concentration of the sludge**. Table 6.3 lists some typical concentrations of various types of sludges. **The specific gravity** of the sludge is another important factor, since the effect of gravity is utilised in the thickening process. The specific gravity of activated sludge increases linearly with the sludge concentration. This corresponds to a specific gravity of 1.08 g/ml for actual solids. However, sludge is normally in sufficiently high concentration to exhibit zone-settling characteristics, which means that laboratory measurement of the settling rate must be carried out in most cases before it is possible to design a thickener.

The rate at which water can be removed from a sludge by such processes as vacuum filtration, centrifugation and sand-bed drying is an important factor, and is expressed by means of the **specific resistance**, R_s, which is calculated from laboratory observations of filtrate production per unit time by constant pressure loss, Δp. The inverse filtration rate per filter area unit, dt/dV, where t is time and V is volume of the filtrate per unit of filter area, yields in accordance to filtration theory a straight line (Hansen and Søltoft, 1988):

$$dt / dV = (\Omega s ñ / \Delta p)(V + V_d) \quad (6.1)$$

where Ω is the specific resistance expressed in m/kg, s is the dry matter in kg per m^3 water, ñ is the viscosity and V_d is the volume of filtrate corresponding to the resistance of the filter. Figure 7.2 shows the applied graph. The slope of the plot, ß, is in accordance with equation (6.1):

$$ß = (\Omega s ñ / \Delta p) \quad (6.2)$$

The slope can be found from the graph, and if s, ñ and Δp are known (they can all be determined by measurements) the specific resistance can be found.

Table 6.3
Typical concentrations of different types of sludge

Type of sludge	Concentration of suspended matter (w/w%)
Primary sludge (fresh)	2.5 - 5.0
Primary sludge (thickened)	7.5 - 10.0
Primary sludge (digested)	9.0 - 15.0
Trickling filter humus (fresh)	5.0 - 10.0
Trickling filter humus (thickened)	6.0 - 10.0
Activated sludge (fresh)	0.5 - 1.2
Activated sludge (thickened)	2.5 - 3.5
Activated sludge (digested)	2.0 - 4.0
Chemical precipitation sludge (fresh)	1.5 - 5.0
Chemical precipitation sludge (digested)	6.0 - 10.0

However, the specific resistance can change during filtration due to compression of the sludge. This is expressed by means of the **coefficient of compressibility**, s, using the following relationship:

$$R_s = R_0 * \Delta P^s \quad (6.3)$$

where R_0 = the cake constant. When s = 0, sludge is incompressible and $R_s = R_0$ = a constant.

The specific resistance of activated sludge has in the order of 100 times greater resistance than digested conditioned sludge after anaerobic treatment. Conditioned digested sludge has a specific resistance which is 2-4 times lower than that of colloidal clay which is difficult to filter. Calcium carbonate which is relatively easy to filter has a specific resistance which is about 80 times lower than digested conditioned sludge.

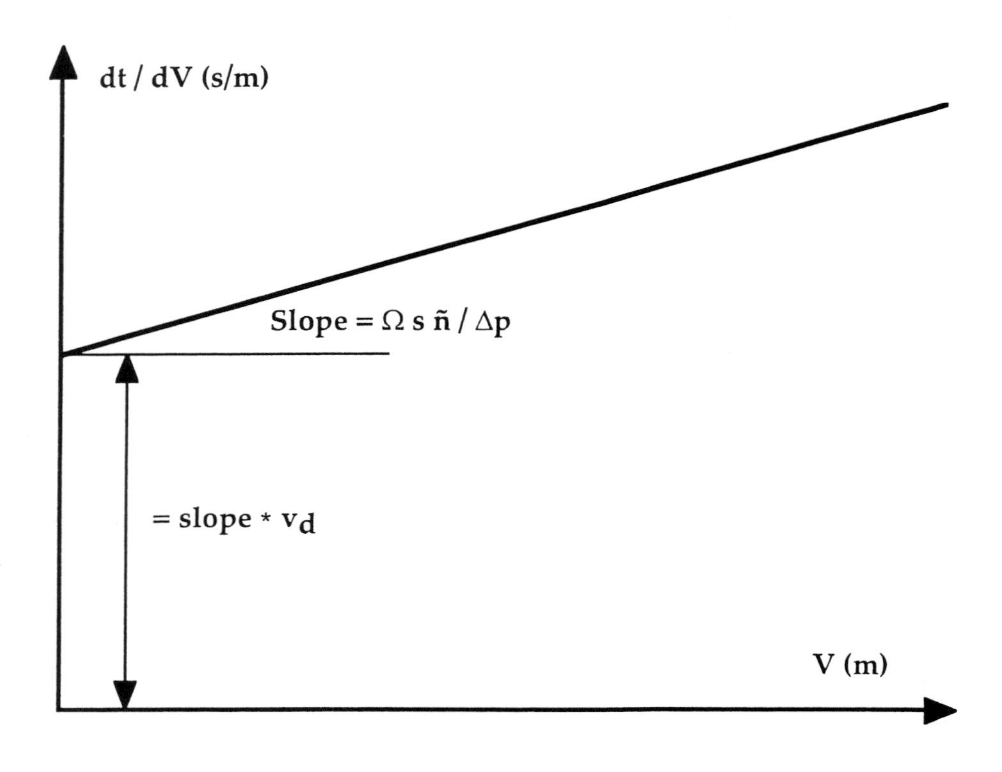

Fig. 6.2 The inverse filtrate rate per unit of filter area is plotted versus the filtrated volume per unit of filter area. In accordance with filtration theory. the plot is a straight line provided that the pressure loss over the filter is constant and the filter cake is incompressible.

When sludge is being considered for use as a soil conditioner, its **chemical properties** are of prime importance. The nutrient content (nitrogen, phosphorus and potassium), in particular, is of interest. Furthermore, knowledge of the heavy metals and organic contaminants in sludge is important because of their toxicity. See also the discussion in Section 4.5, where the standards in Denmark for sludge applicable as soil conditioner are given.

Finally, **the concentration of pathogenic organisms** in the sludge must be considered. Normal waste water treatment processes, such as sedimentation, chemical precipitation and biological treatments, remove considerable amounts of pathogens which are concentrated in the sludge. A significant reduction in the number of pathogenic organisms has been found to occur during anaerobic digestion, but they are not destroyed entirely. Combustion, intensive heat treatment of sludge or treatment with calcium hydroxide would of course eliminate the hazard of pathogenic microorganisms.

Conditioning of sludge.

Sludge conditioning is a process which alters the properties of the sludge to allow the water to be removed more easily. The aim is to transform the amorphous gel-like sludge into a porous material which will release water. Conditioning of the sludge can be accomplished by either chemical or physical means. Chemical treatment usually involves the addition of coagulants or flocculants to the sludge. Inorganic as well as organic coagulants can be used, the difference between typical conditioning by polymers or inorganic chemicals being in the amounts of the chemicals used. Typical doses and by inorganic coagulants, such as aluminium sulphate, iron (III) chloride and calcium hydroxide, are as much as 20% of the weight of the solid, while a typical dose of organic polymer is less than 1% of the weight of the solid. This does not necessarily mean that the cost of using synthetic polymers is lower, since the polymers cost considerably more per kg than the inorganic chemicals used as conditioners.

The amorphous gel-like structure of the sludge is destroyed by heating. Lumb (1951) indicates that the filtration rate of activated sludge is increased more than a thousand-fold after heat treatment. Typically, the heat-treatment conditions are 30-minutes' treatment at 150-200°C under a pressure of 10-15 atmospheres. A great advantage of heat conditioning is, of course, that the pathogens are destroyed.

Conditioning by freezing has also been reported by Klein (1966) and by Burd (1968), but the process seems to be uneconomic.

Thickening of sludge.

The filter area is determined from the specific sludge resistance and the amount to be treated by filtration. As the specific resistance of conditioned sludge may vary considerably, it is recommended to design for a surplus capacity of at least 25-50%

Several types of filter are applicable to achieve a high concentration of dry matter (30-45%) in the sludge which will reduce the cost of transportation significantly.

Vacuum filtration is used to remove water from a sludge by applying a vacuum across a porous medium. The vacuum filter is shown in Fig. 6.3. As the rotary drum passes through the slurry in the slurry tank, a cake of solid is built up on the drum surface and the water is removed by vacuum filtration through the porous medium on the drum surface. As the drum emerges from the slurry the deposited cake is dried further. The cake is removed from the drum by a knife edge. Often the porous filter is washed with water before it is reimmersed in the slurry tank.

Since the specific resistance varies widely with the type of sludge and the waste water treatment used, *it is often best to find the filtration characteristics of the sludge in the laboratory.*

Filter presses (mainly plate and frame type) are more cost effective than vacuum filters and are more widely used. A calcium hydroxide slurry is often used as filter medium. Filter presses operates by charge basis, as the filter cakes have to be removed when the filtration rate is reduced to a certain level.

Filter belts, see Figure 6.4, are also widely used. They have the advantage that they operate continuously, but usually they provide a lower concentration of dry matter than the filter press (about 30-35% versus 40% for filter presses).

The filtrate from all three types of filters has a high ammonium- nitrogen concentration which it may be advantageous to remove by stripping to eliminate the nitrogen loading in the waste water, as the filtrate must always be recycled.; see Section 5.4.

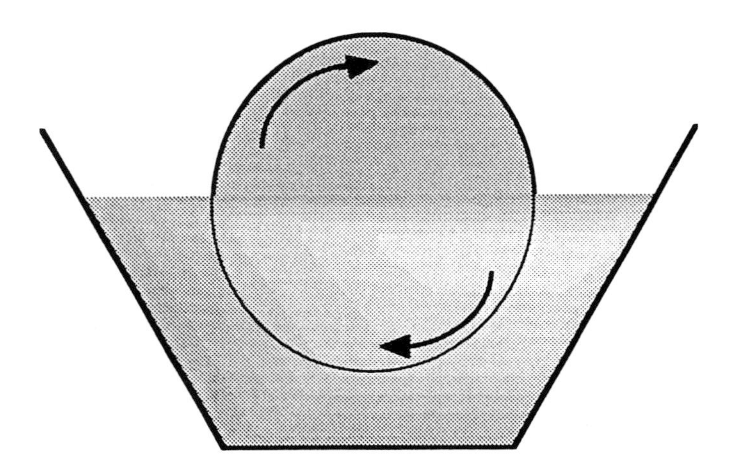

Fig. 6.3. Vacuum filtration.

Centrifugation can also be used in the removal of water from waste water sludge. Of the various types of centrifuge, the solid-bowl centrifuge is considered to offer the best clarification and water-removal properties. It is an important advantage of the centrifugation process that the centrifuge conditions can be adjusted according to the concentration of the volatile material (Albertson and Sherwood, 1968). The disadvantage of using a centrifuge is that the cake concentration is generally slightly less than that obtained by vacuum filtration. *Prediction of the behaviour of sludge in a centrifuge* is largely a matter of experience. However, some general trends can be noted. If the mass flow rate is increased, recovery is reduced. The use of electrolytes will increase the recovery at a given flow rate or increase the flow rate for a given recovery.

Digestion of sludge.

If the sludge contains biodegradable organic material it may be advantageous to treat the sludge by aerobic or anaerobic digestion before thickening. Anaerobic digestion is by far the most common method of treating municipal sludge. It creates good conditions for the growth of micro-organisms. The end products of anaerobic digestion are carbon dioxide and methane. The temperature is commonly set at about 35°C, in order to maintain optimum conditions in the digester. Unfortunately, anaerobic digestion results in considerable quantities of nutrients going into solution (Dalton et al., 1968), which means that a significant amount of nutrient material will be returned to the treatment plant if the supernatant is separated from the sludge.

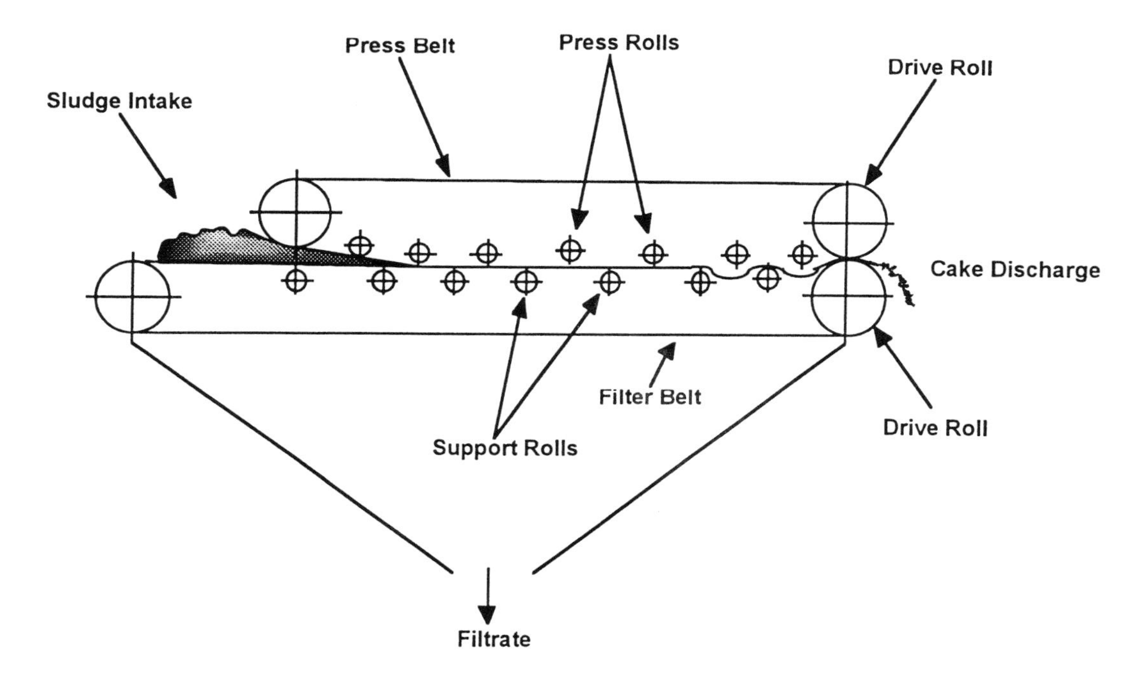

Fig. 6.4 Filter belts are widely used to dewater sludge. Generally, the dry matter concentration is increased from 3-4% to 30-35%.

The principal function of anaerobic digestion is to convert as much as possible of the sludge to end products: liquids and gases. Anaerobic decomposition generally produces less biomass than aerobic processes. As mentioned in Section 5.5 heavy metal ions may slow down the decomposition rate. The presence of ammonia and hydrogen sulphide may have the same effect (Vavilin et al., 1995)

The microorganisms can be divided into 2 broad groups: **acid formers and methane formers. The acid formers** consist of facultative and anaerobic bacteria, and soluble products are formed through hydrolysis. The soluble products are then fermented to acids and alcohols of lower molecular weight. **The methane formers** are strictly anaerobic bacteria that convert the acids and alcohols, along with hydrogen and carbon dioxide, to methane. Design parameters for anaerobic digesters are given in Table 6.4. High-rate digesters are more efficient. The contents are mechanically mixed to ensure better contact between substrate and microorganisms, thus accelerating the digestion process. A standard rate digester (typical retention time 4 weeks) is shown in Figure 6.5. The biogas produced by anaerobic digestion of sludge from waste water treatment is utilised at waste water treatment plants for the production of heat and for the production of electricity (a gas turbine is used). The amount of gas is usually not sufficient to make the sale of gas possible.

TABLE 6.4
Design parameters for anaerobic digesters

Parameter	Normal rate	High rate
Solid retention time	30 - 60	10 - 20
Volatile solid loading (kg/m^3 d)	0.5 - 1.5	1.6 - 6.0
Digested solids conc.%	4 - 6	4 - 6
Volatile solid reduction (%)	35 - 40	45 - 55
Gas production m^3/kg	0.5 - 0.6	0.6 - 0.65
Methane content	65%	65%

The properties of aerobically digested sludge are similar to those of anaerobically digested sludge. An advantage is that some of the operational problems attending anaerobic digesters are avoided, but the disadvantage compared with anaerobic digestion is that the process is more expensive since oxygen must be provided and energy recovery from methane is not possible. Since aerobic digestion is less used in industrial waste water processes than in treatment of municipal waste water, it is not appropriate here to go into further detail. For more extensive coverage of these processes, readers are referred to McCarthy (1964) and to Walker and Drier (1966).

Drying and combustion.

The purpose of drying sludge is to prepare it for use as a soil conditioner or for incineration. Air drying of the sludge on sand beds is often used to reach a moisture content of about 90%. Also, such drying techniques as flash drying and rotary drying are used to remove water from sludge. Often waste heat from the incineration process itself is used in drying. However, it has been reported by Quirk (1964) that the cost of combined drying and combustion is higher than the cost of incineration alone. The economy of sludge drying has recently been reviewed by Burd (1968). He reports that at its present cost, heat drying should only be considered if the product (soil conditioner) can be sold for at least US$ 20.00 a ton (1986-dollars).

Combustion serves as a means for the ultimate disposal of sludge. Two techniques should be mentioned: the atomised suspension technique (Gauvin, 1947), and the and the Zimmerman process (Zimmerman, 1958).In the former process the sludge is atomised at the top of the tower, and droplets pass down the tower where the moisture evaporates. The tower walls are maintained at 600-700°C by hot circulating gas. The solid produced is collected in a cyclone and the heat recovered from the stream and gas.

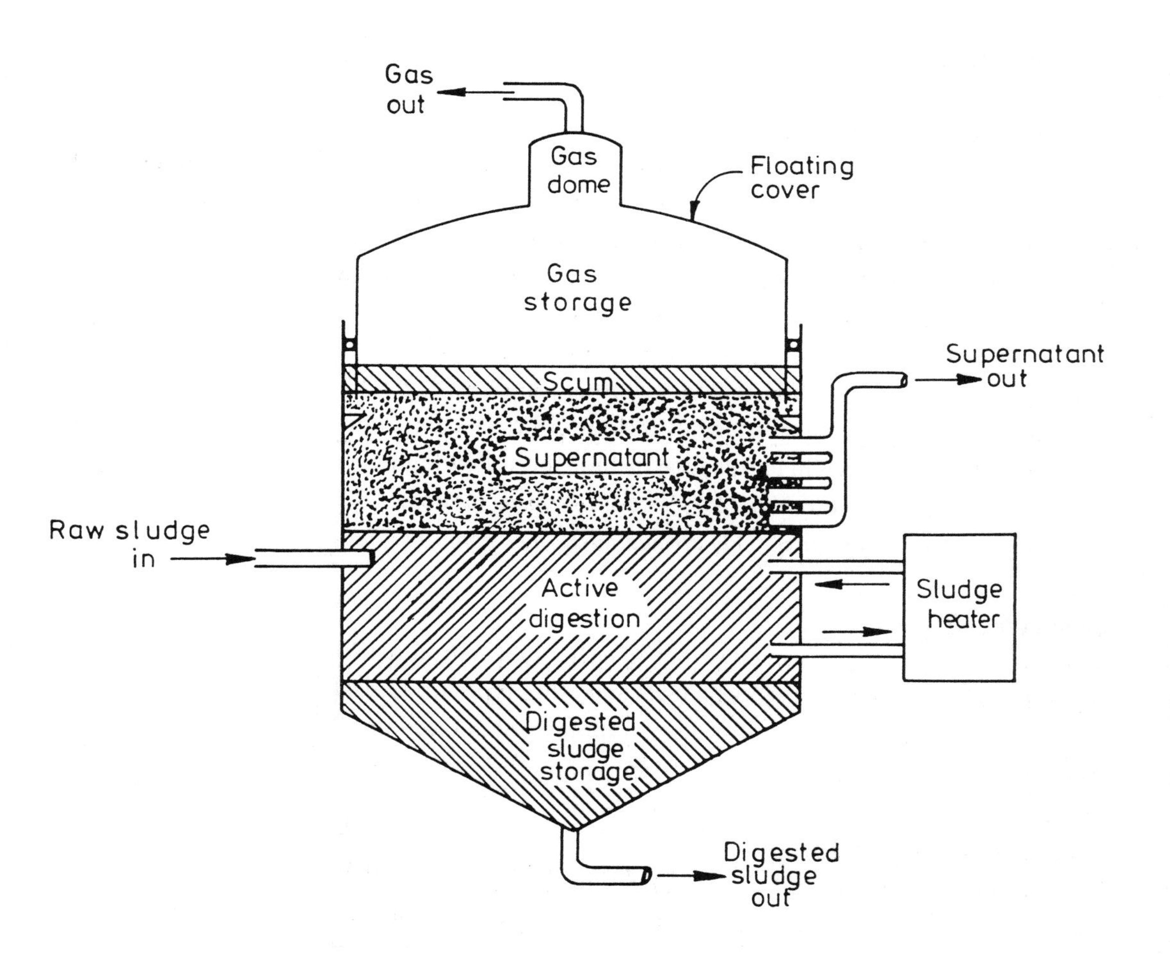

Fig. 6.5. Diagram of standard-rate anaerobic digester.

The Zimmerman process involves wet-air oxidation at high temperature and pressure. Oxidation of organics occurs at 200-300°C and the high pressure is used to prevent evaporation of the water. By means of a heat exchanger the heat developed is used to raise the temperature of the incoming sludge. Tests of wet oxidation of sludge have recently given very promising results (Khan et al., 1999). This process will probably be one of the most attractive solutions to the sludge disposal problems which are expected to accelerate in the coming years due to increased interest in waste water treatment and due to the enhanced requirements imposed on the application of sludge as soil conditioner in agriculture; see for instance the standards in Table 4.6 , Section 4.5.

6.3 DOMESTIC GARBAGE

Characteristics of domestic garbage.

The composition of domestic garbage varies from country to country. To a certain extent the environmental legislation and the economic level of the nation is reflected in this composition. The amount of domestic garbage per inhabitant is increasing. In most developed countries the growth was 2-4% from the mid 1950s to 1963, while it has been lower in the period from the mid seventies to 1999, (1-2%).

It is often advantageous to carry out a comprehensive analysis of the domestic garbage in order to estimate the economic and ecological consequences of various environmental management strategies including the selection of the most relevant treatment methods. The analysis will in this case include the following items: *(1) food waste, (2) paper, (3) textiles, (4) leather and rubber, (5) plastics, (6) wood, (6) iron, (8) aluminium, (9) other metals, (10) glass and ceramic products, (11) ash and dust, (12) stone, (13) garden waste, (14) other types of waste.*

Separation methods.

Separation of solid waste can be achieved either in a central plant or by the organisation of separate collection of paper, glass, metals and other types of domestic waste.

A mass flow diagram for a separation plant in Franklin, Ohio, is shown in Fig. 6.6. After wet grinding, paper fibres, glass, iron and other metals are separated. The remaining part of the solid waste is used for heat production.

Dumping ground (landfills).

This was previously the most common handling method for solid waste. Today it is mainly in use in smaller towns, often after grinding or compression, which reduce the volume by 60-80%.

Deposition of solid waste on a dumping ground is an inexpensive method, but it has **a number of disadvantages:**

1. Possible **contamination of ground water.**

2. Causes inconveniences due to **the smell.**
3. **Attracts noxious animals,** such as flies and rats.

During the deposition several processes take place:

1. *Decomposition* of biodegradable material.
2. *Chemical oxidation* of inorganic compounds.
3. *Dissolution and wash out* of material.
4. *Diffusion processes.*

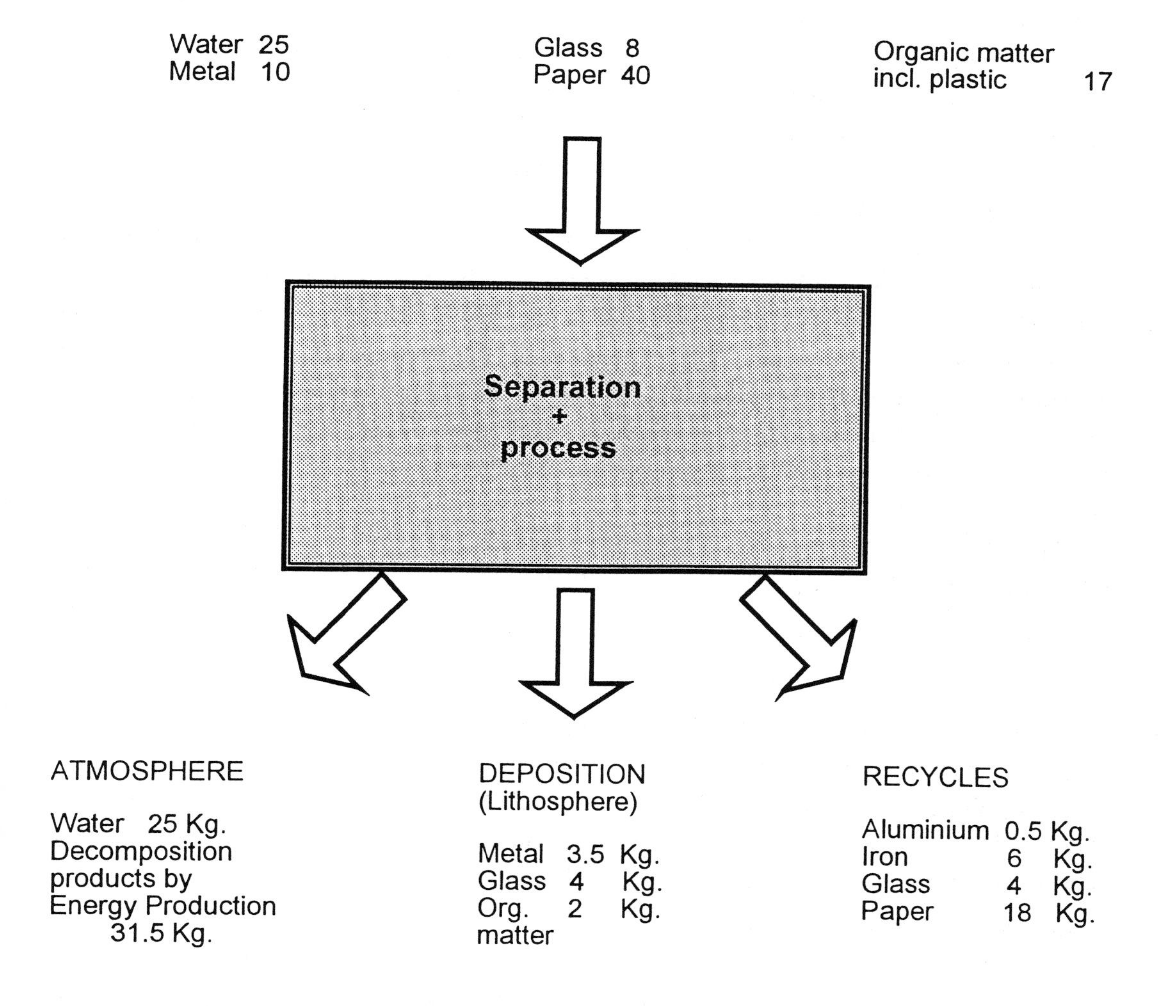

Fig. 6.6. Separation plant, Franklin, Ohio. Capacity: 150 t/d. A mass flow diagram is shown (Basis: 100 kg solid waste).

Where the decomposition takes place in the aerobic layers carbon dioxide, water, nitrates and sulphates are the major products liberated, while decomposition in anaerobic layers leads to the formation of carbon dioxide, methane, ammonia, hydrogen sulphide and organic acids.

Water percolating from dumping grounds has a very high concentration of BOD and nutrients and *can therefore not be discharged into receiving waters.* If it cannot be recycled on the dumping ground, this waste water must be subject to some form of treatment.

Composting.

Composting has been applied as a treatment method in agriculture for thousands of years. The method is still widely applied for treatment of agricultural waste. Organic matter from untreated solid waste cannot be utilised by plants, but it is necessary to let it undergo a degree of biological decomposition, through the action of microorganisms.

Again we can distinguish between **aerobic and anaerobic processes.** A number of factors control these processes:

1. **The ratio of aerobic to anaerobic processes,** which, of course, is determined by the available oxygen (diffusion process).
2. **Temperature.** Different classes of micro-organisms are active within different temperature ranges (see Table 6.5). Heat is produced by the decomposition processes. The thickness of the layer determines to what extent this heat can be utilised to maintain a temperature of 60-65°C, which is considered to be the optimum. Fig. 6.7 shows a typical temperature pattern from composting in stacks. In this context it is also important to mention that composting at a relatively high temperature means a substantial reduction in the number of pathogenic microorganisms and parasites.

TABLE 6.5
Classification of micro-organisms

	Temperature optimum	Temperature range
Psychrophile	15-20°C	0-30°C
Mesophile	25-35°C	10-40°C
Thermophile	50-55°C	25-80°C

3. **The water content** should be 40-60%, as this gives the optimum conditions for the processes of decomposition.
4. **The C/N ratio** should be in accordance with the optimum required by the micro-organisms. Domestic garbage has a C/N ratio of 80 or more due to the high content of paper, while the optimum for composting is 30. It is therefore advantageous to add sludge

from the municipal sewage plant to adjust the ratio to about 30. Sludge usually has a C/N ratio of 10 or even less. During composting the C/N ratio is decreased as a result of respiration, which converts some the organic matter to carbon dioxide and water.

5. The optimum conditions for the micro-organisms include a **pH around 6 (6-8)**. Usually pH increases as a result of the decomposition processes. If the pH is too low calcium hydroxide should be added and if the pH is too high sulphur should be added. Sulphur activates sulphur bacteria, which produce sulphuric acid.

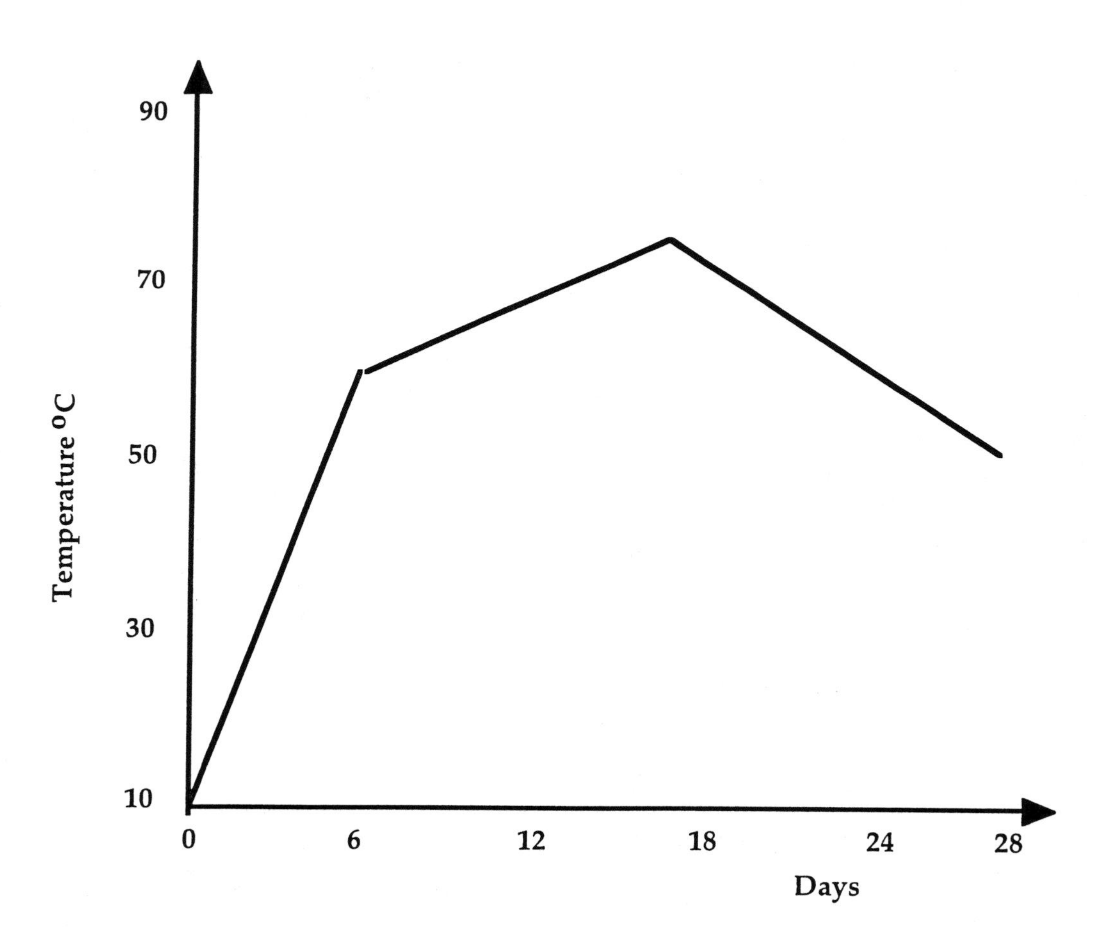

Fig. 6.7 Temperature versus composting time.

Fig. 6.8 shows a flow chart for a composting plant with composting in containers followed by composting in stacks. Composting in containers with addition of *air accelerates* the decomposition processes, but this process might be excluded in smaller plants. The compost is used as soil conditioner when it is stable, i.e. the decomposition processes are almost terminated, which may be determined by a respirometric technique (Lasaridi and

Stentiford, 1998). The standards for sludge are applied also for compost; see Table 4.6; Section 4.5

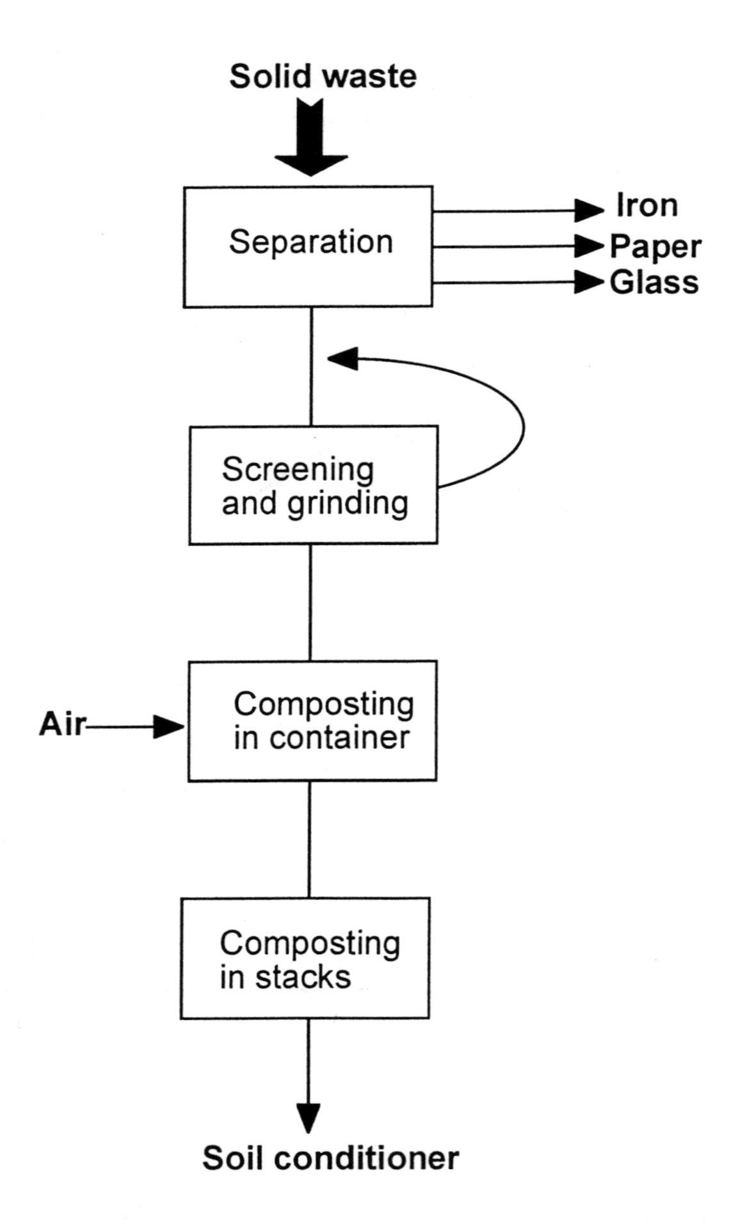

Fig. 6.8 Flow chart of composting plant.

<u>Incineration of domestic garbage.</u>

Incineration is very attractive from a sanitary point of view, but it is a very *expensive method,* which has some environmental disadvantages. Valuable material such as *paper is not recycled, a slag, which must be deposited,* is produced and air pollution problems are involved. Nevertheless, incineration has been the preferred method in many cities and larger towns. This development might be explained by the increasing amount of combustible material in domestic garbage and the possibilities of combining incineration plants with district heating.

The optimum combustion temperature is 800-1050°C. If the temperature is below this range incomplete combustion will result (dioxines may be produced), and at a higher temperature the slag will melt and prevent an even air distribution. The composition of the solid waste determines whether this combustion temperature can be achieved without using additional fossil fuel. If the ash content is less than 60%, the water content is less than 50% and the combustible material more than 25%, no additional fossil fuel is required.

The heating value can be calculated from the following equation:

$$H_u = H^1 \frac{100 - W}{100} - 600 \frac{W}{100} \tag{6.4}$$

where

H^1 = the heat value of dry matter

H_u = the heat value

W = the water content (weight %)

The composition of the slag and the fly ash, which is collected in the filter, is, of course, dependent on the composition of the solid waste. It is notable that a minor amount of the slag and the fly ash always will be unburned material. Slag and fly ash comprise 25-40% of the weight of the solid waste. The slag will have a high concentration of heavy metals in most cases but they are usually chemically bound as very water insoluble oxides.

Although all modern incineration plants have filters, the air pollution problem is not completely solved. Hydrogen chloride, in particular, can cause difficulties, because it is toxic and highly corrosive. It results from incineration of chlorine compounds among which PVC is the most important.

Pyrolysis.

Pyrolysis is the decomposition of organic matter at elevated temperature without the presence of oxygen. For pyrolysis of solid waste a temperature of 850-1000°C is generally used. The process produces gas and a slag, from which metals can easily be separated due to the anoxic atmosphere .

Pyrolysis is a relatively expensive process, although usually slightly less expensive than incineration. The smoke problems are the same as those of incineration, but the easy recovery of metals from the slag is, of course, an advantage. *The gas produced has a composition close to coal gas* and in most cases can be used directly in the gas distribution system without further treatment. Approximately 500 m^3 of gas are produced per ton of solid waste, but

300-400 m^3 are used in the pyrolysis process to maintain the temperature.

Comparison of the methods

Composting is the process which is most acceptable from an ecological point of view for treatment of solid waste including domestic garbage. This is obvious when the 6 principal methods mentioned in Section 1.6 are used to classify the various methods for treatment of solid waste. It can furthermore be seen from Chapter 8 focusing on ecological engineering methods. Composting may be designated an ecological engineering method, as at least to a certain extent it follows some of the 12 ecological principles presented in Chapter 8 to illustrate the characteristics of this approach to pollution abatement. Pyrolysis also gives the opportunity to reuse waste products as raw materials or energy source. When the waste heat from incineration plants is utilised, this method becomes of course more attractive when evaluated according to ecological principles.

Composting is, however, less area intensive than incineration and pyrolysis, although it will require relatively less area than landfills, especially when compressing and grinding are used to reduce the volume of domestic and industrial garbage. Incineration is the most widely used method for treatment of solid waste in bigger towns and cities. The relatively small area required per kg of waste is the dominant factor here.

Table 6.6

Comparison of methods for treatment of solid waste, particularly domestic garbage.

Method	Resource friendly	Ecological Principles	Area needed	Costs per kg waste
1.Dumping ground	unacceptable	unacceptable	large	low
2.Dumping ground with pretreatment	unacceptable	unacceptable	about 50% of 1	slightly higher than 1
3. Composting	acceptable	acceptable	medium	70-85% of 4 and 5
4. Pyrolysis	almost acceptable	almost acceptable	low	high
5. Incineration	unacceptable, unless heat is recovered	unacceptable	low	high

The cost of the treatment of course varies with the size of the plant and the composition of waste. Generally, the treatment costs increases in the following order of treatment methods: dumping ground ----> dumping ground with compressing and grinding of waste ------> composting -------> incineration or pyrolysis.

Comparison of the 5 methods is summarised in Table 6.6.

6.4 INDUSTRIAL, MINING AND HOSPITAL WASTE

Characteristics of the waste.

This type of waste causes particular problems because it may contain toxic matters in relatively high concentrations. Hospital wastes are especially suspect because of contamination by pathogens and the special waste products, such as disposable needles and syringes and radioisotopes used for detection and therapy.

The waste from industry and mining varies considerably from place to place, and it is not possible to provide a general picture of its composition. It the composition permits, it can be used for land filling, but if it contains toxic matter special treatment will be required. The composition may be close to that of domestic garbage in which case the treatment methods mentioned in section 6.3 can be applied.

Hospital solid waste is being studied in only a few locations to provide data on current practices and their implications. Most hospital waste is now incinerated and this might, in many cases, be an acceptable solution. However, if the solid waste contains toxic matter it should be treated along the lines given for industrial waste below. At least, waste from hospital laboratories should be treated as other types of chemical waste.

Treatment methods.

Industrial and mining waste containing toxic substances, such as heavy metals or toxic organic compounds, should be treated separately from other types of solid waste, which means that it cannot be treated by the methods mentioned in Sections 6.2 and 6.3. A number of countries have built special plants to handle this type of waste, which could be called chemical waste. Such plants may include the following treatment methods:

1. Combustion of toxic organic compounds. The heat produced by this process might be used for district heating. Organic solvents, which are not toxic, should be collected for combustion, because discharge to the sewer might overload the municipal treatment plant. For example, acetone is not toxic to biological treatment plants, but 1 kg of acetone uses 2.2 kg of oxygen in accordance with the following process:

$$(CH_3)_2CO + 4O_2 \text{ -------> } 3CO_2 + 3H_2O \qquad (6.5)$$

A different combustion system might be used for pumpable and non-pumpable waste.

2. Compounds containing halogens or sulphur should be treated only in a system which washes the smoke to remove the hydrogen halogenids or sulphuric acid formed.

3. Waste oil can often be purified and the oil reused.

4. Waste containing heavy metals requires deposition under safe conditions after a suitable pretreatment. Recovery of precious metals is often economically viable and is essential for mercury because of its high toxicity. The pretreatment consists of a conversion to the most relevant oxidation state for deposition, e.g. chromates should be reduced to chromium in oxidation state 3. Furthermore metals should be precipitated as the very insoluble hydroxides, (see also Section 6.5) before deposition.

5. Recovery of solvents by distillation becomes increasingly attractive from an economical view-point due to the growing costs of oil products, as most solvents are produced from mineral oil.

It is finally recommended to examine the possibilities of recovering plastics. The main problem with recycling plastics is associated with the wide range of chemically different plastic type. A mixture of different plastic types which may be recovered does not have the same properties as the particular plastic types, and can therefore only be used where the strength and the colour is of minor importance, for instance for plastic bags.

6.5 SOIL REMEDIATION

<u>The soil contamination problem</u>

Until about 25 years ago contaminated soil was largely ignored as an environmental concern. Today, we must acknowledge that millions of cubic meters of soil are contaminated and make normal use of the soil doubtful. The first response to acknowledged soil contamination issues - and such soil is unfortunately still widely used - was removal and placement in a more secure landfill environment. This simply moves contaminated soil from one place to another. Often early landfills were situated adjacent to rivers and lakes which encouraged migration of contaminants. Wastes can be stabilised after removal to further reduce mobility for instance by solidification with concrete or by chemical treatment of certain contaminants. Removal to a safer place coupled with stabilisation may make this otherwise non-ecological method more acceptable.

Today, there is a wide range of available methods to recover contaminated soil. We distinguish between methods based on removal of the contaminated soil and in situ treatment methods. The first group of methods are usually more costly, as they entail relatively expensive transportation which in itself produces pollution. From an ecological point of view, in situ methods must be preferred, but they cannot be used in all possible situations.

Removal and treatment of contaminated soil

Treatment facilities to handle contaminated soil have been constructed in many industrialised countries where the problem is most relevant. They are based on one of three principal treatment options: incineration, bioremediation by a composting like process or extraction.

Incineration or thermal treatment of the contaminated soil can be used to eliminate organic contaminants susceptible to destruction. This process is energy intensive and expensive, as the transportation costs add to the incineration costs. The method is not applicable to inorganic contaminated soil for instance soil contaminated by heavy metals.

Biological degradation of organic contaminants is a cheaper alternative to incineration. Dedicated bioreactors may be used or just open composting sites. The latter possibility is of course more cost effective, but entails that the drainage water must be treated as it will inevitably contain the contaminants in an unacceptably high concentration. It is sometimes necessary to add water to the process to ensure rapid biological decomposition of the organic contaminants, particularly in dry seasons or arid climates. If the contaminant is difficult to decompose, because it has a very slow biodegradation rate (see Section 4.6), addition of adapted microorganisms may offer a solution

Extraction may also be used to reclaim soil, but the method has several technical and ecological disadvantages. The extraction liquid must be treated and recovered which is sometimes technically difficult to accomplish with high efficiency. If even a minor emission of the extraction liquid, takes place, unless water is used, it will cause an additional pollution problem to be solved. Only if the contaminant can be recovered or is soluble in water, will this method offer some perspectives.

In situ treatment of contaminated soil

These methods seem more attractive from both an economic and an ecological view point, because the elimination of the transportation step entails that the cost and the pollution associated with the transportation are eliminated too. In addition, in situ treatment has the advantage that the mixture of contaminant is known which makes it possible to tailor a treatment method to this specific mixture.

If the contaminants are water soluble, it may be advantageous to **dissolve** them and thereby remove them from the soil. Addition of surfactants to the water may increase the solubility of the pollutants. Above the water table, extraction is normally conducted by creating a vacuum to remove vapour - the contaminants in volatilised form - from the unsaturated zone. This method has proven to be an especially useful process for the hazardous components of gasoline and other hydrocarbons. Heavy metal ions may be removed by use of a solution of EDTA or similar ligands which considerably enhance the solubility of heavy metal ions. As the solubility of heavy metal ions in EDTA-solutions is highly dependent on pH, the pH is adjusted to the optimum value (for most heavy metals around pH= 5.0) is applied.

The most desirable method of all treatment processes is **in situ biodegradation** to render the soil harmless. The rate of recovery with these processes should of course be sufficiently high to prevent the contaminants travelling far off-site, where they might pose a risk to humans or ecosystems, before they are decomposed. A promising possibility is to apply airsparging. Air is pumped down pipes to below the water table to accelerate the decomposition in the saturated and unsaturated zone (Walsted and Christensen, 1999).

In situ biodegradation involves a conflict between biodegradation and mobility which obviously is dependent on the polluting compounds. The addition of nutrients or oxygen may provide the enhancement needed to achieve acceptable rates of degradation. Also the use of adapted microorganisms may accelerate the decomposition. The presence of easily degradable compounds can lead to co-metabolism and the destruction of other refractory compounds. This is the case with high molecular weight PAHs in the presence of 2-3 ring polycyclic aromatics.

Recovery of soil by **phytoremediation** is a recently developed technique, but as it is an ecotechnological method, it will be presented in Chapter 8.

Heavy metals may also be removed by application of **electrolysis**. This process is applicable entirely to high concentrations which in addition makes a recovery of the metals possible. A combination of electrolysis and plant bioaccumulators seems to be an attractive solution for soil even highly contaminated with heavy metals.

6.6 QUESTIONS AND PROBLEMS

1. Set up a mass flow diagram for iron and glass in a selected district.

2. Discuss the analytical data in Table 6.5 from an ecological view-point.

3. A selected town or district is considered. Compare from both an ecological and an economical point of view the following solutions to the solid waste problem of domestic garbage and sludge: A. Separation of paper, metals and glass followed by either 1) incineration, 2) composting, or 3) pyrolysis. B. Incineration (no preseparation). C. Pyrolysis followed by separation of metals from the slag.

4. Classify the 5 methods in Table 6.6 in accordance with the 6 principal methods A to F mentioned in Section 1.6

5. Which of the 12 ecological principles presented in Chapter 8 are valid for the 5 methods listed in Table 6.6? Set up a 12x5 table to answer this question.

6. Classify the seven methods for soil remediation presented in Section 6.5 in

accordance with the 6 principal methods A to F mentioned in Section 1.6.

7. Make a detailed list of solid waste components that can be found in domestic garbage. Indicate on the list which components it is possible to recover by which technique.

7. AIR POLLUTION PROBLEMS

7.1. THE PROBLEMS OF AIR POLLUTION - AN OVERVIEW

Air pollution is defined as the presence of components that were not found in a clean atmosphere prior to the industrial activity, or are found in unusually high concentrations compared with the natural level. The concentration is usually indicated in $\mu g/m^3$, or as ppmv (or volume%). The conversion between the two types of units is shown in Section 2.6, and Example 2.2 shows the conversion for carbon dioxide. Air pollution control applies a wide range of remedies, and cleaner technology is playing a more important role in air pollution control than in water pollution control. Cleaner technology is often employed as a result of increasingly stringent legislation, e.g. the setting of lower threshold levels for example for lead in gasoline, sulphur in fuel and carbon monoxide in exhaust gases.

Air pollution problems can be considered in terms of the effect on climate, or of local or regional effects caused by toxicity of particular pollutants. This chapter outlines the environmental technology now available for air pollution control, while the application of cleaner technology is discussed in Chapter 9.

The applied environmental technological methods are classified according to the problems they solve. This chapter covers the control of particulate pollution, carbon dioxide, carbon hydrides and carbon monoxide and sulphur dioxide problems, nitrogenous gas pollution, heavy metals and industrial gaseous pollution. We distinguish between primary pollutants, such as nitrogen oxide and sulphur dioxide, which are emitted directly to the atmosphere and secondary pollutants, resulting from chemical processes in the atmosphere of primary pollutants, for instance sulphur trioxide and nitric acid. The pollution may originate from point sources, linear sources or surface sources. A chimney is a typical point source, a motorway a typical linear source and as an example of a surface source may be mentioned an industrial area.

Emission is the output from a source of pollution. It might be indicated as mass or volume per unit of time or per unit of production. If it is given as a concentration unit, e.g. mg per m^3, it is also necessary to know the number of m^3 discharged per unit of time, or the number of m^3 polluted air produced per unit of production.

Imission is the concentration of pollutants in a given area.It might be indicated as mg per m^3 for particulate matter or as a concentration unit for gaseous pollutants.

Correspondingly, **legislation will distinguish between emission standards and air quality standards respectively.**

Gaseous pollutants are evaluated from 1) toxicity, 2) water solubility 3) absorption band (wave length), which determines their green house effect and 4) photochemical effects. Table 7.1 gives these four evaluation criteria for some pertinent gaseous pollutants.

Appendix 8 shows tables of the global emission of air pollutants from natural and antropogenic sources, imission of various gases and the relationship between pollution scenarios and the most pertinent air pollutants.

TABLE 7.1

Gaseous Pollutants

Gas	Toxicity	Solubility	IR absorption	Photochemical activity
NH3	medium	high	high	none
NO	small	small	small	none
N_2O	small	small	high	medium
NO	high (animals)	medium	high	high (troposphere)
HNO_3	high (animals)	high	high	medium
SO_2	high (plants)	high	high	small
O_3	high	small	medium	high
CO	high (animals)	small	small	small
CO_2	small	high	high	small
CFCs	small	small	high	high (stratosphere)
Halons	small	small	high	high (stratosphere)
CH_4	small	small	high	high (stratosphere)

7.2. APPLICATION OF PLANTS AND VEGETATION AS INDICATORS FOR AIR POLLUTION

Biological monitoring is an evaluation of the environmental quality by observations and determination of effects on living organisms, while physical-chemical monitoring does not give the possibilities to evaluate effects. The scope of biological monitoring is to acknowledge and measure changes of biological systems, interpret the changes and attempt to predict the future changes. Natural factors may also cause the observed changes, for instance variations in climate, soil composition and ecological changes. The man-made changes are the application of the considered ecosystem and / or its adjacent ecosystem and pollution.

The biological monitoring can be performed at different levels:

1) measurement of chemical, biochemical and physiological changes of individuals,
2) measurement of changes in population parameters,
3) determination of changes in distributions and frequencies of species,
4) determination of ecosystem changes.

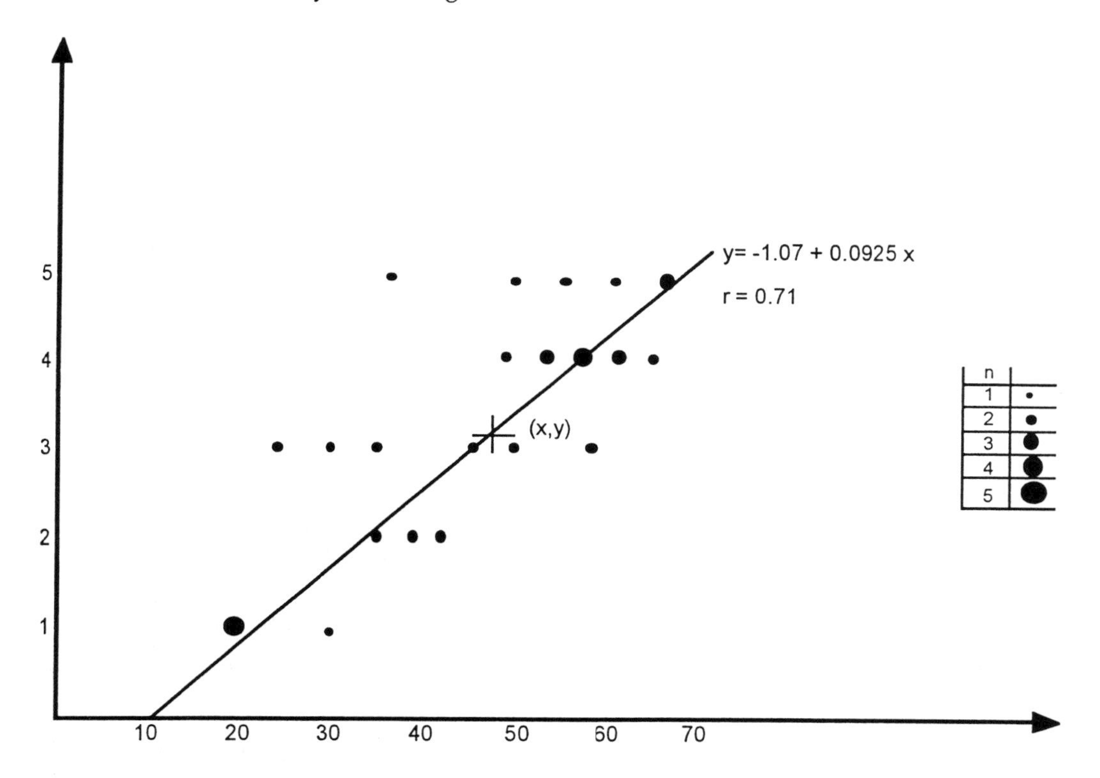

Fig. 7.1. Correlation between levels of sulphur dioxide and extent of damage on thallus, lichen *Hypogymnia physodes*. From various localities in Copenhagen. Exposure 6 months. r = 0 at p = 0.01. From Søchting and Johnsen, 1978.

It is necessary to measure physical and chemical variable in addition to the biological monitoring to find the factors in the environment that are causing the biological changes. The selection of the biological parameters is dependent on the component. If we for instance want to measure the effect of sulphur dioxide, a sulphur dioxide sensitive organism would be selected; see Figure 7.1.

Indicator plants are species that are dependent on well-defined specific conditions, have a well known tolerance or are very sensitive to specific conditions. Monitor species include species that are known to be good accumulators for specific compounds. The following classes of monitor species are used:

1) species that are particularly sensitive to low concentrations of a specific pollutant. The tobacco plant, *Nicotiana tabacum,* is for instance particularly sensitive to ozone,

2) species that show a gradual reaction by increasing exposure, for instance the growth of lichens is inhibited by various pollutants, particularly sulphur dioxide.

3) species that are accumulators of specific compounds, often without any effect. It facilitates the determination of the pollution. Mosses and lichens are for instance able to accumulate heavy metals. See Figure 7.2.

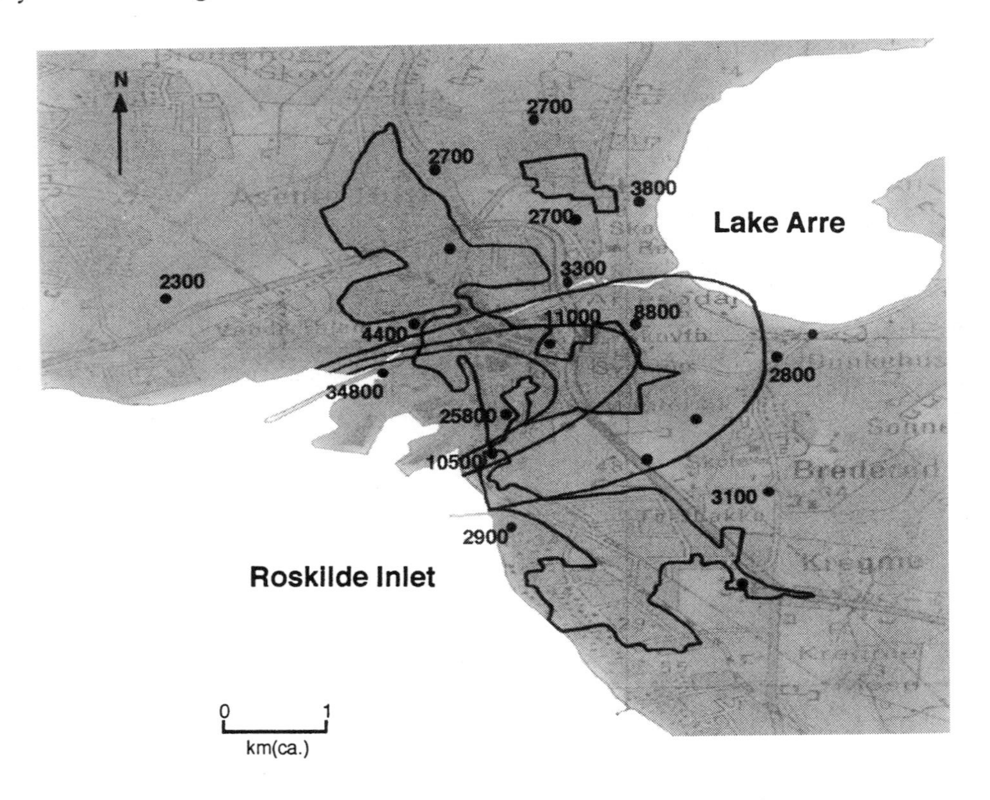

Fig. 7.2. Concentrations of iron (mg per kg dry matter) in ephiphytic lichens, *Lecanora conizaeoides,* in the Frederiksværk area, close to an iron plant.

4) species that are characteristics for areas with a significant pollution. For instance can the lichen, *Lecanora conizaeoides,* often be found in urban areas where the sulphur dioxide concentration is relatively high. See Figure 7.3.

Indicator species can give important information about general environmental problems. If the reactions of the organisms to a specific pollutant is know, they can furthermore assess the effect and distribution of that pollutant.

The advantages of biological indication compared with physical-chemical measurements may be summarised as follow:

1) the effects on living organisms are directly measured,
2) the effect can be expressed as integration over a longer period of exposure,
3) the effect is directly uncovered,

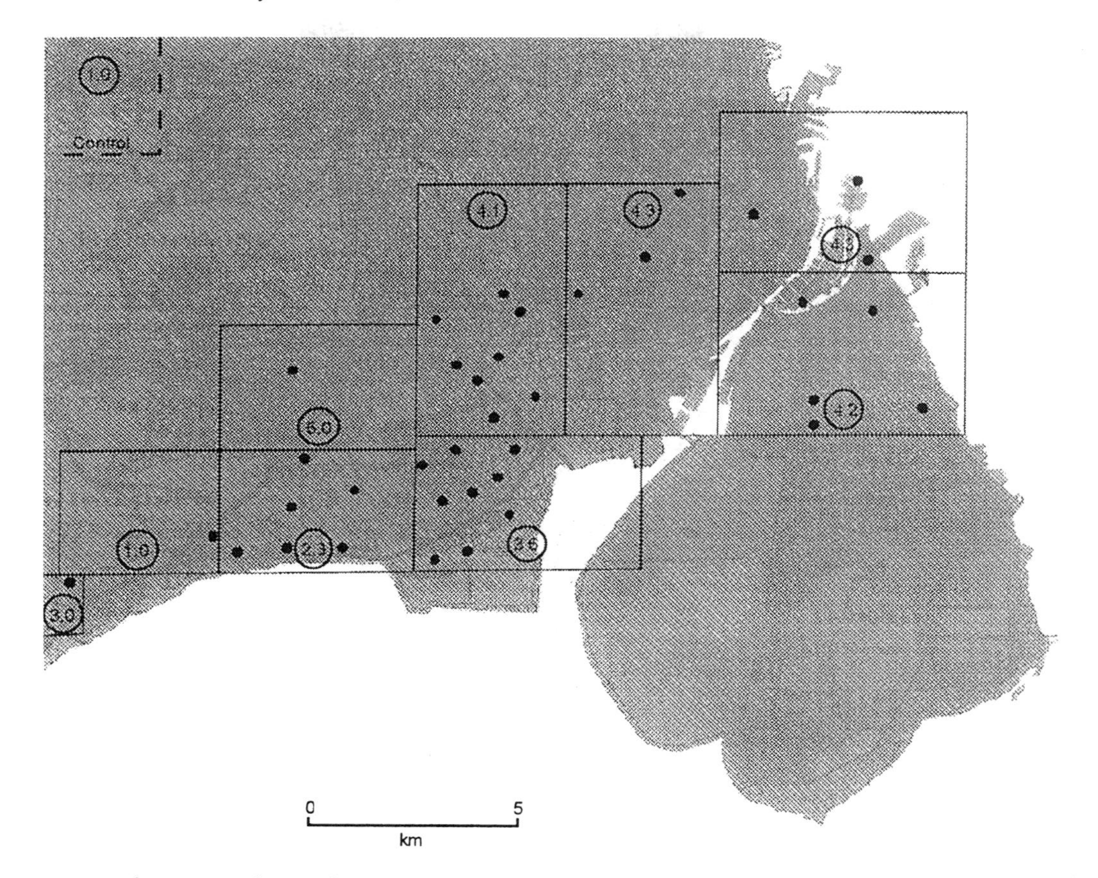

Fig. 7.3. The figure shows the Copenhagen area. The dots indicate transplantation localities. The average number of thallus damages are indicated in circles. Isopleths, $\mu g/m^3$, are shown. Reproduced with permission after Søchting and Johnsen, 1978.

4) information of input to the entire food chain
5) the effect of several components are measured. It is of importance where a synergistic or antagonistic effect will be anticipated.
6) the methods are simple, fast and cost.-moderate.

On the other hand, the disadvantages are as follows:

1) suitable organisms cannot be found (this problem can sometimes be solved by transplantation)
2) information about specific effects of a number pollutants is limited

The need for simple and precise information has provoked development of biological indices, for instancesimilar to the saprobien system in freshwater ecology; see Section 2.7.

"Index of Atmospheric Purity" (IAP) is applied to assess atmospheric pollution (De Sloover and Le Blanch, 1968). A high IAP value means, that there is a well developed ephiphytic lichen community in a unpolluted area.

Many biological indices are based on diversity, although a change in diversity not necessary is related to environmental changes. Biological indices are often successfully applied to assess effects on the ecosystem level; see Section 4.5.

7.3 PARTICULATE POLLUTION

Sources of particulate pollution.

When considering particulate pollution, the source should be categorised with regard to contaminant type. **Inert particulates** are distinctly different from **active solids** in the nature and type of their potentially harmful human health effects. *Inert particulates comprise solid* airborne material, which does not react readily with the environment and does not exhibit any morphological changes as a result of combustion or any other process. Active solid matter is defined as particulate material which can be further oxidised or *which reacts chemically* with the environment or the receptor. Any solid material in this category can, depending on its composition and size, be more harmful than inert matter of similar size.

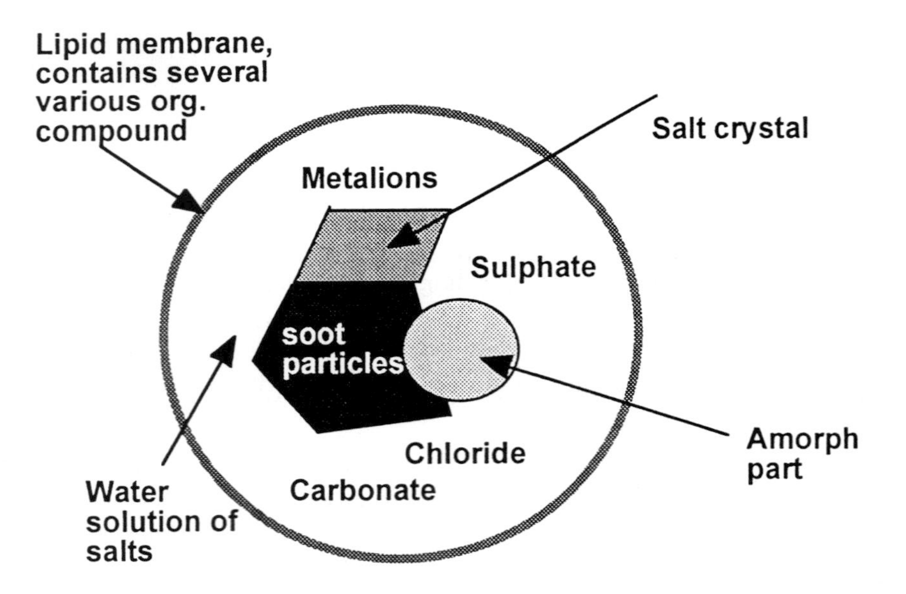

Fig. 7.4. A combination particle is shown. It consists of a solid nucleus with amorph parts and crystals, a solution of salts and a lipid membrane of organic compounds (pesticides and aromatic compounds with the hydrophile part of the molecules toward the nucleus).

A closely related group of emissions are from **aerosols** of liquid droplets, generally below 5 μm. They can be oil or other liquid pollutants (e.g. freon) or may be formed by condensation in the atmosphere. **Fumes** are condensed metals, metal oxides or metal halides, formed by industrial activities, predominantly as a result of pyrometallurgical processes; melting, casting or extruding operations. Products of incomplete combustion are often emitted in the form of particulate matter. The most harmful components in this group are often those of **particulate polycyclic organic matter** (PPOM) which are mainly derivatives of benz-a-pyrene. Natural sources of particulate pollution are *sandstorms, forest fires and volcanic activity.* The major sources in towns are *vehicles, combustion of fossil fuel for heating and production of electricity, and industrial activity.*

The total global emission of particulate matter is in the order of **10^7 t per year.**

Deposition of particles may occur by three processes: 1) sedimentation (Stokes law may be applied, particles > 20 μm), 2) impaction (determined by differences in concentrations by use of Fick's law, particles between 5 and 20 μm) and 3) diffusion (particles < 5 μm). Particles < 20μm are named suspended particulate matter. Particles > 20μm may be denoted dust which id deposited close to the source due to the high sedimentation rate. Dry deposition consists of gases or dry particles. Wet deposition is raindrops containing gases and particles. Particles may consists of a minor concentrations of dissolved salts in water drops, crystals or a combinations of the two. Figure 7.4 shows a typical composition of combination particles. they are frequently found in a polluted troposphere.

The particulate pollution problem.

Particulate pollution is important with regard to health. The toxicity and the size distribution are the most crucial factors. Many particles are highly toxic, such as *asbestos and those of metals such as beryllium, lead, chromium, mercury, nickel and manganese.* In addition, it must be remembered that particulate matter is able to absorb gases, so enhancing the effects of these components. In this context the **particle size distribution** is of particular importance, as particles greater than 10μm are trapped in the human upper respiratory passage and the specific surface (expressed as m^2 per g of particulate matter) increases with l/d, where d is the particle size. The adsorption capacity of particulate matter, expressed as g adsorbed per g of particulate matter, will generally be proportional to the surface area. Table 7.2 lists some typical particle size ranges. However, *size* as well as *shape* and *density* must be considered. Furthermore, particle size is determined by two parameters: the mass median diameter, which is the size that divides the particulate sample into two groups of equal mass, i.e. the 50 percent point on a cumulative frequency versus particle size plot (see the examples in Fig. 7.5); and the geometric standard deviation, which is the slope on the curve in Fig. 7.5. The slope can be found from:

$$ß = \frac{Y_{74.1}}{Y_{50}} = \frac{Y_{50}}{Y_{15.9}} \qquad (7.1)$$

where the subscript numbers refer to the particle diameters at that percentage on the distribution plot.

TABLE 7.2

Typical particle size ranges

	μm
Tobacco smoke	0.01 - 1
Oil smoke	0.05 - 1
Ash	1 - 500
Ammonium chloride smoke	0.1 - 4
Powdered activated carbon	3 - 700
Sulphuric acid aerosols	0.5 - 5

Example 7.1

Find the geometric standard deviation from Figure 7.5.

Solution

From the plot is found, in this case:

$$ß = \frac{7.2}{1.7} = \frac{1.7}{0.4} = 4.25 \qquad (7.2)$$

The particle size also determines the settling out rate of the particulate matters. Particles with a size of 1 μm have a terminal velocity of a few metres per day, while particles with a size of 1 mm will settle out at a rate of about 40 m per day. Particles that are only a fraction of 1 μm will settle out very slowly and can be in suspension in the atmosphere for a very long time.

Control methods applied to particulate pollution.

Particulate pollutants have the ability to adsorb gases included sulphur dioxide, nitrogen oxides, carbon monoxide and so on. The inhalation of these toxic gases is frequently associated with this adsorption, as the gases otherwise would be dissolved in the mouthwash and spittle before entering the lungs.

Particulate pollution may be controlled by modifying the distribution pattern. This method is described in detail below. It represents in principle an obsolete philosophy of pollution abatement: dilution, but it is still widely used to reduce the concentration of pollutants at ground level and thereby minimise the effect of air pollution

Particulate control technology can offer a wide range of methods aimed at the removal of particulate matter from gas. These methods are: settling chambers, cyclones, filters, electrostatic precipitators, wet scrubbers and modification of particulate characteristics.

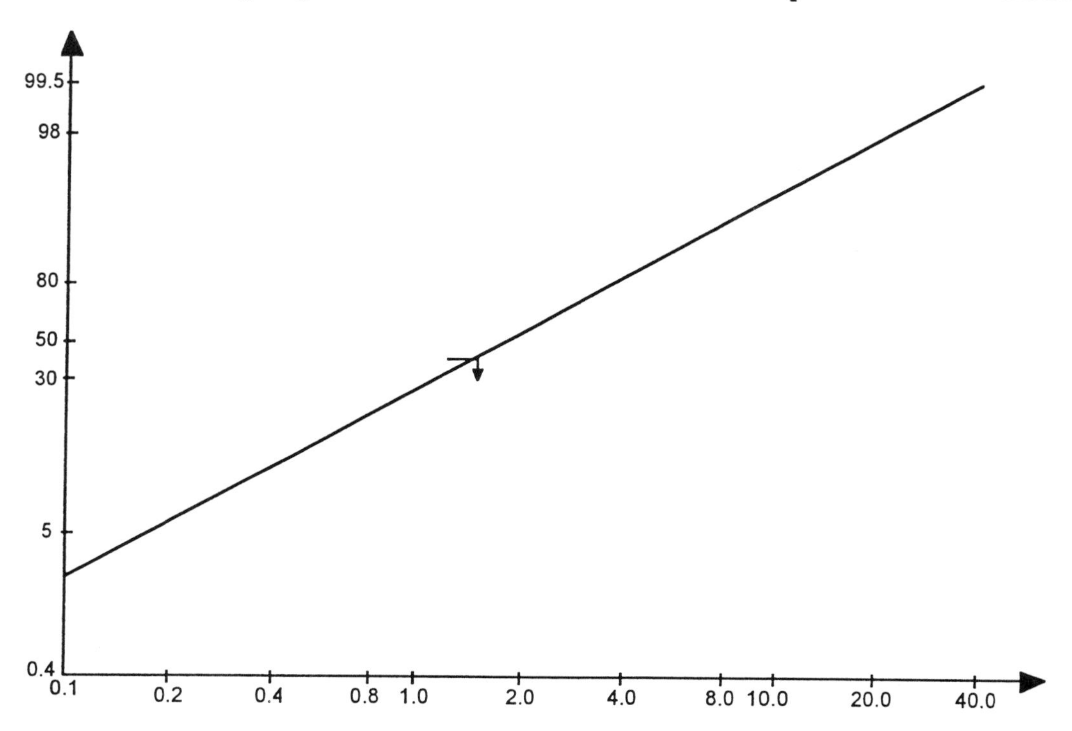

Fig. 7.5 Particle diameter (log scale) plotted against the percentage of particles less than or equal to the indicated size.

Modifying the distribution patterns.

Although emissions, gaseous or particulate, may be controlled by various sorption processes or mechanical collection, the effluent from the control device must still be dispersed into the atmosphere. Atmospheric dispersion depends primarily on horizontal and vertical transport. The horizontal transport depends on the turbulent structure of the wind field. As the wind velocity increases so does the degree of dispersion and there is a corresponding decrease in the ground level concentration of the contaminant at the receptor site.

The emissions are mixed into larger volume, of air and the diluted emission is carried out into essentially unoccupied terrain away from any receptors. Depending on the wind direction, the diluted effluent may be funnelled down a river valley or between mountain ranges. Horizontal transport is sometimes prevented by surrounding hills forming a natural pocket for locally generated pollutants. This particular topographical situation occurs in the

Los Angeles area, which suffers heavily from air pollution.

The vertical transport depends on the rate of change of ambient temperature with altitude. The dry adiabatic lapse rate is defined as a decrease in air temperature of 1°C per 100 m. This is the rate at which, under natural conditions, a rising parcel of unpolluted air will decrease in temperature with elevation into the troposphere up to approximately 10,000 m. Under so-called isothermal conditions the temperature does not change with elevation. Vertical transport can be hindered under stable atmospheric conditions, which occur when the actual environmental lapse rate is less than the dry adiabatic lapse rate. A negative lapse rate is an increase in air temperature with latitude. This effectively prevents vertical mixing and is known as inversion.

The dispersion from a point source - a chimney for instance - may be calculated from the Gaussian plume model (see for instance Reible, 1998).

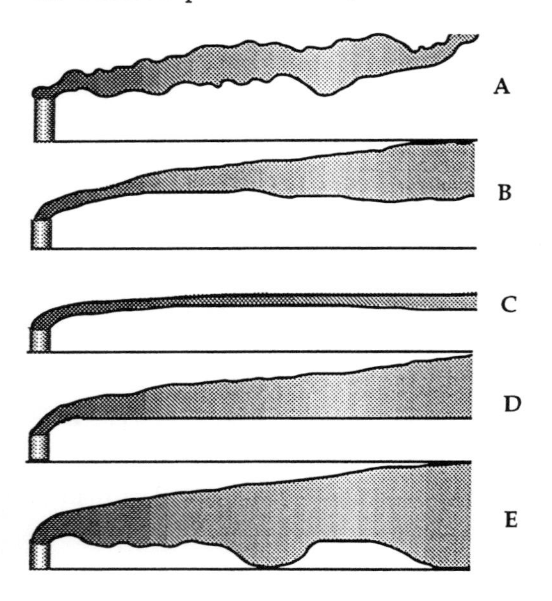

Fig. 7.6 Stack gas behaviour under various conditions. A) Strong lapse (looping), B) Weak lapse (coning), C) Inversion (fanning), D) Inversion below, lapse aloft (lofting), E) Lapse below, inversion aloft (fumigation).

These different atmospheric conditions (U.S.DHEW 1969) are illustrated in Fig. 7.6 where stack gas behaviour under the various conditions is shown. Further explanations are given in Table 7.3 The distribution of particulate material is more effective the higher the stack. **The maximum concentration, C_{max}, at ground level** can be shown to be approximately proportional to the emission and to follow approximately this expression:

$$C_{max} = k\,Q / H^2 \tag{7.3}$$

where Q is the emission (expressed as g particulate matter per unit of time), H is the effective stack height and k is a constant. The effective height is slightly higher than the physical height and can be calculated from information about the temperature, the stack exit velocity and the stack inside diameter.

These equations explain why a lower ground-level concentration is obtained when many small stacks are replaced by one very high stack. In addition to this effect, it is always easier to reduce and control one large emission than many small emissions, and it is more feasible to install and apply the necessary environmental technology in one big installation.

TABLE 7.3
Various atmospheric conditions

Strong lapse (looping)	Environmental lapse rate > adiabatic lapse rate
Weak lapse (coning)	Environmental lapse rate < adiabatic lapse rate
Inversion (fanning)	Increasing temperature with height
Inversion below, lapse aloft (lofting)	Increasing temperature below, app. adiabatic lapse rate aloft
Lapse below, inversion aloft (fumigation)	App. adiabatic lapse rate below, increasing temperature aloft

Example 7.2.
Calculate the concentration ratio of particulate material in cases A and B, given that the total emission is the same.
A: 100 stacks H = 25 m
B: 1 stack H = 200 m

Solution:

$$\text{Ratio} = \frac{200^2}{25^2} = 64$$

Characteristics of particulate pollution control equipment
Environmental technology offers several solutions to the problem of particulate matter removal. The methods their optimum particle size and efficiency are compared in Table 7.4. The cost of the various installations varies of course from country to country and is

dependent on several factors (material applied, standard size or not standard size, automatised and so on). Generally, electrostatic precipitators are the most expensive solution and are mainly applied for large quantities of air. Wet scrubbers also belong among the more expensive installations, while settling chamber and centrifuges are the most cost effective solutions.

TABLE 7.4
Characteristics of particulate pollution control equipment

Device	Optimum particle size (μm)	Optimum concentration (gm^{-3})	Temperature limitations (°C)	Air resistance (mm H_2O)	Efficiency (% by weight)
Settling chambers	> 50	> 100	-30 to 350	< 25	< 50
Centrifuges	> 10	> 30	-30 to 350	< 50-100	< 70
Multiple centrifuges	> 5	> 30	-30 to 350	< 50-100	< 90
Filters	> 0.3	> 3	-30 to 250	> 15-100	> 99
Electrostatic precipitators	> 0.3	> 3	-30 to 500	< 20	< 99
Wet scrubbers	> 2-10	> 3-30	0 to 350	> 5-25	< 95-99

<u>**Settling chambers.**</u>

Simple gravity settling chambers, such as the one shown in Fig. 7.5, depend on gravity or inertia for the collection of particles. Both forces increase in direct proportion to the square of the particle diameter, and the performance limit of these devices is strictly governed by the particle settling velocity. The pressure drop in mechanical collectors is low to moderate, 1-25 cm water in most cases. Most of these systems operate dry but if water is added it performs the secondary function of keeping the surface of the collector clean and washed free of particles.

The settling or terminal velocity can be described by the following expression, which has general applicability:

$$V_t = (\partial_p - \partial)\, g \frac{d_p^2}{17\mu} \tag{7.4}$$

where

V_t = terminal velocity
∂_p = particle density
∂ = gas density
d_p = particle diameter
μ = gas viscosity

This is the equation of Stokes' law, and is applicable to $N_{Re} < 1.9$ where

$$N_{Re} = d_p * V_t * \frac{\partial}{\mu} \tag{7.5}$$

The intermediate equation for settling can be expressed as:

$$V_t = \frac{0.153 * g^{0.71} * d_p^{1.14} (\partial_p - \partial)^{0.71}}{\partial^{0.29} * \mu^{0.43}} \tag{7.6}$$

This equation is valid *for Reynolds numbers between 1.9 and 500,* while the following equation can be applied *for $N_{Re} \geq 500$ and up to 200,000:*

$$V_t = 1.74 \left(d_p * g \frac{(\partial_p - \partial)}{\partial}\right)^{1/2} \tag{7.7}$$

The settling velocity in these chambers is often in the range 0.3-3 m per second. This implies that for large volumes of emission the settling velocity chamber must be very large in order to provide an adequate residence time for the particles to settle. Therefore, the gravity settling chambers are not generally used to remove particles smaller than 100 μm (= 0.1 mm). For particles measuring 2-5 μm the collection efficiency will most probably be as low as 1-2 percent. A variation of the simple gravity chamber is the baffled separation chamber. The baffles produce a shorter settling distance, which means a shorter retention time.

Equations (7.4) - (7.7) can be used to design a settling chamber, and this will be demonstrated by use of equation (7.4).

If it is assumed that equation (7.4) applies, an equation is available for calculating the minimum diameter of a particle collected at 100% theoretically efficiency in a chamber of length L. In practice, some reentrainment will occur and prevent 100% efficiency. We have:

$$\frac{v_t}{H} = \frac{v_h}{L} \qquad (7.8)$$

where H is height of the settling chamber (m), L length of the settling chamber (m) and v_h is the horizontal flow rate (m s^{-1}). Solving for v_t and substituting into equation (7.4) yields:

$$\frac{v_h * H}{L} = \frac{g(\partial_p - \partial)\ d_p^2}{17\mu} \qquad (7.9)$$

$\partial_p >> \partial$ and this equation gives the largest size particle that can be removed with 100% efficiency in a settling chamber:

$$d_p = \left(\frac{17\mu v_h * H}{Lg * \partial_p}\right)^{1/2} \qquad (7.10)$$

A correction factor of 1.5 - 3 is often used in equation (7.10).

Example 7.3

Find the minimum size of particle that can be removed with 100% efficiency from a settling chamber with a length of 10 m and a height of 1.5 m. The horizontal velocity is 1.2 m s^{-1} and the temperature is 75°C. The specific gravity is 1.5 g/ml of the particles. A correction factor of 2 is suggested. At 75°C, μ is $2.1 * 10^{-5}$ kg m^{-1} s^{-1}.

Solution:

$$d_p = \left(2\frac{17\mu v_h * H}{1 * g * \partial_p}\right)^{1/2} = 2\left(\frac{17 * 2.1 * 10^{-5} * 1.2 * 1.5}{9.71 * 10 * 1500}\right)^{1/2}$$

$= \underline{9.62 * 10^{-5}\ m.}$

as $\partial_p = 1.5$ g ml^{-1} = 1500 kg m^{-3} $\qquad$ $\underline{d_p = 96.2\ \mu m}$

Cyclones

Cyclones separate particulate matter from a gas stream by transforming the inlet gas stream into a confined vortex. The mechanism involved in cyclones is the continuous use of inertia to produce a tangential motion of the particles towards the collector walls. The particles enter the boundary layer close to the cyclone wall and lose kinetic energy by mechanical friction, see Fig. 7.7. The forces involved are: the centrifugal force imparted by the rotation of the gas stream and a drag force, which is dependent on the particle density, diameter, shape, etc. A hopper is built at the bottom. If the cyclone is too short, the maximum force will not be exerted on some of the particles, depending on their size and the corresponding drag forces (Leith and Licht, 1975). If, however, the cyclone is too long, the gas stream might reverse its direction and spiral up the centre.

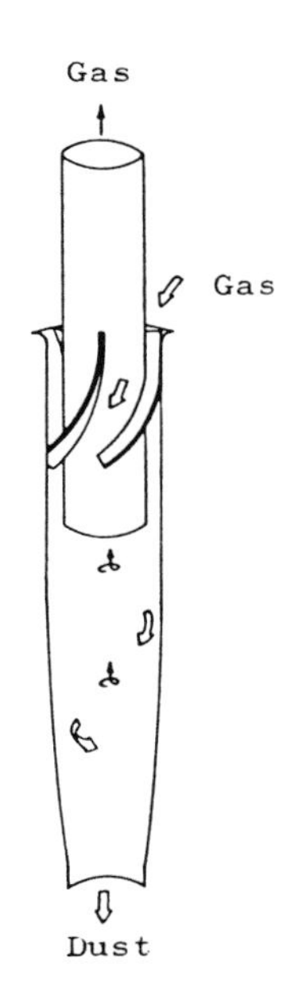

Fig 7.7 Principle of a cyclone.

It is therefore important to design the cyclone properly. The hopper must be deep enough to keep the dust level low. The efficiency of a cyclone is described by a graph similar to Fig. 7.8. which shows the efficiency versus the relative particle diameter, i.e. the actual particle diameter divided by D_{50}, **which is defined as the diameter corresponding to 50 percent efficiency. D_{50} can be found from the following equation:**

$$D_{50} = K * \left(\frac{\mu D_c}{V_c * \partial_p} \right)^{1/2} \quad (7.11)$$

where

D_c = diameter of cyclone

V_c = inlet velocity

∂_p = density of particles
μ = gas viscosity
K = a constant dependent on cyclone performance

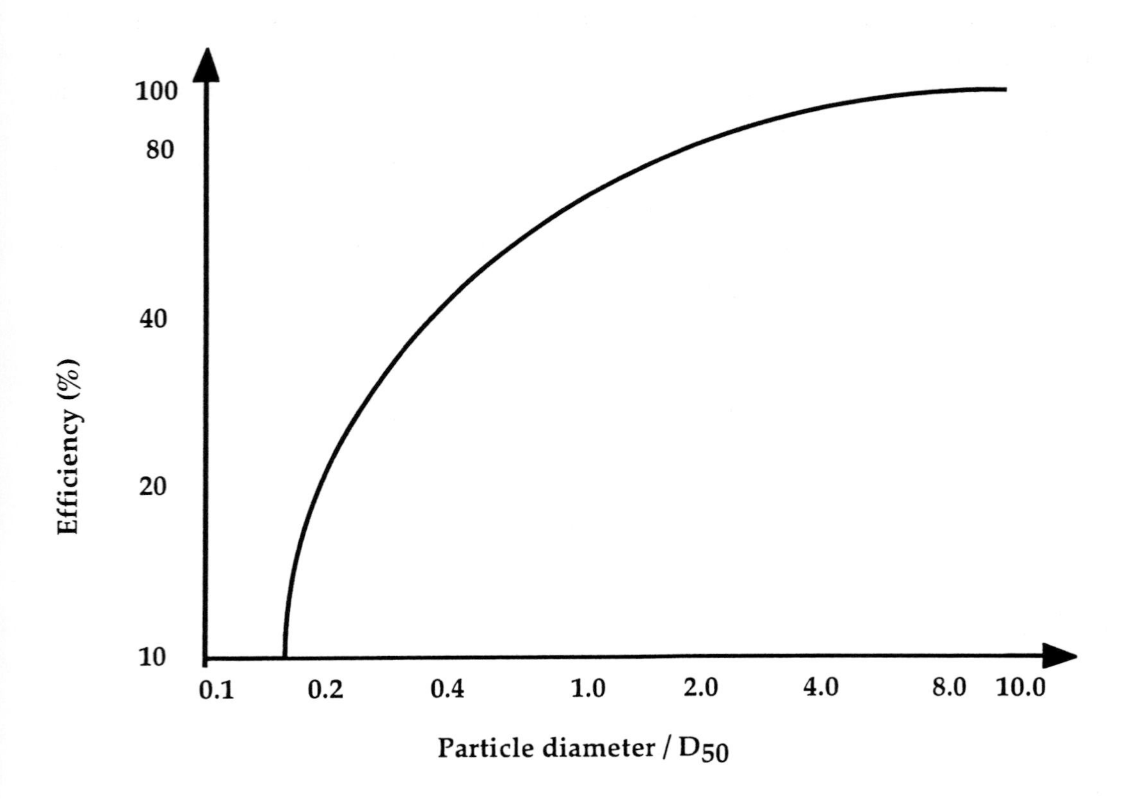

Fig. 7.8 Efficiency plotted against relative particle diameter. Notice that it is a log-log plot.

If the distribution of the particle diameter is known, it is possible from such a graph as in Fig. 7.8 to calculate the total efficiency:

$$eff = \Sigma\ m_i * eff_i \tag{7.12}$$

where

m_i = the weight fraction in the i.th particle size range
eff_i = the corresponding efficiency

The pressure drop for cyclones can be found from:

$$\Delta p = N * \frac{V_c^2}{2g} \tag{7.13}$$

From equations (7.11) and (7.13) it can be concluded that higher efficiency is obtained without increased pressure drop if D_c can be decreased while velocity V_c is maintained. This implies that *a battery of parallel coupled small cyclones will work more effectively than one big cyclone.* Such cyclone batteries are available as blocks, and are known as multiple cyclones. Compared with settling chambers, cyclones offer higher removal efficiency for particles below 50μm and above 2-10 μm, but involve a greater pressure drop.

Example 7.4

Determine D_{50} for a flow stream with a flow rate of 7 m s^{-1}, when a cyclone with a diameter of 2 m is used and a battery of cyclones with diameters of 0.24 m are used. The air temperature is 75°C and the particle density is 1.5 g ml^{-1}. K can be set to 0.2. Find also the efficiencies for particles with a diameter of 5μm.

Solution:

1) $$D_{50} = K \frac{\mu * D_c}{v_c * \partial_p} = 0.2 \frac{2.1 * 10^{-5} * 2}{7 * 1500} = 12.7\ \mu m$$

2) $$D_{50} = K \frac{\mu * D_c}{v_c * \partial_p} = 0.2 \frac{2.1 * 10^{-5} * 0.24}{7 * 1500} = 4.4\ \mu m$$

5 μm corresponds to a relative diameter of

1) $\frac{5}{12.7} = 0.39$, the efficiency will be about 25% (see Fig. 7.8)

2) $\frac{5}{4.4} = 1.14$, the efficiency will be about 55% (see Fig. 7.8)

Filters.

Particulate materials are collected by filters by the following three mechanisms (Wong et al., 1956):

Impaction, where the particles have so much inertia that they cannot follow the streamline round the fibre and thus impact on its surface .

Direct interception where the particles have less inertia and can barely follow the streamlines around the obstruction.

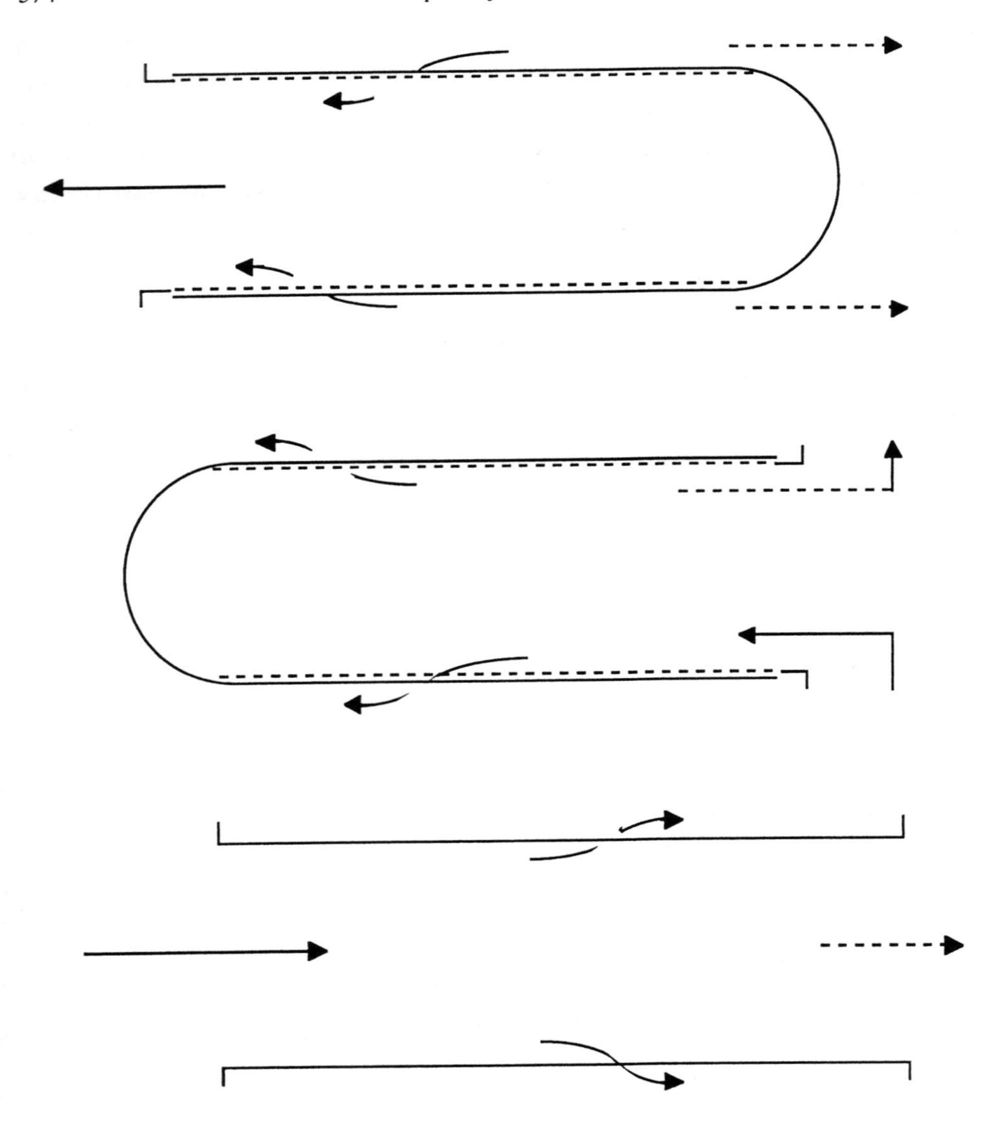

Fig. 7.9 Flow pattern of filters.

Diffusion, where the particles are so small (below 1μm) that their individual motion is affected by collisions on a molecular or atomic level. This implies that the collection of these fine particles is a result of random motion.

Different flow patterns can be used, as demonstrated in Fig. 7.9. The types of fibres used in fabric filters range from natural fibres, such as cotton and wool, to synthetics (mainly polyesters and nylon), glass and stainless steel.

Some properties of common fibres are summarised in Table 7.5. As can be seen, cotton and wool have a low temperature limit and poor alkali and acid resistance, but they are relatively inexpensive. The selection of filter medium must be based on the answer to several questions (Pring, 1972 and Rullman, 1976):

What is the expected operating temperature?

Is there a humidity problem which necessitates the use of a hydrophobic material, such as, e.g. nylon?

How much tensile strength and fabric permeability are required?

How much abrasion resistance is required?

Permeability is defined as the volume of air that can pass through 1 m^2 of the filter medium with a pressure drop of no more than 1 cm of water.

The filter capacity is usually expressed as m^3 air per m^2 filter per minute. A typical capacity ranges between 1 and 5 m^3 per m^2 per minute. The pressure drop is generally larger than for cyclones and will in most cases, be 10-30 cm of water, depending on the nature of the dust, the cleaning frequency and the type of cloth.

TABLE 7.5
Properties of fibres

Fabric	Acid resistance	Alkali resistance	Fluoride strength	Tensile resistance	Abrasion
Cotton	poor	good	poor	medium	very good
Wool	good	poor	poor	poor	fair
Nylon	poor	good	poor	good	excellent
Acrylic	good	fair	poor	medium	good
Polypropylene	good	fair	poor	very good	good
Orlon	good	good	fair	medium	good
Dacron	good	good	fair	good	very good
Teflon	excellent	excellent	good	good	fair

There are several specific methods of filter cleaning. The simplest is *backwash,* where dust is removed from the bags merely by allowing them to collapse. This is done by reverting the air flow through the entire compartment. The method is remarkable for its low consumption of energy. *Shaking* is another low-energy filter-cleaning process, but it cannot be used for sticky dust. The top of the bag is held still and the entire tube sheath at the bottom is shaken. The application of *blow rings* involves reversing the air flow without bag collapse. A ring surrounds the bag; it is hollow and supplied with compressed air to direct a constant steam of air into the bag from the outside.

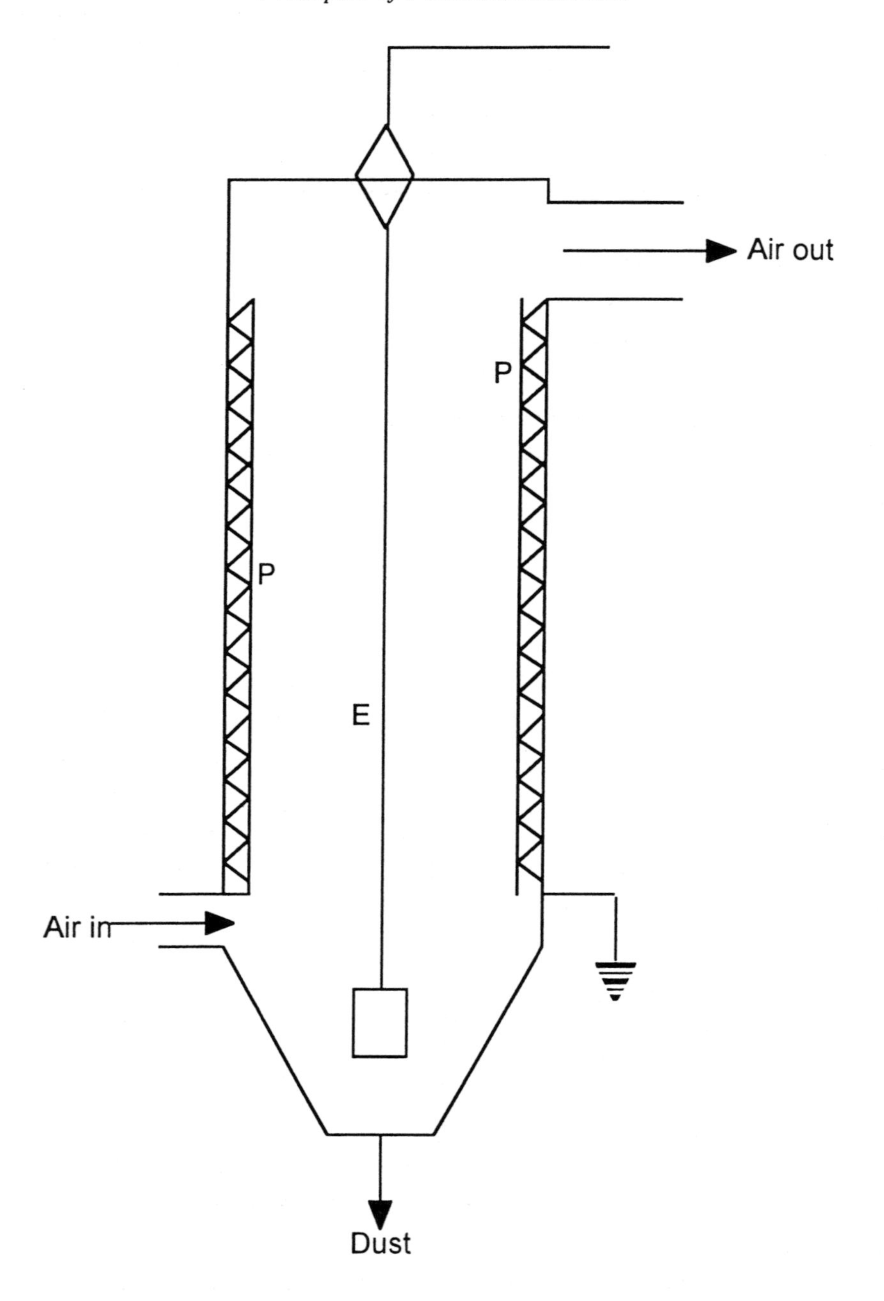

Fig. 7.10. The dust is precipitated on the electrode P. E has a high, usually negative voltage and emits a great number of electrons which give the dust particles a negative charge. The dust particles will therefore be attracted to P.

The pulse and improved jet cleaning mechanism involves the use of a high velocity, high pressure air jet to create a low pressure inside the bag and induce an outward air flow and so clean the bag by sudden expansion and reversal of flow.

In some cases as a result of electrostatic forces, moisture on the surface of the bag and a slight degree of hygroscopicity of the dust itself, the material forms cakes that adhere tightly to the bag. In this case the material must be kept drier and a higher temperature on the incoming dirty air stream is required.

Filters are highly efficient even for smaller particles (0.1 - 2 m), which explains their wide use as particle collection devices.

Electrostatic precipitators.

The electrostatic precipitator consists of four major components:

1. *A gas-tight shell with hoppers* to receive the collected dust, inlet and outlet, and an inlet gas distributor.
2. *Discharge electrodes.*
3. *Collecting electrodes.*
4. *Insulators.*

The principles of electrostatic precipitators are outlined in Fig. 7.10. The dirty air stream enters a filter, where a high, 20-70 kV, usually negative voltage exists between discharge electrodes. The particles accept a negative charge and migrate towards the collecting electrode. The efficiency is usually expressed by use of Deutsch's equation (see the discussion including correction of this equation in Gooch and Francis, 1975). This equation entails that the relationship between migration velocity and particle diameter is as shown in Figure 7.11.

The operation of an electrostatic precipitator can be divided into three steps:

1. The particles *accept a negative charge.*
2. The charged particles *migrate towards the collecting electrode* due to the electrostatic field.
3. The collected dust *is removed from the collecting electrode* by shaking or vibration, and is collected in the hopper.

r, the specific electrical resistance, measured in ohm m, determines the ability of a particle to accept a charge. The practical specific resistance can cover a wide range of about four orders of magnitude, in which varying degrees of collection efficiencies exist for different types of particles. The specific resistance depends on the chemical nature of the dust, the temperature and the humidity.

Electrostatic precipitators have found wide application in industry. As the cost is relatively high, the airflow should be at least 20,000 $m^3 h^{-1}$; volumes as large as 1,500,000 $m^3 h^{-1}$ have been treated in one electrostatic precipitator. *Very high efficiencies* are generally achieved in electrostatic precipitators and emissions as low as 25 mg m^{-3} are quite common. The pressure drop is usually low compared with other devices - 25 mm water at the most. The energy consumption is generally 0.15-0.45 Wh $m^{-3} h^{-1}$.

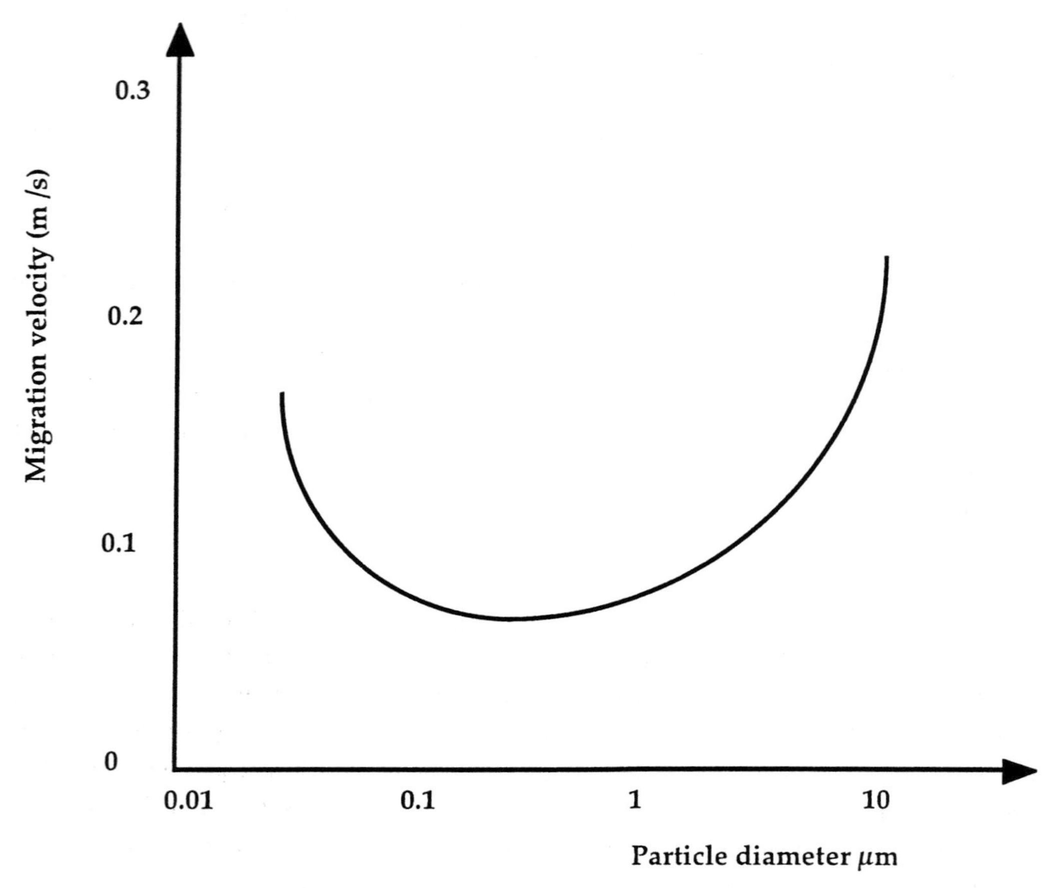

Fig. 7.11. A typical graph of migration velocity versus particle diameter.

<u>Wet scrubbers.</u>

A scrubbing liquid, usually water, is used to assist separation of particles, or a liquid aerosol from the gas phase. The operational range for particle removal includes material less than 0.2 μm in diameter up to the largest particles that can be suspended in air.

Four major steps are involved in collecting particles by wet scrubbing. First, *the particles are moved to the vicinity of the water droplets,* which are 10-1,000 times larger.

Then the particles *must collide with the droplets.* In this step the relative velocity of the gas and the liquid phases is very important: If the particles have an over high velocity in relation to liquid they have so much inertia that they keep moving, even when they meet the front edge of the shock wave, and either impinge on or graze the droplets. A scrubber is no better than its ability to bring the particles directly into contact with the droplets of the scrubber fluid. The next step is *adhesion, which is directly promoted by surface tension.* Particles cannot be retained by the droplets unless they can be wetted and thus incorporated into the droplets. The last step is *the removal of the droplets containing the dust particles from the bulk gas phase.*

Scrubbers are generally very flexible. They are able to operate under peak loads or reduced volumes and within a wide temperature range (Onnen, 1972).

They are smaller and less expensive than dry particulate removal devices, but the operating costs are higher. Another disadvantage is that the pollutants are not collected but transferred into water, which means that the related water pollution problem must also be solved (Hanf, 1970).

Several types of wet scrubbers are available (Wicke, 1971), and their principles are outlined below:

1. **Chamber scrubbers** are spray towers and spray chambers which can be either round or rectangular. Water is injected under pressure though nozzles into the gas phase.
2. **Baffle scrubbers** are similar to a spray chamber but have internal baffles that provide additional impingement surfaces. The dirty gas is forced to make many turns to prevent the particles from following the air stream.
3. **Cyclonic scrubbers** are a cross between a spray chamber and a cyclone. The dirty gas enters tangentially to wet the particles by forcing its way through a swirling water film onto the walls. There the particles are captured by impaction and are washed down the walls to the sump. The saturated gas rises through directional vanes, which are used solely to impart rotational motion to the gas phase. As a result of this motion the gas goes out though a demister for the removal of any included droplets.
4. **Submerged orifice scrubbers** are also called gas-induced scrubbers. The dirty gas is accelerated over an aerodynamic foil to a high velocity and directed into a pool of liquid. The high velocity impact causes the large particles to be removed into the pool and creates a tremendous number of spray droplets with a high amount of turbulence. These effects provide intensive mixing of gas and liquid and thereby a very high interfacial area. As a result reactive gas absorption can be combined with particle removal.
5. The **ejector scrubber** is a water jet pump (see Figure 7.12). The water is pumped through a uniform nozzle and the dirty gas is accelerated by the action of the jet gas. The result is aspiration of the gas into the water by the Bernoulli principle and, accordingly, a lowered pressure. The ejector scrubber can be used to collect soluble gases as well as particulates.
6. The **venturi scrubber** involves the acceleration of the dirty gas to 75-300 m min^{-1} through a mechanical constriction. This high velocity causes any water injected just upstream of, or in, the venturi throat to be sheared off the walls or nozzles and atomised. The droplets are usually 5-20 μm in size and form into clouds from 150-300 μm in diameter, depending on the gas velocity. The scrubber construction is similar to that of the ejector scrubber, but the jet pump is replaced by a venturi constriction.

7. **Mechanical scrubbers** have internal rotating parts, which break up the scrubbing liquid into small droplets and simultaneously create turbulence.

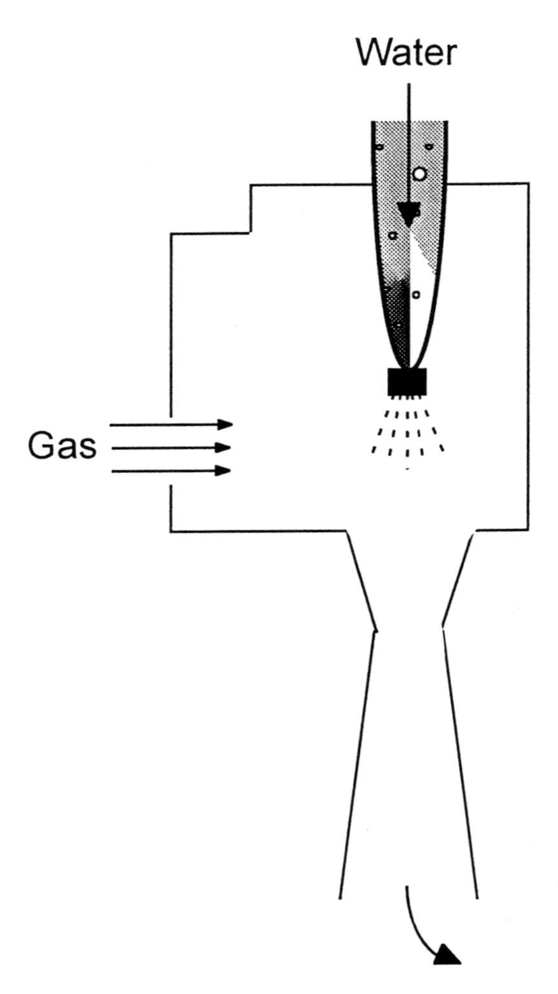

Fig. 7.12. Principle of ejector scrubber

8. **Charged-droplet scrubbers** have a high voltage ionisation section where the corona discharge produces air ions (as in electrostatic precipitators). Water droplets are introduced into the chamber by use of spray nozzles or similar devices. The additional collection mechanism provided by the induction of water droplets increases the collection efficiency.

9. **Packed-bed scrubbers** have a bottom support grid, and an top retaining grid (see Fig. 7.13). The fluid (often water or a solution of alkali or acid) is distributed as shown in the figure over the top of the packed section, while the gas enters below the packing.

The flow is normally counter current. Packed-bed scrubbers offer the possibility of combining gas absorption with removal of particulate material. The pressure drop is often in the order of 3 cm water per m of packing. If the packing consists of expanded fibre, the bed scrubber is known as a fibre-bed scrubber.

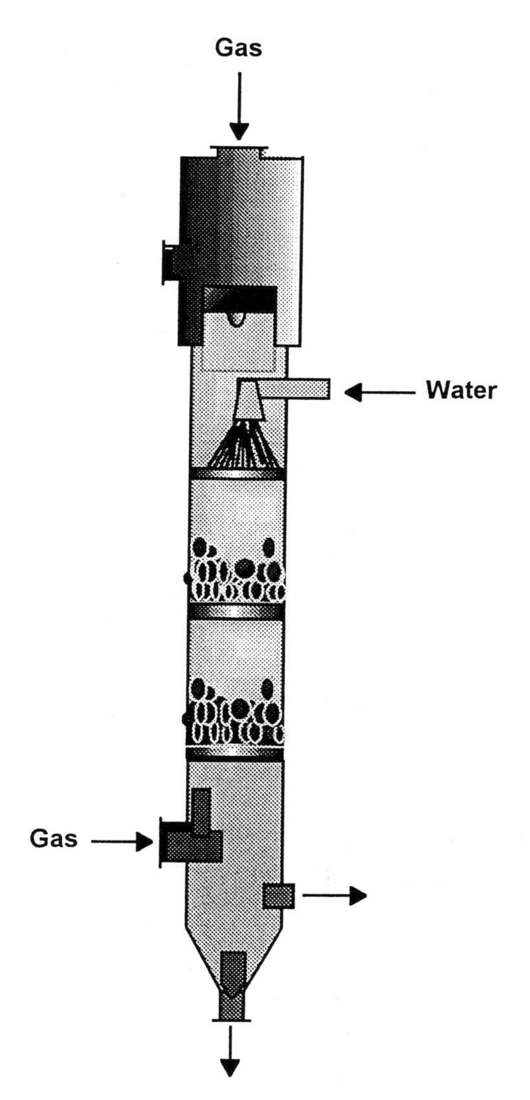

Fig. 7.13. Packed-bed scrubber.

The packed-bed scrubber has a tendency to clog under high particulate loading, which is its major disadvantage.

Common packings used are saddles, rings, etc., like those used in absorption towers. Some important parameters for various scrubbers are plotted in Fig. 7.10, which demonstrates the relationship between pressure drop, energy consumption and D_{50} (the diameter of the particles removed at 50 percent efficiency).

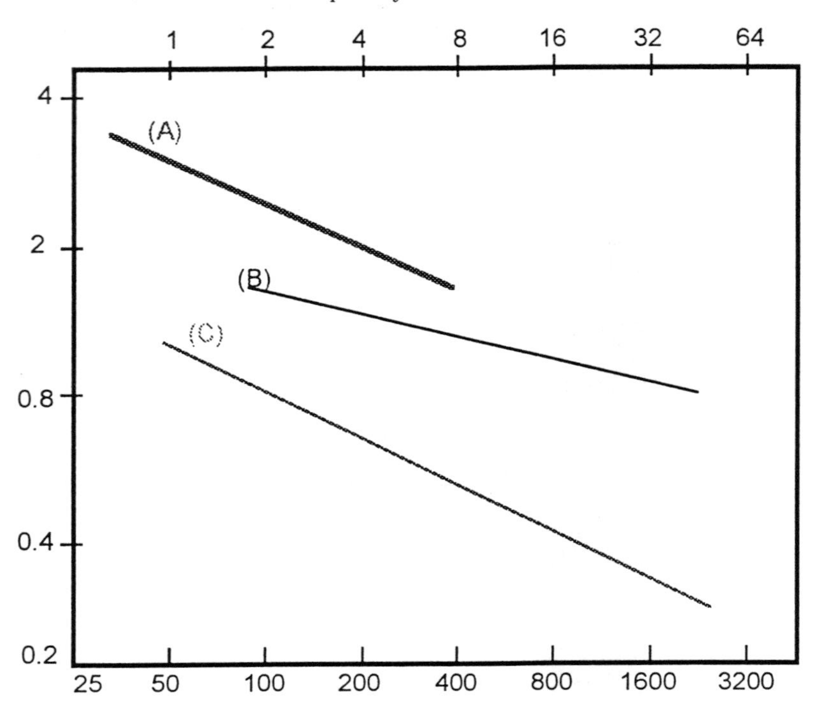

Fig. 7.14 Relationship between D_{50} (μm) pressure drop (mm H_2O, lower axis) and energy consumption (kW/m^3/s; upper axis). A: Packed-bed scrubber. B: Baffled scrubber. C: Venturi scrubber.

7.4. THE AIR POLLUTION PROBLEMS OF CARBON DIOXIDE, CARBON HYDRIDES AND CARBON MONOXIDE.

Sources of pollutants.

All types of fossil fuel will produce carbon dioxide on combustion, which is used in the photosynthetic production of carbohydrates. As such, carbon dioxide is harmless and has no toxic effect, whatever the concentration levels. However, since an increased carbon dioxide concentration in the atmosphere will increase absorption of infrared radiation, the heat balance of the earth will be changed (see Chapter 3 for a detailed discussion).

Carbon hydrides are the major components of oil and gas, and incomplete combustion will always involve their emission. Partly oxidised carbon hydrides, such as aldehydes and organic acids, might also be present.

The major source of carbon hydrides pollution is *motor vehicles.*

In reaction with nitrogen oxides and ozone they form so-called photochemical smog, which consists of several rather oxidative compounds, such as peroxyacyl nitrates and aldehydes.

In areas where solar radiation is strong and the atmospheric circulation weak, the possibility of smog formation increases, as the processes are *initiated by ultraviolet radiation.*

Incomplete combustion produces carbon monoxide. By regulation of the ratio of oxygen to fuel more complete combustion can be obtained, but the emission of carbon monoxide cannot be totally avoided.

Motor vehicles are also the major source of carbon monoxide pollution. On average, 1 litre of gasoline (petrol) will produce 200 litres of carbon monoxide, while it is possible to minimise the production of this pollutant by using diesel instead of gasoline.

The annual production of carbon monoxide is more then 200 million tons, of which 50 percent is produced by the U.S.A. alone.

In most industrial countries more than 75 percent of this pollutant originates from motor vehicles.

The pollution problem of carbon dioxide, carbon hydrides and carbon monoxide.

As mentioned carbon dioxide is not toxic; its problem as a pollutant is related solely to its influence on the global energy balance. As this problem is rather complex it will not be dealt with here. For a comprehensive discussion, see Chapters 2 and 3.

Carbon hydrides, partly oxidised carbon hydrides and the compounds of photochemical smog are all *more or less toxic* to man, animals and plants. Photochemical smog *reduces visibility, irritates the eyes and causes damage to plants* with immense economical consequences, for example for fruit and tobacco plantations. It is also able *to decompose rubber and textiles.*

Carbon monoxide is strongly toxic as it reacts with haemoglobin and thereby reduces the blood's capacity to take up and transport oxygen. Ten percent of the haemoglobin occupied by carbon monoxide will produce such symptoms as headache and vomiting. It should be mentioned here that smoking also causes a higher carboxyhaemoglobin concentration. An examination of policemen in Stockholm has shown that non-smokers had 1.2 percent carboxyhaemoglobin, while smokers had 3.5 percent.

Control methods applied to carbon dioxide, carbon hydrides and carbon monoxide pollution.

Carbon dioxide pollution is inevitable related to the use of fossil fuels. Therefore, it can only be solved by the use of other sources of energy.

Legislation is playing a major role in controlling the emission of carbon hydrides and carbon monoxide. As motor vehicles are the major source of these pollutants, control methods should obviously focus on the possibilities of reducing vehicle emission. The methods available today are:

1. **Motor technical methods.**
2. **Afterburners.**

3. **Alternative energy sources.**

The first method is based upon a motor adjustment according to the relationship between the composition of the exhaust gas and the air/fuel ratio. A higher air/fuel ratio results in a decrease in the carbon hydrides and carbon monoxide concentrations, but to achieve this, a better distribution of the fuel in the cylinder is required, which is only possible through construction of another gasification system. This method may be considered cleaner technology. It will be touched on briefly in Chapter 9.

At present two types of afterburners are in use - **thermal and catalytic afterburners.** In the former type the combustible material is raised above its auto ignition temperature and held there long enough for complete oxidation of carbon hydrides and carbon monoxide to occur. This method is used on an industrial scale (Waid, 1972 and 1974) when low-cost purchased or diverted fuel is available; in vehicles a manifold air injection system is used.

Catalytic oxidation occurs when the contaminant-laden gas stream is passed through a catalyst bed, which initiates and promotes oxidation of the combustible matter at a lower temperature than would be possible in thermal oxidation. The method is used on an industrial scale for the destruction of trace solvents in the chemical coating industry. Vegetable and animal oils can be oxidised at 250-370°C by catalytic oxidation. The exhaust fumes from chemical processes, such as ethylene oxide, methyl methacrylate, propylene, formaldehyde and carbon monoxide, can easily be catalytically incinerated at even lower temperatures.The application of catalytic afterburners in motor vehicles presents some difficulties due to poisoning of the catalyst with lead. With the decreasing lead concentration in gasoline it is becoming easier to solve that problem, and the so-called double catalyst system is now finding wide application. This system is able to reduce nitrogen oxides and oxidise carbon monoxide and carbon hydrides simultaneously.

Lead in gasoline has been replaced by various organic compounds to increase the octane number. Benzene has been applied but it is toxic and causes air pollution problems because of its high vapour pressure. MTBE (methyl tertiary buthyl ether) is another possible compound for increasing the octane number. It is, however, very soluble and has been found as ground water contaminant close to gasoline stations.

Application of alternative energy sources is still at a preliminary stage. The so-called *Sterling motor* is one alternative, as it gives more complete combustion of the fuel. Most interest has, however, been devoted to *electric vehicles.*

7.5 THE AIR POLLUTION PROBLEM OF SULPHUR DIOXIDE

The sources of sulphur dioxide pollution.

Fossil fuel contains approximately 2 to 5 per cent sulphur, which is oxidised by combustion to sulphur dioxide. Although fossil fuel is the major source, several industrial processes produce emissions containing sulphur dioxide, for example mining, the treatment of sulphur containing ores and the production of paper from pulp.

The total global emission of sulphur dioxide has been decreasing during the last 20 years due to the installation of pollution abatement equipment, particularly in North America, the EU and Japan. The concentration of sulphur dioxide in the air is relatively easy to measure, and sulphur dioxide has been used as an indicator component. High values recorded by inversion are typical.

The sulphur dioxide pollution problem.

Sulphur dioxide is oxidised in the atmosphere to sulphur trioxide, which forms sulphuric acid in water. Since sulphuric acid is a strong acid it is easy to understand that sulphur dioxide pollution indirectly causes corrosion of iron and other metals and is able to acidify aquatic ecosystems (these problems have been covered in detail in section 4.8).

The health aspects of sulphur dioxide pollution are closely related to those of particulate pollution. The gas is strongly adsorbed onto particulate matter, which transports the pollutant to the bronchi and lungs. (The relationship between concentration, effect and exposure time has already been discussed in section 4.4).

Control methods applied to sulphur dioxide.

Clean Air Acts have been introduced in all industrialised countries during the seventies and eighties. Table 7.6 illustrates some typical sulphur dioxide emission standards, although these may vary slightly from country to country.

The approaches used to meet the requirements of the acts, as embodied in the standards, can be summarised as follows:

1. Fuel switching from *high to low sulphur fuels.*
2. *Modification of the distribution pattern* - use of tall stacks.
3. *Abandonment of very old power plants,* which have higher emission.
4. *Flue gas cleaning.*

TABLE 7.6
SO_2-emission standards

Duration	Concentration (ppm)	Comments
Month	0.05	
24 h	0.10	might be exceeded once a month
30 min.	0.25	might be exceeded 15 times per month (1% of the time)

Desulphurisation of liquid and gaseous fuel is a well known chemical engineering operation. In gaseous and liquid fuels sulphur either occurs as hydrogen sulphide or can react with hydrogen to form hydrogen sulphide. The hydrogen sulphide is usually removed by absorption in a solution of alkanolamine and then converted to elemental sulphur. The process in general use for this conversion is the so-called Claus process. The hydrogen sulphide gas is fired in a combustion chamber in such a manner that one-third of the volume of hydrogen sulphide is converted to sulphur dioxide. The products of combustion are cooled and then passed through a catalyst-packed converter, in which the following reaction occurs:

$$2H_2S + SO_2 = 3S + 2H_2O \tag{7.14}$$

The elemental sulphur has commercial value and is mainly used for the production of sulphuric acid.

Sulphur occurs in coal both as pyritic sulphur and as organic sulphur. Pyritic sulphur is found in small discrete particles within the coal and can be removed by mechanical means, e.g. by gravity separation methods. However, 20 to 70 per cent of the sulphur content of coal is present as organic sulphur, which can hardly be removed today on an economical basis. Since sulphur recovery from gaseous and liquid fuels is much easier than from solid fuel, which also has other disadvantages, much research has been and is being devoted to *the gasification or liquefaction of coal.* It is expected that this research will lead to an alternative technology that will solve most of the problems related to the application of coal, including sulphur dioxide emission.

Approach (2) listed above has been mentioned earlier in this chapter, while approach (3) needs no further discussion. The next subsection is devoted to flue gas cleaning.

Flue gas cleaning of sulphur dioxide.

When sulphur is not or cannot be economically removed from fuel oil or coal prior to combustion, removal of sulphur oxides from combustion gases will become necessary for compliance with the stricter air pollution control laws.

The chemistry of sulphur dioxide recovery presents a variety of choices and five methods should be considered:

1. Adsorption of sulphur dioxide on active metal oxides with regeneration to produce sulphur.
2. Catalytic oxidation of sulphur dioxide to produce sulphuric acid.
3. Adsorption of sulphur dioxide on charcoal with regeneration to produce concentrated sulphur dioxide.
4. Reaction of dolomite or limestone with sulphur dioxide by direct injection into the combustion chamber. A lime slurry is injected into the flue gas beyond the boilers.
5. Fluidised bed combustion of granular coal in a bed of finely divided limestone or

dolomite maintained in a fluid like condition by air injection. Calcium sulphite is form as a result of these processes .

Particularly the two latter methods have found wide application particularly to large industrial installations. It is possible to recover the sulphur dioxide or elemental sulphur from these processes, making it possible to recycle the spent sorbing material.

7.6 THE AIR POLLUTION PROBLEM OF NITROGENOUS GASES

The sources of nitrogenous gases.

Seven different compounds of oxygen and nitrogen are known: N_2O, NO, NO_2, NO_3, N_2O_3, N_2O_4, N_2O_5 - often summarised as NO_x. From the point of view of air pollution mainly NO (nitrogen oxide) and NO_2 (nitrogen dioxide) are of interest. Nitrogen oxide is colourless and is formed from the elements at high temperatures. It can react further with oxygen to form nitrogen dioxide, which is a brown gas.

The major sources of the two gases are: nitrogen oxide - from combustion of gasoline and oil: nitrogen dioxide - from combustion of oil, including diesel oil. The production of NO is favoured by high temperature. In addition, a relatively small emission of nitrogenous gases originates from the chemical industry.

The total global emission is approximately *10 million tons per year.* This pollution has only local or regional interest, as the natural global formation of nitrogenous gases in the upper atmosphere by the influence of solar radiation is far more significant than the anthropogenous emission.

As mentioned above, nitrogen oxide is oxidised to nitrogen dioxide, although the reaction rate is slow - in the order of 0.007 h^{-1}. However, it can be accelerated by solar radiation.

Nitrogenous gases *take part in the formation of smog,* as the nitrogen in peroxyacyl nitrate originates from nitrogen oxides. They are highly toxic but, as their contribution to global pollution is insignificant, local and regional problems can partially be solved by changing the distribution pattern (see Section 7.3).

The emission from motor vehicles can be reduced by the same methods as mentioned for carbon hydrides and carbon monoxide. The air/fuel ratio determines the concentration of pollutants in the exhaust gas. An increase in the ratio will reduce the emission of carbon hydrides and carbon monoxide, but unfortunately will increase the concentration of nitrogenous gases. Consequently, the selected air/fuel ratio will be a compromise.

A double catalytic afterburner is available. This is able to reduce nitrogenous gases and simultaneously oxidise carbon hydrides and carbon monoxide. The application of alternative energy sources will, as for carbon hydrides and carbon monoxide, be a very useful control method for nitrogenous gases at a later stage.

Between 0.1 and 1.5 ppm of nitrogenous gases, of which 10 to 15 per cent consists of nitrogen dioxide, are measured in urban areas with heavy traffic. On average the emission of nitrogenous gases is approximately 15 g per litre gasoline and 25 g per litre diesel oil.

The nitrogenous gas pollution problem.

Nitrogenous gases in reaction with water form nitrates which are washed away by rain water. In some cases this can be a significant source of eutrophication (see also Section 5.3). For a shallow lake, for example, the increase in nitrogen concentration due to the nitrogen input from rain water will be rather significant. In a lake with a depth of 1.7 m and an annual precipitation of 600 mm, which is normal in many temperate regions, the annual input will be as much as 0.3 mg per litre.

Control methods applied to nitrogenous gases.

The methods used for control of industrial emission of nitrogenous gases, including ammonia, will be discussed in the next paragraph, but as pointed out above industrial emission is of less importance, although it might play a significant role locally. The emission of nitrogenous gases by combustion of oil for heating and the production of electricity can hardly be reduced.

7.7 INDUSTRIAL AIR POLLUTION

Overview.

The rapid growth in industrial production during recent decades has exacerbated the industrial air pollution problem, but, due to increased application of continuous processes, recovery methods, air pollution control, use of closed systems and other technological developments, industrial air pollution has, in general, not increased in proportion to production.

Industry displays a wide range of air pollution problems related to a large number of chemical compounds in a wide range of concentrations.

It is not possible in this context to discuss all industrial air pollution problems, in stead we shall touch on the most important problems and give an overview of the control methods applied today. Only the problems related to the environment will be dealt with in this context.

A distinction should be made between air quality standards, which indicate that the concentration of a pollutant in the atmosphere at the point of measurement should not be greater than a given amount, and emission standards, which require that the amount of pollutant emitted from a specific source should not be greater than a specific amount (see also section 7.1).

The standards reflect, to a certain extent, the toxicity of the particular component, but also the possibility for its uptake.

Here the distribution coefficient for air/water (blood) plays a role. The more soluble the component is in water, the greater the possibility for uptake. For example, the air quality standard for acetic acid, which is very soluble in water, is relatively lower than the toxicity of aniline, which is almost insoluble in water.

Control methods applied to industrial air pollution.

Since industrial air pollution covers a wide range of problems, it is not surprising *that all three classes of pollution control methods* mentioned in Section 7.1 have found application: modification of the distribution pattern, alternative (cleaner) production methods and particulate and gas/vapour control technology.

All the methods mentioned in Sections 7.2 and 7.4 are valid for industrial air pollution control.

In gas and vapour technology a distinction has to be made between condensable and non-condensable gaseous pollutants. The latter must usually be destroyed by incineration, while the condensable gases can be removed from industrial effluents by absorption, adsorption, condensation or combustion. Recovery is feasible by the first three methods.

Gas absorption.

Absorption is a diffusion process that involves the mass transfer of molecules from the gas state to the liquid state along a concentration gradient between the two phases. Absorption is a unit operation which is enhanced by all the factors generally affecting mass transfer; i.e. *high inter facial area, high solubility, high diffusion coefficient, low liquid viscosity, increased residence time, turbulent contact between the two phases and possibilities for reaction of the gas in the liquid phase.*

TABLE 7.7
Absorber reagents

Reagents	Applications
$KMnO_4$	Rendering, polycyclic organic matter
NaOCl	Protein adhesives
Cl_2	Phenolics, rendering
Na_2SO_3	Aldehydes
NaOH	CO_2, H_2S, phenol, Cl_2, pesticides
$Ca(OH)_2$	Paper sizing and finishing
H_2SO_4	NH_3, nitrogen bases

This last factor is often very significant and almost 100 per cent removal of the contaminant is the result of such a reaction. Acidic components can easily be removed from gaseous effluents by absorption in alkaline solutions, and, correspondingly, alkaline gases can easily be removed from effluent by absorption in acidic solutions. Carbon dioxide, phenol and hydrogen sulphide are readily absorbed in alkaline solutions in accordance with the following processes:

$$CO_2 + 2NaOH \text{ -------> } 2Na^+ + CO_3^{2-} \quad (7.15)$$

$$H_2S + 2NaOH \text{ ------> } 2Na^+ + S^{2-} + 2H_2O \quad (7.16)$$

$$C_6H_5OH + NaOH \text{ ------> } C_6H_5O^- + Na^+ + H_2O \quad (7.17)$$

Ammonia is readily absorbed in acidic solutions:

$$2NH_3 + H_2SO_4 \text{ ------ >} 2NH_4^+ \; SO_4^{2-} \quad (7.18)$$

Gas adsorption.

Adsorption is the capture and retention of a component (adsorbate) from the gas phase by the total surface of the adsorbing solid (adsorbent). In principle the process is the same as that mentioned in Section 5.3, which deals with waste water treatment; the theory is equally valid for gas adsorption.

Adsorption is used to concentrate (often 20 to 100 times) or store contaminants until they can be recovered or destroyed in the most economical way. Fig. 7.15 illustrates some adsorption isotherms applicable to practical gas adsorption problems. These are (as mentioned in Section 5.3) often described *as either Langmuir's or Freundlich's adsorption isotherms*. Adsorption is dependent on temperature: increased temperature means that the molecules move faster and therefore it is more difficult to adsorb them. There are four major types of gas adsorbents, the most important of which is activated carbon, but also aluminium oxide (activated aluminium), silica gel and zeolites are used.

The selection of adsorbent is made according to the following criteria:

1. *High selectivity* for the component of interest.
2. *Easy and economical to regenerate.*
3. *Availability of the necessary quantity at a reasonable price.*
4. *High capacity for the particular application,* so that the unit size will be economical. Factors affecting capacity include total surface area involved, molecular weight, polarity activity, size, shape and concentration.
5. *Pressure drop,* which is dependent on the superficial velocity.
6. *Mechanical stability in the resistance of the adsorbent particles to attrition.* Any wear and

abrasion during use or regeneration will lead to an increase in bed pressure drop.

7. *Micro structure* of the adsorbent should, if at all possible, be *matched to the pollutant* that has to be collected.

8. *The temperature,* which, has a profound influence on the adsorption process, as already mentioned.

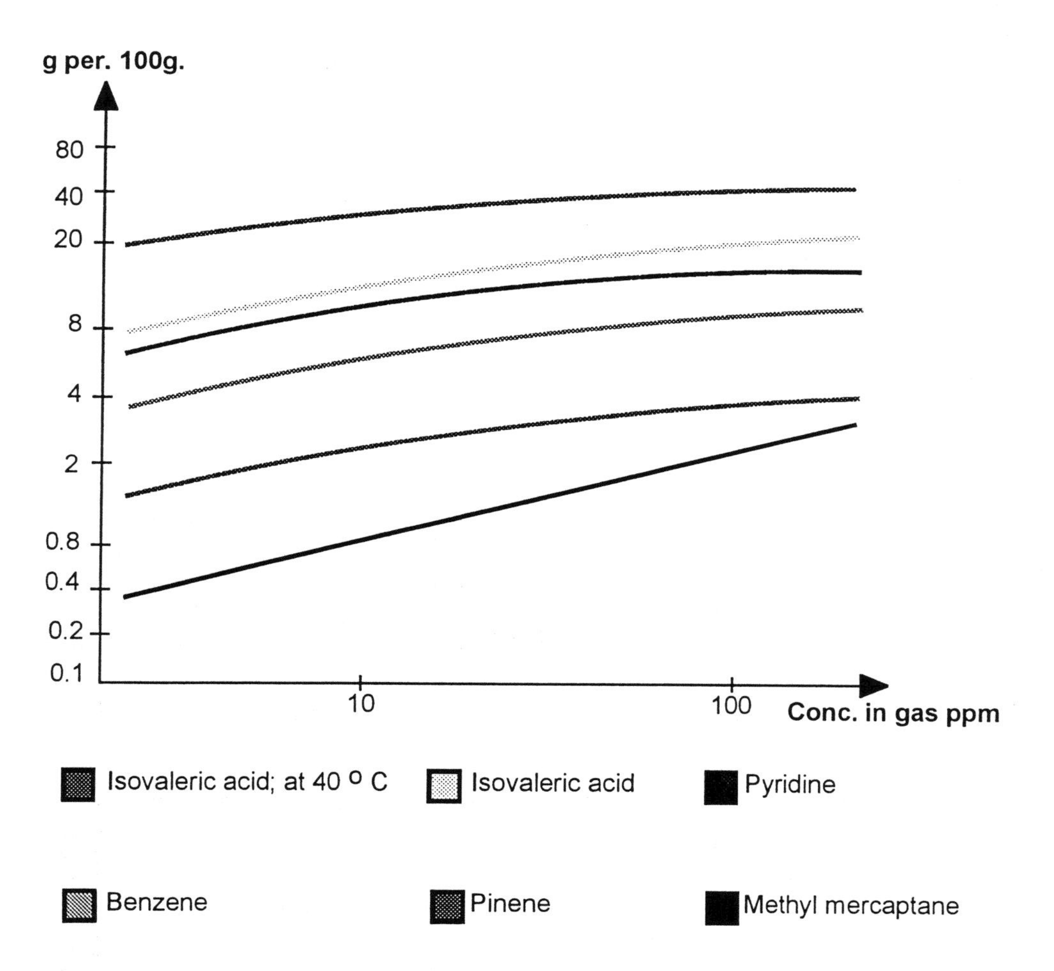

Fig. 7.15. Adsorption isotherms, at 20°C.

As already mentioned **regeneration of the adsorbents** is an important part of the total process. A few procedures are available for regeneration:

1. **Stripping** by use of steam or hot air.

2. **Thermal desorption** by raising the temperature high enough to boil off all the adsorbed material.

3. **Vacuum desorption** by reducing the pressure enough to boil off all the adsorbed

material.

4. **Purge gas stripping** by using a non-adsorbed gas to reverse the concentration gradient. The purge gas may be condensable or non-condensable. In the latter case it might be recycled, while the use of a condensable gas has the advantage that it can be removed in a liquid state.

5. **In situ oxidation**, based on the oxidation of the adsorbate on the surface of the adsorbent.

6. **Displacement** by use of a preferentially adsorbed gas for the desorption of the adsorbate. The component now adsorbed must, of course, also be removed from the adsorbent, but its removal might be easier that that of the originally adsorbed gas, for instance, because it has a lower boiling point.

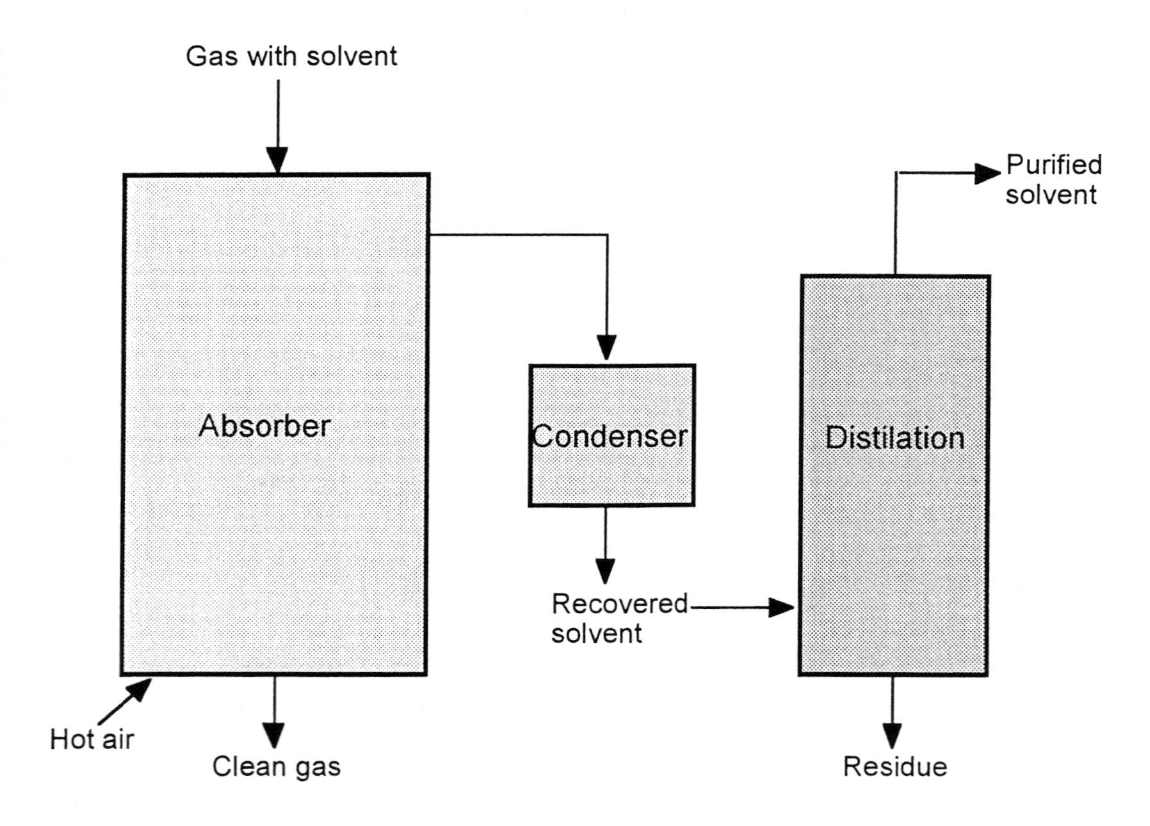

Fig. 7.16 Flow-chart of solvent recovery by the use of activated carbon.

Although the regeneration is 100 per cent the capacity of the adsorbent may be reduced 10 to 25 per cent after several regeneration cycles, due to the presence of fine particulates and/or high molecular weight substances, which cannot be removed in the regeneration step. A flow/chart of solvent recovery using activated carbon as an adsorbent is shown in Fig. 7.16 as an illustration of a plant design.

Combustion.

Combustion is defined as rapid, high temperature gas-phase oxidation. The goal is the complete oxidation of the contaminants to carbon dioxide and water, sulphur dioxide and nitrogen dioxide.

The process is often applied to control odours in rendering plants, paint and varnish factories, rubber tyre curing and petrochemical factories. It is also used to reduce or prevent an explosion hazard by burning any highly flammable gases for which no ultimate use is feasible. The efficiency of the process is highly dependent on temperature and reaction time, but also on turbulence or the mechanically induced mixing of oxygen and combustible material. The relationship between the reaction rate, r and the temperature can be expressed by Arrhenius' equation:

$$r = A * e^{-E/RT} \quad (7.19)$$

where A = a constant, E = the activation energy, R = the gas constant and T = the absolute temperature. A distinction is made between combustion, thermal oxidation and catalytic oxidation, the latter two being the same in principle as the vehicles afterburners.

7.8 HEAVY METAL DEPOSITION AND POLLUTION

Heavy metals, which may be defined as the metals with a specific gravity > 5.00 kg/l., comprise 70 elements. Most of them are, however, only rarely found as pollutants. The heavy metals of environmental interest form very heavy soluble compounds with sulphide and phosphate and form very stable complexes with many ligands present in the environment. It means, fortunately, that most of the heavy metals are not very bioavailable in most environment, as already discussed in Chapter 4,

A number of enzymes activated by metal ions and metallo-enzymes are known. The first mentioned group comprises iron, cobalt, chromium, vanadium and selenium, Copper, zinc, iron, cobalt, molybdenum are able with a stronger bond to form metallo-enzymes: metallo-proteins, metallo-porphyrines and metallo-flavines.

As pollutants are particularly lead, cadmium and mercury in focus, because of their extremely high toxicity. A general overview of the biological activity of elements can be found in Appendix 6.

Heavy metals are emitted to the atmosphere by energy production and a number of technological processes (see Table 7.8). It makes the atmospheric deposition of heavy metals, originated from human activities, the dominant pollution source for the vegetation of natural ecosystems - forests, wetlands, peat lands and so on. The heavy metal content in sludge and fertilisers plays a more important role for agricultural land where also the inputs of heavy metals by irrigation, natural fertilisers and application of chemicals including pesticides may add to the overall pollution level. The atmosphere and hydrosphere have both a well developed ability for "self purification" - for heavy metals by removal processes, for instance sedimentation. The lithosphere has a high buffer capacity toward the effects of most pollutants, and also has an ability of self purification, for instance by run off and uptake by plants, although the rates usually are much lower than in the two other spheres. Table 7.9 illustrates the removal rates.

TABLE 7.8
Important atmospheric pollution sources of heavy metals

Source	Heavy metals
Incineration of oil	V,Ni
Incineration of coal	Hg, V, Cr, Zn, As
Gasoline	Pb (leaded gasoline)
Metal industry	Fe, Cu, Mn, Zn, Cr, Pb, Ni, Cd and other
Application of pesticides	Hg, Cr, Cu, As
Incineration of solid waste	Hg, Zn, Cd, and other

TABLE 7.9
Removal of heavy metals by run-off and drainage from a typical cultivate clay soil (Hovmand, 1980)

Metal	Removal (mg /m^2y)	Removal % of pool
Pb	0.5	1-3
Cu	1.2	2-3
Zn	15.9	30-50
Cd	0.07	15-30

Heavy metals are bound to clay particles due to their ion exchange capacity and to hydratised metal oxides, such as iron sesquioxide (As, Cr, Mo, P, Se, and V) and manganese sequioxides (Co, Ba, Ni and lanthanides).Calcium phosphate is further more able to bind As, Ba, Cd and Pb in alkaline soil. Fulvic acids (molecular weight about 1000) and humic acid (molecular weight about 150, 000) are able to form complexes with a number of heavy metals, Hg(II), Cu(II), Pb(II) and Sn(II).

The mobility of heavy metals is dependent on a number of factors. The soil pore water contains soluble organic compounds (acetic acid, citric acid, oxalic acid and other organic acids), partly excreted by the roots. These small organic molecules form chelated, soluble compounds with metals ions such as Al, Fe and Cu. Activity of living organisms in soil may also enhance the mobility of heavy metal ions. Fungi and bacteria may utilise phosphate and thereby release cations. Formation of insoluble metal sulphide under anaerobic conditions from sulphate implies a reduced mobility. The lower oxidation stages of heavy metals are generally more soluble than the higher oxidation stages which implies increased mobility.

The many possibilities of binding heavy metals in soil explain the long residence time. Cadmium, calcium, magnesium and sodium have the most mobile metal ions with a residence time of about 100 years. Mercury has a residence time of about 750 years, while copper, lead, nickel, arsenic, selenium and zinc have residence times of more than 2000 years under temperate conditions. Tropic residence times are typically lower- for all heavy metals about 40 years.

The biological effect of the heavy metal pollution occurs in accordance with Sections 4.4 and 4.5 on two levels: on organisms level and on the higher level, first of all ecosystem level.

The plant toxicity is very dependent on the presence of other metal ions. For instance are Rb and Sr very toxic to many plants, but the presence of the biochemically more useful K and Ca is able to reduce or eliminate the toxicity. The toxicity of arsenate and selenate can be reduced in the same manner by sulphate and phosphate.

Formation of complexes by reaction with organic ligands reduces also the toxicity due to reduced bioavailability. The plant toxicity of heavy metals in soil is consequently also correlated with the concentration of heavy metal ions in the soil solution.

The heavy metals that are most toxic to plants are silver, beryllium, copper, mercury, tin, cobalt, nickel, lead and chromium. With exception of silver and chromium, the divalent form is most toxic. For silver it is Ag^+ and for chromium it is chromate and dichromate, that are most toxic. Silver and mercury ions are very toxic to fungus spores, and copper and tin ions are very toxic to green algae; lethal concentrations may be as low as 0.002-0.01 mg /l.

One of the key processes on ecosystem level is the mineralisation process, because they determine the cycling of nutrients. Heavy metals can inhibit the mineralisation due to blocking of enzymes. The effect is known not only for the enzymes produced in the organisms but also for extra cellular enzymes - exo-enzymes - originated from dead cells or excreted from roots and living microorganisms. As the various processes forming the cycling of nutrients are coupled, the entire mineralisation cycle is disturbed if only one process is reduced. It is therefore possible to determine the change of the mineralisation cycle by measuring the respiration, the transformation of nitrogen and the release of phosphorus. As low a concentration of copper as 3-4 times the background concentration may imply a reduced soil respiration. A few hundreds of mg copper per kg soil is furthermore able to diminish the nitrogen release rate by one half. The composition of uncontaminated soil is shown in Appendix 1.

The most sensitive mineralisation process is the phosphorus cycling. Biological material binds phosphorus as esters of phosphoric acid. The phosphate is released by hydrolysis of the ester bond, a process catalysed by phosphatase. This process is inhibited by the presence of heavy metals. The inhibition is decreasing in the following sequence: Molybdenum (VI) > wolframate (VI) > vanadate (V) > nickel (II) > cadmium > mercury (II) > copper (II) > chromate (VI) > arsenate (V) > lead (II) > chromium (III).

The inhibition of exo-enzymes by heavy metals doesn't form a clear pattern. It is therefore difficult to generalise. Most experiments give, however, a clear picture of the influence of heavy metals on the mineralisation: the rate of mineralisation may be reduced significantly with a consequent reduction of the productivity of the entire ecosystem.

The atmospheric deposition causes in Denmark, representing a country with relatively little heavy industry and a good pollution control, an average annual increase of the total content of heavy metal in soil between 0.4 % and 0.6%, but it varies very much from location to location. In accordance with the many possibilities for side reactions of heavy metals in soil, including adsorption to the soil particles, the amount of heavy metals ions that are available to plants is only a fraction of the total content. If only the bioavailable heavy metals are used as the basis, the annual percentage increase in the soil concentration due to atmospheric deposition is probably higher.

Most lead in soil is not mobile, and cannot be transported via the root system to the leaves and stems. This is in contrast to cadmium that is very mobile. About 50% of the cadmium in soil will be found in the plants after the growth season, although the concentration may be very different in different parts of the plants. The cadmium in grains has for instance not increased parallel to the increased atmospheric deposition of cadmium.

The heavy metal pollution of soil is one of the major challenges in environmental management in industrialised countries. Due to the many diffuse sources of heavy metal pollution, the solution of the problem requires a wide spectrum of methods, but first of all application of cleaner technology (see Chapter 9). It is with other words necessary to reduce the total emission of heavy metals. Dilution (for instance higher chimneys) is not an applicable solution. Moreover, as pollution, particularly air pollution, has no borders, it is necessary to take international initiatives and agree on international standards, particularly for the most problematic heavy metals, i.e, cadmium, mercury, nickel, chromium and vanadium. A three point program must be adopted:

- a national and international environmental strategy is accepted
- agreed international standards and long-term goals

- monitoring program to assess the pollution level and compare the measured concentrations with standards

7.9 QUESTIONS AND PROBLEMS

1. Calculate the ratio in concentration of particulate matter in cases A and B provided that the total emission is the same in the two cases:

   A: 2000 stacks $H = 10$ m

   B: 1 stack $H = 250$ m

2. Indicate a method to remove the following as air pollutants:
   1) ammonia, 2) hydrogen sulphide, 3) phenol, 4) hydrogen cyanide.

3. What is the energy consumption for a packed-bed scrubber treating 650 m^3 min^{-1} with a $D_{50} = 2\mu m$?

4. Compare the influence of SO_2 on the sulphur cycle and CO_2 on the carbon cycle.

5. Would the total SO_2 emission on earth be able to change the pH of the sea over a period of 100 years, if: a) the total SO_2 emission is constant at the present level, b) the total SO_2 emission increases 2% per year?

6. Calculate the cost of activated carbon per day used for a 90% removal of 10 ppm (v/v) pyridine from 100 m^3 air per hour. The temperature is 20°C. The cost of one kg activated carbon is US$ 1.15.

7. Henry's constant for sulphur dioxide is 1 atm/M. pK_{a1} for sulphurous acid is 1.75. Oxidation of sulphur dioxide is not considered. Calculate the pH of rainwater in equilibrium with sulphur dioxide in a polluted air with a concentration of 0.5 ppm.

   Calculate the concentration of sulphur dioxide that must be reached in polluted air if the air- water equilibrium should produce a pH of 4.0 in raindrops.

8. Calculate the mass of fine particles inhaled by an adult each year, assuming that 15 m^3 is inhaled per 24h, that the concentration of particulate matter is $10\mu g/l$, that the particle diameter is 1 μm and that the specific gravity of the particles is 0.6 g/ml. Calculate also the surface area of this annual load of particles.

8. Ecological Engineering

8. 1 WHAT IS ECOLOGICAL ENGINEERING?

H.T. Odum was among the first to define ecological engineering (Odum, 1962 and Odum et al., 1963) as the "environmental manipulation by man using small amounts of supplementary energy to control systems in which the main energy drives are coming from natural sources". Odum further developed the concept (Odum, 1983) of ecological engineering as follows: ecological engineering, the engineering of new ecosystems designs, is a field that uses systems that are mainly self-organising.

Straskraba (1984 and 1985) has defined ecological engineering, or as he calls it ecotechnology, more broadly, as the use of technological means for ecosystem management, based on a deep ecological understanding, to minimise the costs of measures and their harm to the environment. Ecological engineering and ecotechnology may be considered synonymous.

Mitsch and Jørgensen (1989) gives a slightly different definition which, however, covers the same basic concept as the definition given by Straskraba and also encompasses the definition given by H.T. Odum. They define ecological engineering and ecotechnology as the design of human society with its natural environment for the benefit of both. It is engineering in the sense that it involves the design of man-made or natural ecosystems or parts of ecosystems. Like all engineering disciplines it is based on basic science, in this case ecology and system ecology. The biological species are the components applied in ecological engineering. Ecological engineering represents therefore a clear application of ecosystem theory.

Ecotechnic is another often applied word, but it encompasses in addition to ecotechnology or ecological engineering also the development of technology applied in society, based upon ecological principles, for instance all types of cleaner technology, particularly if they are applied to solve an environmental problem.

Ecological engineering should furthermore not be confused with bioengineering or biotechnology. Biotechnology involves the manipulation of the genetic structure of the cells to produce new organisms capable of performing certain functions. Ecotechnology does not manipulate at the genetic level, but at several steps higher in the ecological hierarchy. The manipulation takes place on an assemblage of species and/or their abiotic environment as a self-designing system that can adapt to changes brought about by outside forces, controlled by human or by natural forcing functions.

Ecological engineering is also not the same as environmental engineering which is involved in cleaning up processes to prevent pollution problems. It uses settling tanks, filters, scrubbers and man-made components which have nothing to do with the biological and ecological components that are applied in ecological engineering, although the use of environmental engineering aims towards reducing man-made forcing functions on ecosystems. Ecotechnic mentioned above may be considered to include in addition to ecological engineering, environmental technology based on ecological principles such as recirculation. The tool boxes of the two types of engineering are completely different where ecological engineering uses ecosystems, communities, organisms and their immediate abiotic environment.

All applications of technologies are based on quantification. Ecosystems are very complex systems and the quantification of their reactions to impacts or manipulations therefore becomes complex. Fortunately, ecological modelling represents a well developed tool to survey ecosystems, their reactions and the linkage of their components. Ecological modelling is able to synthesise our knowledge about an ecosystem and makes it possible to a certain extent to quantify any changes in ecosystems resulting from the use of both environmental engineering and ecological engineering. Ecological engineering may also be used directly to design constructed ecosystems. Consequently, ecological modelling and ecological engineering are two closely cooperating fields. Research in ecological engineering was originally covered by the Journal of Ecological Modelling which was initially named "Ecological Modelling - International Journal on Ecological Modelling and Engineering and Systems Ecology" to emphasise the close relationship between the three fields: ecological modelling, ecological engineering and systems ecology. Ecological Engineering was launched as an independent journal in 1992, and the name of Ecological Modelling was changed to "Ecological Modelling - An International Journal on Ecological Modelling and Systems Ecology". Meanwhile, the journal "Ecological Engineering" has been successful in covering the field of ecological engineering which has grown rapidly during the nineties due to the increasing acknowledgement of the need for other technologies than environmental technology in our efforts to solve pollution problems. This development does not imply that ecological modelling and ecological engineering are moving in different directions. On the contrary, ecological engineering is increasingly using models to perform design of constructed ecosystems or to quantify the results of application of specific ecological engineering methods for comparison with alternative, applicable methods.

8.2 EXAMPLES AND CLASSIFICATION OF ECOTECHNOLOGY

Ecotechnology may be based on one or more of the following four classes of ecotechnology:

1. Ecosystems are used to reduce or solve a pollution problem that otherwise would

be (more) harmful to other ecosystems. A typical example is the use of wetlands for waste water treatment.

2. Ecosystems are imitated or copied to reduce or solve a pollution problem, leading to constructed ecosystems. Examples are fishponds and constructed wetlands for treatment of waste water or diffuse pollution sources.

3. The recovery of ecosystems after significant disturbances. Examples are coal mine reclamation and restoration of lakes and rivers.

4. The use of ecosystems for the benefit of humankind without destroying the ecological balance, i.e. utilisation of ecosystems on an ecologically sound basis. Typical examples are the use of integrated agriculture and development of organic agriculture. This type of ecotechnology finds wide application in the ecological management of renewable resources.

TABLE 8.1

Ecological Engineering Examples. Alternative Environmental Engineering methods are given, if possible.

Type of Ecological Engineering	Example of Ecological Engineering		Environmental Engineering alternative
	Without Env. Eng. Alternative	With Env. Eng. Alternative	
1	Wetlands utilised to reduce diffuse pollution	Sludge disposal on agricultural land	Sludge incineration
2	Constructed wetland to reduce diffuse pollution	Root zone plant	Traditional waste water treatment
3	Recovery of lakes	Recovery of contamined land in situ	Transport and treatment of contaminated soil
4	Agroforestry	Ecologically sound planning of harvest rates of resources	

The idea behind these four classes of ecotechnology is illustrated in Fig. 8.1. Notice that ecotechnology operates in the environment and the ecosystems. It is here that ecological engineering has its toolbox.

Illustrative examples of all four classes of ecological engineering may be found in situations where ecological engineering is applied to replace environmental engineering because the ecological engineering methods offer an ecologically more acceptable solution and where ecological engineering is the only method that can offer a proper solution to the problem. Examples are shown in Table 8.1 where also the alternative environmental technological solution is also indicated.

This does not imply that ecological engineering consequently should replace environmental engineering. On the contrary, the two technologies should work hand in hand to solve environmental management problems better than they could do alone. This is illustrated in Figure 8.2, where control of lake eutrophication requires both ecological engineering and environmental technology.

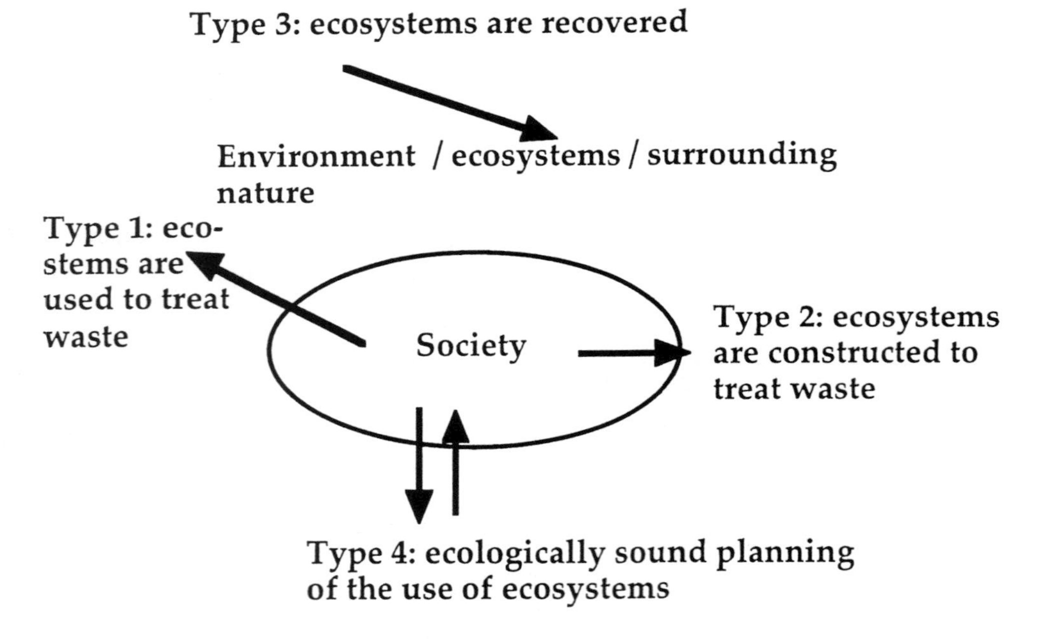

Fig. 8.1. An illustration of the four types of ecological engineering.

Type 1 ecological engineering, application of ecosystems to reduce or solve pollution problems, may be illustrated by wetlands utilised to reduce the diffuse nutrient loadings of lakes. This problem could not be solved by environmental technology. Treatment of sludge could be solved by environmental technology, namely by incineration, but the ecological engineering solution, sludge disposal on agricultural land which involves a utilisation of the organic material and nutrients in the sludge, is a considerably more sound method from ecological perspectives.

The application of constructed wetlands to cope with the diffuse pollution is a good example of ecological engineering type 2. Again, this problem cannot be solved by environmental technology. The application of root zone plants for treatment of small amounts of waste water is an example of ecological engineering, type 2, where the environmental engineering alternative, a mechanical-biological-chemical treatment cannot compete, because it will have excessive costs relative to the amount of waste water (sewage system, pumping stations and so on). A solution requiring less resources will always be ecologically more sound.

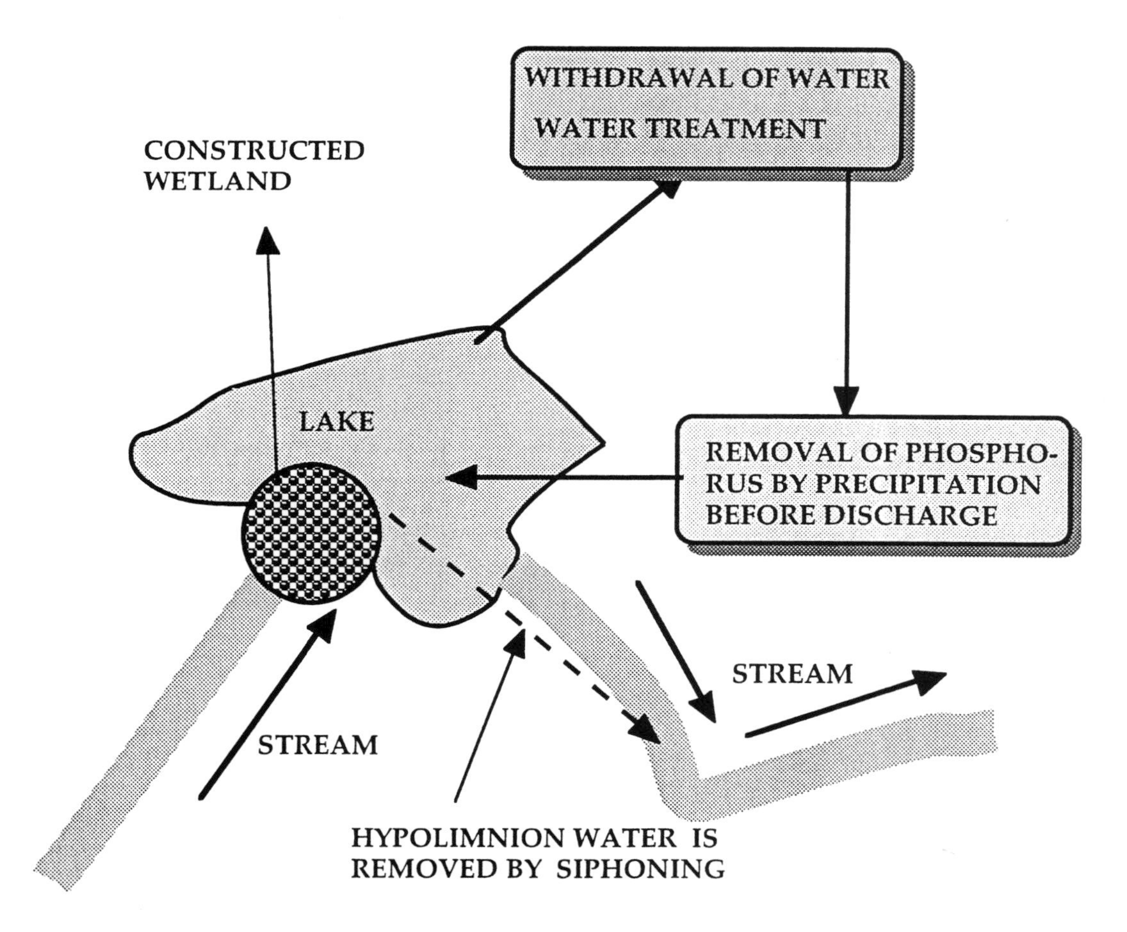

Fig. 8.2. Control of lake eutrophication with a combination of chemical precipitation for phosphorus removal from waste water (environmental technology), a wetland to remove nutrients from the inflow (ecotechnology, type 1 or 2), and siphoning off of hypolimnetic water, rich in nutrients, downstream (ecotechnology type 3).

Recovery of land contaminated by toxic chemicals is possible using environmental technology, but it will require transportation of soil to a soil treatment plant, where biological biodegradation of the contaminants takes place. Ecological engineering will propose a treatment in situ with adapted microorganisms or plants. The latter method will be much more cost effective and the pollution related to the transport of soil will be omitted. Restoration of lakes by biomanipulation, installation of an impoundment, sediment removal or coverage, siphoning off hypolimnetic water, rich in nutrients, downstream or by several other proposed ecological engineering techniques are type 3 examples of ecological engineering. It is hardly possible to obtain the same results by environmental engineering because this requires activities in the lake and/or the vicinity of the lake.

Type 4 of ecological engineering is to a great extent based on prevention of pollution by utilisation of ecosystems on an ecologically sound basis. It is hardly possible to find environmental engineering alternatives in this case, but it is clear that a prudent harvest rate of renewable resources, whether of timber or fish, is the best long term strategy from an ecological and economic point of view. An ecologically sound planning of the landscape is another example of the use of type 4 ecological engineering.

8.3 ECOSYSTEM THEORY APPLIED TO ECOTECHNOLOGY

Ecological engineering has been presented in the two previous sections of this chapter as a useful technological discipline. This section is devoted to draw the relations to ecosystem theory by presentation of a number of principles applied to understand the ecological engineering methods in practice and extracted from ecosystem theory.

Mitsch and Jørgensen (1989) apply twelve system ecological principles to elucidate the basic concepts of ecotechnology and to ensure proper application of the practical use of this approach. All 12 principles are based on system concepts applied in ecology. The principles are presented below with at least two examples of the application of each of the 12 principles in the practical use of ecological engineering. The two examples are taken from terrestrial ecosystem, often from agriculture, and from aquatic ecosystems respectively.

Principle 1: Ecosystem structure and functions are determined by the forcing functions of the system. Ecosystems are open systems, which implies that they exchange mass and energy with the environment. There is a close relation between the anthropogenic forcing functions and the state of the agricultural ecosystems, but due to the openness of all ecosystems, the adjacent ecosystems are also affected. Intensive agriculture leads to drainage of the surplus nutrients and pesticides to the adjacent ecosystems. This is the so-called non-point or diffuse pollution. The abatement of this source of pollution requires a wide range of ecotechnological methods.

The use of constructed wetlands and impoundments to reduce the concentrations of nutrients in streams entering a lake ecosystem illustrate the application of this system ecological principle in ecotechnology. The forcing functions, the nutrient loadings, are reduced and a corresponding reduction of the eutrophication should be expected.

Principle 2: Homeostasis of ecosystems requires accordance between biological function and chemical composition. The biochemical functions of living organisms define their composition, although these are not to be considered as fixed concentrations, but as ranges. The application of the principle implies that the flows of elements through agricultural systems should be according to the biochemical stoichiometry. If this is not the case the elements in surplus will be exported to adjacent ecosystems and make an impact on the natural balances and processes there. An investigation (R. Skaarup and T. Sørensen, 1994) on a well-managed Danish farm has shown that much can be gained by a complete material flow analysis of a farm. The results will not only lead to reduction in the emission level of pollutants but will often also imply cost reductions.

Several restoration projects on lakes have considered this principle. The limiting nutrient determining the eutrophication of a lake is found and the selected restoration method will reduce the limiting nutrient further. When sediment is removed or covered in lakes, it is to prevent the otherwise limiting nutrient, phosphorus, reaching the water phase.

Principle 3: It is necessary in environmental management to match recycling pathways and rates to ecosystems to reduce the effect of pollution. The application of sludge as a soil conditioner illustrates this principle very clearly. The recycling rate of nutrients in agriculture has to be accounted for in any use of sludge. If the sludge is applied faster than it can be utilised by the plants, a significant amount of the nutrients might contaminate the streams, lakes and / or ground water adjacent to the agricultural ecosystem. If, however, the influence of the temperature on nitrification and denitrification, the hydraulic conductivity of the soil, the slope of field and the rate of plant growth are all considered in an application plan for manure, the loss of nutrients to the environment will be maintained at a very low and probably acceptable level. This can be achieved through ecological models to develop a plan for the application of the sludge.

Elements are recycled in agro-systems but to a far less extent than occurs in nature. For the last couples of decades animal husbandry has to great extent become separated from plant production. This has made the internal cycling more difficult to achieve, of less economical value and thus less attractive. We have come to accept losses of the relatively inexpensive, easy-to-apply artificial fertilisers and to compensate for this by increasing their application.

So, the message is: know the ecological processes of farming and their rates and manage the system accordingly, i.e., recycle as much as possible at the right rates and don't use more fertilisers than can be recycled.The restoration of a eutrophied lake by application of shading is an illustrative example of the same principle for aquatic ecosystems.

Ecosystems with pulsing patterns often have greater biological activity and chemical cycling than systems with relatively constant patterns. A specific case, an estuary in Brazil, named the Cannaneia study will illustrate the recognition of pulsing forces and how it is possible to take advantage of them in ecological engineering. The shores of the islands in the estuary and the coast are very productive mangrove wetlands and the entire estuary is an important nesting area for fish and shrimps. A channel was built to avoid flooding upstream, where productive agricultural land is situated. The construction of the channel has caused a conflict between farmers, who want the channel open, and fishermen, who want it closed due to its reduction of the salinity in the estuary (the right salinity is of great importance for the mangrove wetlands). The estuary is exposed to tide which is important for maintenance of a good water quality with a certain minimum of salinity. The conflict can be solved by use of an ecological engineering approach that takes advantages of the pulsing force (the tide). A sluice in the channel could be constructed to discharge the fresh water when it is most appropriate. The tide would in this case be used to transport the fresh water as rapidly as possible to the sea. The sluice should be closed when the tide is on its way into the estuary. The tidal pulse is frequently selectively filtered to produce an optimal management situation.

Principle 4: Ecosystems are self-designing systems. The more one works with the self-designing ability of nature, the lower the costs of energy to maintain that system. Many of our actions are undertaken to circumvent or to counteract the process of self-design. For example, the biodiversity of agricultural fields would be significantly higher if pesticides were not used and nature left to rule on its own. While the self-designing systems are able to implement sophisticated regulations before violent fluctuations or even chaotic events occur, agriculture attempts to regulate chemically, for instance undesired organisms are regulated by the use of pesticides. This very coarse regulation sometimes causes more harm than anticipated, for instance when the insect-predators are affected more than the insects. The conclusion seems clear: don't eliminate the well-working natural regulation mechanisms, i.e., maintain a pattern of nature within agricultural systems.

The application of green in contrast to bare fields during the winter in Northern Europe is consistent with this principle, as the self-designing ability is maintained at this time of the year. Bare soil should generally also be avoided due to possible erosion.

The closer the agricultural system is to a natural ecosystem, the more self-designing capacity the system has. Integrated agriculture is therefore less vulnerable than modern industrialised agriculture, because it offers more components for self-designing regulation and it has a wider range of flows that facilitates possibilities for recycling.

The use of constructed wetlands in lake restoration is an example of the application of this principle of self-design taken from aquatic ecosystems. If we design a wetland to remove partially the nutrients from streams entering the lake, the lake can itself do the self-design and reduce the level of eutrophication accordingly. The constructed wetland will also use self-design. The diversity (complexity) and nutrients removal efficiency will increase gradually, provided that the wetland is not disturbed.

Principle 5: Processes of ecosystems have characteristic time and space scales, that should be accounted for in environmental management. Environmental management should consider the role of a certain spatial pattern for the maintenance of biodiversity. Violence of this principle by drainage of wetlands and deforestation on too large a scale has caused desertification. Wetlands and forests maintain high soil humidity and regulate precipitation. When the vegetation is removed, the soil is exposed to direct solar radiation and dries, causing organic matter to be burned off. Application of excessively large fields prevents wild animals and plants from finding their ecological niches as important components in the pattern of agriculture and more or less untouched nature. The solution is to maintain ditches and hedgerows as corridors in the landscape or as ecotones between agricultural and other ecosystems. Fallow fields also should be planned as contributors to the pattern of the landscape. The example mentioned above on the use of the tide to transport fresh water as rapidly as possible to the sea may also be used to illustrate this principle of using the right time and space scales in ecological engineering.

Principle 6: Chemical and biological diversity contributes to the buffering capacity and self-designing ability of ecosystems. A wide variety of chemical and biological components should be introduced or maintained for the ecosystem's self-designing ability to choose from. Thereby a wide spectrum of buffer capacities is available to meet the impacts from anthropogenic pollution.

Biodiversity plays an important role in the buffer capacity, and the ability of the system to meet a wide range of possible disturbances by the use of the ecosystem's self designing ability. There are many different buffer capacities, corresponding to any combination of a forcing function and a state variable. It has been shown that vegetables cultivated to a great extent in mixed cultures give a higher yield and are less vulnerable to disturbances for instance attacks by herbivorous insects. In agricultural practice this implies that it is advisable to use small fields with different crops.

This principle also means that integrated agriculture is less vulnerable than modern industrialised agriculture simply due to its higher biological and chemical diversity.

Restoration of lakes by use of biomanipulation usually increases the biodiversity and some buffer capacities.

Principle 7: Ecotones, transition zones, are as important for the ecosystems as the membranes are for cells. Agricultural management should therefore consider the importance of transition zones. Nature has developed transition zones, denoted ecotones, to make a soft transition between two ecosystems. Ecotones may be considered as buffer zones, that are able to absorb undesirable changes imposed on an ecosystem from adjacent ecosystems. We must learn from nature and use the same concepts when we design interfaces between manmade ecosystems (agriculture, human settlements) and nature. Some countries have required a buffer zone between human settlements and the coast of lakes or marine ecosystems (for instance in Denmark it is 50m). Without buffer zones between agriculture and natural ecosystems emissions will be transferred directly to the ecosystem, while a buffer zone such as a wetland would at least partly absorb the emissions and thereby prevent their negative influences on natural ecosystems. Some countries require a buffer zone (for instance in Denmark of 2m and it is discussed to increase it to 5m) between arable land and streams or lakes.

A pattern of wetlands in the landscape will be able to prevent the emission of particular nitrate from agriculture which is one of the hot issues in the environmental management of the diffuse pollution originating from agriculture.

The role of the littoral zone in lake management is another obvious example. A sound littoral zone with a dense vegetation of macrophytes will be able to absorb contamination before it reaches the lake and will thereby be for a lake like a membrane for the cell.

Principle 8: The coupling between ecosystems should be utilised to the benefit of the ecosystems in the application of ecotechnology and in environmental management of agricultural systems. An ecosystem cannot be isolated - it must be an open system, because it needs an input of energy to maintain the system (see also principle 1). The coupling of agricultural systems to natural systems leads to transfer of pesticides and nutrients from agriculture to nature and measures should be taken to (almost) complete utilisation of the pesticides and the nutrients in the agricultural system, for instance by implementation of a proper fertilisation plans accounting for these transfer processes. Ecological management should always consider all ecosystems as interconnected systems, not as isolated subsystems. This means that not only local but also regional and global effects have to be considered. For instance, the methane emitted from rice fields may increase the greenhouse effect and thereby the global climate which again will feed back upon the cultivation of rice.

Lake management can only be successful if it is based upon this principle, as all the inputs to the open ecosystem, a lake, should be considered. Recovery of a eutrophied lake will require that all nutrient sources are quantified and that a plan is made which takes all the sources into account is realised.

Principle 9: It is important that the application of ecotechnology and environmental management considers that the components of an ecosystem are interconnected, interrelated and form a network, which implies that direct as well as indirect effects are of importance. An ecosystem is an entity; everything is linked to everything in the ecosystem. Any effect on any component in an ecosystem is therefore bound to have an effect on *all* components in the ecosystem either directly or indirectly, i.e., the entire ecosystem will be changed. It can be shown that the indirect effect is often more important than the direct one (Patten, 1991). Application of ecotechnology attempts to take the indirect effect into account, while management considering only the direct effects often fails. There are numerous example of the use of pesticides against herbivorous insects, that also might have a pronounced effect on the carnivorous insects and therefore will result in the opposite effect on the herbivorous insects than intended. Therefore pesticides should not be used in a vacuum, but sufficient knowledge about the insect populations and their predators should be the basis for decision about the application of pesticides. Preferably, a model should be developed to ensure that the pesticides reach the right target organisms and do not cause an inverse effect.

The use of ecotechnology in the abatement of toxic substances in aquatic ecosystem requires that this principle is considered. The biomagnification of toxic substances through the food chain is a result of the interconnectedness of ecological components. Due to the biomagnification, it is necessary to aim at a far lower concentration of toxic substances in aquatic ecosystems to avoid undesirable high concentration of the toxic substance in fish for human consumption.

Principle 10: It is important to realise that an ecosystem has a history in the application of ecotechnology and environmental management in general. The components of ecosystems have been selected to cope with the problems that nature has imposed on the ecosystems for million of years. The high biodiversity of old ecosystems compared with the immature ecosystems is another important feature of this principle. The structure of mature ecosystems should therefore be imitated in the application of ecological engineering. An ecosystem with a long history is better able to cope with the emissions from its environment than an ecosystem with no history. This again emphasises the importance of establishing a pattern of agriculture and natural terrestrial and aquatic ecosystems to ensure that the history is preserved and the right solution can therefore be offered to emerging environmental problems.

Many lakes store considerable amounts of nutrients in the sediment due to a sad history of discharge of insufficiently treated waste water. Restoration of such lakes often requires removal of the sediment, an extremely expensive restoration methods for large and deep lakes. Ecologically sound management will take the history into account and try to prevent arising problem in good time.

Principle 11: Ecosystems are most vulnerable at their geographical edges. Therefore ecological management should take advantage of ecosystems and their biota in their optimal geographical range. When ecological engineering involves ecosystem manipulation, the system will have enhanced buffer capacity if the species are in the middle range of their environmental tolerance. Ecosystem manipulation should therefore consider a careful selection of the involved species in accordance with this principle. For agriculture this means that the crops should be selected following this principle. The cultivation of tomatoes and other subtropical vegetables in Northern Europe demonstrates how this principle is easily violated in modern agriculture. These products may compete in price due to good management or subsidies, but they cannot compete in quality.

Ecologically sound planning will use this principle and avoid the use of biological components which are at their geographical edges. This rule is of course important for both terrestrial and aquatic ecosystems.

Principle 12: Ecosystems are hierarchical systems and all the components forming the various levels of the hierarchy make up a structure, that is important for the function of the ecosystem. It is important for instance to maintain the components, that make up the landscape diversity, such as hedges, wetlands, shorelines, ecotones, ecological niches etc. They will all contribute to the buffer capacity of the entire landscape. Clearly, the integrated agriculture can more easily follow this principle than can industrialised agriculture, because it has more components to use for construction of a hierarchical structure.

It is equally important in our management of lakes to consider the benthic zone, the littoral zone, the epilimnion and hypolimnion. All the zones require the right conditions with respect to oxygen, pH, temperature and so on to maintain the various organisms (the next lower level in the hierarchy) fitted to these zones. Selection of lake restoration methods requires consideration of this issue. In which zone does the problem of the right ecological balance occur? What can we do to solve the problem? Which restoration methods should then be selected? The answers to these questions require extensive use of ecological modelling, applied on ecotechnology.

8. 4 ECOSYSTEM CONCEPTS IN ECOTECHNOLOGY

Straskraba (1993) has presented 7 principles of ecosystems which he uses to set up 17 rules on how to use ecotechnology. The rules may, however, be expanded to be applied in environmental management in general. They are based upon basic scientific principles of systems ecology and should therefore be respected by all means whenever an environmental management decision is to be taken. The principles (taken from Straskraba, 1993) are presented below with a short explanation, followed by the rules (Straskraba, 1993) with comments on their application in ecological technology and environmental management in general.

Principle 1: Energy inputs to the ecosystems and available storage of matter are limited. This principle is based on the conservation of matter and energy (Patten et al., 1997); nuclear processes may convert matter to energy, but these processes are insignificant in ecosystems. The dominant energy input to life on the Earth is solar energy. We may be able to supplement this energy source with other forms of energy, for instance fossil fuel, but the only sustainable energy source is solar energy.

Favourable conditions for vegetation growth are represented by temperature, humidity (water) and sufficient concentrations of nutrients. Mass conservation leads to the concept of limiting factors for growth and the Michaelis-Menten equation: growth = v* S / (k

S), where S represents the concentration of the substrate or limiting nutrient, and v and k are constants. A maximum growth rate is attained when S >> k which implies that another nutrient or factor becomes limiting.

Principle 2: Information is stored in ecological systems in structures. Structures are a result of the input of energy which is utilised to move away from thermodynamic equilibrium, gain exergy (see Chapters 2-4) or build structures. Such structures include not only organisms but also the physical structure of the landscape. Size is an important characteristic of structures. Organism size determines many important features of life, like the rate of development, speed of movement and the range of areas they inhabit. A certain minimum size of structures surrounding the organisms is necessary to satisfy their needs.

Principle 3: The genetic code of single organisms stores information about the development of the Earth's environment, and the process of adaptation to the changing conditions of existence. For organisms information about the past evolution is stored in the genes. The genetic code reflects not only the past of an individual organism, but also its surroundings, included other coexisting organisms. It is the consequence of tight coupling between each organism and its environment. The evolution of the organisms is highly dependent on the conditions of life (Straskraba, 1993).

In Chapters 3-4 use of exergy calculations, denoting an exergy index, was proposed to describe the propensity for the development of an ecosystem and to assess ecosystem health. These proposals are in accordance with the concepts behind principle 3.

Principle 4: Ecosystems are open and dissipative systems. They are dependent on a steady input of energy from outside. This principle is an application to ecosystems of the second law of thermodynamics. As ecosystems of course must obey the conservation principle, they must also obey basic scientific laws, including the basic laws of thermodynamics. The energy input is utilised to cover the energy needs for maintenance, respiration and evapotranspiration. If the energy (exergy) input exceeds these needs, the surplus energy (exergy) is utilised to build more structure.

Principle 5: Ecosystems are multi-mediated feedback systems. Feedback systems are characterised by influences on one part of the system being fed back to other components of the system. Feedbacks create many completely unexpected effects and the entire network describing the interrelations of the organisms may also work as feedbacks. Patten (1991) has shown that indirect effects may often exceed direct effects due to feedbacks and the relationships determined by the network.

Principle 6: Ecosystems have homeostatic capability which results in smoothing out and depressing the effects of strongly variable inputs. However, this capability is limited. Once it is exceeded, the system breaks down. Several homeostatic mechanisms are known in biology, for instance the maintenance of pH in our blood and the maintenance of the body temperature of warm-blooded animals. The homeostatic capability may be expressed by means of the concept of ecological buffer capacity (see Chapter 3). As the homeostatic capability is limited, so are the buffer capacities. It is obviously important in environmental management to know and respect the buffer capacities, as the ecosystem otherwise may change radically and even collapse.

Principle 7: Ecosystems are adaptive and self-organising systems. We may distinguish between biochemical and biological adaptation. Biochemical adaptation takes place when the individual organism adapt to changed conditions. Phytoplankton for instance can adjust its content of chlorophyll a to the intensity of solar radiation. Biological adaptation implies adaptation at the ecosystem level. The properties of the species in the ecosystem is changed currently in accordance with the prevailing conditions, either for the next generation by dominant heritage of the properties best fitted to the prevailing conditions, or by complete or partial replacement of species with different tolerances. Other species waiting in the wings take over, if they have a combination of properties better fitted to the (new) emergent conditions.

Self-organisation of the ecosystem may be understood as a directional change of the species composition of the ecosystem when its surroundings change, e.g. as a result of human activities (Straskraba, 1993).

These principles give occasion to introduce 17 rules (Straskraba, 1993) for the application of ecotechnology or environmental management in general. The 17 rules with reference to the principles are presented below.

Rule 1: Minimise energy waste. This rule is based on the conservation principle; see principle 1. Energy is limited and therefore a resource and in addition all uses of energy cause pollution: carbon dioxide for fossil fuel, nuclear waste for nuclear energy, noise and landscape pollution for windmills etc. The selection of environmental technology should also be based on this rule and rule number two, so not only are we reducing the use of resources, but we are also reducing the pollution correspondingly, as energy not wasted and matter recycled are not discharged to the environment.

Rule 2: Recycle. This rule is also based on the conservation principle, principle 1. This rule is a core principle in cleaner technology (see Chapter 6) and is an imitation of ecosystems. For all essential elements, ecosystems use recycling which will ensure that the resources can be used again and again. The development of green audit is based upon the conservation principle. The purpose is to increase recycling; see Chapters 5 and 6.

Rule 3: Retain all kinds of structures. This rule is based on principle 2. It is in accordance with the use of ecotechnology which prescribe that hedges, littoral zones, a landscape pattern of agriculture, wetlands, forests, hedges and hills are essential to maintain.

Rule 4: Consider long-term horizons. Sustainable development, or sustainable life, may be realised only when we switch from short-term evaluation of our economic practices to long-term horizons (Straskraba, 1993). This implies that we need to consider the long-term consequences of changes in the ecological structures and extinction of species due to our activities (principles 2 and 3).

Rule 5: Do not neglect that mankind is dependent on many organisms and that their loss may lessen our ability to survive in the steadily changing environment. This rule is based on principle 3. Our dependence on many species is not necessarily known today. We may find medicine for curing cancer in a rain forest plant which has not yet been discovered. The presence of important indirect effects makes it very difficult to evaluate the importance of many species. It seems therefore to be a prudent strategy to maintain as many species as possible and maintain the networks in which they are operating to the greatest possible extent.

Rule 6: Consider ecosystem dynamics. This rule is directly translated from principle 4. Open and dissipative systems have complex dynamics which should not be neglected in environmental practices. Dynamic and structurally dynamic models appear to be tools needed for decisions concerning ecosystem management, as they consider these important dynamics of ecosystems.

Rule 7: Understand that nature is a teaching ground for handling complex systems. In particularly, the ability to survive by adapting to changing conditions can be learned only from nature. This rule is based on principles 4 and 5. The study of adaptation and evolution of individual organisms and ecosystems seems to be the only way we can recognise our own possibilities and limitations of survival under conditions of the environment subject to unexpected major changes created by our activity. The recent development of cleaner technology and organic agriculture is based on imitation of nature or learning from nature.

Rule 8: People are part of the ecosystems dependent on solar energy directly: we need it in the form of light to see and in the form of heat for warmth. We are also indirectly dependent on energy stored in fossil fuel as well as in vegetable and animal food - we are dissipative structures, too. This rule acknowledges that principle 4 is also valid for us - we are ourselves open and dissipative systems and we are part of open and dissipative systems. We are therefore also victims of principle 4.

Rule 9: Understand that we are dependent on and sensitive to external inputs, in particularly solar radiation, but also to the supply of minerals from the Earth. This rule is based upon principles 1, 2 and 4. It is also associated with rules 1, 2 and 8. Our activity should not alter the inputs by solar radiation by destruction of the ozone layer as that may cause skin cancer. On the other hand, the modern post-industrialised society needs still supply of minerals and energy - it demonstrates that we are indeed a part of the global dissipative structure. Therefore we should minimise energy waste and recycle.

Rule 10: Manage the environment as an interconnected system, not as isolated subsystems. This rule is based on principle 4. You cannot close an ecosystem. It has to be open which means that we have to consider inputs and outputs, but outputs from one ecosystem may be inputs for another ecosystem. Pollution problems can therefore not be solved by discharge / emission / dilution, but only by degradation and recycling.

Rule 11: Evaluate available management options simultaneously. Any technological goal can be achieved in different ways, using different options. The options are, however, different in expenditure, quality of product, material needs, waste etc. The options available to solve an environmental problem have to be recognised and evaluated with respect to best satisfying the goal while respecting all the principles and rules presented in this section.

Rule 12: Include secondary effects elsewhere. This rule is based on principle 5 and partly principle 6. It is very difficult to survey the consequences of our activities for ecosystems due to the complexity and openness of these systems and due to the often surprising indirect effects. As wide an evaluation as possible - if possible a global system evaluation - of the consequences of our activities is therefore needed.

Rule 13: Do not exceed the ecosystem homeostatic capacity. This rule is directly derived from principle 6. It is important to know the buffer capacities of focal ecosystems, possibly by use of models to consider the indirect effects and the ability of ecological networks to absorb disturbances.

Rule 14: Consider self-adaptation in management strategies. This rule is directly based on principle 7. The self adaptation ability of natural components in ecosystems should be reflected in our environmental management strategy, The best illustration may be taken from pest control. The use of pesticides is problematic because the organisms become adapted to the pesticide, while biological methods do not violate principle 7. There are unfortunately numerous examples of environmental management where we have omitted to use this rule and therefore failed. To mention a few: the introduction of rabbits to Australia and introduction of the Nile perch to Lake Victoria.

Rule 15: Evaluate the socioeconomic environment. This rule is an acknowledgement of the self-organisation (principle 7) of human society, which is another complex, open, dissipative system with many properties similar to an ecosystem's. As society and ecosystems are both open systems; they are also interchanging matter and energy. The ultimate environmental management should therefore consider both types of system which makes the problem even more complex. It has been attempted to develop ecological-sociological-economic models but this type of model is in its very early infancy.

Rule 16: Evaluate all possible human uses of the environment. The same environmental area or object, like a body of water or a forest, can be used for different purposes. When decisions are made, all of these possible purposes have to be taken into account to find the most environmentally friendly use.

Rule 17: Base measures on ecosystem principles. In other words: use the principles presented in this section and in Section 8.3.

8.5 RESTORATION OF LAKES

It has been pointed out throughout this book that a close to optimum solution of an environmental problem can only be found by quantification of the problem combined with correct selection among the wide spectrum of applicable methods. This principle is even more important to use when the solution requires interference with the ecosystem. Without quantification of the problem, we are unable to measure and compare the effects of the various alternatives. Ecological engineering is obviously more applicable when the pollutants are threshold rather than non-threshold agents, although it has also been applied for non-threshold pollutants. In this section an overview of ecological engineering methods applied to lake restoration will be given. Those who seek a more comprehensive introduction into the application of these methods are referred to Mitsch and Jørgensen (1989). The most important methods are listed below and a brief description of the application, advantages and disadvantages is given.

1. *Diversion* of waste water has been extensively used, often to replace waste-water treatment. Discharge of effluents into an ecosystem which is less susceptible than the one used at present is, as such, a sound principle, which under all circumstances should be considered, but quantification of all the consequences has often been omitted. Diversion might reduce the number of steps in the treatment, but **cannot replace** waste water treatment totally, as discharge of effluents, even to the sea, always should require at least mechanical treatment to eliminate suspended matter. Diversion has often been used with a positive effect when eutrophication of a lake has been the dominant problem. Canalisation, either to the sea or to the lake outlet, has been used as solution in many cases of eutrophication. However, effluents must be considered as a fresh-water resource. If it is discharged into the sea, effluent cannot be recovered; if it is stored in a lake, after sufficient treatment of course, it is still a potential water resource, and it is far cheaper to purify eutrophic lake-water to an acceptable drinking-water standard than to desalinate seawater. Diversion is often the only possibility when a *massive* discharge of effluents goes into a susceptible aquatic ecosystem (a lake, a river, a fjord or a bay). The general trend has been towards the construction of larger and larger waste water plants, but this is quite often an ecologically unsound solution. Even though the waste water has received multi step treatment, it will still have a large amount of pollutants relative to the ecosystem, and the more massive the discharge is at one point, the greater the environmental impact will be. If it is considered that the canalisation is often a significant part of the overall cost of handling waste water, it might often turn out to be a both better and cheaper solution to have smaller treatment units with individual discharge points.

2. *Removal of superficial sediment* can be used to support the recovery process of very eutrophic lakes and of areas contaminated by toxic substances (for instance, harbours). This method can only be applied with great care in small ecosystems. Sediments have a high

concentration of nutrients and many toxic substances, including trace metals. If a waste water treatment scheme is initiated, the storage of nutrients and toxic substances in the sediment might prevent recovery of the ecosystem due to exchange processes between sediment and water. Anaerobic conditions might even accelerate these exchange processes; this is often observed for phosphorus, as iron(III) phosphate reacts with sulphide and forms iron(II)-sulphide by release of phosphate. The amount of pollutants stored in the sediment is often very significant, as it reflects the discharge of untreated waste water for the period prior to the introduction of a treatment scheme. Thus, even though the retention time of the water is moderate, it might still take a very long time for the ecosystem to recover.

The method is, however, costly to implement, and has therefore been limited in use to smaller systems. Maybe the best known case of removal of superficial sediment is Lake Trummen in Sweden. The transparency of the lake was improved considerably, but decreased again due to the phosphorus in overflows from rainwater basins. Probably, treatment of the overflow after the removal of superficial sediment would have given a better result.

3. *Uprooting and removal of macrophytes* has been widely used in streams and also to a certain extent in reservoirs, where macrophytes have caused problems in the turbines. The method can, in principle, be used wherever macrophytes are a significant result of eutrophication. A mass balance should always be set up to evaluate the significance of the method compared with the total nutrient input. A simultaneous removal of nutrients from the effluent should also be considered.

4. *Coverage of sediment by an inert material* is an alternative to removal of superficial sediment. The idea is to prevent the exchange of nutrients (or maybe toxic substances) between sediment and water. The general applicability of the method is limited due to the high costs, even though it might be more moderate in cost than removal of superficial sediment. It has only been used in a few cases and a more general evaluation of the method is still lacking.

5. *Siphoning of hypolimnetic water* is more moderate in cost than methods 2 and 4. It can be used over a longer period and thereby gives a pronounced overall effect. However, the effect is dependent on a significant difference between the nutrient concentrations in the epilimnion and the hypolimnion, which, however, is often the case if the lake or the reservoir has a pronounced thermocline. This implies, on the other hand, that the method will only have an effect during the period of the year when a thermocline is present (in many temperate lakes from May to October/November), but as the hypolimnetic water might have a concentration 5-fold or higher than the epilimnetic water, it might have a significant influence on the nutrients budget to apply the method anyhow.

As the hypolimnetic water is colder and poorer in oxygen, the thermocline will move downwards and the possibility of anaerobic zones will be reduced. This might have an indirect effect on the release of nutrient from the sediment.

If there are lakes or reservoirs downstream, the method cannot be used, as it only removes, but does not solve the problem. A possibility in such cases would be to remove phosphorus from the hypolimnetic water before it is discharged downstream. The low concentration of phosphorus in hypolimnetic water (maybe 0.5 - 1.0 mg l-1) compared with waste- water makes it almost impossible to apply chemical precipitation, see section 6.3. However, it will be feasible to use ion exchange, because the capacity of an ion exchanger is more dependent on the total amount of phosphorus removed and the flow than on the total volume of water treated (see section 6.3). Figure 8.3 illustrates the use of siphoning and ion exchange of hypolimnetic water.

6. *Flocculation* of phosphorus in a lake or reservoir is another alternative. Either aluminium sulphate or iron(III)-chloride can be used. Calcium hydroxide cannot be used, even though it is an excellent precipitant for waste water, as its effect is pH-dependent and a pH of 9.5 or higher is required. The method is not generally recommended as 1) it is not certain that all flocs will settle and thereby incorporate the phosphorus in the sediment, and 2) the phosphorus might be released from the sediment again at a later stage.

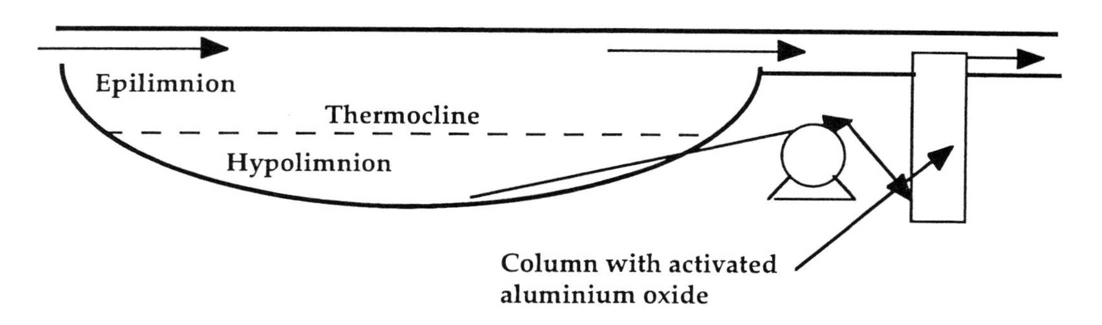

Fig. 8.3 Application of siphoning and ion exchange of hypolimnetic water. The dotted line indicates the thermocline. The hypolimnetic water is treated by activated aluminium oxide to remove phosphorus..

7. *Circulation* of water can be used to break down the thermocline. This might prevent the formation of anaerobic zones, and thereby the release of phosphorus from sediment.

8. *Aeration* of lakes and reservoirs is a more direct method to prevent anaerobic conditions from occurring. Aeration of highly polluted rivers and streams has also been used to avoid anaerobic conditions.

9. *Regulation of hydrology* has been extensively used to prevent floods. Lately, it has also been considered as a workable method to change the ecology of lakes, reservoirs and wetlands. If the retention time in a lake or a reservoir is reduced with the same annual input of nutrients, eutrophication will decrease due to decreased nutrient concentrations. The role of the depth, which can be regulated by use of a dam, is more complex. Increased depth has a positive effect on the reduction of eutrophication, but if the retention time is increased

simultaneously, the overall effect cannot generally be quantified without the use of a model. The productivity of wetlands is highly dependent on the water level, which makes it highly feasible to control a wetland ecosystem by this method.

10. *Fertiliser control* can be used in agriculture and forestry to reduce nutrient loss to the environment. Utilisation of nutrients by plants is dependent on a number of factors (temperature, humidity of soil, soil composition, growth rate of plant (which again is dependent on a number of factors), chemical speciation of nutrients, etc.). Models of all these processes are available today on computers, and in the near future it is foreseen that the fertilisation scheme will be worked out by a computer on the basis of all the above-mentioned information. This will make it feasible to come closer to the optimum fertilisation from an economic-ecological point of view.

11. *Application of wetlands or impoundments as nutrient traps* could be considered as an applicable method, wherever the non-point sources are significant. The use of wetlands has also been applied as a direct waste-water treatment method, for instance in Florida, but it will probably be most effective in dealing with nutrient losses from agricultural areas. Inputs of nutrients into a wetland will to a great extent be denitrified, adsorbed on the sediment or used for growth of algae and macrophytes (phragmites, etc.). Management of a wetland or an impoundment as a nutrient trap obviously requires that a major part of the nutrient input is removed by denitrification (nitrogen only) and stored in sediment and plants. The storage capacity is often large, but of course limited. This implies that nutrients must be removed by harvest of macrophytes, which is quite feasible mechanically. However, if the retention time of a lake or reservoir in the temperate zone is short, it might be favourable to let the in flowing water pass through a wetland or an impoundment.

In the wintertime, when almost no growth takes place in the wetland, the nutrients are washed out, but due to the short retention time, the water with high nutrient concentration will have passed the lake or reservoir before the spring bloom starts. The nutrients of the water which passes the wetland during the spring and summer will, on the other hand, be used to a large extent in the wetland or the impoundment for growth of algae and macrophytes. The water flowing to the lake or reservoir will therefore have a significantly lower nutrient concentration at the time of the year when eutrophication may appear.

The role of wetlands in maintaining an ecological balance is still not fully understood, although our knowledge of the topic is far better today than 10 or 20 years ago. Lately, many large land-reclamation projects have been questioned due to increased experience in the field. Indications show that wetlands are important not only as nutrient traps but also for maintenance of species diversity and for an ecologically sound hydrology in the region.

12. *Calcium hydroxide* is widely used to neutralise low pH-values in streams and lakes in those areas where acidic rain has a significant impact. Sweden is using in the order of 100 mill. $ per year to neutralise acid streams and lakes.

13. *Algicides,* for instance various copper salts, have previously been used widely for relatively small lakes, but are now hardly in use due to the general toxicity of copper which is accumulated in the sediment and thereby can contaminate the lake for a very long time.
14. *Shading by use of trees* at the shore line is a cost effective method which, however, only can give an acceptable result for small lakes due to the low area / circumference ratio.

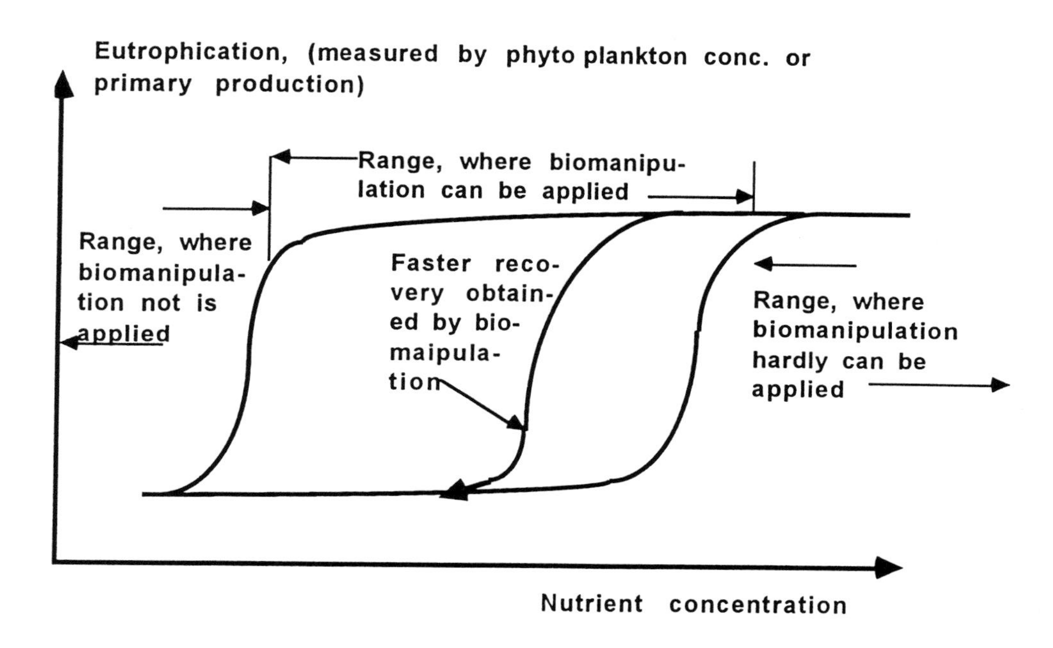

Fig. 8.4 The hysteresis relation between nutrient level and eutrophication measured by the phytoplankton concentration is shown. The possible effect of biomanipulation is shown. An effect of biomanipulation can only be expected in the range approximately 50 -150 μg P /l. Biomanipulation can hardly be applied successfully above 150 μg /l (see also de Bernardi and Giussani, 1995 and Jørgensen and de Bernardi, 1998).

15. *Biomanipulation* can only be used in the phosphorus concentration range from about 50 μg/ l to about 150 μg/ l, dependent on the lake. In this range two ecological structures are possible. This is illustrated in Figure 8.4. When the phosphorus concentration initially is low and increases, zooplankton is able to maintain a relatively low phytoplankton concentration by grazing. Carnivorous fish is also able to maintain a low concentration of planktivorous fish which implies relatively low predation on zooplankton. At a certain phosphorus concentration (about 120 - 150 μg /l), zooplankton is not any longer able to control the phytoplankton concentration by grazing and as the carnivorous fish (for instance Nile perch or pike) is hunting by the sight, and the turbidity increases, the planktivorous fish become more abundant which involves more pronounced predation on zooplankton. In other words the structure changed from control by zooplankton and carnivorous fish to control by phytoplankton and planktivorous fish. When the phosphorus concentration

decreases from a high concentration the ecological structure is initially dominated by phytoplankton and planktivorous fish. This structure can, however, be maintained until the phosphorus concentration is reduced to about 50 μg /l. There are therefore two possible ecological structures in the phosphorus range of approximately 50 - 150 μg/l. Biomanipulation (de Bernardi and Giussani, 1995) can be used in this range - and only in this range - to make a "short cut" by removal of planktivorous fish and release carnivorous fish. If biomanipulation is used above 150 μg P/l, some intermediate improvement of the water quality will usually be observed, but the lake will later get the ecological structure corresponding to the high phosphorus concentration, i.e, a structure controlled by phytoplankton and planktivorous fish. Biomanipulation is a relatively cheap and effective method provided that it is applied in the phosphorus range where two ecological structures are possible. De Bernardi and Giussani, 1995 give a comprehensive presentation of various aspects of biomanipulation. Simultaneously, biomanipulation makes it possible to maintain relatively high biodiversity which does not change the stability of the system, but a higher biodiversity gives the ecosystem a greater ability to meet future, unforeseen changes without changes in the ecosystem function (May, 1977).

From this brief survey of lake restoration methods, it is possible to conclude that an examination of the cost-benefit by application of methods 4, 5, 6 and 9 and perhaps 8 in some cases should be carried out, but of course only in situations where these methods can work hand in hand with methods which are simultaneously reducing the loading.

8.6 AGRICULTURAL WASTE

Characteristics of agricultural waste

Particular animal waste causes great problems in intensive farming. The productions of chickens, pigs and cattle are in many industrialised countries concentrated in rather large units, which implies that the waste from such production units requires hundreds of hectares for suitable distribution and feasible use of its value as fertiliser. Animal waste has a high nitrogen concentration, 5-10% based on dry matter (2-4% dry matter). If the nitrogen is not used as fertiliser it may

1) either evaporate as ammonia
2) be lost to deeper layers, where it contaminates the ground water, or
3) be lost by surface run off to lakes and streams.

Agricultural waste has therefore become a crucial pollution problem in many countries with intensive agriculture.

Treatment methods

Many of the methods described in the previous chapters may be used for treatment of animal waste. The following possibilities give a summary of the available methods, based on ecological engineering or interface between environmental technology and ecotechnology, to reduce the pollution originating from agricultural waste.

1. Storage capacity for animal waste to avoid spreading on bare fields.
2. Green fields in winter to ensure use of the fertilising value of agricultural waste.
3. Composting of agricultural waste assure conditioning before it is used. The composting heat may be utilised.
4. Anaerobic treatment of agricultural waste for production of biogas and conditioning before use as fertiliser. The biogas can be used to heat the farm buildings. It is preferably to mix cattle and swine manure to obtain the highest efficiency with this process, as an excessive free ammonia concentration (more than 1.1 g/l) results in a decreased rate of gas formation (Hansen et al., 1998).
5. Chemical precipitation of agricultural (animal) waste with activated benthonite is applied to bind ammonium and obtain a solid concentration of 6-10%. This process gives 2 advantages:

A. The required storage capacity is reduced by a factor 2-4 (the solid concentration is increased from 2-4% to 6-10%)

B. The loss of nitrogen by evaporation of ammonia is reduced by a factor of 3-8 corresponding to an adsorption of about 60-90% of the ammonium on the added benthonite.

6. In China animal waste is applied in fish ponds. Zooplankton eats the detritus and fish feed on the zooplankton.

8.7 WASTE STABILISATION PONDS (WSPs)

Traditionally waste stabilisation ponds (WSPs) are built as flow-through systems with an anaerobic, a facultative and one or more maturation ponds. The following processes are utilised in the pond system: settling (mainly in the first ponds), anaerobic decomposition of organic matter (mainly in the first ponds), aerobic decomposition of organic matter (mainly in the last ponds, where algae are present and produce oxygen), uptake of phosphorus and nitrogen by algae (facultative and maturation ponds), evaporation of ammonia (mainly where pH is high, i.e. in the last ponds), settling of algae, denitrification (in the anaerobic zones). High removal efficiencies of BOD, COD, microorganisms, nitrogen and phosphorus may be obtained provided that the guidelines for design and maintenance are followed.

The results from several WSPs operating in a tropical climate where this method is mostly applied have been obtained. Figure 8.5 illustrate a regression of the expected removal results with indication of the standard deviation. A first order reaction scheme (Ellis and Rodrigues, 1995) is often assumed in the design of WSPs which is consistent with Figure 8.5. A reasonable removal efficiency is obtained for linear alkyl sulphonates and sulphophenyl carboxylates (Moreno et al., 1994). WSPs may be supplemented by a trickling filter as presented by Meiring and Oellermann (1995), but the use of constructed wetlands seems a more attractive solution in most developing countries where sufficient areas for erection of a wetland are available.

The removal % of E. coli, EC%, and Streptococcus faecalis, SF%, in anaerobic and facultative ponds may be derived from the following empirical equations according to Almasi and Pescod (1996b):

$$EC\% = 95.58 - 0.0021L^2 + 0.0069L*t \quad (8.1)$$

$$SF\% = 95.57 - 0.0014L^2 + 0.0007L*t \quad (8.2)$$

where L and t are defined as g BOD5 /(m^3 24h) and t is temperature in ^oC, respectively.

One of the major maintenance problems is associated with removal of sludge. It is absolutely necessary to remove sludge at least once a year because the sludge reduces the volume of particularly the anaerobic ponds and causes an undesired internal loading. Generally, an increase in the sediment depth of 0.2 cm per month can be expected (Saqqar and Pescod, 1996).

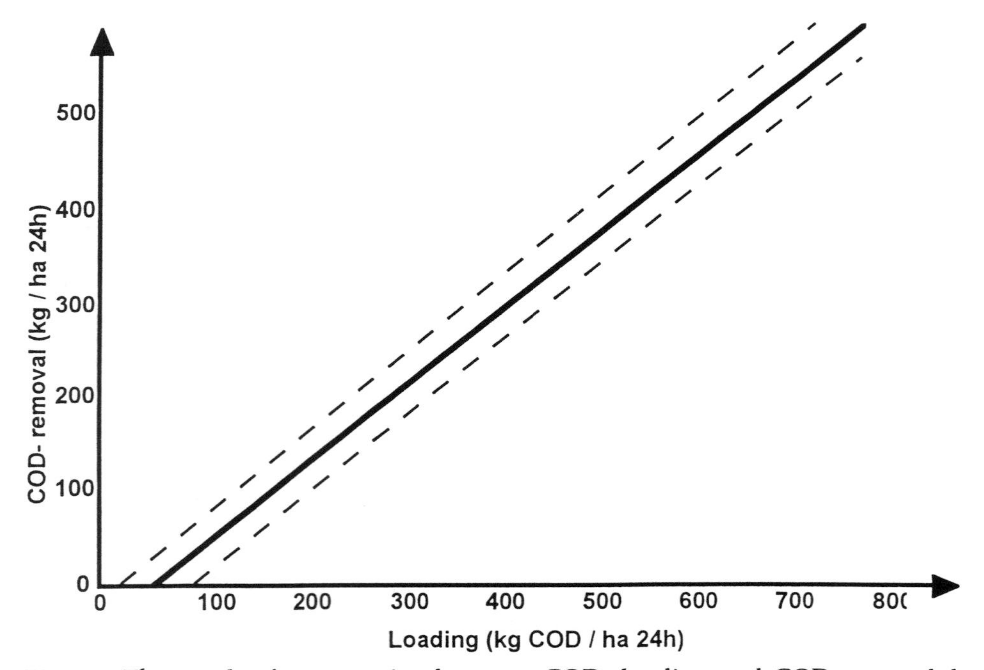

Fig. 8.5 The result of a regression between COD- loading and COD-removal, based upon experience from several tropical waste stabilisation ponds. When the loading (in the unit kg /ha 24h) is known, it is possible to read the removal of COD expressed in kg / ha 24h. As seen the efficiency increases with the loading.

Ammonia volatilisation is the main mechanism of nitrogen removal and the following equation may be used to estimate the removal by this process (derived from Soares et al., 1996 and Middlebrooks et al., 1982:

$$NH_3\text{-inf} / NH_3\text{-eff} = 1 + 0.0388\ (A/Q) \exp(1.585\ (pH - 7.5)) \qquad (8.3)$$

Nitrate if present in waste water will be completely removed in practically all WSPs by denitrification in the anaerobic zones.

Removal of phosphorus by WSPs is in the range of 20-50% but depends highly on the possibility of removing algae from the effluent before discharge. Enhanced removal of phosphorus which is often required for discharge of waste water to lakes and reservoirs can be achieved by addition of precipitants such as calcium compounds and clay minerals. It is recommended to examine the possibilities to use addition of cheap local and natural compounds. Addition of the chemicals to the influent may increase the removal efficiency increases to about 80-95%; Nurdogan and Oswald (1995). Overall high efficiency is, however, best obtained in most cases by the use of a constructed wetland after the WSPs (see next section), preferably with harvest of the wetland 2-4 times per year.

Algal ponds.

The uptake of nutrients by algae can be utilised as a waste water treatment process. Per 100 g of dry matter phytoplankton will, on average, contain 4-10 g nitrogen and 0.5-2.5 g phosphorus. The process is known from the eutrophication problem, see section 2.8. It can be considered a form of biological nutrient removal. Algal ponds are, under favourable conditions be able to produce 100 g phytoplankton per m^3 during 24 hours, which means that as much as 25% of the nitrogen and phosphorus can be removed from solution into phytoplankton (i.e. into suspension) during 24 hours. If a retention time of 3-5 days is applied, as much as 40-60% of the nutrient in municipal waste water will be removed from solution, but it is hardly possible to increase this efficiency as the uptake rate decreases with decreasing concentration as presented in Section 2.3. If chemical precipitation follows the treatment in algal ponds the suspended phytoplankton can easily be removed. The cost of the method is moderate, and the removal of nitrogen especially is often a great advantage. The method is used in the plant in Windhoek (see Fig. 5.22) prior to chemical precipitation with aluminium sulphate. Particularly for the last ponds in the series, waste stabilisation ponds will react as algae ponds.

The method has, however, some pronounced disadvantages that limit its application:

1. Only in tropical or subtropical regions is solar radiation sufficient throughout the year to ensure an acceptable efficiency.
2. The areas needed are large - retention time 3-5 days.
3. The growth of phytoplankton is sensitive to many organic compounds, e.g. mineral oil.

8.8 CONSTRUCTED WETLANDS

The ecotones between lakes and terrestrial ecosystems are crucial for protection of the lake ecosystem against anthropogenic impacts. The transition area has the same function for a lake as the membrane has for a cell: it prevents to a certain extent undesirable components penetrating into the lake. It is therefore crucial to preserve the shore ecotones round a lake and the wetlands in the catchment area, independent of the management strategy implemented. Any man-made construction should be omitted in a zone 50-100 m from the lake shore line to keep the ecotone intact.

Non-point or diffuse pollutants from the environment will inevitably flow toward the lake, but the transition zone is able to transform and/or adsorb the pollutants entirely or partially. It will thereby reduce significantly the overall irreversible effects on the lake ecosystems. The most important processes may be summarised as follows:

1. Nitrate is denitrified by the anaerobic conditions in the wetlands. Organic matter accumulated in the wetland converts nitrate to free dinitrogen (see for instance Ingersoll and Baker, 1998).
2. Clay mineral is able to adsorb ammonium and metal ions.
3. Organic matter is able to adsorb metal ions, pesticides and phosphorous compounds. Metal ions form complexes with humic acids and other polymer organic substances , which reduce the toxicity of these ions significantly.
4. Biodegradable organic matter is decomposed aerobically or anaerobically by the microorganisms in the transition zone.
5. Pathogens are out-competed by the natural microorganisms in the transition zone.
6. Macrophytes are able to take up heavy metals with high efficiency. Other toxic substances may also be removed by macrophytes, but it is not possible to indicate any general rule for the efficiency.
7. Toxic organic compounds will to a certain extent be decomposed by anaerobic processes in wetlands, dependent on the biodegradability of the compounds and the retention time in the wetland.
8. P-soprtion by soil with a high total metal content. Among the four major metal ions (magnesium, calcium, iron and aluminium) calcium has the strongest correlation to P-sorption capacity. High pH implies also increasing sorption capacity to soil with high metal content, Zhu et al., 1997. This relationship between high calcium content and high pH on the one side and high P- sorption capacity on the other side may be utilised in construction of artificial wetlands. It is often beneficial to transport a soil with high P-sorption capacity to the wetland under construction. It may increase the P-removal capacity by more than one magnitude. As much as 1-3 g P/kg soil can be achieved by a 200-600 g of total metal/kg soil.

The denitrification potential of wetlands is often surprisingly high. As much as 2-3000 kg of nitrate-nitrogen can be denitrified per ha of wetland per year, dependent on the hydraulic conditions (see Jørgensen, 1994 and Jørgensen et al., 1995b). It is of great importance for the protection of lakes, as significant amounts of nitrate are released from agriculture. As much as 100 kg nitrate-N / ha may be found in the drainage water from intensive agriculture. As the denitrification is accompanied by a stoichiometric oxidation of organic matter, significant amounts of organic matter are also removed by this process. The phosphorus bound as organic matter or adsorbed to the organic matter may, however, be released by these processes (Jørgensen et al., 1995b). These processes should be examined carefully and quantitatively in each case, including whether the released phosphorus will flow towards the lake or towards the ground water. These possibilities should be included in the development of management strategies.

The adsorption capacity of the transition zone offers significant protection against pollution by toxic substances, both heavy metals and toxic organic substances (primarily pesticides originating from agriculture). The ratio of the concentration of heavy metals or pesticides in organic matter to the concentration in water at equilibrium is strongly dependent on the composition of the organic matter, but is usually between 50 and 5000, which indicates that the transition zone has an enormous binding capacity for these pollutants.

Ecotones serve as a buffer zone not only for pollutants, but also for the species present in the adjacent ecosystems. Preservation of wetlands at the lake shore line may therefore be crucial for maintenance of the biodiversity in the lake ecosystem - a function, which the manager must not overlook in development of an appropriate lake management strategy.

Table 8.2 gives an overview of the different types of wetlands, that are found adjacent to lakes: wet meadows, forested wetlands, marshes, bogs and shore wetlands. The table gives the characteristics of the seven types of wetlands and their different ability to cope with the non-point pollution problems.

The importance of wetlands adjacent to lake ecosystems has caused drainage of wetlands to cease in many countries, and the restoration of previously drained wetlands (Jørgensen and Löffler, 1990). The construction of artificial wetlands offers a solution to diffuse pollution originating from agriculture, septic tanks and other sources (examples are given in Mitsch 1993, 1996 and 1998, Mitsch and Gosselink, 1993 and Jørgensen et al., 1994). U.S. legislation it does not allow wetlands to be drained, unless another wetland of the same size is installed somewhere else.

Construction of artificial wetlands is an attractive and cost-effective solution to pollution by diffuse sources and even waste water. Wetlands are first of all able to cope with the nitrogen and heavy metal pollution from these sources, but it is essential to make a proper planning on where to place the artificial wetlands, as their effects are dependent on the hydrology (they should be covered by water most of the year and should have sufficient retention time to allow them to solve the key pollution problems) and on the landscape pattern (they should protect the most vulnerable ecosystems which often are lakes and reservoirs). It is furthermore important to ensure that the wetlands are not releasing other components such as phosphorus as mentioned above.

The following emergent types of plants are proposed for use in constructed wetlands: cattails, bulrush, reeds, rushes. papyrus and sedges. Submerges species can be applied in deep water zones. Species that have been used for this purpose include coon tail or horn wart, redhead grass, widgeon grass, wild celery and watermil foil

It should also be kept in mind that a wetland in most cases will reduce the water budget due to evapotranspiration (Mitsch and Gosselink, 1993). Wetlands, however, reduce the wind speed at the water surface and may thereby also reduce the evaporation. It is important to consider these factors in the planning of artificial wetlands. Finally, it should not be forgotten that an artificial wetland is not fully developed overnight. In most cases it will require 2-4 years for an artificial wetland to obtain sufficient plant coverage and biodiversity to be fully operational. It is, however, clear from the experience gained in the relatively few constructed wetlands that the application of models for the wetland, encompassing all the processes reviewed above, as well as for the lake is mandatory, if positive results are to be expected.

Wetlands encompassing the so-called root zone plants may also be utilised as waste water treatment facilities. This application of soft technology seems particularly advantageous for developing countries due to its moderate cost.

The self-purification ability of wetlands has found wide application as a waste water treatment method in several developing countries (China, Philippines, Burma, India and Thailand). Generally, the constructed wetlands in the tropic regions show higher efficiency than in the temperate regions due to higher photosynthetic and microbiological activity. In addition, the seasonal variations of the removal capacity of constructed wetlands are in the tropic regions much smaller than in the temperate regions. Constructed wetlands in the northern part of Europe have in the winter period (December - February) a capacity expressed for instance as kg N or kg COD removed / ha 24h which may be less than 30% of the summer capacity. This implies that constructed wetlands in the temperate zone have to be constructed with an overcapacity relative to the average situation to be able to meet the required efficiency in the winter period. These differences between constructed wetlands in tropic regions and in temperate regions explain why it is usually more expensive to use constructed wetlands in the temperate zone.

Much attention has for instance been devoted to the stocking of fish cultivated in biological sewage stabilisation ponds. These studies show that nutrients such as nitrogen and phosphorus are retrieved from the sewage by fish, through the intermediate activities of bacteria, algae and other types of plankton. In order to mobilise the self-purification of a body of water for sewage treatment, the main principle is to utilise coordination among four chief components:

1) producers, aquatic vegetation,
2) consumers including fish,
3) decomposers (bacteria mainly) and

4) abiotic factors such as solar radiation and water exchange.

Bacteria decompose organic materials, including organo-phosphates and purify the water in the preliminary phase. Carbon dioxide, ammonia, phosphates are produced and utilised by algae and other aquatic vegetation. The oxygen released by photosynthesis is supplied for the needs of oxidising organic matter by bacteria. A great quantity of algae and other types of plankton are utilised by rearing fries. The energy fixed by algae appears finally in the form of fish harvesting and the water is thereby further purified.

Waste water treatment using floating species has also been proposed. Different types of duck weed and water hyacinths (*Eichhornia crassipes*) have been applied as an alternative to waste stabilisation ponds. Use of water hyacinths requires, however, strict control, as they easily become widely spread as a weed which may get completely out of control.

The inorganic nitrogen and phosphorus brought in by sewage and decomposed from organic pollutants by microorganisms are absorbed by water hyacinths.

Table 8.2

Characteristic of Wetlands adjacent to Lakes (Patten et al., 1990)
(Classification of wetlands varies according to authors and regions.)

Type of wetland	Characteristics	Ability to retain non-point pollutants
wet meadows	grassland with waterlogged soil. Standing water for a part of the year.	denitrification only in standing water. Removal of N and P by harvest
fresh water	marshes, reed-grass dominated, often with peat accumulation	high potential for denitrification, which is limited by the hydraulic conductivity
forested wetlands	dominated by trees shrubs. Standing water not always for the entire year	high potential for denitrification and accumulation of pollutants, provided that standing water is present
salt water marshes	herbaceous vegetation usually with mineral soil	medium potential for denitrification. Harvest possible.
bogs	a peat-accumulating wetland with minor flows	high potential for denitrification but limited by small hydraulic conductivity
shore wetlands	littoral vegetation, often of great importance for the lake	high potential for denitrification and accumulation of pollutants

Experiments in China show that an average yield of up to about 10 kg $/m^2$ may be obtained during the growing period from May to November. Such production is able to absorb 1500-3600 kg of nitrogen, 150-500 kg of phosphorus and 100-250 kg of sulphur per ha (see Greenway, 1997). The water hyacinths with microorganisms and organic pollutants attached or coagulated on root surfaces are harvested to serve as feed in fish culture ponds, duck farms, pig farms and oxen farms. From May to November the concentrations of COD, total nitrogen, ammonium, total phosphorus and ortho phosphate were less than half the concentrations in the inlet. On the other side, when the water hyacinths were absent from December to April the differences in concentrations of COD, nitrogen and phosphorus between the inlet and outlet were very small. In tropical areas it may be possible to obtain the removal efficiencies indicated above for a wetland with floating plants. It can recommended to estimate the nitrogen removal from the following equation for aquatic treatment systems based on water hyacinths in tropical regions:

$$N / No = \exp(-k\,t) \qquad (8.4)$$

where N is the nitrogen concentration in the effluent and No in the in fluent. k is the rate constant which is 0.2-0.6 1/24h dependent on the plant density (0.2 at 3 t/ha, 0.5 at 10 t/ ha and 0.6 at 20 t/ ha) and t is the retention time in days. Equation (8.4) assumes harvest of the plants on at least an annual basis. The BOD5 removal can be estimated at 50 kg - 250 kg / ha, dependent on the plant density (from 3 - 20 t/ha; see Wolverton, 1982). The phosphorus removal is about 5-10 times lower than the nitrogen removal, based on at least an annual harvest.

Many heavy metals are concentrated and accumulated in water hyacinths from very low concentrations in water, but the heavy metal enrichment in this plant varies with its aquatic habitat. Obviously, water hyacinths with high residual amounts of heavy metals cannot be used as fodder, which limited the application of this ecological engineering approach for treatment of water polluted by organic pollutants (mainly municipal waste water).

The root zone plant has found its application for treatment of small volumes of municipal waste water in industrialised countries, particularly, where construction of a sewage system to an adjacent waste water treatment plant would be prohibitively expensive. The decomposition of organic matter and denitrification usually do not cause any problem, provided that the plant is 5-10 m^2 per person equivalent, dependent on the climatic conditions. Phosphorus is mostly not removed by an efficiency of more than 10-20% by a root zone plant, but with addition of iron chloride the efficiency may be increased to 80% or more due to precipitation of iron phosphate. The application of this method seems attractive to recreational areas, where the density of population is low, but where the loading from the waste water has a significant impact.

Constructed wetlands or root zone plants are designed for BOD-removal in accordance with the following equations (Jørgensen and Johnsen, 1989):

$$C / Co = \exp(- K^*V / Q), \quad (8.5)$$

$$Q_{w+p} = b^*d^*HC^* i \quad (8.6)$$

Where C and Co the BOD5 concentration or COD concentration of the inflowing water and the treated water respectively, while K is a rate constant (1/24h), V is the volume (m^3) of the plant (depth 0.4- 1.0 m), Q is the waste flow expressed in m^3 / 24h. Q_{w+p} is the flow rate of waste water plus rain often estimated to be 3-4 times Q, b is the width of the plant, d is the depth (0.5-1.0 m) , HC is the hydraulic conductivity of the soil and i is the slope in cm/ m. K is estimated to be 0.2-0.5 for root zone plants and constructed wetlands at approximately 20°C. K at temperature t can be found from the following equation:

$$Kt = K20 * 1.06^{(t-20)} \quad (8.7)$$

Example 8.1
Design a root zone plant for the treatment of the waste water from 1000 inhabitants (water consumption 0.2 m^3 / 24h). The BOD_5 of the untreated water is 200 mg/l and it must be reduced to 20 mg/l by the planned treatment. 20° C is presumed. The plot to be applied for construction of the root zone plant has a slope of 5 cm/m and the soil has a hydraulic conductivity of 4 m/ 24h.

Solution
K = the average = 0.35 (range 0.2 - 0.5). It is presumed that 0.5 m depth can be applied. By use of equation (8.5), we get:

$$20 / 200 = \exp(- 0.35^* V / 200)$$

This gives V = 1315. We choose an area of 2800 m^2 (0.5 m depth).
Equation (8.6) is used to control the hydraulic loading:

$$800 = 4x200 = bx0.5 x4x5$$

This leads to b = 80 m. The length is therefore 35 m, which seems to be a reasonable dimension.

The recommended management of natural or artificial wetlands and of the lake shore ecotones may be summarised in the following points:

1. Maintenance and preservation of the transition zone between the terrestrial ecosystems and the lake ecosystem should be considered mandatory. The ecotone between the two types of ecosystems functions as a buffer zone for preservation of species diversity (May, 1977) in both ecosystems.
2. The different types of wetlands forming the transition zone have a high adsorption capacity for many pollutants such as heavy metals and toxic organic compounds.
3. The different types of wetlands forming the transition zone are able to denitrify up to several tons of nitrate-nitrogen per ha and year.
4. Recovery and construction of artificial wetlands are crucial abatement methods for diffuse pollutants originating from agriculture.
5. Application of artificial wetlands, including waste stabilisation ponds and root zone plants may be an attractive waste water treatment method for developing countries, for recreational areas adjacent a lake ecosystems and generally for areas with a low population density.
6. The application of quantitatively based management of all the mentioned types of wetlands, including development of models for these ecosystems, is recommended.

The combination of WSPs and constructed or reclaimed wetlands are particularly attractive (Gschlöbl et al., 1998), because:

- wetlands offer by use of gravel media a significant reduction of suspended matter from maturation pond effluent.
- wetlands buffer the pH value of the effluent from WSPs
- effluents from WSPs need most often a polishing step as a post treatment. wetlands offer an excellent and cost effective solution to this problem.
- the two steps, WSPs and wetlands, are to a great extent independent, and it therefore provides a much greater certainty to play on both steps.

Constructed wetlands may also be used to treat dairy farm waste waters, mine water pollutants, textile waste water and pulp mill waste water.

8.9 IN SITU REMOVAL OF CONTAMINANTS BY PHYTO-REMEDIATION

We have already touched on the possibilities of recovering soil in situ in Chapter 6. Use of aeration to enhance the decomposition of organic contaminants is a technology which is on the interface between environmental technology and ecotechnology, but it was chosen to treat it under the heading of environmental technology, while the use of plants to remove contaminants from soil clearly must be considered ecotechnology and is therefore presented in this chapter.

Recently, it has been shown that a few plant species (for instance the alpine pennycress) can be used as **bioaccumulators**, (hyperaccumulators). When they are growing in contaminated soil, they are able to achieve a many times higher (10 - 100 times) heavy metal concentration than normal plant species. If the plants are harvested, the corresponding amount of heavy metal is removed from the soil. If the soil is only slightly contaminated with heavy metal, this method may be able to remediate the soil in a few years which makes the method very attractive due to its low costs. For very contaminated soil, the number of years needed for complete recovery of the soil will probably be too high. By watering the plants carefully with a diluted EDTA solution, it is possible to enhance the uptake of heavy metals in the plants. Jørgensen (1993) reports a removal of 11.5 % lead per harvest of plants watered by 0.02 M EDTA solution at an initial concentration of 380 mg Pb/kg soil (dry matter). Several plants species release chelating ligands and enzymes into the soil and thereby accelerate the removal rate. Chelating agents will also reduce the toxicity of heavy metal ions; see Section 4.4..

The economics of the various alternatives offered by this method determines the practical applicability of the method.

Hyperaccumulators are usually plants that are slow to grow and therefore slow to remove heavy metals as mg/ m^2 year. Fast growing poplar trees are not such good hyperaccumulators as many other species, but due to their fast growth they are promising phytoremidators.

Aquatic plants, *Salvinia* and *Spriodela,* have been used to remove chromium and nickel from waste water. A significant removal efficiency in the concentration range 1- 8 ppm, although with fluctuations, has been reported by Srivastav et al., (1994). Fungal biomass has recently been proposed for removal of heavy metals. The amounts of metal adsorbed per unit weight are several mg/g, for cadmium, copper, lead and nickel (Kapoor and Viraraghavan, 1998).

Plant can efficiently take up organic substances that are moderately hydrophobic, with a K_{OW} from about 0.5- 3.0. More hydrophobic substances exceeding a K_{OW} of 3.0 approximately bind so strongly to the soil and the roots that they are not easily taken up within the plants.

8. 10 QUESTIONS AND PROBLEMS

1. A wetland has a hydraulic conductivity of 0.25 m /24h. It is presumed that the upper 1 m layer can be applied for denitrification of 2000 m^3 of agricultural drainage water which contains 12 mg/l nitrogen compounds. How much wetland would you recommend to use

a removal of 90% of the nitrogen, when it can be anticipated that the plant density is 20 t/ ha? What is the expected BOD5 removal?

2. Design a root zone plant for the treatment of the waste water from 500 inhabitants (water consumption 200 l/24h). The BOD5 of the untreated water is 250 mg/l and it must be reduced to 10 mg/l by the planned treatment. 15° C is presumed. The plot to be applied for construction of the root zone plant has a slope of 6 cm/m and the soil has a hydraulic conductivity of 2.8 m/ 24h. It has been recommended to use a depth of 0.8 m.

3. Which of the 12 ecological principles and the 17 rules given in the text are violated in conventional agriculture? In organic agriculture? In integrated agriculture?

4. How many ha waste stabilisation ponds are needed for the reduction of COD from 700 mg/l to 100 mg/l in domestic waste water mixed with industrial waste water. The volume to be treated is 1000 m^3 / 24 h.

9. Environmental Management by Changes in Products and Production Methods

9.1 INTRODUCTION

All human activities cause an impact on the environment. Particularly agriculture, industry and their products have caused many pollution problems. We can, in principle, solve these problems by environmental technology for point sources and ecotechnology for non-point sources. This, however, obviously raises also the question: couldn't we change the production or the product in such a way that we would be able to reduce the impact on the environment? Such a change in production for instance by recycling inside the production unit may even be much cheaper to realise than the treatment of emissions. It may even give a cost reduction as recycling means less use of resources. There are numerous examples of simultaneous reductions in costs and environmental impact due to changes in a production method.

Figure 1.3 shows how the various technological approaches can work hand in hand with the basic idea to find the optimum solution by playing on the entire spectrum of available methods. It is difficult to find the optimum solution to an environmental problem, because of the many possible solutions including those based on combinations of methods, which often in practice give the optimum solution. In many industries it has at least been close to the optimum solution to enhance recycling, reduce water consumption, introduce many minor changes in production which would together reduce emissions, combined with some treatment of waste water and smoke.

Improved environmental management can be achieved by mainly four initiatives:

1) Introduction of an environmental management system where the first step would be to examine all the production steps from an environmental view point: couldn't we reduce the emission here and there? How do we do it? This step often leads to environmental certification of the industry which may be utilised in negotiation with customers. Many global and regional firms require that their sub-suppliers have environmental certification in order to ensure that their final product is "green". The following ISO standards support the enterprises in their effort to improve their green image and reduce emissions by internal changes in production: ISO 14001-14004 and partly ISO 140014. This is the topic of the next section.

2) Environmental auditing requires that the enterprise set up a mass balance for all components. The quantities of consumed raw materials, produced products, and material lost as pollutants to the environment at various points of production must be indicated. This was described in Section 2.11 as "green accounting", but may also be denoted as environmental auditing. As the principles have been presented in Section 2.11, they will not be repeated here, but we shall discuss how the results can be used in environmental management context. The ISO standards 14010-14013 give guidance for environmental auditing. This the topic of Section 9.3.

3) Cleaner technology covers the actual changes in the production frequently resulting from 1 and 2. It is an attempt to give a full answer to the pertinent question: how can we produce our products by a more environmentally friendly method? Which changes in the production should be made? The ISO standards mentioned under 1 and 2 are supplied in ISO 14014, 14015 and 14031. Section 9.4 is devoted to cleaner technology.

4) Not only the production but also the product may be changed. Environmental risk assessment which was introduced in Sections 4.1-4.2 may be used to change from a product with a high environmental risk to a product with a low environmental risk. A life cycle analysis, LCA, examines all the emissions associated with the use of the product from the cradle to the grave. The idea is to examine the possibilities for reductions in emissions by minor or major changes in the product, including possibilities of recycling the final discarded waste product. This topic will be discussed further in Section 9.5.

9.2 ENVIRONMENTAL MANAGEMENT SYSTEMS

The control of the impact of industrial activities on health and the environment has undergone a significant transition since the late seventies and early eighties in the EU, North America and Japan. Since the mid eighties industries have taken a more proactive stance in recognising that sound environmental management on a voluntary basis can enhance corporate image, increase profits, reduce costs and obviate the need for further environmental legislation in the area. The commercial values of green products have been fully acknowledged. In some countries, the state has even set up what may be named a green purchase policy which is followed not only by the state but also by counties and communities. It is therefore not surprising that there has been an enormous interest among many enterprises

Introduction of an environmental management system can help companies to approach environmental issues systematically in order to integrate environmental care as a normal part of their operation and business strategy. The major triggers for the companies to initiate such environmental changes are rooted in legislation, in pressure from the stake-holders, in the importance of image and reputation, in the competition (environmental certification is required for sub-suppliers and the commercial value of a green product) and in economies (sales potential, recycling means lower consumption of resources, green taxes).

ISO 14001 uses the following definition of the term environmental management system, EMS: that part of the overall management system which includes organisational structure, planning activities, responsibility, practices, procedures, processes and resources for developing, implementing, achieving, reviewing and maintaining the environmental policy.

EMS is intended to:

- identify and control the environmental aspects, impacts and risks relevant to the organisation
- develop and achieve its environmental policy, objectives and targets, including compliance with environmental legislation
- establish short-, medium- and long-term goals for environmental performance, ensuring a balance of costs and benefits, for the organisation and its stakeholders
- determine what resources are needed to achieve these goals, assign responsibility for them and commit the necessary resources
- define and document specific tasks, responsibilities, authorities and procedures to ensure that every employee acts in the course of their daily work to help to achieve the defined goals
- communicate these throughout the organisation and train people to effectively fulfil their responsibilities effectively.
- measure performance against pre-agreed standards and goals, and modify the approach if necessary

EMS has benefited mainly from two developments:

A. The rising cost of environmental liabilities led companies to develop environmental auditing as a management tool.

B. Total quality management concepts were adopted to reduce defects = noncompliance with specifications.

The development of EMS may be described in several phases (Dyndgaard and Kryger, 1999) :

1. Set up an environmental policy, expressing the commitments of the enterprise to specific environmental goals. These are most clear if they are quantified for instance to reduce the emission of a specific pollutant by 90%.
2. An environmental program of action plan describing the measures to be taken over a given period of time. The action plan should identify the steps to be taken by each department to reduce emissions,. commit the necessary resources and provide monitoring and coordination of progress towards the achievement of the defined goals.
3. Organisational structures delegating authority and assigning responsibility for actions.
4. Integration of environmental management into business operations and internal communication including risk assessment to identify potential accidents.

5. Monitoring and measurements and record keeping procedures to document and monitor the results as well as the overall effects of environmental improvements.
6. Corrective actions to eliminate potential nonconformances to objectives, targets and specifications.
7. EMS audits to check the adequacy of efficiency of the implementation.
8. Management reviews to assess the status and adequacy of the EMS in light of changing circumstances.
9. Internal information and training to ensure that all employees understand to fulfil their environmental responsibility, and why they should.
10. External communications and community relations to communicate the enterprise's environmental goals and performance to interested persons outside the enterprise to obtain the advantage of a green image.

All ten steps are covered in detail by ISO 14001 and are consistent with the quality management in ISO 9000 which discusses four phases: the planning phase, action phase, evaluation phase and correction phase.

Performance measures are needed in EMS to provide indication of how well the goals have been reached and to serve as an incentive for the employers as their efforts are recognised and appreciated through the results. The reward to the enterprise can be substantial because even a limited EMS can often demonstrate that the economic gains can outweigh the costs and that inexpensive measures may yield important environmental and financial returns.

Pedersen and Nielsen (1999) have examined the effect of EMS introduced into 18 different enterprises including several industries with high pollution potential (manufacturers of dye stuff, cables, enzymes and printed matter). They summarise the results of the examination in the following list of advantages including indirect advantages gained by introduction of EMS:

1. A direct measurable reduction of emissions with all the benefits that may entail
2. A general increased engagement by all employees not only with EMS but with all the enterprise's areas of activities.
3. Improved (green) image.
4. The economic gains outweigh all environmental investments and associated costs from within a few months to a maximum 3.8 years.
5. In several cases new orders have been obtained directly due to the introduction of EMS.
6. Internal reductions in noise, dust and odours were obtained.
7. Optimisation of processes and better utilisation of resources.
8. Better cooperation with environmental agencies.

9.3 ENVIRONMENTAL AUDIT

Green or environmental audit covers a mass and energy balance of the raw materials and energy used in a production unit. Section 2.11 has presented the concept and a specific example, namely the nitrogen balance in Denmark, was used to illustrate the principles. Through a quantification of the mass and energy flows in a production unit, often presented in the form of one or more flow charts, it becomes possible to identify how much materials and energy are lost and where this takes place. These results enable us to answer the following pertinent questions:

- Could the losses (= emissions) be reduced?
- Could the loss of energy eventually be recycled?
- How would that influence the mass and energy balances and the quality of the product?
- How could the mass and energy flows be optimised with relation to the total costs, including energy, raw material, environmental, labour and miscellaneous costs

Mass and energy balances are in this context called environmental audits, as they are comparable with accounts of income (useful outputs), expenses (inputs), profit (value of useful outputs - inputs), loss (non-useful outputs). An account is settled by looking at the money flow in the same manner, using the principles of conservation: the input - the output corresponds to the accumulation in the firm = an increased asset. This doesn't entail that the environmental audit is made by an auditor which has been the tendency. Mass balances, particularly in the chemical industry, are often based on stoichiometric calculations which requires a good knowledge of chemistry and the involved chemical processes. Similarly, energy balances can hardly be set up without use of thermodynamics and energy transfer processes.

Sørensen et al., 1999, have examined more than 500 environmental audits. They conclude that most enterprises give sufficient information about consumption of energy, water and raw materials, but often the information about emissions, noise, dust and odours is not sufficiently detailed. The environmental issues are, however, in most environmental audits covered at least partially, while 8-10% of the audits give very incomplete information about the environmental impact. This examination has also uncovered the economy behind the use of green audit. 1200 environmental audits have costed approximately 4 mill. $ or 3300 $ on average. About 50% of the enterprises show gains of about the same order of magnitudes, while 15% of the enterprises can show gain on average 12,000 $ from the introduction of green audit. 3% of the enterprises even have a profit of 35,000 $ or more from the results of environmental audit.

It is often surprising how little enterprises know about their losses (emissions to the environment) before they introduced green audit. Most enterprises are focusing almost entirely on the two Ps: Product (quality and price) and Profit. By introduction of green audit many enterprises have, however, acknowledged they can increase profit by reduction of the losses (emissions) and decrease the price and increase the quality of the product from a commercial point of view by developing a green image. A few examples are presented below to illustrate how green audit has been able to elucidate the value of waste.

A well managed farm (Skaarup and Sørensen, 1994) bought waste proteins used as animal feed from a fermentation industry. The manure was utilised as natural fertiliser. The composition of the waste proteins was, however, not the optimum for animal feed. The potassium content was too high. The manure therefore also had too high a potassium content which meant that the fertiliser effect of the potassium content was not utilised, but was emitted to the environment. By an adjustment of the composition of the animal feed obtained mainly by use of a mixture of soya beans, barley and the waste proteins, it was possible to get an optimum composition which brought the manure closer to a fertilisation optimum. The result of the changes was that the farmer reduced the emission to the environment caused by the manure and could simultaneously increase his profit.

Fibrous material is lost to the waste water in paper and pulp production. During the seventies it was realised on the basis of mass balance that a significant amount of fibrous material was being lost to the environment. It is, however, possible using chemical precipitation with benthonite and a polyflocculant as precipitant to collect the fibres as a sludge (5-10% dry matter) which can be used to produce lower quality paper.

Mass balances have also revealed 25-40 years ago that the amounts of fish oil lost by fish filleting plants and the amounts of chromium lost by tanneries were very significant particularly from an economic point of view. Today, a fish filleting plant which is not recovering fish oil or a tannery which is not recycling chromium cannot survive the tight competition.

9.4 CLEANER PRODUCTION

Cleaner production is in accordance with UNEP defined as the continuous application of an integrated preventative environmental strategy to processes, products and services so as to increase efficiency and reduce the risks to humans and the environment. The main emphasis is on manufacturing processes and products. Cleaner production therefore includes the efficient use of raw materials and energy, the elimination of toxic and dangerous materials and the reduction of emissions and wastes at the source. In addition, it focuses on reduction of the impacts along the entire life cycle of the products and services - from design to use and to ultimate disposal (for further details; see Section 9.5).

Table 9.1 shows a comparison of pollution control based mainly on environmental technology and cleaner production based on all the approaches presented in this chapter, EMS, environmental audit, changes of processes and technologies applied in production and LCA.

Cleaner production is especially important to developing countries (it also explains UNEP interest in this topic), because it provides industries in these countries for the first time with an opportunity to bypass the more established industries which are still stocked with the costly environmental technology to solve the problems at the end of the pipeline. Table 9.2 mainly based on Kryger and Dyndgaard (1999), but partially on the examples given above in the context of environmental audit and UNEP, (1994), gives several examples of how sometimes relatively simple changes or introduction of low cost equipment can provide cleaner production, with reduced emissions and lower costs.

Table 9.1.

Comparison of pollution control approaches and cleaner production approaches

Pollution control approaches	Cleaner production approaches
Pollutants are controlled with filters and waste water treatment methods	Pollutants are prevented at their sources through integrated measures
Pollution control is evaluated when processes and products have been developed and when the problems arise	Pollution prevention is an integrated part of product and process development
Pollution control is considered to be a cost factor	Pollutants and wastes are considered to be potential resources
Environmental challenges are to be adressed by environmental experts	Environmental challenges are the responsibility of people throughout the company
Environmental improvements are to be accomplished with technology	Environmental improvements include non-technical and technical approaches
Environmental improvements should full fill authorised standards	Environmental improvement measures involve a process of working continuously to achieve higher standards
Quality is defined as meeting the customer's requirements	Quality means the production of products that meet the customer's needs and have minimal impacts upon human health and the environment

Kryger and Dyndgaard (1999) attempt to review the barriers to introduction of cleaner production. They are in short: bureaucratic resistance, conservatism, uncoordinated legislation, misinformation, scarce money and lack of technical reliable information. (Huising (1995).

Table 9.2

Examples of Cleaner Production

Industry	Country	Method	Waste and emission reduction	Pay back period
Fish filleting	Denmark	Removal of fish oil from waste water	75% reduction of BOD5	3 months
Tannery	Denmark	Recycling of chromium	95% reduction of chromium in waste water	< 1 year
Paper	Switzerland	Precipitation of fibres	70% reduction of BOD5	2-3 years
Fruit juice	The Philippines	Collection of fruit drops	55 l organic waste / h	9 months
Wood finishing	Malaysia	Waste segregation	54,000 kg of hazardous waste + 5700 m^3 w.w.	3 months
Lead oxide	India	New insulation in furnace	fuel consumption reduced 50%, increase production 3%	3 months
Cement	Indonesia	Control system to ensure optimum temperature	3% reduced energy use, 9% higher production, reduced emission of NO_x	< 1 year

Introduction of cleaner production often follows the procedure presented in Figure 9.1. An option generating process is often used extensively from the planning stage to the implementation. It is often beneficial to consider the following elements in this process (Kryger and Dyndgaard, 1999):

1. **Change in raw materials** either by changing to another (more pure) grade of the same raw material, eventually by purification of the raw material or by substituting the present raw material with a less hazardous material.

Example: detergents containing polyphosphates are replaced with detergents based on zeolites

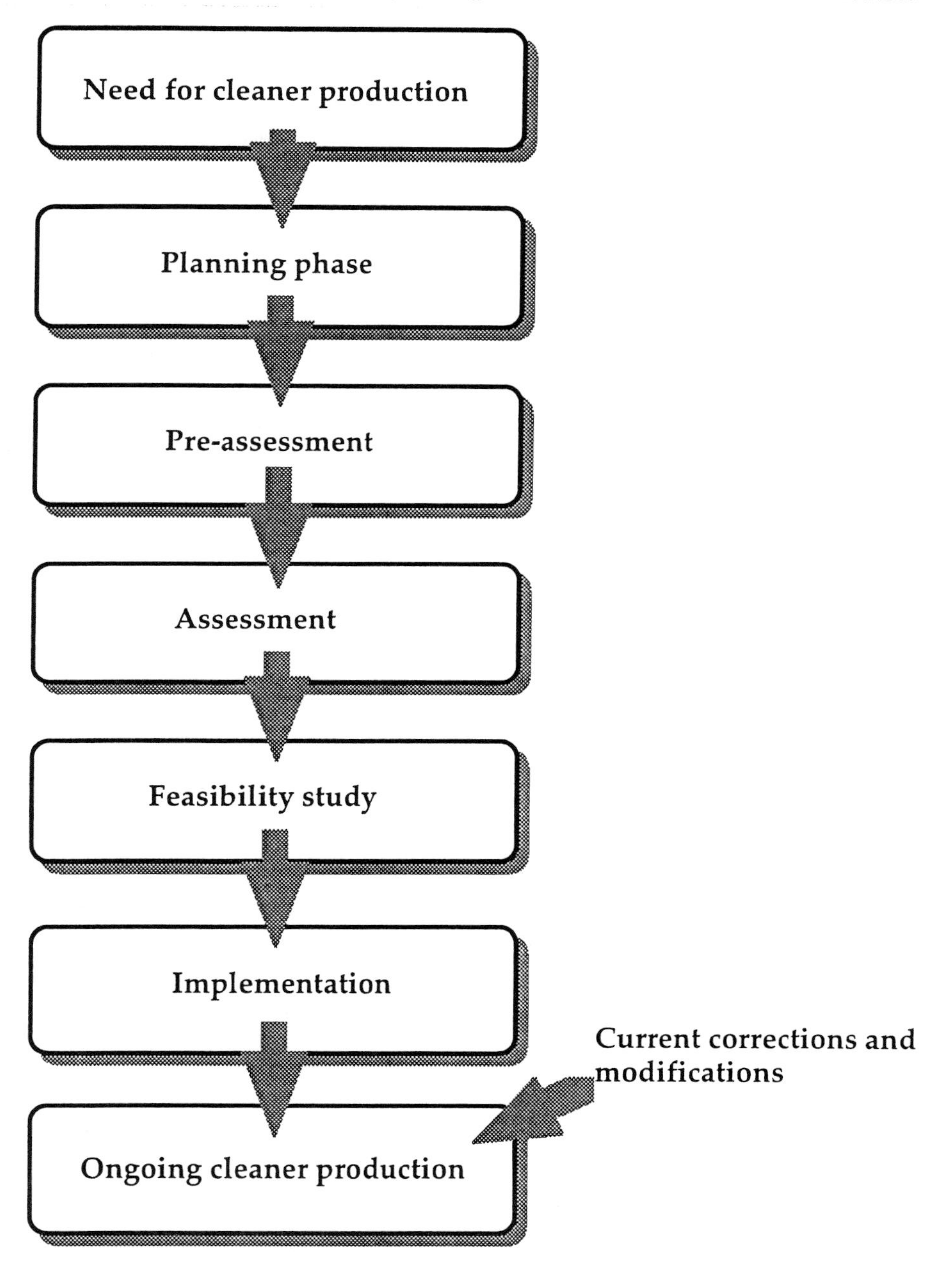

Figure 9.1. Introduction of cleaner production by a seven step procedure.

2. **Technological changes** covering changes in the production processes, modification of equipment, use of automation and changes of process conditions (flow rates, temperatures, pressures, residence times and so on). The aim with these changes is to reduce the emissions and / or improve the quality of the product.

Example: use of mechanical cleaning to replace cleaning by chemicals.

3. **Good operating practices** can often be implemented with little cost. They include management and personnel practices, material handling and inventory practices, training of employees, loss prevention, waste segregation, cost accounting practice and production scheduling. Example: reduce loss due to leaks.

4. **Product changes** includes changes in quality standards, product composition, and product durability or product substitution..

Example: replace CFCs with ammonia in refrigerators

5. **Reuse and recycling** involves the return of a waste material to the originating or another process to substitute input material.

Examples: recover nickel plating solution with an ion exchanger, recover dye stuffs from waste water in the textile industry by ultrafiltration.

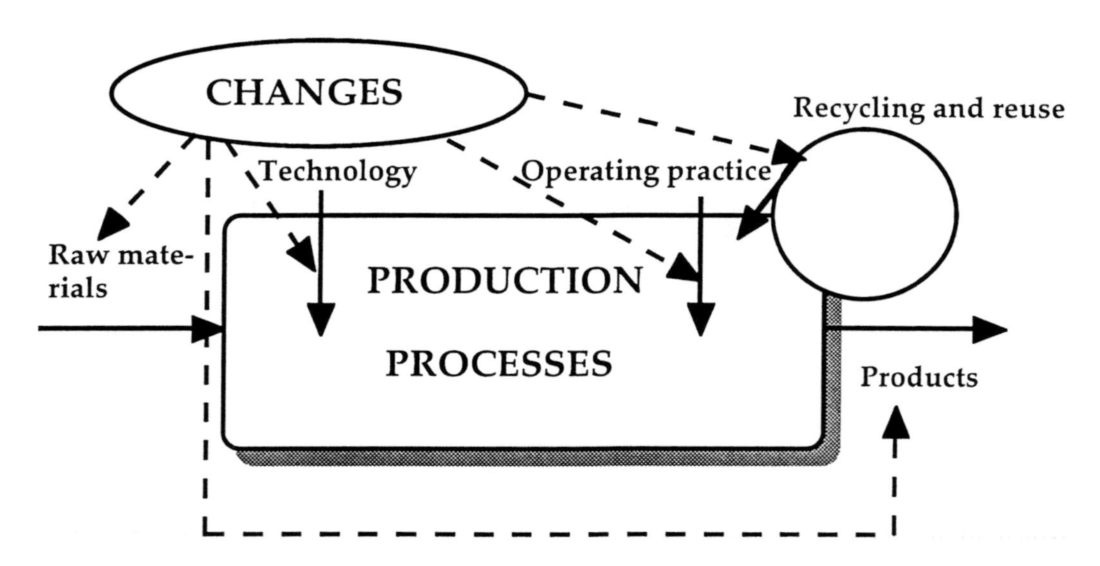

Figure 9.2. Cleaner production can be introduced by changes (dotted arrows) in raw materials, technology, operating practice, products and reuse or recycling.

These five possibilities for cleaner production are summarised in Figure 9.2.

The range of tools to catalyse industry to adopt cleaner production is large (Kryger and Dyndgaard, 1999). Governmental agencies may use the following instruments:

A. Regulations, standards, legislations

B. Green taxes

C. Economic support

D. Providing external assistance.

We presented six principal methods to solve pollution problems in Section 1.6. A (to reduce emissions by cleaner production) and B (reuse and recycling) work hand in hand and can hardly be separated. B is however not always a consequence of cleaner technology but may be an obvious solution after environmental technology has been introduced .

On the other hand cleaner technology may apply environmental technology as a tool to achieve the goals set up in a cleaner production scheme. This illustrates very clearly that integrated environmental management is needed to achieve an optimum solution to a pollution problem. Up-to-date integrated environmental management requires that the entire spectrum of considerations and tools presented in this volume is applied.

9.5 LIFE CYCLE ANALYSIS, LCA

LCA is a methodology for assessing the environmental impacts and resource consumption associated with a product throughout its entire life - from cradle to grave, or from raw material to production to product to use of product to disposal. It is a holistic approach, using system analysis.

. Several industrial countries are discussing or have already implemented LCA; see for instance DEPA, 1996 and 1998. The following initiatives have enhanced the interest by industry for introduction of LCA:

1) Ecolabelling or environmental declarations about products based on LCA.

2) Green purchase guidance taking environmental considerations into account.

3) Take-back responsibility for certain products. e.g. cars and electronics.

A LCA consists of four steps (Hautschild and Wenzel, 1999) :

1. **Definition of the goals of the assessment** and the scope of the study. The goals are often formulated as questions:

Which product on the market causes the least impact on the environment?

Which are the most important environmental impacts of the products during its life cycle?

What would happen if the product was not produced at all?

How does the impact of a specific product during its life cycle compare with alternative products (comparative studies which should be based on functional units)?

What influence would a specific proposed change in the product or production processes have on the results of the LCA?

Which properties of the product cause the undesirable impacts? Can we change them? How?

2. **Preparation of an inventory of inputs and outputs** for the processes that occur during the life cycle of the product. Traditionally, emissions are divided into air pollutants, waste water and solid waste. The emissions are expressed relative to the function of the product in order to make a proper comparison.

3. **Impact assessment,** where the results under point 2 are translated into environmental impacts. We distinguish the impact on human health, ecosystem health and the resource base. Various techniques are used to compare two or more emissions with the

same or different types of effects. When the emissions have the same type of effects a normalisation may be applied, while different effects require the introduction of weighting factors. The selection of weighting factors is at least to certain extent a political question, as it is impossible by objective criteria to compare effects on human health and on ecosystems or the depletion of two different resources. Table 9.3 shows some of the weighting factors based on Danish Environmental Policy Targets from the period 1990-2000.

Table 9.3
Weighting factors based on Danish Environmental Policy Targets, applied during the decade 1990-2000. Source: Wenzel et al., 1997.

Impact	Weighting factor
Global warming	1.3
Ozone depletion	23
Photo chemical smog formation	1.2
Acidification	1.3
Nutrient enrichment	1.2
Persistent toxicants	2.5
Human toxicity	2.8
Ecotoxicity	2.3

4. **Interpretation of the impact assessment according to the defined goals and scope of the study** including sensitivity analysis of key elements of the assessment. This phase attempts to answer the questions raised in the first phase. The outcome may be a series of recommendations to the decision maker, but may also lead to revision of the scope of the study. The result of the sensitivity analysis is used to identify which product properties and processes are most significant for the total result. The significance of the uncertainty of the key figures is examined by letting them vary within their estimated range and investigating how this affects the total result of the LCA and the conclusions that can be drawn.

The following stages of the life cycle are considered in a complete LCA:

A. Extraction of raw materials.

B. Transportation of raw materials to the production site.

C. Manufacture of the product.

D. Transportation from production to the user. This may consists of several stages.

E. Use of product.

F. Disposal.

In all these stages emissions may occur and the following treatment processes may be considered as the most obvious alternatives: reuse (after a suitable treatment), recycling, incineration, composting, waste water treatment and dumping. The result of the LCA is changed in accordance with any decision on treatment of the waste.

9.6 QUESTIONS AND PROBLEMS

1. Explain how an LCA can reduce the waste originating from disposal of a used car.

2. Give examples on the use of cleaner technology in dairies and abattoirs.

3. A specific resource A is used at a rate of 1000 t / y, increasing by 2% per year. The known amount of resource A is 25 000 t, increasing with new discoveries linearly by 1000 t/ y. Calculate the lifetime of the resource under these circumstances. How much longer would the life time of resource A be with introduction of A. 50% and B. 75% recycling?

APPENDIX 1

Composition of the Spheres

A. Composition of the earth (total)

	Weight (%)
Fe	34.63
O	29.53
Si	15.20
Mg	12.70
Ni	2.39
S	1.93
Ca	1.13
Al	1.09
Na	0.57
Cr	0.29
Mn	0.22
Co	0.13
P	0.10
K	0.07
Ti	0.05
	100.00

B. Background concentration: Abundance in average crustal rock

Item	Value (% by weight)
Ag	0.000007
Al	8.1
Au	0.000004
Co	0.0025
Cr	0.010
Cu	0.0055
Fe	5.0
Hg	0.000008
Mg	2.1
Mn	0.10
Mo	0.00015
Ni	0.0075
Pb	0.0013
Pt	0.000001
Sb	0.00002
Sn	0.00020
Ti	0.44
U	0.00018
V	0.014
W	0.00015
Zn	0.0070

C. Background concentration: Biosphere

Item	Value (moles/hectare)
Al	4.12
C	6560
Ca	18.9
Cl	2.80
Fe	1.38
H	13,100
K	11.7
Mg	8.18
Mn	0.765
N	71.6
Na	1.65
O	6540
P	3.35
S	4.44
Si	8.62

D. Composition of the hydrosphere

element	mg/l	tons	element	mg/l	tons
Cl	19,000.0	$29.3 * 10^{15}$	V	0.002	$3.0 * 10^{9}$
Na	10,500.0	16.3 * -	Mn	0.002	3.0 * -
Mg	1,350.00	2.1 * -	Ti	0.001	1.5 * -
S	885.0	1.4 * -	Sb	0.0005	0.8 * -
Ca	400.0	0.6 * -	Co	0.0005	0.8 * -
K	380.0	0.6 * -	Cs	0.0005	0.8 * -
Br	65.0	0.1 * -	Ce	0.0004	0.6 * -
C	28.0	0.04 * -	Y	0.0003	0.5 * -
Sr	8.0	$12.0 * 10^{12}$	Ag	0.0003	0.5 * -
B	4.6	7.1 * -	La	0.0003	0.5 * -
Si	3.0	4.7 * -	Kr	0.0003	0.5 * -
F	1.3	2.0 * -	No	0.0001	$15.0 * 10^{7}$
A	0.6	0.93 * -	Cd	0.0001	15.0 * -
N	0.5	0.78 * -	W	0.0001	15.0 * -
Li	0.17	0.26 * -	Xe	0.0001	15.0 * -
Rb	0.12	0.19 * -	Ge	0.00007	11.0 * -
P	0.07	0.11 * -	Cr	0.00005	7.8 * -
I	0.06	$93.0 * 10^{9}$	Th	0.00005	7.8 * -
Ba	0.03	47.0 * -	Sc	0.00004	6.2 * -
In	0.02	31.0 * -	Pb	0.00003	4.6 * -
Zn	0.01	16.0 * -	Hg	0.00003	4.6 * -
Fe	0.01	16.0 * -	Ga	0.00003	4.6 * -
Al	0.01	16.0 * -	Bi	0.00002	3.1 * -
Mo	0.01	16.0 * -	Nb	0.00001	1.5 * -
Se	0.004	6.0 * -	Tl	0.00001	1.5 * -
Sn	0.003	5.0 * -	He	0.000005	0.8 * -
Cu	0.003	5.0 * -	Au	0.000004	0.6 * -
As	0.003	5.0 * -	Pa	$2 * 10^{-9}$	3000
U	0.003	5.0 * -	Ra	$1 * 10^{-10}$	150
Ni	0.002	3.0 * -	Rn	$0.6 * 10^{-15}$	$1 * 10^{-3}$

E. Composition of the atmosphere

	Volume (ppm)	Weight (ppm)	Mass * 10^{20} g
N_2	780,900	755,100	38,648
O_2	209,500	231,500	11,841
A	9,300	12,800	0.665
CO_2	300	460	0.0233
Ne	18	12.5	0.000636
He	5.2	0.72	0.000037
CH_4	1.5	0.9	0.000043
Kr	1	2.9	0.000146
N_2O	0.5	0.8	0.000040
H_2	0.5	0.03	0.000002
O_3	0.4	0.6	0.000031
Xe	0.08	0.36	0.000018

F. The atomic composition of the four spheres

Element	Atoms % in (v.l. = very low)			
Element	Biosphere	Lithosphere	Hydrosphere	Atmosphere
H	49.8	2.92	66.4	v.l.
O	24.9	60.4	33	21
C	24.9	0.16	0.0014	0.03
N	0.27	v.l.	v.l.	
Ca	0.073	1.88	0.006	v.l.
K	0.046	1.37	0.006	v.l.
Si	0.033	20.5	v.l.	v.l.
Mg	0.031	1.77	0.034	v.l.
P	0.030	0.08	v.l.	v.l.
S	0.017	0.04	0.017	v.l.
Al	0.016	6.2	v.l.	v.l.
Na	v.l.	2.49	0.28	v.l.
Fe	v.l.	1.90	v.l.	v.l.
Ti	v.l	0.27	v.l.	v.l.
Cl	v.l.	v.l.	0.33	v.l.
B	v.l.	v.l.	0.0002	v.l.
Ar	v.l.	v.l.	v.l.	0.93
Ne	v.l.	v.l.	v.l.	0.0018

G. Background concentration in atmosphere, at stations in unpolluted area.

Element	$\mu g\ m^{-3}$
Ag	0.05 - 0.1
Al	1 - 12
As	0.1 - 1.0
Ca	70 - 250
Cd	0.05 - 1.5
Cu	1 - 4
Fe	3 - 20
H_2	$(250 - 365) . 10^3$
He	$5.2 . 10^3$
K	60 - 180
Kr	$1.1 . 10^3$
Mg	170 - 600
Mn	0.05 - 0.33
Na	1500 - 5500
Ne	$18 . 10^3$
Pb	0.4 - 0.8
S	$(3 - 50) . 10^3$
Sr	0.9 - 4.0
V	0.05 - 0.27
Xe	86

H. Composition of the sea

Element	$mg\ l^{-1}$	Present as	Retention time (y)
H	108,000	H_2O	
He	0.000005	He(g)	
Li	0.17	Li^+	$2.0 * 10^7$
Be	0.0000006		$1.5 * 10^2$
B	4.6	$B(OH)_3$, $B(OH)_2O^-$	
C	28	HCO_3^-, H_2CO_3, CO_3^{2-}	
N	0.5	NO_3^-, NO_2^-, NH_4^+, $N_2(g)$	
O	857,000	H_2O, $O_2(g)$, SO_4^{2-}	
F	1.3	F^-	
Ne	0.0001	Ne (g)	
Na	10,500	Na^+	$2.6 * 10^8$
Mg	1350	Mg^{2+}, $MgSO_4$	$4.5 * 10^7$
Al	0.01		$1.0 * 10^2$
Si	3	$Si(OH)_4$, $Si(OH)_3O^-$	$8.0 * 10^3$
P	0.07	HPO_4^{2-}, $H_2PO_4^-$ HPO_4^{3-}, H_3PO_4	
S	885	SO_4^{2-}	
Cl	19,000	Cl^-	

H. continued

Element	mg l^{-1}	Present as	Retention time (y)
A	0.6	A (g)	
K	380	K^+	$1.1 * 10^7$
Ca	400	Ca^{2+}, $CaSO_4$	$8.0 * 10^6$
Sc	0.00004		$5.6 * 10^3$
Ti	0.001		$1.6 * 10^2$
V	0.002	$VO_2(OH)_2^{2-}$	$1.0 * 10^4$
Cr	0.00005		$3.5 * 10^2$
Mn	0.002	Mn^{2+}, $MnSO_4$	$1.4 * 10^3$
Fe	0.01	$Fe(OH)_3$ (s)	$1.4 * 10^2$
Co	0.0005	Co^{2+}, $CoSO_4$	$1.8 * 10^4$
Ni	0.002	Ni^{2+}, $NiSO_4$	$1.8 * 10^4$
Cu	0.003	Cu^{2+}, $CuSO_4$	$5.0 * 10^4$
Zn	0.01	Zn^{2+}, $ZnSO_4$	$1.8 * 10^5$
Ga	0.00003		$1.4 * 10^3$
Ge	0.00007	$Ge(OH)_4$, $Ge(OH)_3O^-$	$7.0 * 10^3$
As	0.003	$HAsO_4^{2-}$, $H_2AsO_4^-$ H_3AsO_4, H_3AsO_3	
Se	0.004	SeO_4^{2-}	
Br	65	Br^-	
Kr	0.0003	Kr (g)	
Rb	0.12	Rb^+	$2.7 * 10^5$
Sr	8	Sr^{2+}, $SrSO_4$	$1.9 * 10^7$
Y	0.0003		$7.5 * 10^3$
Nb	0.00001		$3.0 * 10^2$
Mo	0.01	MoO_4^{2-}	$5.0 * 10^5$
Ag	0.0003	$AgCl^{2-}$, $AgCl_3^{2-}$	$2.1 * 10^6$
Cd	0.00011	Cd_{2+}, $CdSO_4$	$5.0 * 10^5$
In	< 0.02		
Sn	0.003		$5.0 * 10^5$
Sb	0.0005		$3.5 * 10^5$
I	0.06	IO_3^-, I^-	
Xe	0.0001	Xe (g)	
Cs	0.0005	Cs^+	$4.0 * 10^4$
Ba	0.03	Ba^{2+}, $BaSO_4$	$8.4 * 10^4$
La	0.0003		$1.1 * 10^4$
Ce	0.0004		$6.1 * 10^3$
W	0.0001	WO_4^{2-}	$1.0 * 10^3$
Au	0.000004	$AuCl_4^-$	$5.6 * 10^5$
Hg	0.00003	$HgCl_3$, $HgCl_4^{2-}$	$4.2 * 10^4$
Tl	< 0.00001	Tl^+	
Pb	0.00003	Pb^{2+}, $PbSO_4$	$2.0 * 10^3$
Bi	0.00002		$4.5 * 10^5$
Rn	$0.6*10^{-15}$	Rn (g)	
Ra	$1.0*10^{-10}$	Ra^{2+}, $RaSO_4$	
Th	0.00005		$3.5 * 10^2$
Pa	$2.0*10^{-9}$		
U	0.003	$UP_2(CO_3)_3^{4-}$	$5.0 * 10^5$

I. Characteristic background concentration range in uncontaminated soil.

Element	mg (kg dry matter)$^{-1}$		
Ag	0.01	-	0.1
A	10,000	-	300,000
As	0.1	-	40
B	2	-	100
Ba	100	-	3000
Be	0.1	-	40
Br	1	-	10
Ce	1	-	50
Cl	1	-	150
Co	1	-	40
Cr	5	-	3000
Cs	0.3	-	25
Cu	2	-	100
F	2	-	300
Fe	7000	-	550,000
Ga	0.4	-	300
Ge	1	-	50
Hg	0.01	-	0.8
I	1	-	5
K	400	-	30,000
La	1	-	5000
Li	7	-	200
Mg	600	-	6000
Mo	0.2	-	5
N	200	-	2500
Na	750	-	7500
Ni	10	-	1000
P	100	-	4000
Pb	2	-	200
Ra	(3-20) * 10^{-7}		
Rb	20	-	600
S	30	-	900
Sc	1	-	10
Se	0.01	-	2
Si	0.2	-	5
Sn	2	-	200
Sr	50	-	1000
Th	0.1	-	12
Ti	1000	-	10,000
Te	0.01	-	0.5
U	0.9	-	9
V	20	-	500
Y	2.5	-	250
Zn	10	-	300
Zr	30	-	2000

APPENDIX 2

Decomposition Rates

Pesticides: Half-life time in soil

Item		Conditions
Akton	1.5 years	10 lbs/acre, spray, corn, sultan silt loam
Akton	1 year	2 lbs/acre, granules, corn, sultan silt loam
Akton	1.2 years	2 lbs/acre, spray, corn, sultan silt loam
Akton	32 weeks	LAB
Aldicarb	7-17 days	Oxidation, value depending of soil
Aldrin	10 days	In irrigated soil, cotton, Gezira, Sudan
Aldrin	40% remaining after 14 years	Max value
Aldrin	28% remaining after 15 years	Max value
Azinphosmethyl	484 days	Sterile soil, dry, lag period included, 279 K
Azinphosmethyl	135 days	Sterile soil, dry, lag period included,298 K
Azinphosmethyl	36 days	Sterile soil, dry, lag period included, 313 K
Benomyl	3-6 months	On turf, Delaware
Benomyl	6.12 months	Bare soil, Delaware
BHC	10% remaining after 14 years	Max. value
Bux	1-2 weeks	Hydrolysis, lab.
Carbaryl	8 days	Agricultural soil
Carbaryl	64 days	Sandy loam, Application = 25.4 kg/ha
Carbofuran	0.60 years	In soil, better drained, application 2.5 ppm, clay-muck, 1973
Carbofuran	0.92 years	In soil, poorly drained, Application 2.5 ppm, clay, 1973

Pesticides: Half-life time in soil (continued)

Item		Conditions
Carbofuran	30 days	Hydrolysis, formation of phenol
Chlordane	40% remaining after 14 years	Max. value
Chlorobenzilate	21 days	In Leon and lakeland sands
DDT	8 months	P,P'-DDT, subtropical soil, fall and winter, Application = 5 kg/ha
DDT	14 days	In irrigated soil, cotton, Gezira, Sudan
DDT	39% remaning after 17 years	Max. value
Dicamba	3 weeks	On litter, Texas
Dicamba	2 weeks	On native grasses, Texas
Dieldrin	2.7 days	Loss by volatilization, grass, first 5 d
Dieldrin	7.5 months	P,P′-DDT, subtropical soil, fall and winter, Application = 5 kg/ha
Dieldrin	11 days	In irrigated soil, cotton, Gezira, Sudan
Dieldrin	31% remaining after 15 years	Max. value
Dioxathion	55 days	Soil dust 2, pH = 7.3, org. matter = 2.1%
Dioxathion	45 days	Soil dust 3, pH = 7.3, org. matter = 2.3%
Dioxathion	30 days	Soil dust 4, pH = 7.6, org. matter = 1.8%
Endosulfan	7 days	In irrigated soil, cotton, Gezira, Sudan
Endrin	41% remaining after 14 years	Max. value
Ethion	420 days	None
Heptachlor	1.7 days	Loss by volatilization, grass, first 5 d
Heptachlor	16% remaining after 14 years	Max. value
Malathion	3 days	Basic silty loam, Illinois, Application = 10 ppm, pH = 6.2

Pesticides: Half-life time in soil (continued)

Item		Conditions
Malathion	7 days	Basic silty loam, Illinois, Application = 10 ppm, pH = 8.2
Malathion	4.3 days	Basic silty loam, Illinois, Application = 10 ppm, pH = 7.2
Meobal	7 days	Field condition
Methomyl	30-42 days	None
Naproamide	54 days	Soil moisture content = 10.0%, 301 K, initial = 4.5 kg/ha
Naproamide	63 days	Soil moisture content = 7.5%, 301 K, initial = 4.5 kg/ha
Carbaryl	1.0 day	pH = 8.0, 301 K, seawater
Carbaryl	99 min	pH = 10.0, 285 K, init. concentration=3 * 10^{-3} M, dark
Carbaryl	20 min	pH = 10.0, 298 K, init. concentration=3 * 10^{-3} M, dark
Carbaryl	8 min	pH = 10.0, 308 K, init. concentration=3 * 10^{-3} M, dark
Carbaryl	27 min	pH = 9.8, 298 K, init. concentration= 3 * 10^{-3} M, dark
Carbaryl	58 min	pH = 9.5, 298 K, init. concentration=3 * 10^{-3} M, dark
Carbaryl	116 min	pH = 9.2, 298 K, init. concentration=3 * 10^{-3} M, dark
Carbaryl	173 min	pH = 9.0, 298 K, init. concentration=3 * 10^{-3} M, dark
Carbaryl	1 month	pH = 8.0, 276.5 K, seawater
Carbaryl	4.8 days	pH = 8.0, 290 K, seawater
Carbaryl	3.5 days	pH = 8.0, 293 K, seawater
Carbaryl	3.2 hours	Hydrolysis, pH = 9.0, 300 K
Carbaryl	1.3 days	Hydrolysis, pH = 8.0, 300 K
Carbaryl	13 days	Hydrolysis, pH = 7.0, 300 K
Carbaryl	4.4 months	Hydrolysis, pH = 6.0, 300 K
Carbaryl	3.6 years	Hydrolysis, pH = 5.0, 300 K
Diazinon	0.49 days	Hydrolysis, pH = 3.1
Diazinon	31 days	Hydrolysis, pH = 5.0
Diazinon	185 days	Hydrolysis, pH = 7.5
Diazinon	136 days	Hydrolysis, pH = 9.0
Diazinon	0.017 days	Hydrolysis, pH = 3.1
Diazinon	1.27 days	Hydrolysis, pH = 5.0
Diazinon	29 days	Hydrolysis, pH = 7.5
Diazinon	18 days	Hydrolysis, pH = 9.0
Diazinon	0.42 days	Hydrolysis, pH = 10.4

Pesticides: Half-life time in soil (continued)

Item	Conditions	
Dimilin	22.9 days	pH = 7.7, 283 K, initial = 0.1 ppm
Dimilin	80.5 days	pH = 10.0, 283 K, initial = 0.1 ppm
Dimilin	28.7 days	pH = 7.7, 297 K, initial = 0.1 ppm
Dimilin	8.31 days	pH = 10.0, 297 K, initial = 0.1 ppm
Dimilin	8 days	pH = 7.7, 311 K, initial = 0.09 ppm
Dimilin	3.45 days	pH = 10.0, 311 K, initial = 0.1 ppm
Heptachlor	23.1 hour	299.8 K, for hydrolysis, in distilled water
Methoprene	30 hours	Freshwater pound, initial = 0.01 ppm, February, in sunlight, California
Methoxychlor	58 hours	Conc. in water = 10^{-7} M, NO H_2O_2 added, 338 K
Methoxychlor	58 hours	Hydls. water conc. = 10^{-6} M, 5% aceto-nitrile, NO H_2O_2 added, 338 K
Methoxychlor	<1 hour	Hydls. water conc. = 10^{-6} M, 5% aceto-nitrile, H_2O_2-conc. = 0.1 M, 338 K
Methoxychlor	<1.7 hours	Hydls. water conc. = 10^{-6} M, 5% aceto-nitrile, H_2O_2-conc. = 8 * 10^{-2} M, 338 K
Methoxychlor	2 hours	Hydls. water conc. = 10^{-6} M, 5% aceto-nitrile, H_2O_2-conc. = 8 * 10^{-3} M, 338 K

Rate of degradation: biological degradation in water

Item		Conditions
Diethanolamine	19.5 mg COD/ gram-hour	Init. COD = 200 mg/l, no other source of C, aerob, 293 ± 3 K
Diethylene glycol	13.7 mg COD/ gram-hour	Init. COD = 200 mg/l, no other source of C, aerob, 293 ± 3 K
Dimethyl-cyclohexanol	21.6 mg COD/ gram-hour	Init. COD = 200 mg/l, no other source of C, aerob, 293 ± 3 K
Ethylene diamine	9.8 mg COD/ gram-hour	Init. COD = 200 mg/l, no other source of C, aerob, 293 ± 3 K

Rate of degradation: biological degradation in water (continued)

Item		Conditions
Ethylene glycol	41.7 mg COD/ gram-hour	Init. COD = 200 mg/l, no other source of C, aerob, 293 ± 3 K
Furfuryl alcohol	41.1 mg COD/ gram-hour	Init. COD = 200 mg/l, no other source of C, aerob, 293 ± 3 K
Furfuryl-aldehyde	37.0 mg COD/ gram-hour	Init. COD = 200 mg/l, no other source of C, aerob, 293 ± 3 K
Gallic acid	20.0 mg COD/ gram-hour	Init. COD = 200 mg/l, no other source of C, aerob, 293 ± 3 K
Gentisic acid	80.0 mg COD/ gram-hour	Init. COD = 200 mg/l, no other source of C, aerob, 293 ± 3 K
Glucose	180.0 mg COD/ gram-hour	Init. COD = 200 mg/l, no other source of C, aerob, 293 ± 3 K
Glucose	8 to > 26 days	Half-life time, 278 K, 5 stations at Southampton
Glucose	3 to 10 days	Half-life time, 295 K, 5 stations at Southampton
Glucose	21 to >30 days	Half-life time, 300 K, Porto Novo 2 stations from River Vellar
Glucose	11 days	Half-life time, 300 K, Porto Novo Kille Backwater
Glucose	9 to 10 days	Half-life time, 300 K, 2 stations from Kille Backwater and Mangrove Swamp
Glucose	> 17 days	Half-life time, 300 K, Porto Novo Bay of Bengal
Glucose	45 days	Half-life time, 300 K, distilled water
Glycerol	85.0 mg COD/ gram-hour	Init. COD = 200 mg/l, no other sources of C, aerob, 293 ± 3 K
Hydroquinone	54.2 mg COD/ gram-hour	Init. COD = 200 mg/l, no other sources of C, aerob, 293 ± 3 K
Iso-propanol	52.0 mg COD/ gram-hour	Init. COD = 200 mg/l, no other sources of C, aerob, 293 ± 3 K
Isophthalic acid	85.0 mg COD/ gram-hour	Init. COD = 200 mg/l, no other sources of C, aerob, 293 ± 3 K

Rate of degradation: biological degradation in water (continued)

Item		Conditions
Lineal alkyl benzen sulphonate in presence of:		
Activated sludge	0.79 mg surfact./l day	21 days, batch culture, initial conc. = 20 mg surfactant/l
Anabaena cylindrica	0.72 mg surfact./l day	21 days, batch culture, initial conc. = 20 mg surfactant/l
Anabaena variabilis	0.92 mg surfact./l day	21 days, batch culture, initial conc. = 20 mg surfactant/l
Anacystis nidulans	0.40 mg surfact./l day	21 days, batch culture, initial conc. = 20 mg surfactant/l
Ankistrodesmus braunii	0.35 mg surfact./l day	21 days, batch culture, initial conc. = 20 mg surfactant/l
Calothrix parietina	0.93 mg surfact./l day	21 days, batch culture, initial conc. = 20 mg surfactant/l
Chlorella pyrenoidosa	0.58 mg surfact./l day	21 days, batch culture, initial conc. = 20 mg surfactant/l
Chlorella vulgaris	0.42 mg surfact./l day	21 days, batch culture, initial conc. = 20 mg surfactant/l
Cylindrospernum sp.	0.84 mg surfact./l day	21 days, batch culture, initial conc. = 20 mg surfactant/l
Gloeocapsa alpicola	0.92 mg surfact./l day	21 days, batch culture, initial conc. = 20 mg surfactant/l
Nostoc Muscorum	0.46 mg surfact./l day	21 days, batch culture, initial conc. = 20 mg surfactant/l
Oscillatoria borneti	0.58 mg surfact./l day	21 days, batch culture, initial conc. = 20 mg surfactant/l
P-chloroaniline	5.7 mg COD/gram-hour	Init. COD = 200 mg/l, no other sources of C, aerob, 293 ± 3 K
P-chlorophenol	11.0 mg COD/gram-hour	Init. COD = 200 mg/l, no other sources of C, aerob, 293 ± 3 K
P-cresol	55.0 mg COD/gram-hour	Init. COD = 200 mg/l, no other sources of C, aerob, 293 ± 3 K
P-hydroxybenzoic acid	100.0 mg COD/gram-hour	Init. COD = 200 mg/l, no other sources of C, aerob, 293 ± 3 K
P-nitroacetophenone	5.2 mg COD/gram-hour	Init. COD = 200 mg/l, no other sources of C, aerob, 293 ± 3 K
P-nitro aniline	No degradation	Init. COD = 200 mg/l, no other sources of C, aerob, 293 ± 3 K
P-nitrobenzaldehyde	13.8 mg COD/gram-hour	Init. COD = 200 mg/l, no other sources of C, aerob, 293 ± 3 K
P-nitrobenzoic acid	19.7 mg COD/gram-hour	Init. COD = 200 mg/l, no other sources of C, aerob, 293 ± 3 K

Rate of degradation: biological degradation in water (continued)

Item		Conditions
P-nitrophenol	17.5 mg COD/ gram-hour	Init. COD = 200 mg/l, no other sources of C, aerob, 293 ± 3 K
P-nitro toluene	32.5 mg COD/ gram-hour	Init. COD = 200 mg/l, no other sources of C, aerob, 293 ± 3 K
P-phenylen-diamine	More degrad-able	Init. COD = 200 mg/l, no other sources of C, aerob, 293 ± 3 K
P-toluenesul-phonic acid	8.4 mg COD/ gram-hour	Init. COD = 200 mg/l, no other sources of C, aerob, 293 ± 3 K
Phenol	80.0 mg COD/ gram-hour	Init. COD = 200 mg/l, no other sources of C, aerob, 293 ± 3 K
Phloroglucinol	22.1 mg COD/ gram-hour	Init. COD = 200 mg/l, no other sources of C, aerob, 293 ± 3 K
Phosphorus	0.14 l/day	Potomac (Estuary), 293 K, org. P
Phosphorus	0.40 l/day	Lake Erie, 293 K, org. P
Phosphorus	0.14 l/day	Lake Ontario, 293 K, org. P
Phthalic acid	78.4 mg COD/ gram-hour	Init. COD = 200 mg/l, no other sources of C, aerob, 293 ± 3 K
Phthalimide	20.8 mg COD/ gram-hour	Init. COD = 200 mg/l, no other sources of C, aerob, 293 ± 3 K
Pyrocatechol	55.5 mg COD/ gram-hour	Init. COD = 200 mg/l, no other sources of C, aerob, 293 ± 3 K
Pyrogallol	No degradation	Init. COD = 200 mg/l, no other sources of C, aerob, 293 ± 3 K
Resorcnol	57.5 mg COD/ gram-hour	Init. COD = 200 mg/l, no other sources of C, aerob, 293 ± 3 K
Ribose	8 to > 30 days	Half-life time, 278 K, 5 stations at Southampton
Ribose	3 to 9 days	Half-life time, 295 K, 5 stations at Southampton
Ribose	> 36 days	Half-life time, 300 K, Porto Novo, River Vellar
Salicyloic acid	95.0 mg COD/ gram-hour	Init. COD = 200 mg/l, no other sources of C, aerob, 293 ± 3 K
Sec. butanol	55.0 mg COD/ gram-hour	Init. COD = 200 mg/l, no other sources of C, aerob, 293 ± 3 K
Si	0.0015 l/day	Detritus Si to dissolved Si

Rate of degradation: biological degradation in water (continued)

Item		Conditions
Sulphanilic acid	4.0 mg COD/ gram-hour	Init. COD = 200 mg/l, no other sources of C, aerob, 293 ± 3 K
Sulphosalicylic acid	11.3 mg COD/ gram-hour	Init. COD = 200 mg/l, no other sources of C, aerob, 293 ± 3 K
Tert. butanol	30.0 mg COD/ gram-hour	Init. COD = 200 mg/l, no other sources of C, aerob, 293 ± 3 K
Tetrahydrofurfuryl alcohol	40.0 mg COD/ gram-hour	Init. COD = 200 mg/l, no other sources of C, aerob, 293 ± 3 K
Tetrahydrophthalic acid	No degradation	Init. COD = 200 mg/l, no other sources of C, aerob, 293 ± 3 K
Tetrahydrophthalimide	No degradation	Init. COD = 200 mg/l, no other sources of C, aerob, 293 ± 3 K
Thymol	15.6 mg COD/ gram-hour	" " "
Triethylene	27.5 mg COD/ gram-hour	" " "
1-Napthalenesulfonic acid	18.0 mg COD/ gram-hour	" " "
1-Naphtol	38.4 mg COD/ gram-hour	" " "
1-Naphthol-2-sulfonic acid	18.0 mg COD/ gram-hour	" " "
1-Naphthylamine	No degradation	" " "
1-Naphthylamine 6-sulfonic acid	No degradation	" " "
1,2-Cyclohexanediol	66.0 mg COD/ gram-hour	" " "
1,3-Dinitrobenzene	No degradation	" " "
1,4-Butanediol	40.0 mg COD/ gram-hour	" " "
1,4-Dinitrobenzene	No degradation	" " "
2-Chloro-4-nitrophenol	5.3 mg COD/ gram-hour	" " "

Rate of degradation: biological degradation in water (continued)

Item		Conditions		
2-Naphthol	39.2 mg COD/ gram-hour	"	"	"
2,3-Dimethyl-aniline	12.7 mg COD/ gram-hour	"	"	"
2,3-Dimethyl-phenol	35.0 mg COD/ gram-hour	"	"	"
2,4-Diamino-phenol	12.0 mg COD/ gram-hour	"	"	"
2,4-Dichloro-phenol	10.5 mg COD/ gram-hour	"	"	"
2,4-Dimethyl-phenol	28.2 mg COD/ gram-hour	"	"	"
2,4-Dinitri-phenol	6.0 mg COD/ gram-hour	"	"	"
2,4-Trinitro-phenol	No degradation	"	"	"
2,5-Dimethyl-aniline	3.6 mg COD/ gram-hour	"	"	"
2,5-Dinitri-phenol	10.6 mg COD/ gram-hour	"	"	"
2,5-Dinitro-phenol	No degradation	Init. COD = 200 mg/l, no other sources of C, aerob, 293 ± 3 K		
2,6-Dimethyl-phenol	9.0 mg COD/ gram-hour	"	"	"
2,6-Dinitro-phenol	No degradation	"	"	"
3,4-Dimethyl-aniline	30.0 mg COD/ gram-hour	"	"	"
3,4-Dimethyl-phenol	13.4 mg COD/ gram-hour	"	"	"
3,5-Dimethyl-phenol	11.1 mg COD/ gram-hour	"	"	"
3,5-Dinitro-benzoic acid	No degradation	"	"	"
4-Methylcyclo-hexanol	40.0 mg COD/ gram-hour	"	"	"

Rate of degradation: biological degradation in water (continued)

Item		Conditions
Sulphanilic acid	4.0 mg COD/ gram-hour	Init. COD = 200 mg/l, no other sources of C, aerob, 293 ± 3 K
4-Methylcyclo-hexanone	62.5 mg COD/ gram-hour	" " "

APPENDIX 3

Concentration Factors

Concentration factors (463), ww in brackets means that CF is based upon wet weight

Component	Species	CF	Concentration in water	Conditions
Ag	Daphnia magna	26 (ww)	0.5 mg l^{-1}	-
Ag	Phytoplankton	620-15,000 (ww)	wide range	-
Aldrin	Buffalo fish	30,000 (ww)	0.007 $\mu g\ l^{-1}$	-
Aldrin	Catfish	1590 (ww)	0.044 $\mu g\ l^{-1}$	-
Aldirn	Oyster	10 (ww)	0.05 $\mu g\ l^{-1}$	-
As	Salmo gardneri egg	18.5 (ww)	0.05 mg l^{-1}	279.5°K 33 days
Au	Brown algae	270 (ww)	wide range	-
Cd	Brown algae	890 (ww)	wide range	-
Cd	Zooplankton	6000 (ww)	10^{-4} mg m^{-3}	-
Cd	32 Freshwater plant species	1620	wide range	-
Chlordane	Algae	302 (ww)	6.6 ng l^{-1}	-
Chlorinated naphthalene	Chloroccum sp.	120 (ww)	100 $\mu g\ l^{-1}$	24 h
Co	32 freshwater plant species	4425	wide range	-
Cr	Fish species	10 (ww)	wide range	Freshwater
Cr	Molluscs	21,800 (ww)	wide range	Marine sp.
Cs	Salmo trutta	1020 (ww)	wide range	Soft water 6.6 g fish
Cu	Chorda filum	560	$2.5*10^{-7}$ g l^{-1}	Seawater
Cu	Ulva sp.	47,000-56,000	low	Seawater
DDT	Algae	500 (ww)	0.016 ng l^{-1}	-
DDT	Crab	144 (ww)	50 $\mu g\ l^{-1}$	Seawater
DDT	Crayfish	97 (ww)	0.1 $\mu g\ l^{-1}$	-
DDT	Oyster	70,000	0.1 $\mu g\ l^{-1}$	-
DDT	Sea squirt	160,000 (ww)	0.1 $\mu g\ l^{-1}$	Seawater
DDT	Snail	480	50 $\mu g\ l^{-1}$	-
DDT	Trout	200 (ww)	20 $\mu g\ l^{-1}$	-

Concentration factors (CF), ww in brackets means that CF is based upon wet weight (continued)

Component	Species	CF	Concentration n water	Conditions
Dieldrin	Algae	4091	0.011 ng l^{-1}	-
Dieldrin	Catfish	4444 (ww)	0.009 μg l^{-1}	-
Dieldrin	Trout	3300 (ww)	2.3 μg l^{-1}	-
Fe	Brown algae	17,000 (ww)	-	-
Fe	Zooplankton	144,000 (ww)	0.01 mg m^{-3}	Seawater
Heptachlor	Bluegill	1130 (ww)	50 μg l^{-1}	-
Hexabro-mobiphenyl	Salmo salar	1.73 (ww)	all	5.3 g fish 48 h
Hexachlo-robenzene	Salmo salar	690 (ww)	all	288°K 6 g fish
Hg	Daphnia magna	50 (ww)	2 mg m^{-3}	10 weeks
Hg	Zooplankton	650 (ww)	0.02 mg m^{-3}	Freshwater
Trimethyl-naphthalene	Rangia cuneata	26.7 (ww)	0.03 mg l^{-1}	24 h
Pb	Brown algae	70,000 (ww)	all	-
Pb	Fucus vesiculosus	870	Very low	Seawater
Pb	Zooplankton	1500 (ww)	2 mg m^{-3}	-
PCB	Salmo salar	282 (ww)	wide range	5.29 g fish, 24 h
PCB	Yellow perch	17,000 (ww)	1.0 μg l^{-1}	Freshwater
Ra	Brown algae	370 (ww)	wide range	-
Se	Phytoplankton	900-5500 (ww)	wide range	-
W	Brown algae	87 (ww)	wide range	-
Zn	Phytoplankton	8900-75,000 (ww)	all	-
Zn	Pike	1250 (ww)	low	Freshwater

APPENDIX 4 Composition of Plants, Animals and amount of elements in diet of adult mammals

A - mg per kg dry matter, elements in dry plant tissue

Element	Plankton*)	Brown algae	Ferns	Bacteria	Fungi
Ag	0.25	0.28	0.23		0.15
Al	1,000	62		210	29
As		30			
Au		0.012			
B		120	77	5.5	5
Ba	15	31	8		
Be					<0.1
Br		740			20
C	225,000	345,000	450,000	538,000	494,000
Ca	8,000	11,500	3,700	5,100	1,700
Cd	0.4	0.4	0.5		4
Cl		4,700	6,000	2,300	10,000
Co	5	0.7	0.8		0.5
Cr	3.5	1.3	0.8		1.5
Cs		0.067			
Cu	200	11	15	42	15
F		4.5			
Fe	3,500	690	300	250	130
Ga	1.5	0.5	0.23		1.5
H	46,000	41,000	55,000	74,000	55,000
Hg		0.03			
I	300	1,500			
K		52,000	18,000	115,000	22,300
La		10			
Li		5.4			
Mg	3,200	5,200	1,800	7,000	1,500
Mn	75	53	250	30	25
Mo	1	0.45	0.8		1.5
N	38,000	15,000	20,500	96,000	51,000
Na	6,000	33,000	1,400	4,600	1,500
Ni	36	3	1.5		1.5
O	440,000	470,000	430,000	230,000	340,000
P	4,250	2,800	2,000	30,000	14,000
Pb	5	8.4	2.3		50
Ra	$4 * 10^{-7}$	$9 * 10^{-8}$			
Rb		7.4			
Re		0.014			
S	6,000	12,000	1,000	5,300	4,000
Se		0.84			2
Si	200,000	1,500	5,500	180	
Sn	35	1.1	2.3		5
Sr	260	1,400	13		320
Ti	80	12	5.3		
U					0.25
V	5	2	0.13		0.67
W		0.035			
Y			0.77		0.5
Zn	2,600	150	77		150
Zr	20		2.3		5

*) Mainly diatoms

B - mg per kg dry matter elements in dry animal tissues *)

Element	Coelenterata	Annelida	Mollusca	Crustacea	Insecta	Pisces	Mammalia
Ag	5?				≤0.07	11?	0.006
Al		340	50	15	100	10	<3
As	30	6	0.005	0.08		0.3	0.2
Au	0.007		0.008	0.0005		0.0003	<0.0009
B		2.1?	20	15		20	<2
Ba			3	0.2			2.3
Bi	0.3?					0.04?	
Br	1,000	100?	1,000	400		400	4
C	436,000	402,000	399,000	401,000	446,000	475,000	484,000
Ca	1,300	11,000	1,500	10,000	500	20,000	85,000
Cd	1		3	0.15		3	
Cl	90,000		5,000	6,000	12,000	6,000	3,200
Co	4?	5?	2	0.8	<0.7	0.5	0.3
Cr	1.3					0.2	<0.3
Cs							0.06
Cu	50	4?	20	50	50	8	2.4
F			2	2		1,400	500
Fe	400	630	200	20	200	30	160
Ga	0.5?					0.15?	
Ge	1.5?					0.3?	
H	45,000	59,000	60,000	60,000	73,000	68,000	66,000
Hg			1?			0.3?	0.05
I	15	160	4	1	0.9	1	0.43
K	3,000	16,000	19,000	13,000	11,000	12,000	7,500
La							0.09
Li			1?		≤7		<0.02
Mg	5,500	6,000	5,000	2,000	750	1,200	1,000
Mn	30?	0.06?	10	2?	10	0.8	0.2
Mo	0.7		2	0.6	0.6	1	<1
N	63,000	99,000	85,000	84,000	123,000	114,000	87,000
Na	48,000		16,000	4,000	3,000	8,000	7,300
Ni	26?	11?	4	0.4	9	1	<1
O	271,000	340,000	390,000	400,000	323,000	290,000	186,000
P	14,000	8,100	6,000	9,000	17,000	18,000	43,000
Pb	35?		0.7	0.3	≤7	0.5	4
Ra			$1.5*1^{-7}$	$7*10^{-9}$		$1.5*10^{-8}$	$7*10^{-9}$
Rb			20				18
Re			0.006	0.0005		0.0008	
S	19,000	14,000	16,000	7,500	4,400	7,000	5,400
Sb	0.2					0.2	0.14
Sc							0.006
Se							1.7
Si		150	1,000	300	6,000	70	120
Sn	23?		15?	0.2		3?	<0.16
Sr		20	60	500			21
Th	0,03						
Ti	7		20	17	160	0.2	<0.7
U						≤0.06	0.023
V	2.3	1.2	0.7	0.4	0.15	0.14	<0.4
W			0.05	0.0005		0.0014	
Zn	1500?	6?	200	200	400	80	160

*) Most of the figures for marine animals were derived from the compilation by Vinogradov (1953)

C - mg elements per kg dry mammalian tissues

Element	Brain	Heart	Kidney	Liver	Muscle	Skin	Hair
Ag	0.04	0.01	<0.005	0.03	<0.004	0.022	
Al	0.92	0.8	1.1	1.7	0.67	4.4	30
As	0.08	0.01	0.34	0.5	0.16	0.36	1.1
Au	<0.5	0.00013	<0.5	<0.0001	<0.4	<0.2	
B	<0.6	0.2	<0.5	0.48	0.31	<0.2	
Ba	0.012	0.08	0.06	<0.007	0.013	0.15	
Be	<0.002	<0.002	0.002	0.0009	<0.003	<0.04	
Bi	<0.1	<0.08	<0.09	<0.07	<0.08	<0.03	
Br	3	8	16	10	4	10	6
Ca	320	150	390	140	105	360	200
Cd	<3	0.05	130	6.7	<0.06	<1	
Ce		0.0064			0.00003		
Cl	8,000	6,000	9,000	4,800	2,800	11,000	20,000
Co	0.0055	0.05	0.05	0.23	0.016	<0.03	15
Cr	0.12	0.025	0.05	0.026	0.042	0.29	2
Cs	0.03	0.05	0.03	0.05	0.09	<0.04	
Cu	22	14	12	196	3.1	1.7	80
Eu					0.00012		
F	2	2	3.2	4	5		
Fe	200	190	290	520	140	29	130
Ga	<0.04	<0.04	<0.04	<0.04	<0.04	<0.02	
Hf					<0.04		
Hg		0.17	0.25	0.022	0.02		
I	4		0.09	0.0015	0.12	1.7	
In					0.016		
Ir					0.00002		
K	11,600	9,200	7,800	7,400	10,500	1,900	
La		0.00012					
Li	<0.03	<0.03	<0.03	<0.02	<0.02	0.084	
Lu					0.00012		
Mg	550	640	550	480	630	150	
Mn	1.1	0.8	3.8	3.7	0.21	0.22	1
Mo	<0.2	0.2	1.4	2.8	<0.2	<0.07	
N	99,000	132,000	115,000	112,000	108,000	161,000	
Na	10,000	4,500	800	5,500	4,000	9,300	
Ni	<0.3	<0.2	<0.2	<0.2	0.008	0.8	6
P	12,200	6,000	6,900	7,400	6,300	680	800
Pb	0.24	0.2	4.5	4.8	<0.2	0.78	35
Pd					0.002		
Pt					0.002		
Ra			$4*10^{-9}$	$8*10^{-9}$	10^{-10}		
Rb	15	13	17	30	24	8	
Ru	<0.5	<0.4	<0.4	<0.4	0.002	<0.2	
S	6,700	9,500	6,600	8,400	6,800	3,200	38,000
Sb		0.006					
Sc		0.00006			0.008		
Se	2.1	0.7	2.1	2.1	2.5		0.3-13
Si	80	100	95	70	130	450	
Sm		0.01					

C - mg elements per kg dry mammalian tissues (continued)

Element	Brain	Heart	Kidney	Liver	Muscle	Skin	Hair
Sn	<2	0.2	0.74	0.85	<0.2	0.36	
Sr	0.085	0.1	0.24	0.06	0.05	0.15	
Te					0.02		
Ti	<0.3	<0.2	<0.2	<0.2	<0.2	0.54	3
Tl	<0.5	<0.4	<0.4	<0.4	<0.4	<0.2	
Tm					0.0004		
U		0.03	0.03	0.04	0.03		0.13
V	<0.3	<0.04	<0.05	<0.04	<0.04		0.02
W		0.005					
Yb					0.00012		
Zn	46	110	210	130	180	13	170
Zr	<5	<4	<4	<4	<0.3	<2	

References for tables A, B and C:

Aten et al., 1961; Arrhenius, 1963; Beharrell, 1942; Baumeister, 1958; Bowen and Dymond, 1955; Bertrand, 1950; Bowen, 1963; Boirie et al., 1962; Brooksbank and Leddicotte, 1953; Black and Mitchell, 1952;Bowen and Cawse, 1963; Bowen, 1966; Bertrand, 1942; H.J.M. Bowen, unpublished; Bowen, 1960; Bertrand and Levy, 1931; Cannon, 1960; Cannon, 1963; Chau and Riley, 1965; Chilean Iodine Educational Bureau, 1952; Dye et al., 1963; Fukai and Meinke, 1959 and 1962; Force and Morton, 1952; Forbes et al., 1954; Hunter, 1953; Hunter, 1942 and 1953; Ferguson and Armitage, 1944; Moon and Pall, 1944; Hamaguchi et al., 1960; Henderson et al., 1962; Harrison et al., 1963; International Commission on Radiological Protection, 1964; Johnson and Butler, 1957; Jervis et al., 1961; Koczy and Titze, 1958; Kehoe et al., 1940; King, 1957; King and Belt, 1938; Koch and Roesmer, 1962; Kringsley, 1959; Long, 1961; Lounamaa, 1956; Leddicotte, 1959; ; Low, 1949; Lux, 1938; Mayer and Gorham, 1957; Matsumura et al., 1955; McCance and Widowson, 1960; McConnell, 1961; Muth et al., 1960; Mitchell, 1944; Moiseenko, 1959; Mackle et al., 1939; Monier-Williams, 1950; Mullin and Riley, 1956; Neufeld, 1936; Newman, 1949; Porter, 1946; Pavlova,1956; Parr and Taylor, 1963, 1964; Samsahl and Soremark, 1961; Schofield and Hackin, 1964; Shacklette, 1965; Schwartz and Foltz, 1958; Shimp et al., 1957; Shibuya and Nakai, 1963; Smales and Pate, 1952; Smales and Salmon, 1955; Soremark and Bergman, 1962; Sowden and Stitch, 1957; Soremark, 1964; Spector, 1956; ; Stamm and Fernandez, 1958; Stitch, 1957; Suzuki and Hamada, 1956; Tipton and Cook, 1963; Turner et al., 1958; Thompson and Chow, 1956; Thomas, Hendricks and Hill, 1950; Tyutina et al., 1959; Vinogradow, 1953; Vinogradova and Kobalsky, 1962; Wester, 1965; Wakita and Kigoshi, 1964;; Yamagata, 1950, 1962; Yamagata, Murata and Toril, 1962.

D. Amounts of elements *) in the diet of adult mammals in mg day^{-1}

Species Mean weight Wt. of dry diet		Man (Homo sapiens) 70 kg 750 g/day					Rat (Ratus norvegicus) 0.3 kg 10 g/day	
Element/ state	Defi-cient	Normal	Toxic	Lethal	Defi-cient	Normal	Toxic	Lethal
Ag^{+}		0.06-0.08	60	1300				
Al^{3+}		10-100			0.001		200	220
AsIII or V		0.1-0.3	5-50	100-300		0.002	0.6	1.3-5
B Borate		10-20	4000		0.0006	0.15	130-270	
Ba^{2+} soluble		(1-5)	200				70-100	
Bi^{3+}		(0.06)					1.5	
Br^{-}		1-10	3000		0.005		800	
Ca^{2+}		400-1500			1	45-60		>400
Cd^{2+}		(0.6)	3				0.5	16
Cl^{-}	70	2400-4000			0.4	5-30		>900
Co^{2+}	0.0002	500				0.7		
Cr^{VI} Chromate		(0.05)	200	3000			5	
Cu^{2+}		2-5	250-500			0.05-0.2		20
F^{-}		0.5	20	2000	0.0007	0.001	0.1	30
$Fe^{II\ or\ III}$		12-15				0.1-0.5		>60
Ga^{3+}		(0.02)					10	
Hg^{II}		0.005-0.02		150-300				8
I^{-}	0.015	0.2	10,000		0.001-0.002			
In^{3+}		(0.01)					30	200-300
K^{+}		1400-3700		6000	0.3	50		>400
Li^{+}		2		200				
Mg^{2+}		220-400			0.1	2-5		
Mn^{2+}		3-9			0.003	0.03-0.2		
Mo^{VI} Molybdate		(0.7)			0.0005	0.0005-0.001		5 50
N Organic		8000-22,000						
Na^{+}	45	1600-2700		0.2	5-50			
Ni^{2+}		0.3-0.5					50	
P Phosphate		1200-2700					35-45	
Pb^{2+}		0.3-0.4		10,000				270
Rb^{+}		(10)				0.5	10	
S Sulphate, etc.		420-3000						
$Sb^{III\ or\ V}$		(0.1)	100					11-75
Se^{IV} Selenite		(0.2)	5		0.0007		0.06	1-2
Si Silicate		600						
Sn^{2+}		17-45	2000					
Sr^{2+}		1.5-5					8	900
Ta^{V} Tantalate		(1)						300
Te^{VI} Tellurate		(0.02)		2000			0.25	1-9
Ti TiO_2		(1-10)						
Tl^{+}		(0.1)		600				7.5
U^{VI}, UO_2^{2+}		(0.05)						36
V^{V} Vanadate		(0.3)					0.5	1.5
W^{VI} Tungstate		(0.05)						30-50
Zn^{2+}		10-15			0.016	0.02-0.04	50	150
Zr^{IV}		(0.1)						250-700

*) For comparative purpose, and for order of magnitude estimates for other species of mammals, the amounts in mg per kg body weight are more useful than the absolute amounts given above. Figures given in parantheses are provisional.

APPENDIX 5

Definition of expressions, concepts and indices

Acceptable daily intake - ADI - is the estimate of the amount of a substance in food or drinking water, which can be ingested per day over a lifetime by humans without appreciable health risk. ADI is normally used for food additives. The applied unit is usually mg/(kg body weightx24h)

Adaptation is the response of an organism to changing environmental conditions.

Adverse effect is the change in morphology, physiology, growth, biochemistry, and / or development of lifespan of an organisms which results in impairment of the functional capacity or impairment of the capacity to compensate for additional harmful effects of other environmental influences.

Bioconcentration factor - BCF - is the ratio of the examined chemical substance concentration in the test organisms to the concentration in the test medium, water (or air), at steady state

Biodegradability - indicates the ability of microorganisms to decompose a specific compound or a specific water sample to inorganic components. Biodegradability may be expressed as biological half life, as a first order kinetic coefficient or as the biological oxygen demand relative to the theoretical oxygen demand understood as the oxygen demand needed for a complete decomposition.

Biological half life ($t_{1/2}$) - the time needed to reduce the concentration of a chemical in environmental compartments or organisms to half the initial concentration by various biological processes (biodegradation, metabolism or growth).

Biological oxygen demand - measures the amount of organic matter in a water sample as mg oxygen consumed per litre by microorganisms over a period of time often indicated as days.

Biomagnification factor - BMF - measure of the tendency of a compound to be taken up through the food. It is the concentration of a chemical compound in a living organism divided by the concentration of the chemical compound in the food at steady state.

Body burden - concentration of a toxic substance in an organisms, usually expressed as mg/kg dry matter.

Carcinogenicity is the development of cancer. Any chemical which can cause cancer is said to be carcinogenic.

CF - is the ratio of the examined chemical substance concentration in the test organisms to the concentration in the test medium, water (or air), at steady state, assuming that only the medium (and not for instance the food) is contaminated. In practice, it is very close to the same value as BCF. In addition it is difficult to have a medium that is contaminated without contaminating the food.

COD - chemical oxygen demand expressed as the amount of oxygen in mg/l necessary to decompose a specific compound or a specific water sample by a chemical, most often dichromate. Permanganate is also used as oxidator, and the result is in this case indicated as the permanganate number, expressed either as the normal COD as mg oxygen /l or as potassium permanganate / l.

Critical body burden - the concentration in an organism which is critical for the organism. It is suggested to find it as the critical concentration in the medium (for instance LC_{50}) x BCF.

Critical range - the range of concentrations in mg/l below which all organisms lived for 24 hr and above which all died. Mortality is expressed as a fraction indicating the death rate (e.g. 3/4).

DOC - is an abbreviation for dissolved organic carbon.

Endocrine disrupters - components that are able to imitate the reaction of hormones or block the hormone receptors

HCp is the harzadous concentration for p% of the species, derived by a statistical extrapolation procedure.

ICp - is the inhibiting concentration to produce an inhibiting effect of p %.

LC50 (lethal concentration fifty) - a calculated concentration which, when administered by the respiratory route, is expected to kill 50% of a population of experimental animals during an exposure of a specified duration. Ambient concentration is expressed in mg per litre.

LCn (lethal concentration n) - a calculated concentration which, when administered by the respiratory route, is expected to kill n% of a population of experimental animals during an exposure of a specified duration. Ambient concentration is expressed in mg per litre.

LD50 (lethal dose n) - a calculated dose of a chemical substance which is expected to kill 50% of a population of experimental animals exposed through a route other than respiration. Dose concentration is expressed in mg per kg of body weight. the LD50 has often been used to classify toxicity between chemical compounds. The following classification may be used - oral LD50 to rat, expressed in mg / kg body weight: highly toxic < 25 ; toxic > 25 and < 200; harmful > 200 and < 2000.

LDn (lethal dose fifty) - a calculated dose of a chemical substance which is expected to kill n % of a population of experimental animals exposed through a route other than respiration. Dose concentration is expressed in mg per kg of body weight.

LOEC - is an abbreviation for lowest observed effect concentration. The LOEC is generally reserved for sublethal effects but can in principle also be used for mortality which is usually the most sensitive effect observed.

MAC (maximum allowable concentration) - a value in accordance with environmental legislation. Often dependent on time. This relation may be expressed as follows: $\log C = 1.8 - 0.7 \log t + 0.068 \log t2$, where C is the MAC (in mg/m3) and t is the time (in hours).

NC (narcotic concentration) - NC50 - median narcotic concentration.

No effect level - NEL - implies that test animals have remained in good condition. In most experiments, blood and urine tests are performed. The urine tests include specific gravity, pH, reducing sugars, bilirubin, and protein. The blood tests include haemoglobin concentration, packed cell volume, mean corpuscular Hb content, a white and differential cell count, clotting function, and the concentration of urea, sodium, and potassium. Control tests for hematological examination were made on the group of animals before exposure. No effect level in this context means: no toxic signs, autopsy - organs normal, blood and urine tests - (if made) normal.

No observed effect concentration (level) - NOEC (NOEL) - is defined as the highest concentration (level) of a test chemical substance to which the organisms are exposed that does not cause any observed and statistically significant adverse effects on the organisms compared with controls.

POM is an abbreviation for particulate organic matter.

ppb - parts per billion. For water this is presumed to be $\mu g/l$, for air $\mu l/m^3$ and for soil $\mu g/kg$ (dry matter), unless another unit is indicated.

ppm - parts per million. For water this is presumed to be mg/l, for air $\mu l/l$ and for soil mg/kg (dry matter), unless another unit is indicated.

Predicted environmental concentrations - PEC - is the concentration of a chemical in the environment, calculated primarily by use of models on the basis of available information about its properties and application pattern.

Predicted no effect concentration - PNEC - is the environmental concentration below which it is probable that an unacceptable effect will not occur according to predictions.

Predicted no effect level - PNEL - is the maximum level expressed as dose or concentration which on the basis of our present knowledge is likely to be tolerated by a particualr organism without producing any adverse effect.

Principal component analysis - PCA - is a multi variate technique to derive a set of orthogonal parameters from a large number of properties.

Quantitative structure activity relationship - QSAR - covers the relationships between physical and / or chemical properties of substances and their ability to cause a particular effect or enter into certain processes.

Retrospective risk assessment - is a risk assessment performed for hazards that began in the past.

Risk-benefit analysis - is the process of setting up the balance of risks and benefits of a proposed risk-reducing action.

Risk quotient is the PEC / PNEC ratio.

Teratogenesis is the capacity of a substance to cause defects in embryonic and foetal development. Any chemical which can cause these defects is said to be teratogenic

The median tolerance (TLm) - this term has been accepted by most biologists to designate the concentration of toxicant or substance at which m% of the test organisms survive. In some cases and for certain special reasons, the TL10 or TL90 might be used. The TL90 might be requested by a conservation agency negotiating with an industry in an area where an important fishery exists, and where the agency wants to establish waste concentrations that will definitely not harm the fish. The TL10 might be requested by a conservation agency which is buying toxicants designed to remove undesirable species of fish from fishing lakes.

***Threshold-effect concentration* - *TEC* -** is the concentration calculated as the geometric mean of NOEC and LOEC. It is equivalent to MATC - maximum acceptable toxicant concentration.

***Threshold limit value* - *TLV* -** is the concentration in air of a chemical to which most workers can be exposed daily without adverse effects according to current knowledge.

***TOC* -** is defined as the total organic carbon expressed for instance as kg organic carbon / kg solid. Organic carbon can often be estimated as 50-60% of all organic matter.

***Tolerable daily intake* - *TDI* -** is the acceptable daily intake established by the European Committee of Food. It is expressed in mg/(person x24h), assuming a body weight of 60 kg. TDI is normally used for food contaminants, unlike ADI. See also ADI.

***Toxicity equivalency factor* - *TEF* -** is a factor used in risk assessment to estimate the toxicity of a complex mixture of compounds.

***Ultimate median tolerance limit* - *UMTL* -** is the concentration of a chemical at which acute toxicity ceases.

***Xenobiotic* -** is a man-made chemical not produced in nature and not considered a normal constituent component of a specified biological system.

APPENDIX 6

Elements: Abundance and Biological Activity

Symbols used:

a = elements formed by radioactive decay of uranium and thorium. Have short physical half-lives and their crustal abundance are too low to be measured accurately.
b = very low, unmeasureable
ra = radioactive
cs = carcinogenic, suspected only.
s = stimulatory
cp = carcinogenic, proven
en = essential nutrient, established
ep = essential nutrient, probably or required under special conditions
t1 = toxic
t2 = very toxic

Element	Symbol	Atomic number	Crustal weight (%)	Abundance abundance in hydros-phere (mg/l)	Abundance in atmos-phere (vol ppm)	Biological activity	Threshold limit (mg/ m^3 in air in 8 hours)
Actinium	Ac	89	manmade	manmade		ra	
Aluminium	Al	13	8	0.01		cs	
Americum	Am	93	manmade	manmade		ra	
Antimony	Sb	51	0.00002	0.0005		s t2	
Argon	Ar	18	0	0.6	9300		
Arsenic	As	33	0.00020	0.003		cs s t2	
Astatine	At	85	manmade	manmade		ra	
Barium	Ba	56	0.0380	0.03		s t1	0.5
Berkelium	Bk	97	manmade	manmade		ra	
Beryllium	Be	4	0.0002	b		cp (s) t2	
Bismuth	Bi	83	4E-7	2E-5		t1	
Boron	B	5	0.0007	4.6			
Bromine	Br	35	0.00040	65		t1=Br2	
Cadmium	Cd	48	0.000018	0.001			0.2
Calcium	Ca	20	5.06	400		en	
Californium	Cf	98	manmade	manmade		ra	
Carbon	C	6	0.02	28	CO_2=330	en	
Cerium	Ce	58	0.0083	0.0004		s	
Cesium	Cs	55	0.00016	0.0005			
Chlorine	Cl	17	0.019	18,980		Cl(-)=en Cl_2=t1	
Chromium	Cr	24	0.0096	5E-5		en cp s t1	0.1(CrO_3)
Cobalt	Co	27	0.0028	0.0005		en cp	
Copper	Cu	29	0.0056	0.003		en s t1	
Curium	Cm	96	manmade	manmade		ra	
Dysprosium	Dy	66	0.00085	b		s	
Einsteinium	Es	99	manmade	manmade		ra	
Erbium	Er	68	0.00036	b		s	
Europium	Eu	63	0.00022	b			
Fermium	Fm	100	manmade	manmade		ra	
Fluorine	F	9	0.0460	1.3		ep s	

Elements: Abundance and Biological Activity (continued)

Element	Symbol	Atomic number	Crustal weight (%)	Abundance abundance in hydros- phere (mg/l)	Abundance in atmos- phere (vol ppm)	Biological activity	Threshold limit (mg/ m^3 in air in 8 hours)
Francium	Fr	87	manmade	manmade		ra	
Gadolinium	Gd	64	0.00063	b			
Gallium	Ga	31	0.00063	b			
Germanium	Ge	32	0.00013	7E-5		s	
Gold	Au	79	2E-7	4E-6		s t1	
Hafnium	Hf	72	0.0004	b			
Helium	He	2	0	5E-6	5.2		
Holmium	Ho	67	0.00016	b		s	
Hydrogen	H	1	0.14	H2O	CH_4=1.5 H_2=0.5	en	
Indium	In	49	0.00002	0.02		s t2	
Iodine	I	53	0.00005	0.06		I(-)=en I_2=t1	
Iridium	Ir	77	2E-8	b		t1	
Iron	Fe	26	5.80	0.01		en	
Krypton	Kr	36	0	0.0003	1		
Lanthanium	La	57	0.0050	0.0003			
Lead	Pb	82	0.0010	3E-5		t2 cp s	0.2
Lithium	Li	3	0.0020	0.17		s	
Lutetium	Lu	71	8E-5	b			
Magnesium	Mg	12	2.77	1350		en	
Manganese	Mn	25	0.100	0.002		en cs	5
Mendelevium	Md	101	manmade	manmade		ra	
Mercury	Hg	50	2E-6	3E-5		s t2	
Molybdenum	Mo	42	0.00012	0.01		en	5-15
Neodymium	Nd	60	0.0044	b			
Neon	Ne	10	0	0.0001	18		
Neptunium	Np	93	manmade	manmade		ra	
Nickel	Ni	28	0.0072	0.002		ep cp s	
Niobium	Nb	28	0.0072	0.002		ep cp s	
Nitrogen	N	7	0.0020	0.5	780,900	en	
Nobelium	No	102	manmade	manmade		ra	
Osmium	Os	76	2E-8	b		ra	
Oxygen	O	8	45.2	H_2O	209,500	en	
Palladium	Pd	46	3E-7	b		cs	
Phosphorus	P	15	0.1010	0.07		en	
Platinum	Pt	78	5E-7	b		t1	0.002
Plutonium	Pu	94	manmade	manmade		ra	
Polonium	Po	84	a	b		ra	
Potassium	K	19	1.68	380		en	
Praseody- mium	Pr	59	0.0015				
Promethium	Pm	61	manmade				Protach- tinium
Radium	Ra	88	a	1E-10		ra	
Radon	Rn	86	a	0.6E-15		ra	

Elements: Abundance and Biological Activity (continued)

Element	Symbol	Atomic number	Crustal abundance weight (%)	Abundance in hydrosphere (mg/l)	Abundance in atmosphere (vol ppm)	Biological activity	Threshold limit (mg/ m^3 in air in 8 hours)
Rhenium	Re	75	4E-8	b			
Rhodium	Rh	45	1E-8	b		cs	0.001
Rubidium	Rb	37	0.0070	0.12		s	
Rutherium	Ru	44	1E-8	b			
Samarium	Sm	62	0.00077	b			
Scandium	Sc	21	0.0022	4E-5		cs	
Selenium	Se	34	5E-6	0.004		cp s en t2	
Silicon	Si	14	27.2	3		ep cs	
Silver	Ag	47	8E-7	0.0003		cs t1	0.01
Sodium	Na	11	2.32	10,556		en	
Strontium	Sr	38	0.045	8			
Sulphur	S	16	0.030	885		en	
Tantalum	Ta	73	0.00024	b			5
Technetium	Tc	43	manmade	manmade		ra	
Uranium	U	92	0.00018	0.003	b	ra,cp	
Vanadium	V	23	0.009	0.002	b	ep,t1	
Wolfram	W	74	0.00015	0.0001	b	cs,t1	
Zinc	Zn	30	0.03	0.01		en	

APPENDIX 7

Density and dynamic viscosity of water and air

Density and dynamic viscosity of water and pure air (1 atm.) as function of temperature

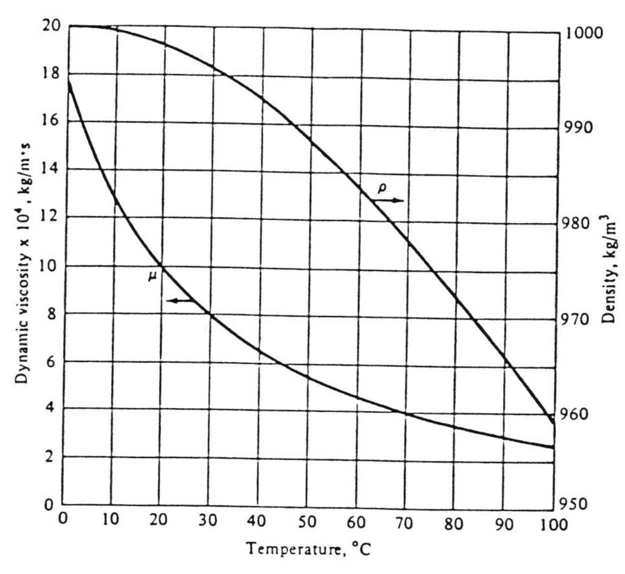

Density and dynamic viscosity of liquid water as a function of temperature

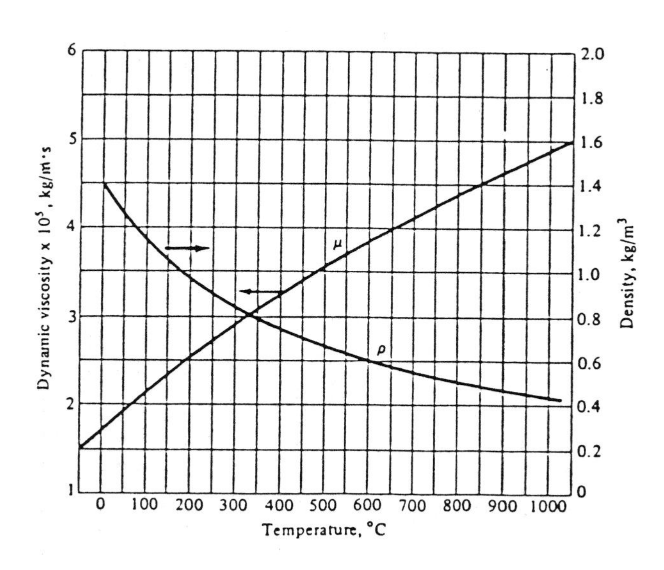

Density and dynamic viscosity of pure air at 1.0 atm pressure as a function of temperature

APPENDIX 8

Air Pollution: Emission, Imission and Scenarios

TABLE 1 (modified from Fenger, J & Tjell, 1994).

Global emissions of gasses and particles

	Natural	Antropogenic	Antropogenic
	Million t C, N ell. S pr år		%
Carbon dioxide	212000	6000	3
Carbon oxide	1600	1100	41
Carbon hydrides	1000	90	8
Methan	150	360	71
Nitrogenoxid and dioxid	30	30	50
Dinitrogenoxide	7	1.5	18
Sulphate	180	-	-
Sulphur dioxide	10	80	89
Hydrogen sulphide	15	-	-
Dimethylsulphide	40	-	-
Other S-compounds	10	-	-

TABLE 2

	Mean concentration (imission)		
	%	ppmv	Range
Nitrogen	78.08		
Oxygen	20.95		
Argon	0.93		
Carbon dioxide		360	
Neon		18.2	
Helium		5.2	
Methan		1.5	
Krypton		1.1	
Hydrogen		0.500	
Dinitrogenoxide		0.340	
Carbon oxide		0.100	
Ozone		0.020	0.01-0.1
Sulphur dioxide		0.010	0 - 0.100
Hydrogen sulphur		8	0 -0.100
Ammonia		5	0 - 0.100
Nitrogenoxide and dioxide		5	0 -0.010
Halogenated carbonhydrides		2	0 -0.010

TABLE 3

Pollution scenarios and air pollutants

(+ indicates that the pollutant contibutes to the pollution scenario, brackets mean that the contribution is minor or unknown.)

Component	Acidification	Eutrophication	Photochemistry	Decomposition of ozone layer	Greenshouse effect	Impact on biodiversity
Sulphur dioxide	+	(-)			(+)	+
Sulphur trioxide	+	(-)			(+)	
Sulphuric acid	+	(-)				
Dinitrogen oxide				(+)	(+)	
Nitrogen oxide	+	+	+		(+)	
Nitrogen dioxide	+	+	+	(+)	(+)	+
Nitric acid	+	+			(+)	(+)
Ammonia	(-)	+			(+)	(+)
Ammonium	+	+				+
Ozone			+		(+)	+
PAN			+			+
Trace elements	(+)					(+)
POP						(+)
PAH						(+)
NMHC/VOC			+		(+)	
Methane				+	(+)	
Pesticides		(+)				+
HFC, PFC, SF_6					+	
Freons				+	+	
Halons				+	+	
Carbon dioxide	+			(+)	+	(+)
Aerosols	(+)	(+)	(+)	(+)	(+/-)	(+)

References

Ahl, T. and Weiderholm, T., 1977. Svenska vattenkvalitetskriterier. Eurofierande ämnen. SNV PM (Swed)., 918.

Alabaster, J.S. and Lloyd, R., 1980. Water quality criteria for freshwater fish. pp. 21-45. FAO, Butterworths, London.

Albertson, O.E. and Sherwood, R.J., 1968. Improved sludge digestion is possible. Water Wastes Eng., 5: 8, 43.

Alkan, U., Anderson, G.K. and Ince, O. 1996. Toxicity of trivalent chromium in the anaerobic digestion process. Wat. Res. **30** No. 3: 731-741.

Almasi, A. and Pescod. 1996a. Waste water treatment mechanism in anoxic stabilisation ponds. Wat. Sci. Tech. Vol. 33, No. 7: 125-132.

Almasi, A. and Pescod, M.B. 1996b. Pathogen removal mechanisms in anoxic waste water stabilisation ponds. Wat. Sci. Tech. Vol. 33, No. 7: 133-140.

Almer, B.U., 1972. Inf. from Freshwater Lab., Drottningholm, Sweden, No. 12.

Ambühl, H., 1969. Die neueste Entwicklung der Vierwaldstättersees. Inst. Verein. theor. angew. Limnologie, 17: 210-230.

Ames, L.L., 1969. Evaluation of operating parameters of alumina columns for the selective removal of phosphorus from waste water and the ultimate disposal of phosphorus as calcium phosphate. Final report FWPCA Contract No. 14-12: 413.

Ames, Lloyd L. jr. and Dean, Robert B., 1970. Phosphorus removal from effluents in alumina columns. J. Wat. Pol. Contr. Fed., 42: 161-172.

Andersen, H., R. 1998. Master thesis in Danish at DFH, Environmental Chemistry: *Examination of Endocrine Disrupters.*

Anderson, P.D. and D' Apollonia, S., 1978. Aquatic Animals. In: G.C. Butler (ed.), Principles of Ecotoxicology, SCOPE 12. John Wiley % Sons, New York. p. 187.

Andreozzi, R., Insola, A., Caprio, V. and d'Amore, M.G. 1992. Quinoline ozonation in aqueous solution. Wat. Res. Vol. 26, No. 5: 639-643.

Andreozzi, R., Caprio, V., Insola, A., Marotta, R. and Tufano, V. 1998. The ozonation of pyruvic acid in aqueous solutions catalysed by suspended and dissolved manganese. Wat. Res. vol. 32, No. 5: 1492-1696.

Anon, 1958. Ozone counter waste cyanide's lethal punch. Chem. Eng., 65: 63.

Arcangeli, J.P., Arvin, E., Mejlhede, M. and Lauritsen, F.R. 1996. Biodegradation of *CIS*-1,2-dichloro-ethylene at low concentrations with methane-oxidizing bacteria in a biofilm reactor. Wat. Ras. Vol. 30, No. 8: 1885-1893.

Arrhenius, G., 1963. The Sea, vol. 3, p. 655. (N.M. Hill, ed.), Interscience, New York.

Aston, R.N., 1947. Chlorine dioxide use in plants on the Niagara boarder. J. Amer. Water Works Assoc., 39: 687.

Aten, A.H.W., 1961. Health Phys., 6: 114.

Baird, C. 1998. Environmental Chemistry. W.H. Greeman and Company, New York. 557 pp.

Baker, J. R., Milke, M.W. and Mihelcic, J.R. 1999. Relationship between chemical and theoretical oxygen demand for specific classes of organic chemicals. Wat. Res. Vol. 33, No. 2: 327-334.

Balmer, Peter, Blomqvist, Magnus and Lindholm, Margareta, 1968. Simultanfällning i en högbelastet aktivslamproces. Vatten, 24: No. 2, 112-116. (Simultaneous precipitation in high loaded activated sludge treatment).

Barry, R.G., 1978. Cryospheric responses to a global temperature increase. In: Carbon Dioxide, Climate and Society - Proceedings of an IIASA Workshop, Febr. 21-24, 1978 (Ed. Jim Williams). Pergamon Press, Oxford.

Bartell, S.M., Gardner, R.H. and O'Neill, R.V. 1992. Ecological Risk Estimation. Lewis Publishers, Boca Raton.

Baumester, W., 1958. Encyclopaedia of Plant Physiology, vol. 4, pp. 5, 482. (W. Ruhland, Ed.), Springer, Berlin.

Beccari, M., Di Pinto, A.C., Ramadori, R. and Tomei, M.C. 1992. Effects of dissolved oxygen and diffusion resistances on nitrification kinetics. Wat. Res. Vol. 26, No. 8: 1099-1104.

Beharrell, J., 1942. Nature, London, 149: 306.

Beltran, F.J., Garcia-Araya, J.F. and Acedo, B. 1994a. Advanced oxidation of atrazine in water-II. Ozonation combined with ultraviolet radiation. Wat. Res. Vol. 28, No. 10: 2165-2174.

Beltran, F.J., Garcia-Araya, J.F. and Acedo, B. 1994b. Advanced oxidation of atrazine in water-I. Ozonation. Wat. Res. Vol. 28, No. 10: 2153-2164.

Beltran, F.J., Gómez-Serrano, V. and Durán, A. 1992. Degradation kinetics of *p*-nitrophenol ozonation in water. Wat. Res. Vol. 26, No. 1: 9-17.

Bernhart, E.L., 1975. Nitrification in industrial treatment works. 2nd International Congress on Industrial Waste Water and Wastes, Stockholm, Febr. 4-7, 1975.

Berry, James W. et al., 1974. Chemical Villains: A Biology of Pollution. St. Louis: Mosby.

Bertrand, D., 1942. Annls. Inst. Pasteur, Paris, 68: 58.

Bertrand, D., 1950. Bull. Am. Mus. Nat. Hist., 94: 409.

Bertrand, G. and Levy, G., 1931. C. r. hebd. Sénanc. Acad. Sci., Paris, 192: 525.

Black, A.P., 1960. Basic mechanism of coagulation. J. Am. Water Works Ass., 52: No. 4, 492-501.

Black, W.A.P. and Mitchell, R.L., 1952. J. mar. biol. Ass. U.K., 30: 575.

Boeghlin, 1972. Les polyelectrolytes wt leur utilitation dans l'amelioration des traitements de clarification des eaux. Residuaire et dans les procedes de deshydratation mecanique des bones. Chemie et Industrie. Génie Chimique, 105: 527-534.

Bloor, J.C., Anderson, G.K. and Willey, A.R. 1995. High rate aerobic treatment of brewery waste water using the jet loop reactor. Wat.Res. Vol. 29, No. 5: 1217-1223.

Boirie, C., Boss, D., Hugot, G. and Platzer, R., 1962. Acta chim., hung., 33: 281.

Boliden Information FKK 120S (30/10-1967).

Boliden AVR, September 1969.

Bolin, B., Degens, E.T., Kempe S. and Ketner, P., 1979. The global carbon cycle. SCOPE Report 13. John Wiley and Sons, New York.

Boltzmann,L., 1905. The Second Law of Thermodynamics. (Populare Schriften, Essay No.3, address to Imperial Academy of Science in1886).

Bowen, H.J.M., 1960. Biochem. J., 77: 79.

Bowen, H.J.M., 1963. U.K. Atomic Energy Authority Report AERE-R 4196.

Bowen, H.J.M., 1966. Trace Elements in Biochemistry. A.P. New York, 241 pp.

Bowen, H.J.M., and Cawse, P.A., 1963. Analyst, London, 88: 721.

Bowen, H.J.M., 1966. Trace Elements in Biochemistry. A.P. New York, 241 pp.

Bowen, H.J.M. and Dymond, J.A., 1955. Proc. R. Soc., B. 144, 355.

Bowman, K.O., Hutcheson, K., Odum, E.P. and Shenton, L.R., 1970. Comments on the distribution.

Bridges, B.A., 1980. Introduction of Enzymes Involved in DNA Repair and Mutagenesis. In: Ciba Foundation Symposium 76, Environmental Chemicals,

Enzyme Function and Human Disease. Excerpta Medica, Amsterdam, pp. 67-81.

Brooksbank, W.A. and Leddicotte, G.W., 1953. J. Phys. Chem., 57: 819.

Brosset, C., 1973. Air-borne acid. Ambio, 2: 1-9.

Brown, C.C., 1978. The Statistical Analysis of Dose-Effect Relationships. In: G.C.

Brown,D.S. and Flagg,E.W., 1981. Empirical prediction of organic pollutant sorption in natural sediments. J. Environ. Qual., 10: 382-386.

Burd, R.S., 1968. Study of sludge handling and disposal. Water Poll. Contr. Res. Ser., Fed. Water Poll. Contr. Adm. Publ. WP-20-4, Washington, D.C.

Burton, I.D, Milbourne, G.M. and Russel, R.S., 1960. Relationship between the rate of fall-out and the concentration of strontium-90 in human diet in the U.K. Nature, 185: 498.

Butler, G.C.., (ed.), Principles of Ecotoxicology, SCOPE 12. John Wiley & Sons, New York.

Butler, G.C., 1972. Retention and excretion equations for different patterns of uptake. In: Assessment of Radioactive Contamination in Man, IAEA, Vienna, STI/PUB/290, p. 495.

Buzzell, James C. and Sawyer, Clair N., 1967. Removal of algal nutrients from raw waste with lime. Wat. Poll. Contr. Fed., 39: R 16.

Cairns Jr., J., Dickson, K.L., and Maki, A.W.. 1987. *Estimating Hazards of Chemicals to Aquatic Life*. STP. 675, American Society for Testing and Materials, Philadelphia.

Calow, P. 1998. Ecological Risk Assessment: Risk for What? How Do We Decide? Ecotoxicology and Environmental Safety, 40: 15-18.

Camp, T.R., 1946. Sedimentation and design of settling tanks. Trans. Amer. Soc.

Canham, R.A., 1955. Some problems encountered in spray irrigation of canning plant waste. Proc. k10th Ind. Wast. Conf.

Cannon, H.L., 1960. Science, N.Y., 132: 591.

Cannon, H.L., 1963. Soil Sci., 96: 196.

Chamberlin. T.A. et al., 1975. Colour removal from bleached kraft. TAPPI Env. Conf., May.

Chang, W.C., Ouyang, C.F., Chiang, W.L. and Hou, C.W. 1998. Sludge precycle control of dynamic enhanced biological phosphorus removal system: an application of on-line fuzzy controller. Wat. Res. Vol. 32, No. 3: 727-736.

Chau, Y.K. and Riley, J.P., 1965. Analytica chem. Acta, 33: 36.

Chem. Eng., 111: 895.

Chemical Week, 1976. Cyanide tamer looks promsin. 118: 44.

Chemistry, 1968. The lead we breathe. Chemistry, 41: 7.

Chichilnisky G, Heal, G., and Vercelli, A. 1998 Sustainability: Dynamics and Uncertainty. Kluwer Academic Publ. Dordrecht, Boston, London.
Chilean Iodine Educational Bureau, 1952. Iodine Content of Foods, London.

Chiou, C.T., Freed, V.H, Schmedding, D.W. and Kohnert, R.L., 1977. Partition coefficient and bioaccumulation of selected organic chemicals. Environmental Science and Technology 11: 475-478.

Clark, W.E., 1962. Prediction of ultrafiltration membrane performance. Science, 138: 148.

Cloud, Preston E., Jr., 1971. Resources, Population, and Quality of Life. In: S. Fred Singer, ed., Is There an Optimum Level of Population? New York, McGraw-Hill.

Cole, C.A., Paul, P.E. and Brewer, H.P., 1976. Odour control by hydrogen peroxide. J. Wat. Poll. Contr. Fed., 48: 297.

Commoner, B, 1971. The Closing Circle. Bantam Books, Inc. New York.

Connell, Des W. 1997. Basic Concepts of Environmental Chemistry. Lewis Publ., Bocca Raton, New York. 506 pp.

Council on Environmental Quality: Twelfth Annual Report of the Council on Environmental Quality, Washington D.C., 1982.

Cox, J.L., 1970. Accumulation of DDT residues in Triphoturus mexicanus from the Gulf of California. Nature, 227: 192-193.

Crittenden, J.C., Hu, S., Hand, D.W. and Green, S.A. 1999. A kinetic model for H_2O_2/UV process in a completely mixed batch reactor. Wat. Ras. Vol. 33, No. 10: 2315-2328.

Cullen, J. and Forsberg, C. , 1988. Swedish Report on the Examination of 1000 lakes in Sweden. In Swedish. Naturvårdsverket, Stockholm.

Culp, Gordon, 1967a. Chemical treatment of raw sewage 1. Water Wastes Eng., July 61-63, vol. 4.

Culp, Gordon, 1967b. Chemical treatment of raw sewage 2. Water Wastes Eng., Oct. 54-57, vol. 4.

Culp, G., Hansen, S. and Richardson, G., 1968. High-rate sedimentation in water treatment works. J. Am. Water Works Ass., 60: 681.

Dalton, F.E., Stein and Lynam, H.T., 1968. Land reclamation - a complete solution to the sludge and solids disposal problem. J. Water Poll. Contr. Fed., 40: 789.

Dancraft Management. 1997. Information about Ionex, a commercial ion exchanger.

Davidson, G. and Ullman, P., 1971. Use of prickling acid for precipitation of municipal waste water. Vatten, 27: No. 1, pp. 95-106.

Davis, G.E., Foster, J. and Warren, C.E., 1963. The influence of oxygen concentration on the swimming performance of juvenile Pacific salmon at various temperatures. Trans. Am. Fish Soc. 92, 111.

Davis, Wayne, H., 1970. Overpopulated America. New Republic, January 10.

de Bernardi, R., and Giussani, G., (editors), 1995. Biomanipulation in Lakes and Reservoirs Management. ILEC, UNEP. 208 pp.

de Freitas, A.S.W., and Hart, J.S., 1975. Effect of body weight on uptake of methyl mercury in fish. Water Quality Parameters, ASTM STP 573, Amer. Soc. Testing Materials, p. 356.

DeSloover, J. & LeBlanc, F. (1970): Pollutions atmosphériques et fertilité chez les mousses et chez les lichens épiphytiques. Acad. Soc. Lorr. Sci. Bull. 9:82-90

DEPA, Danish Environmental Protection Agency, 1996. A Strengthened Product-Oriented Environmental Effort, Copenhagen (in Danish).

DEPA, Danish Environmental Protection Agency, 1998. Account of the Danish Environmental Protection Agency on the Product-Oriented Environmental Effort, Copenhagen (in Danish).

Diab, S., Kochba, M. and Avnimelech. Y. 1993. Nitrification Pattern in a fluctuating anaerobic-aerobic pond environment. Wat. Research. vol. 27, No. 9: 1469-1475.

Dilek, F.B., Gokcay, C.F. and Yetis. U. 1998. Combined effects of Ni(II) and Cr(VI) on activated sludge. Wat. Res. Vol. 32, No 2: 303-312.

Dharmappa, H.B., Vering, J., Fujiwara, O. and Vigneswaran, S. 1993. Technical Note. Optimal Design of a Flocculator. Wat. Res. Vol. 27, No.3: 513-519.

Dillon, P.J. and Rigler, F.H., 1974. A test of a simple nutrient budget model predicting the phosphorus concentration in lake water. J. Fisk. Res. Board Can., 31: 1771-1778.

Dillon. P.J. and Kirchner, W.B., 1975. The effects of geology and land use on the export of phosphorus from watersheds. Water Res., 9:, 135-148.

Downing, A., 1966. Advances in water quality improvement, 1. University of Texas Press, Austin, Texas, April, 1966.

Drijvers, D., Langenhoven van H. and Beckers, M. , 1999. Decomposition of phenol and trichloroethylene by the ultrasound/H2O2/CuO process. Wat. Res. Vol. 33,

No.5:1187-1194.

Dryden, F.D. and Stern, G., 1968. Renovated waste water creates recreational environ. Science Tech., 2: 1079.

Dye, W.B., Bretthauer, E., Seim, H.J. and Blincoe, C., 1963. Analyt. Chem., 35: 1687.

Dyndgaard, R. and Kryger, J. 1999. Environmental Management System. In: S.E. Jørgensen: A Systems Approach to the Environmental Analysis of Pollution Minimization."Lewis Publishers, pp. 67-87.

Eberle, S.H., Donnert, D. and Stöber, H., 1976. Möglichkeiten des Einsatzes von Aluminium Oxide zur Reinigung organisch belasteter Abwässer. Chem. - Ing. - Tech., 48: 731.

Eckenfelder, W.W., Lawler, J.P. and Walsh, J.T., 1958. Study of fruit and vegetable processing waste disposal methods in the eastern region. U.S. Department of Agriculture, Final Report, September.

Edwards, J.D., Ph.D., P.E. 1995. Industrial Waste water Treatment. A Guidebook. Lewis Publishers, CRC, Boca Raton, New York, London, Tokyo. 170 pp.

Ehrlich, Paul R. and Raven, Peter H., 1969. Differentiation of populations. Science, 165: 1228-1232.

Eisenhauer, H.R., 1968. The ozonation of phenolic wastes. J. W. P. Contr. Fed., 40: 1887, 1896.

Ekholm, P. and Krogerus, K. 1998. Bioavailability of phosphorus in purified municipal waste waters. Wat. Res. Vol. 32, No. 2: 343-351.

Ellis, K.V., Cotton, A.P. and Khowaja, M.A. 1993. Iodine disinfection of poor quality waters. Wat. Res. Vol. 27, No. 3: 369-375.

Ellis, K.V. and Rodrigues, P.C. 1995. Developments to the first-order, complete-mix design approach for stabilisation ponds. Wat. Res. Vol. 29, No. 5: 1343-1351.

Ericsson, Bernt and Westberg, Nils, 1968. Use of flocculants by reduction of phosphorus in waste water. Vatten, 24: No. 2, 125-31.

Eriksson, C. and Mortimer, D.C., 1975. Mercury uptake in rooted higher plants - laboratory studies. Verh. Internat. Verein. Limnol., 19: 2087.

Ettala, M., Koskela, J. and Kiesilä, A. 1992. Removal of chlorophenols in a municipal sewage treatment plant using activated sludge. Wat. Res. Vol. 26, No. 6: 797-804.

Fair, G.M., Geyer J.C. and Okun, D.A., 1968. Water and Waste Engineering. J. Wiley and Sons, Inc., New York.

Fenger, J and Tjell, J.C., (Editors.), (1994). Luftforurening (Air Pollution, in Danish) Polyteknisk Forlag, 479 pp.

Ferguson, W.S. and Armitage, E.R., 1944. J. Agric. Sci., 34: 165.

Forbes, R.M., Cooper, A.R. and Mitchell, H.H., 1954. J. Biol. Chem., 209: 857

Force, H. and Morton, R.A., 1952. Biochem. J., 51: 598

Fukai, R. and Meinki, W.W., 1959. Nature, London, 184: 815.

Fukai, R. and Meinki, W.W., 1962. Limnol, Oceanorg., 7: 186.

Gardner, M.R. and Ashby, W.R., 1970. Connectance of large synamical (cybernetic) systems: critical values for stability. Nature, 288: 784.

Gellman, I. and Blosser, R.O., 1959. Proc. 14th Ind. Waste Conf., Purdue University.

Geyer, H., Sheehan, P., Kotzias, D., Freitag, D. and Korte,F., 1982. Chemosphere, 11: 1121.

Glass, C. and Silverstein, J. 1999. Denitrification of high-nitrate, high-salinity waste water. 1999. Wat. Res. Vol. 33, No. 1: 223-229.

Gleason, G.H. and Loonam, A.C., 1933. The development of a chemical process for treatment of sewage. Sewage Works Journal p. 61.

Golley, Frank B., 1960. Energy dynamics of a food chain of an old-field community. Ecol. Monogr., 30: 187-206.

Gooch, N.P. and Francis, N.L., 1975. A theoretically based mathematical model for calculation of electrostatic precipitator performance. J. Air Pollut. Contr. Assoc., 25: 108-113.

Gould, J.P. and Weber, W.J., 1976. Oxidation of phenols by ozone. J. Wat. Poll. Contr. Fed., 48.

Grahn, O., Hultberg, H. and Landner, L., 1974. Oligotrophication - a self-Accelerating Process in Lakes Subjected to Excessive Supply of Acid Substances. Ambio vol. 3, No. 2, pp. 93-94.

Granstrom, M.L. and Lee, G.F., 1958. Generation and use of chlorine dioxide. J. Amer. Water Works Assoc., 50: 1453.

Greenway, M. 1997. Nutrient content of wetland plants in constructed wetlands receiving municipal effluent in tropical Australia. Wat. Sci. Tech. Vol. 35, No. 5: 135-142.

Groeneweg, J., Sellner, B. and Tappe, W. 1994. Ammonia oxidation in nitrosomonas at NH3 concentrations near *Km*: effects of pH and temperature. Wat. Res. Vol. 28, No. 12: 2561-2566.

Gromiec, M.J., 1983. Biochemical Oxygen Demand - Dissolved Oxygen. River Models. p. 131-226. In: Application of Ecological Modelling in Environmental Management. Part A. Elsevier, Amsterdam.

Gschlöbl, T., Steinmann, C., Schleypen, P. and Melzer, A. 1998. Constructed wetlands for effluent polishing of lagoons. Wat. Res. Vol. 32, No. 9: 2639-2645.

Gustafson, Bengt and Westberg, Nils, 1968. Information från Statens Naturvårdsverk, vol. 3, p. 13-24.

Halling-Sørensen, B. and Jørgensen, S.E. 1993. Studies in Environmental Science 54. The Removal of Nitrogen Compounds From Waste water. Elsevier, Amsterdam, London, New York, Tokyo. 444 pp.

Hamaguchi, H., Kuroda, R. and Hosohara, K., 1960. J. atom. Energy Soc., Japan, 2: 317.

Hanf, E.B., 1970. A guide to scrubber selection. Environ. Sci. Technol., 4: 110-115.

Hansen, A. and Søltoft, P., 1988. Chemical Unit Operation. 3. edition. Polyteknisk Forlag. Copenhagen. 524 pp.

Hansen, K.H., Angelidaki, I. and Ahring, B.K. 1998. Anaerobic digestion of swine manure: inhibition by ammonia. Wat. Res. Vol. 32, No. 1: 5-12.

Hansen, S.E. and Jørgensen, S.E. 1988. Introduction to Environmental Management. Elsevier. Amsterdam. 390 pp.

Harrison, G.E. and Sutton, A., 1963. Nature, London, 197: 809.

Hasebe, S. and Yamamoto, K., 1970. Studies of the removal of cadmium ions from mine water by utilizing xanthate as selective precipitate. Int. congr. on Ind. Waste Water, Stockholm.

Hauck, A.R. and Sourirajan, S., 1972. Reverse osmosis treatment of diluted nickel plating solution. J. Wat. Poll. Contr. Fed., 44: 372.

Hauschild, M. and Wenzel, H. 1999. Life Cycle Assessment - Environmental Assessment of Products. In: A Systems Approach to the Environmental Analysis of Pollution Minimization. Lewis Publishers, pp. 155-191.

Henderson, E.H., Parker, A. and Webb, M.S.W., 1962. U.K. Atomic Energy Authority Report AERE-R 4035.

Holluta, t., 1963. Das Ozon in der Wasserchemie. GWF, 104: 1261.

Hovmand, M.F., 1981. Circulation of lead, cadmium, cobber, zinc and nickel in Danish Agriculture. Im : The Application of Sludge in Agriculture volume II. Polyteknisk Forlag p. 85-118. In Danish

Howard, P.H. et al., 1991. Handbook of Environmental Degradation Rates. Lewis

Publishers. New York.

Howard. P.H. Handbook of Environmental fate and Exposure Data. Lewis Publishers. New York.
Volume I. Large Production and Priority Pollutants. 1989
Volume II. Solvents. 1990.
Volume III. Pesticides. 1991.
Volume IV. Solvents 2. 1993.
Volume V. Solvents 3. 1998.

Howard, R., 1971. Paper presented to the American Assoc. for Advancement of Science, 1971 meeting.

Huang, C-P., Wang, H-W. and Chiu, P-C. 1998. Nitrate reduction by metallic iron. Wat. Res. Vol. 32, No. 8: 2257-2264.

Hummel, J.R. and Reck, R.A., 1978. Development of a global surface Albedo Model, presented at the Meteorology Section, American Geophysical Union, San Francisco, December 1977. GMR-2607, General Motors Research Laboratories, Warren, Michigan.

Hunter, J.G., 1942. Nature, London, 150: 578.

Hunter, J.G., 1953. J. Sci. Fd. Agric., 4: 10.

Hutcheson, K., 1970. A test for comparing diversities based on the Shannon formula. J. Theor. Biol., 29: 151-154.

Hwang, Y., Matsuo, T., Hanaki, K. and Suzuki, N. 1994. Removal of odorous compounds in waste water by using activated carbon, ozonation and aerated biofilter. Wat. Res. Vol. 28, No. 11: 2309-2319.

Hynes, H.B.N., 1971. Ecology of Running Water. Liverpool University Press, Liverpool, England.

Hägg, G., 1979. Kemisk Reaktionslære. Almquist and Wiksell, Stockholm.

ICRP, 1977. Principles and Methods for Use in Radiation Protection Assessments.

Ingols, R.S. and Fetner, R.H., 1957. Proc. Soc. Water Treatment Exam., 6: 8.

International Commission on Radiological Protection, Publication 6. Recommendations of the committee on permissible doses for internal radiation, 1964. Pergamon Press, Oxford.

Islam, S. and Suidan, M.T. 1998. Electrolytic denitrification: long term performance and effect of current intensity. 1998. Wat. Res. Vol 32, No. 2: 528-536.

Jenkins, David, Ferguson, John F. and Menar, Arnolds B., 1971. Chemical processes for phosphate removal. Water Research, vol. 5, No. 7, p. 369-390.

Jensen, J., Nielsen, S., and Halling-Sørensen. Report (in Danish) on the environmental risk assessment of drugs. (The Danish Environmental Protection Agency, 1998).

Jensen, K.S., Moth Iversen, T., and Lindegaard, C., 1988. Freshwater Ecology, in Danish, published by Freshwater-Biological Laboratory, Hillerød, Copenhagen University.

Jensen, K.W. and Snevik, E., 1972. Low pH levels wipe out salmon and trout populations in southernmost Norway. AMBIO, vol. 1, No. 6, pp. 223-225.

Jervis, R.E., Perkons, A.K., Mackintosh, W.D. and Kerr, K.F., 1961. Modern trends in activation analysis. Proc. Int. Conf., Texas, p. 107.

Johnsen, I. & Søchting U (1973): Influence of air pollution on the epiphytic lichen vegetation and bark properties of deciduous trees in the Copenhagen area. Oikos 24 (1973) 344-351.

Johnson, J.M. and Butler, G.W., 1957. Physiologia Pl., 10: 100.

Jones, L.R., Owen, S.A., Horrell, P. and Burns, R.G. 1998. Research Note. Bacterial inoculation of granular activated carbon filters for the removal of atrazine from surface water. Wat. Res. Vol. 32, No. 8: 2542-2549.

Jonge de, R.J., Breure, A.M. and van Andel, J.G. 1996. Bioregeneration of powdered activated carbon (PAC) loaded with aromatic compounds. Wat. Res. Vol. 30, No. 4: 875-882.

Judd, S.J. and Jeffrey, J.A. 1995. Research Note. Trihalomethane formation during swimming pool water disinfection using hypobromous and hypochlorous acids. Wat. Res. Vol. 29, No. 4: 1203-1206.

Juhola, A.J. and Tupper, F., 1969. Laboratory investigation of the regeneration spent granular activated carbon. FWPCA Report, No. TWRC-7.

Jørgensen, L.A., Jørgensen, S.E. and Nielsen, S.N., 2000. Ecotox, a CD with 120,000 ecological and ecotoxicological data. Elsevier, Amsterdam, Oxford, New York, Tokyo.

Jørgensen, S.E., 1968. The purification of waste water containing protein and carbohydrate. Vatten 24(4): 332-338.

Jørgensen, S.E., 1969. Purification of highly polluted waste water by protein precipitation and ion exchange. Vatten, 25(3): 278-288.

Jørgensen, S.E., 1970. Ion exchange of waste water from the food industry. Vatten, 26(4): 350-357.

Jørgensen, S.E., 1971. Precipitation of proteins in waste water. Vatten, 27(1): 58-72.

Jørgensen, S.E., 1973a. A new advanced waste water treatment plant in Ätran, Falkenberg, Sweden. Vatten, 29(1): 36-40.

Jørgensen, S.E., 1973b. The combination precipitation - ion exchange for waste water from the food industry. Vatten, 29(1): 40-42.

Jørgensen, S.E., 1973. Industrial Waste Water Treatment by Precipitation and Ion Exchange. In: Environmental Engineering, Eds. G. Lindner and K. Nyberg. D. Reidel Publ. Co., Holland, U.S.A., 239 pp.

Jørgensen, S.E., 1974. Recirculation of waste water from the textile industry. Vatten, 29: 364.

Jørgensen, S.E., 1975. Recovery of ammonia from industrial waste water. Water Res., 9: 1187.

Jørgensen, S.E., 1976a. A eutrophication model for a lake. J. Ecol. Model., 2: 147-165.

Jørgensen, S.E., 1976b. An ecological model for heavy metal contamination of crops and ground water. Ecol. Modelling, 2: 59-67.

Jørgensen, S.E., 1976c. Reinigung häuslicher Abwässer durch Kombination eines chemischen Fällungs- und Ionenaustausch Verfahrens. (Doctor thesis, Karlsruhe).

Jørgensen, S.E., 1976d. Recovery of phenols from industrial waste water. Prog. Wat. Tech., 8: 2/3, 65-79.

Jørgensen, S.E., 1978. The application of cellulose ion exchanger to industrial waste water management. Water Supply Management, 12.

Jørgensen, S.E., 1979a. A holistic approach to ecological modelling. Ecol. Model., 7: 169-189.

Jørgensen, S.E., 1979b.. Modelling the distriburion and effect of heavy metals in an aquatic ecosystem. Ecol. Model., 6: 199-222.

Jørgensen, S.E., 1979c. Removal of phosphorus by ion exchange. Vatten, 34: 179-182.

Jørgensen, S.E., 1979d. Industrial Waste Water Management. Elsevier Amsterdam, New York, Oxford. 388pp.

Jørgensen, S.E., 1980. Ecological submodels for Upper Nile Lake System. WMO Report.

Jørgensen, S.E., 1981. Application of Ecological Modelling in Environmental Management. Elsevier.

Jørgensen, S.E., 1986. Structural dynamic model. Ecol. Modelling, 31: 1-9.

Jørgensen, S.E. 1993. Removal of heavy metals from compost and soil by ecotechnological methods. Ecological Engineering 2: 89-100.

Jørgensen, S.E., 1994. Fundamentals of Ecological Modelling (2nd Edition), Developments in Environmental Modelling, 19. Elsevier, Amsterdam, 628 pp.

Jørgensen, S.E., 1997. Integration of Ecosystem Theories: A Pattern. Second revised edition. Kluwer Acad. Publ. Dordrecht, Boston, London. 388 pp.

Jørgensen, S.E. 1999. A Systems approach to the Environmental Analysis of Pollution Minimisation. Lewis Publishers, 352 pp.

Jørgensen, S.E., Jacobsen, O.S. and Hoi, I., 1973. A prognosis for a lake. Vatten, 29: 382-404.

Jørgensen, S.E. and Mejer, J.F., 1977. Ecological buffer capacity. Ecol. Model., 3: 39-61.

Jørgensen, S.E., and Gromiec, M.J., 1985. Mathematical Models in Biological Waste Water Treatment. Editors. Elsevier, Amsterdam, Oxford, New York, Tokyo. 802 pp.

Jørgensen, S.E. and Vollenweider, R.A. 1988. Guidelines of Lake Management, I: Principles of Lake Management, ILEC/UNEP, 199 pp.

Jørgensen, S.E., and Johnsen, I. 1989. Principles of Environmental Science and Technology. Studies in Environmental Science 33. Elsevier, Amsterdam, Oxford, New York, Tokyo. 628 pp.

Jørgensen, S.E. and H. Löffler (eds.). 1990. Guidelines of Lake Management. Volume 3: Lake Shore Management. 175 pp.

Jørgensen, S.E., Nors Nielsen, S., and Jørgensen, L.A. 1991. Handbook of Ecological Parameters and Ecotoxicology, Elsevier, Amsterdam.

Jørgensen, S.E., Halling-Sørensen, B. and Nielsen, S.N., 1995a. Handbook of Environmental and Ecological Modeling. CRC Lewis Publishers, Boca Raton, New York, London, Tokyo, 672 pp.

Jørgensen, S.E., Nielsen, L.K., Ipsen, L.G.S. and Nicolaisen, P. 1995b. Lake restoration using a reed swamp to remove nutrients from non-point sources. Wetland Ecology and Management vol.3 no. 2: 87-95.

Jørgensen, S.E., Halling-Sørensen, B. and H. Mahler. 1997. Handbook of Estimation Methods in Ecotoxicology and Environmental Chemistry. Lewis Publishers, Boca Raton, Boston, London, New York, Washinton, D.C. 230 pp.

Jørgensen, S.E. and de Bernardi, R., 1998. The use of structural dynamic models to explain the success and failure of biomanipulation. Hydrobiologia 379: 147-158.

Jørgensen, S.E. and Mitsch, W. 1999. Ecological Engineering. In: A Systems Approach to the Environmental Analysis of Pollution Minimisation. Lewis Publishers, pp. 225-241.

Jørgensen, S.E., Patten, B.C. and Straskraba, M., 1999. Ecosystems emerging: 3. Openness. Ecological Modelling, 117: 41-64.

Kamp Nielsen, L., 1986. Modelling of eutrophication processes. In: Frangipane, E.F. (editor): Lakes Pollution and Recovery. Proc. Internat. EWPCA Congr., 15-18. April 1985, Rome, p.61-101.

Kamp Nielsen, L., 1989. The relation between external P-load and in-lake P-concentration. Chapter III in: H. Sas (editor): Lake Restoration by Reduction of Nutrient Loading. Academia Verlag. Sankt Augustin, Germany.

Kapoor, A. and Viraraghavan, T. 1998. Research note. Removal of heavy metals from aqueous solutions using immobilized fungal biomass in continuous mode. Wat. Res. Vol. 32, No. 6: 1968-1977.

Kehoe, R.A., Cholak, J. and Story, R.V., 1940. J. Nutr., 19: 579.

Khan, Y., Anderson, G.K. and Elliott, D.J. 1999. Wet oxidation of activated sludge. Wat. Res. Vol. 33, No. 7: 1681-1687.

Khandelwal, K.K., Barduhn, A.J. and Grove, C.S., Jr., 1959. Kinetics of ozonation of cyanides. Ozone Chemistry and Technology, Adv. Chem. Ser., 21, 78. Amer. Chem. Soc., Washington, D.C.

King, E.J. and Belt, T.H., 1938. Physiol. Rev., 18: 329.

King, R.C., 1957. Am. Nat., 41: 319.

Kirk-Othmer, 1967. Encyclopedia of Chemical Technology. John Wiley, New York. Sec. Ed., 14: 410-43.

Klein, L., 1966. River Pollution 3. Control Butterworth and Co. Ltd., Washington, D.C.

Klostermann, J.E.M., and Tukker, 1998. A. Product Innovation and Eco-Efficiency. Kluwer Academic Publ. Dordrecht, Boston, London.

Koch, R.C. and Roesmer, J., 1962. J. Fd. Sci., 27: 309.

Koch, R.C. and Keisch, B., 1963. U.S. Atomic Energy Commission Report, NSEC-100.

Koch, G.W. and Mooney, H.A., 1996. Carbon Dioxide and Terrestrial Ecosystems. Academic Press. New York. 443 pp.

Koczy, F.F. and Titze, H., 1958. J. Mar. Res., 17: 302.

Kraus, K.A., Shore, A.J. and Johnson, J.S., 1967. Hyperfiltration studies. Desalination, 2: 243.

Kringsley, D., 1959. Nature, London, 183: 770.

Kristensen, P., Jensen, .P., and Jeppesen E., 1990. Eutrophication Models for Lakes. Research Report C9. DEPA. Copenhagen, 120pp.

Kryger, J. and Dyndgaard, R. 1999. Introduction to Cleaner Production. In: A

Systems Approach to the Environmental Analysis of Pollution Minimisation. Lewis Publishers, pp. 87-101.

Ku, Y., Chang, J-L., Shen, Y-S. and Lin, S-Y. 1998. Decomposition of diazinon in aqueous solution by ozonation. Wat. Res. Vol. 32, No. 6: 1957-1963.

Lahav, O. and Green, M. 1998. Ammonium removal using ion exchange and biological regeneration. Wat. Res. Vol. 32, No. 7: 2019-2028.

Lasaridi, K.E. and Stentiford, E.I. 1998. A simple respirometric technique for assessing compost stability. Wat. Res. vol. 32, No. 12: 3717-3723.

Lawrence, A. and McCarty, P.L., 1967. Kinetics of methane fermentation in anaerobic waste treatment. Tech. Rep. 75, Department of Civil Engineering, Stanford University, Stanford, California, Feb. 1967.

Leddicotte, G.W., 1959. International Commission on Radiological Protection. Report of committee on permissible doses for internal radiation. Pergamon Press, New York and Oxford.

Lee, G.F. and Kluesener, J.W., 1971. Nutrient sources for Lake Wingra, Madison, Wisconsin. Rep. Univ. Wisconsin Water Chemistry Program, December 1971.

Leeuwen, van C.J., and Hermens, J.L.M., 1995 (editors). Risk assessment of Cehmicals: An Intorduction. Kluwer Acad. Publ. Dordrecht, Boston and London. 374 pp.

Leith, D. And Licht, W., 1972. The collection of cyclone type particle collectors - a new theoretical approach. AICHE Symp. Ser. 68(126): 196-206.

Lemons, J., Westra, L., and Goodland, R. 1998. Ecological Sustainability and Integrity: Concepts and Approaches. Kluwer Academic Publ. Dordrecht, Boston, London.

Levy, N., Magdassi, S. and Bar-Or, Y. 1992. Physico-chemical aspects in flocculation of bentonite suspensions by a cyanobacterial bioflocculant. Wat. Res. Vol. 26, No. 2: 249-254.

Lieth, H. and Whittaker, R.H., 1975. Primary Productivity of the Biosphere. Springer-Verlag. pp. 203-215.

Lindegaard, C., Dall, P.C., and Jacobsen, D., 1999. Biological Evaluation of Streams contaminated by Organic Matter. Compendium. The Freshwater Biological Laboratory, Copenhagen University. 60 pp. In Danish

Loehr, R.C., 1974. Characteristics and comparative magnitude of nonpoint sources. J. Wat. Poll. Contr. Fed., 46: 1849-1872.

Logsdon, G.S. and Symons, J.M., 1973. Mercury removal by concentional water treatment techniques. J. AWWA: 554-562.

Long, C., 1961. Biochemists Handbook. Spon. London.

Lounamaa, J., 1956. Suomal. eläin-ja kasvit. Seur. van Julk, 29: No. 4.

Low, E.M., 1949. J. Mar. Res., 8: 97.

Lu, J.C.S. and Chen, K.Y., 1977. Migration of Trace Metals in Interface of Seawater and Polluted Superficial Sediments. Env. Science and Technology. Vol. 1. pp. 174-182.
Luley, H.G., 1963. Spray irrigation of vegetable and fruit processing wastes. J. Wat. Contr. Fed., 35: 1252.

Lumb, C., 1951. Heat treatment as an aid to sludge dewatering - ten years full-scale operation. J. Proc. Inst. Sew. Purif. part 1, 5.

Lundelius, E.F., 1920. Adsorption and solubility. Kolloid Z., 26: 145.

Luo, Bin., Patterson, J.W. and Anderson, P.R. 1992. Kinetics of cadmium hydroxide precipitation. Wat. Res. Vol. 26, No. 6: 745-751.

Lux, H., 1938. Z. anorg. allg. Chem., 240: 21.

Lyman, W.J., Reehl, W.F. and Rosenblatt, D.H. 1990. Handbook of Chemical Property Estimation Methods. Environmental Behaviour of Organic Compounds. American Chemical Society.

Lønholdt, Jens, 1973. The BOD_5, P and N content in raw waste water. Stads- og Havneingeniøren, 7: 1-6.

Lønholdt, Jens, 1976. Nutrient engineering WMO Training Course on Coastal Pollution (DANIDA): 244-261.

Mackay, D., 1991. Multimedia Environmental Models. The fugacity Approach. 257 pp. Lewis Publishers. Boca Raton, Ann Arbor, London and Tokyo.

Mackay, D. and Paterson, S., (1982). Fugacity revisited. Environ. Sci. Technol. 16: 654A-660A.

Mackay, D., Shiu, W.Y., and Ma, K.C. . Illustrated Handbook of Physical-Chemical Properties and Environmental Fate for organic Chemicals. Lewis Publishers.
Volume I. Mono-aromatic Hydrocarbons. Chloro-benzens and PCBs. 1991.
Volume II. Polynuclear Aromatic Hydrocarbons, Polychlorinated Dioxines, and Dibenzofurans. 1992.
Volume III. Volatile Organic Chemicals. 1992.

Mackle, W., Scott, E.W. and Treon, J., 1939. Am. J. Hyg., 29A: 139.

Manabe, S. and Wetherald, R.T., 1975. The effect of doubling the CO_2 concentration on the climate of a general circulation model. J. Atmos. Sci., 32: 3-15.

Margalef, R., 1963. Successions of populations. Adv. Frontiers of Plant Sci. (Instit. Adv. Sci. and Culture, New Delhi, India), 2: 137-188.

Margalef, R., 1968. Perspectives in Ecological Theory. University of Chicago Press,

Chicago. 122 pp.

Margalef, R., 1969. Diversity and stability: A practical proposal and a model for interdependence. In: Diversity and Stability in Ecological Systems. (Woodwell & Smith, eds.).

Matsumura, S., Kokubu, N., Watanabe, S. and Sameshima, Y., 1955. Mem. Fac. Sci., Kyushu University Ser. C2: 81.

Matter-Müller, C., Gujer, W., Giger, W., and Stumm, W., 1980. Non-biological elimination mechanisms in biological sewage treatment plant. Prog. Wat. Tech., 12: 299-314.

May, R.M., 1974. Stability in ecosystems: Some comments. Proc. First Int. Congr. Ecol., p. 67.

May, R.M., 1975. Patterns of species abundance and diversity. Chapter 4 (pp. 81-120). In: M.L. Cody and J.M. Diamond, eds. Ecology and evolution of communities. Harvard Univ. Press, Cambridge, Mass.

May, R.M. 1977. Stability and Complexity in Model Ecosystems. 3. edition. Princeton University Press, Princeton, N.J. 530 pp.

Mayer, A.M. and Gorham, E., 1957. Ann. Bot., 15: 247.

Mazet, M., Farkhani, B. and Baudu, M. 1994. Influence of heat or chemical treatment of activated carbon onto the adsorption of organic compounds. Wat. Res. Vol. 28, No. 7: 1609-1617.

Mazierski, J. 1994. Effect of chromium (CrVI) on the growth rate of denitrifying bacteria. Wat. Res. Vol. 28, No. 9: 1981-1985.

McCance, R.A. and Widdowson, E.P., 1960. The Composition of Foods. H.M. Stationary Office.

McCarty, P.L., 1964. Anaerobic waste treatment fundamentals. Publ. Works, 95: 9, 107; 95: 10, 123; 95: 11, 19; 12, 95.

McCarty, L.S., Mackay, D., Smith, A.D., Ozburn, G.W. and Dixon, D.G. 1992. Residue interpolation of toxicity and bioconcentration QSARs from aquatic bioassay: neutral narcotic organics. Environmental Toxicology and Chemistry, Vol. 11: 917-930.

McConnell, K.P., 1961. Modern Trends in Activation Analysis. Proc. Int. Conf., Texas, p. 137.

Meadows, Donella H. et al., 1972. The Limits to Growth. New York: Universe Books for Potomac Assosiates.

Meiring, P.G.J. and Oellermann, R.A. 1995. Biological removal of algae in an integrated pond system. Wat. Sci. Tech. Vol. 31, No. 12: 21-31.

Middlebrooks, E.J., Middlebrooks, C.H., Reynolds, J.H., Watters, G.Z., Reed, S.D.

and George, D.B. 1982. Waste water Stabilisation Lagoon Design, Performance and Upgrading. Macmillan Publishing Co., NeMayer, A.M. and Gorham, E., 1957. Ann. Bot., 15: 247.

Miljøstyrelsen No. 5, 1998. Biological Evaluation of Danish Streams. 39 pp. In Danish.

Milne, G.W.A. 1994. CRC Handbook of Pesticides. CRC.

Mishara, V.S., Joshi, J.B. and Mahajani, V.V. 1994. Kinetics of wet air oxidation of diethanolamine and morpholine. Wat. Res. Vol. 28, No. 7: 1601-1608.

Mitchell, R.L., 1944. Proc. Nutr. Soc., 1: 183.
Mitsch, W.J. 1993. Ecological engineering - a cooperative role with the planetary life-support systems. Environmental Science & Technology, 27: 438-445.

Mitsch, W.J. 1993. Ecological engineering - a cooperative role with the planetary life-support systems. Environmental Science & Technology, 27: 438-445.

Mitsch, W.J. 1996. Ecological engineering: A new paradigm for engineers and ecologists. In: P.C. Schulze (ed.), Engineering within Ecological Constraints. Washington, DC. National Academy Press, pp. 111-128.

Mitsch, W.J. 1998. Ecological engineering - the seven-year itch. Ecological Engineering, 10: 119-138.

Mitsch, W.J. and Jørgensen, S.E. (eds.). 1989. Ecological Engineering. An Introduction to Ecotechnology. John Wiley & Sons, New York, Chichester, Brisbane, Toronto, Singapore, pp. 430.

Mitsch, W.J. and Gosselink, J.G. 1993. Wetlands. Second Edition. Van Nostrand Reinhold, New York, NY. 722 pp.

Miyaji, I. and Cato, K., 1975. Biological treatment of industrial waste water by using nitrate as an oxygen source. Wat. Research, 9: 95.

Mohr, Eugen, 1969. Konventionelle oder Ionenaustauschanlage. Galvanotechnik, 8: 60.

Moiseenko, U.I., 1959. Geochemistry, 117.

Monier-Williams, G.W., 1950. Trace Elements in Food. Wiley and Sons, New York.

Moody, G.J. and Thomas, J.D.R., 1968. Analyst, 93: 557.

Moon, F.E. and Pall, A.K., 1944. J. Agric. Sci., Camb., 34: 165.

Moore, E.W., 1951. Fundamentals of chlorination of sewage and wastes. Water and Sewage Works, 98(3): 130.

Moreno, A., Ferrer, J., Bevia, F.R., Prats, D., Vazquez, B. and Zarzo, D. 1994.
Las monitoring in a lagoon treatment plant. Wat. Res. Vol. 28, No. 10: 2183-2189.

Moriarty, Frank, 1972. Pollutants and food chains. New Scientist, March 16, pp. 594-596.

Morowitz, H.J., 1968. Energy flow in biology. Biological Organisation as a Problem in Thermal Physics. Academic Press, N.Y. 179 pp. (See review by H.T. Odum, Science, 164: 683-84 (1969)).

Mortimer, D.C. and Kundo, A., 1975. Interaction between aquatic plants and bed sediments in mercury uptake from flowing water. J. Environ. Qual., 4: 491.

Moth Iversen, T. and Lindegaard, C., 1983. Biological assessment of water courses, in Danish, published by Freshwater-Biological Laboratory, Hillerød, Copenhagen University.

Moth Iversen, T., Lindegaard, C., Sand Jensen, K., and Thorup J., 1989. Water Course Ecology, in Danish, published by Freshwater-Biological Laboratory, Hillerød, Copenhagen University.

Mulligan, T. J. and Fox, R. D., 1976. Removal of dye stuff. Chem. Eng.: 49-66.

Mullin, J.B. and Riley, J.P., 1956. J. mar. Res., 15: 103.

Muth, H., Rajewsky, B., Hantke, H.J. and Aurand, K., 1960. Health Phys., 2: 239.

Narbaitz, R.M. and Cen, J. 1994. Electrochemical regeneration of granular activated carbon. Wat. Res. Vol. 28, No. 8: 1771-1778.

Neufeld, A.H., 1936. Can. J. Res., 14B: 160.

Neufeld, R.D. and Thodos, g., 1969. Removal of orthophosphates from aqueous solutions with aluminia. Environ. Sci. Tech., vol. 3, p. 661.

Newman, W.F., 1949. Pharmacology and Toxicology of Uranium Compounds. McGraw-Hill, New York.

Niemczynowicz, J. 1994. New aspects of urban drainage and pollution reduction towards sustainability. In: Wat. Sci. Tech., vol. 30, No. 5, London.

Nilsson, Rolf, 1969. Precipitation of phosphates at municipal waste water plants. Stads- og Havneingeniøren, 4: 1-8.

Nordström, R.J., McKinnon, A.E. and de Freitas, A.S.W., 1975. A bioenergetics-band model for pollutant accumulation by fish. Simulation of PCB and methyl mercury levels in Ottawa River perch (Perca flevescens). J. Fish. Res. Bd. Canada, 33: 248.

Nurdogan, Yakup. and Oswald, W.J. 1995. Enhanced nutrient removal in high-rate ponds. Wat. Sci. Tech. Vol. 31, No. 12: 33-43.

O' Connor, D.J., and Dobbings, W.E., 1958. Mechanisms of Re-aeration in Natural

Streams. American Society of Civil Engineers Trans., 123: 641.

Odum, E.P., 1950. Bird populations of the Highlands (North Carolina) Plateau in relation to plant succession and avian invasion. Ecology, 31: 587-605.

Odum, E.P., 1969. The strategy of ecosystem development. Science, 164: 262-270.

Odum. E.P., 1971. Fundamentals of Ecology. W.B. Saunders Co., Philadelphia.

Odum, H.T., 1956. Primary production in flowing waters. Limnol. Oceanogr., 1: 102-117.

Odum, H.T. 1962. Man in the ecosystem. In: Proceedings Lockwood Conference on the Suburban Forest and Ecology. Bull. Conn. Agr. Station 652. Storrs, CT, pp. 57-75.

Odum, H.T. 1983. System Ecology. Wiley Interscience, New York. 510 pp.

Odum, H.T., Siler, W.L., Beyers, R.J. and Armstrong, N. 1963. Experiments with engineering of marine ecosystems. Publ. Inst. Marine Sci. Uni. Texas, 9: 374-403.

Olson, T.M. and Barbier, P.F. 1994. Oxidation kinetics of natural organic matter by sonolysis and ozone. Wat. Res. Vol. 28, No. 6: 1383-1391.

O'Neill, R.V., 1976. Ecosystem persistence and heterotrophic regulation. Ecology, 57: 1244-1253.

O'Neill, R.V., DeAngelis, D.L., Waide, J.B and Allen,T.F.H., 1986. A Hierachical Concept of Ecosystems. Princeton University Press., Princeton, N.J. 253 pp.

Onnen, J.H., 1972. Wet scrubbers tackle pollution. Environ. Sci. Technol., 6: 994-998.

Orlob, G., 1981. State of the Art of Water Quality Modelling. IIASA, Laxenburg, Austria.

Park, Richard A. et al., 1978. The Aquatic Ecosystem Model MS. CLEANER. Proc. Int. conf. on Ecol. Modelling, 28 Aug.-2 Sep. 1978, Copenhagen (ISEM), p. 579.

Parr, R.M. and Taylor, D.M., 1963. Physics Med. Biol., 8: 43.

Parr, R.M. and Taylor, D.M., 1964. Biochem. J., 91: 424.

Patten, B.C. 1991. Network ecology: indirect determination of the life-environment relationship in ecosystems. In: M. Higashi and T.P. Burns (editors). Theoretical Studies of Ecosystems: The Network Perspective. Cambridge University Pres, pp. 288-351.

Patten, B.C., Jørgensen, S.E. and Dumont, H. 1990. Wetlands and Continental Shallow Water bodies, Volume I, SPB Academic Publishers, The Hague, 760 pp.

Patten, B.C., Strakraba, M., and Jørgensen, S.E., 1997.Ecosystems emerging 1: Conservation. Ecol. Modelling, 96: 221.

Pavlova, A.K., 1956. Vestsi Akad. Navuk. BSSR, No. 3: 83.

Peters,, R.H., 1983. The Ecological Implications of Body Size. Cambridge University Press. Cambridge. 329 pp.

Pielou, E.C., 1966. Species-diversity and pattern diversity in the study of ecological succession. J. Theoret. Biol., 10: 370-83.

Pelekani, C. and Snoeyink, V.L. 1999. Competitive adsorption in natural water: role of activated carbon pore size. Wat. Res. Voll. 33, No. 5: 1209-1219.

Porter, J.R., 1946. Bacterial Chemistry and Physiology. John Wiley and Sons, New York.

Posselt, H.S., 1966. Sodium and potassium ferrate (VI). A Bibliography and Literature Review. Unpublished Research Report, Carus Chemical Company, Inc., LaSalle, I11.

Pring, R.T., 1972. Specification considerations for fabric collectors. Pollution Eng., 4(12): 22-24.

Push, C.H. and Walcha, W., 1975. Fractionation of metal salts by hyperfiltration. 2nd Int. Congr. on Industrial Waste Water, Stockholm.
Pytkowicz, R.M., 1967. Geochim cosmochim Acta, 31: 63.

Quirk, T.P., 1964. Economic aspects of incineration vs. incineration-drying. J. Wat. Poll. contr. Fed., 36: 11, 1355.

Ramanathan, V., 1975. Greenhouse effect due to chlorofluorocarbons: Climatic Implications. Science, 190: 50-52.

Rasmussen K. and Lindegaard, C., 1988. Effects of pollution with iron compounds (ochre) on the macroinvertebrate fauna of the river Vidå in South-West Jutland, Denmark. Wat Res. 22: 1101-1108.

Reck, R.A., 1978. Global Temperature Changes: Relative Importance of different Parameters as calculated with a Radioactive-Convective Model. In: Carbon Dioxide, Climate and Society. Proceedings of an IIASA Workshop, Febr. 21-24, 1978. (Jim Williams, ed.). Pergamon Press, Oxford. p. 193-200.

Reck, R.A. and Fry, D.L.,1977. The Direct Effect of Chlorofluoromethanes on the Atmospheric Temperature. GMR-2564. General Motors Research Laboratories, Warren, Michigan.

Reible D.D. 1998. Fundamentals of Environmental Engineering. Lewis Publishers, Springer Verlag, Boca Raton. London. New York , Washington D.C. 526 pp.

Rinaldi, S., Soncini-Sessa, R., Stehfest, H. and Tamura, H., 1979. Modelling and Control of River Quality. McGraw-Hill, Great Britain.

Rizzo, J.C. and Shepherd, A.R., 1977. Chem. Eng.: 95-100.

Rosendahl, A., 1970. Undersökelser i halvteknisk målestok vedrörende fjerning av fosfor från mekanisk och biologisk renset kommunalt avlöpsvann ved kjemisk rensning. 6. Nordiske Symposiet om Vattenforskning, Scandicon 21-23 April 1970.

Rovel, J.M., 1972. Chemical regeneration of activated carbon. Proc. Wat. Techn., 1: 187-190.

Rosenzweig, M.L., 1971. Paradox of enrichment: destabilization of exploitation ecosystems in ecological time. Science, 171: 385-387.

Rozema, J., Lambers, H., van de Geijn, S.C. and Cambridge, M.L., CO_2 and biosphere. Kluwer Acad. Publ. Dordrecht, Boston, London. 489 pp.

Rullman, D.H., 1976. Baghouse technology: a perspective. J. Air Pollut. Control Assoc., 26: 16-18.

Rüb, Friedmond, 1969. Aufbereitung von spielwässern der Metalindustrie nach dem Ionenaustauscherverfahren. Wasser, Luft und Betreib, 8: 292-296.

Samsahl, K. and Soremark, R., 1961. Modern Trends in Activation Analysis. Proc. Int. Conf., Texas.

Sanborn, N.P., 1953. Disposal of food processing water by spray irrigation. Sewage and Int. Wastes, 25: 1034.

Saqqar, M.M. and Pescod, M.B. 1996. Performance evaluation of anoxic and facultative waste water stabilisation ponds. Wat. Sci. Tech. Vol. 33, No. 7: 141145.

Scaramelli, A.B. and Dibiano, F.A., 1973. Upgrading the activated sludge system by addition of powdered carbon. Water and Sewage Works: 90-94.

Schaufler, Gerhard, 1969. Ion exchangers in industrial waste water clarification. Chem. Tech. Ind.: 787-790.

Schindler, D.W. and Nighswander, J.E., 1970. Nutrient supply and primary production in Clear Lake, Eastern Ontario. J. Fish. Res. North Canada, 27: 2009-2036.

Schjødtz-Hansen, P., 1968. Dansk Teknisk Tidsskrift, 5.

Schjødtz-Hansen, P. and Krogh, O., 1968. Nordisk Mejeritidsskrift, 34: 194.

Schofield, A. and Haskin, L., 1964. Geochim. cosmochim. Acta, 28: 437.

Schrøder, H. 1999. A Systems Approach to the Environmental Analysis of Pollution Minimisation. Lewis Publishers, pp. 101-131.

Schüürmann, G. and Markert, B., 1998. Ecotoxicology. Wiley-Interscience. New York, Chichester, Weinheim, Brisbane, Singapore, Toronto. 900 pp.

Schwarz, K. and Foltz, C.M., 1958. Fed. Proc. Fedn. Am. Socs. exp. Biol., 17: 492.

Seip, Knut Lehre, 1979. Mathematical model for uptake of heavy metals in benthic algae. J. Ecol. Model., 6: 183-198.

SETAC. 1995. The Multi-Media Fate Model: A vital tool for predicting the fate of chemicals.

Shacklette, H.T., 1965. Bull. U.S. geol. Surv. 1198D.

Shannon, C.E. and Weaver, W., 1963. The mathematical theory of communication. University of Illinois Press, Urbana. 117 pp.

Shibuya, M. and Nakai, T., 1963. Proc. 5th Conf. Radioisotopes, Japan.

Shimp, N.F., Connor, J., Prince, A.L. and Bear, F.E., 1957. Soil Sci., 83: 51.

Sillen, I.G. and Martell, A.J., 1964. Stability Constants of Metal - Ion Complexes. The Chemical Society, London.

Sillen, L.G., 1961. Oceanography Publ. No. 67 AAAS, Washington D.C.: 549-581.

Sillen, L.G., 1967. Adv. Chem. Serv., 67: 45-57.

Simon,' H.A., 1973. The organisation of complex systems. In: Pattee, Hierarchy Theory. Braziller, New York. pp. 3-27.

Simpson, E.H., 1949. Measurement of diversity. Nature, 163: 688.

Simpson, R.L., Good, R.E., Leck, M.A. and Whigham, D.F. 1983. The ecology of freshwater tidal wetlands. Bioscience 33: 255-59.

Sison, N.f., Hanaki, K. and Matsuo, T. 1996. Denitrification with external carbon source utilising adsorption and desorption capability of activated carbon. Wat. Res. Vol. 30, No. 1: 217-227.

Skaarup, T. and Sørensen C., 1994. Master's thesis at the Royal Danish School of Agriculture and Veterinary. Green Audit of Agriculture (in Danish with English summary).

Skinner, Brian J., 1969. Earth Resources. Englewood Cliffs, N.J.: Prentice-Hall.

Smales, A.A. and Pate, B.D., 1952. Analyst, London, 77: 196.

Smales, A.A. and Salmon, L., 1955. Analyst, London, 80: 37.

Smidth, F.L./MT, 1973. Report on the eutrophication of Lake Lyngby.

Soare, J., Silva, S.A., de Oliveira, R., Araujo, A.L.C., Mara, D.D. and Pearson, H.W.

1996. Wat. Sci. Tech. Vol. 33, No. 7: 165-171.

Soremark, R. and Bergman, B., 1962. Acta Isotopica, 2: 1.

Sorial G.A., Smith, F.L., Suidan, M.T., Pandit, A., Biswas, P. and Brenner, R.C. 1998. Evaluation of trickle-bed air biofilter performance for styrene removal. Wat. Res. Vol. 32, No. 5: 1593-1603.

Sowden, E.M. and Stitch, S.R., 1957. Biochem. J., 67: 104.

Spanier, G., 1969. Ionenaustauscher für die Wasser un Abwasseraufbereitung in der Metallveredlung. Oberfläche, 10: 365-373.

Spector, W.S., 1956. Handbook of Biological Data. Saunders, Philadelphia.

Sprague, J.B., 1970. Water Res., 4(1): 3-32.

Srivastav, R.K., Gupta, S.K., Nigam, K.D.P. and Vasudevan, P. 1994. Treatment of chromium and nickel in waste water by using aquatic plants. Wat. Res. Vol. 28, No. 7: 1631-1638.

Stamm, M.D. and Fernandez, F., 1958. Revta. esp. Fisiol., 14: 177,185.

Statens Naturvårdsverk Publikationer, 1969. Experiences of Chemical Purification, 10: 12-13.

Stevenson, D.G. 1997. Flow and filtration through granular media - the effect of grain and particle size dispersion. Wat. Res. Vol. 31, No. 2: 310-322.

Stitch, S.R., 1957. Biochem. Z., 304: 73.

Straskraba, M. 1984. New ways of eutrophication abatement. In: M. Straskraba, Z. Brandl and P. Procalova (eds.). Hydrobiology and Water Quality of Reservoirs. Acad. Sci., Ceské Budejovice, Czechoslovakia, pp. 37-45.

Straskraba, M. 1985. Simulation Models as Tools in Ecotechnology Systems. Analysis and Simulation. Vol. II. Academic Verlag, Berlin, pp. 362.

Straskraba, M. 1993. Ecotechnology as a new means for environmental management. In: Ecological Engineering, Vol. 2, No. 4, pp. 311-332.

Straskraba, M., Jørgensen, S.E. and Patten, B.C. 1999. Ecosystems emerging: 2. Dissipation. Ecological Modelling 117: 3-39.

Stumm, W., 1958. Ozone as a disinfectant for water sewage. J. Boston Soc. Civ. Eng., 45(1): 68.

Stumm, W. and Leckie, J.O., 1970. Phosphate exchange with sediments; Its role in the productivity of surface waters. Advances in water pollution research. Proc. of 5th Int. Conf. in San Francisco and Hawaii, III 26/1 to 16.

Stumm, W. and Morgan, J,J., 1970. Aquatic Chemistry. Wiley Interscience, New

York.

Stumm, W. and Morgan, J.J. 1996. Aquatic Chemistry. Chemical Equilibria and Rates in natural Waters. John Wiley & Sons, Inc., New York, Chichester, Brisbane, Toronto, Singapore. 1022 pp.

Suter, G.W. 1993. Ecological Risk Assessment. Lewis Publishers, Chelsea, MI.

Suzuki, T. and Hamada, I., 1956. J. Chem. Soc., Japan, Pure Chem. Sect., 77: 125.

Svensson, B.H. and Söderlund, R. (eds.), 1976. Ecological Bulletin, No. 22. Nitrogen, phosphorus and sulphur - Global Cycles. SCOPE Report 7, Örsundsbro, Sweden.

Särkkä, Mirja, 1970. Praktiska erfarenheter av närsaltreduktion. 6. Nordiske Symposiet om Vattenforskning, Scandicon 21-23 April 1970: 177-187.

Thomas, M.D., Hendricks, R.H. and Hill, G.R., 1950. Soil Sci., 70: 9.

Thomas, D.N., Judd, S.J. and Fawcett, N. 1999. Review Paper. Flocculation modelling: a review. Wat. Res. Vol. 33, No. 7: 1579-1592.

Thompson, L.M., 1975. Weather variability, climatic change, and grain production. Science, 188: 553-541.

Thompson, T.G. and Chow, T.J., 1956. Univ. Washington Publs. Oceanog. No. 184: 20.

Thomsen, A.B. 1998. Degradation of quinoline by wet oxidation-kinetic aspects and reaction mechanisms. Wat. Res. Vol. 32, No. 1: 136-146.

Tipton, I.H. and Cook, M.J., 1963. Health Phys., 9: 103.

Turner, R.C., Radley, J.M. and Mayneord, W.V., 1958. Health Phys. 1: 268.

Turner, R.C., Radley, J.M. and Mayneord, W.V., 1958. Br. J. Radiol., 31: 397.

Tyutina, N.A., Aleskovsky, V.B. and Vasilev, P.I., 1959. Geochemistry: 668.

Tönseth, E.I. and Berridge, H.B., 1968. Removal of proteins from industrial waste water. Effluent and Water Treatment J., 8: 124-128.

UNEP, 1994. Government Strategies and Policies for Cleaner Production, Industry and Environmental Program Activity Centre. Paris.

U.S. Department of Health, Education and Welfare, 1969. Tall Stacks Various Atmospheric Phenomena, and Related Aspects. Report APTD 69-12, pp. 1-12, Washington D.C.

Vavilin, V.A., Vasiliev, V.B. , Rytov, S.V. and Ponomarev, A.V. 1995. Modelling ammonia and hydrogen sulphide inhibition in anaerobic digestion. Wat. Res. Vol. 29, No. 3: 827-835.

Verschueren, K.,1983. Handbook of Environmental Data on Organic Chemicals. Van Nostrand Reinhold.

Vinogradov, A.P., 1953. The elementary chemical composition of marine organisms. Sears Foundation, New Haven, Conn.

Vinogradova, Z.A. and Kobalsky, V.V., 1962. Dokl. Acad. Nauk SSSR, 147: 1458.

Vollenweider, R.A., 1968. The scientific basis of Lake and Stream Eutrophication with particular reference to phosphorus and nitrogen as eutrophication factors. Tech. Rep. OECD, Paris. DAS/DSI/68, 27: 1-182.

Vollenweider, R.A., 1969. Möglichkeiten und Grenzen elementarer Modelle der Stoffbilanz von Seen. Arch. Hydrobiol., 66: 1-136.

Vollenweider, R.A., 1975. Input-output models with special reference to the phosphorus loading concept in limnology. Schweiz. Z. Hydrol., 37: 53-83.

Waid, D.E., 1972. Controlling pollutants via thermal incineration. Chem. Eng. Prog., 68(8): 57-58.

Waid, D.E., 1974. Thermal oxidation or incineration. Proc. Special - Pollut. Control Assoc., Pittsburgh, PA: 62-79.

Wakita, H. and Kigoshi, K., 1964. J. Chem. Sic., Japan, Pure Chem. Sect., 85: 476.

Walker, J.D. and Drier, D.E., 1966. Aerobic digestion of sewage sludge solids. Paper presented at 35th Annual Meeting of Georgia Water Pollution Association, Walker Process Equipment, Inc., Publ. 26.130.

Wang, W.C., Young, Y.L., Lacis, A.A, Mo, T. and Hansen, J.E., 1976. Greenhouse effects due to manmade perturbations of trace gases. Science, 194: 685.

WCDE, 1987. Our common future. Oxford University Press. Oxford.

Weber, W.J. Jr. and Morris, J.C., 1963. Kinetics of adsorption on carbon from solution. J. Sanit. Eng. Div. Amer. Sic. Civ. Eng., 89: SA 2, 31-59.

Weber, W.J. Jr. and Morris, J.C., 1964. Equilibria and capacities for adsorption of carbon. J. Sanit. Eng. Div. Amer. Soc. Civ. Eng., 90: SA 3, 79-107.

Weber, W.J. Jr. and Morris, J.C., 1965. Adsorption of biochemically resistant materials from solution. J. Wat. Poll. Contr. Fed., 37: 425.

Webster, J.R., 1979. Hierarchical organizartion of ecosystems. In E. Halfon (editor) Theoretical Systems Ecology. Academic Press, New York, 119-131.

Weibel, S.R. et al., 1964. Urban land runoff as a factor in stream pollution. J. Wat. Poll. Contr. Fed., 36(7): 914.

Weiderholm, T., 1980. Use of benthos in lake monitoring. J. Water Pollut. Control Fed. 52, 537.

Wenzel, H., Damborg, A., and Jacobsen, B.N., 1990. Danish Emissions of Industrial Waste water, Environmental Project No.153. DEPA, Copenhagen, (in Danish, with English summary).

Wenzel, H., Hauschild, M., and Alting,L., 1997. Environmental Assessment of Products, Volume 1, Methodology, Tools and Case Studies in Product Development. Chapman and Hall., London 378 pp.

Wester, P.O., 1965. Biochem. Biophys. Acta, 109: 268.

Wetzel, R.G., 1983. Limnology. Second edition. Saunders College Publ. Philadelphia, 760 pp.

Whealer, G.L., 1976. Chlorine dioxide a selective oxidant for industrial waste water treatment. 4th Annual Ind. Poll. Conf. April 1976.

Wicke, M., 1971. Collection efficiency and operation behaviour of wet scrubbers. Paper EN 16H, pp. 713-718, in Proceedings of the 2nd Int. Clean Air Congress, H.M. Englund and W.T. Beery, eds. Academic Press, Inc., New York.

Willers, B. 1994. Sustainable Development: A New World Deception. Conservation Biology 8: 1146-1148.

Williamson, G., 1959. Proc. 6th Ind. Waste Conf., Ontario Canada, p. 17.

Wing, A. Bruce and Steinfeld, William M., 1970. Comparison of stone-packed and plastic-paced trickling filters. J.W. Poll. Contr. Fed., 42: 255.

Winther, l., Henze, M., Linde J.J., Jensen, H.T. 1998. Spildevandsteknik. 355 pp.

Wisniewski, T.F., Wiley, A.J. and Lueck, B.F., 1956. TAPPI, 39: 2,65.

Wisniewski, J. and Lugo A.E., Natural Sinks of CO_2. Kluwer Acad. Publ. Dordrecht, Boston, London. 462 pp.

Wolverton, B.C. 1982. Hybrid Waste water Treatment Systems Using Anaerobic Microorganisms and Reeds. Econ. Bot. 36: 378-388.

Wong, J.B., Ranz, W.E. and Johnstone, H.F., 1956. Collection of aerosols by fiber mats. Technical Report No. 11. Engineering Experiment Station, University of Illinois.

Woodwell, G.M. et al., 1967. DDT residues in an East Coast estuary: A case of biological concentration of a persistent insecticide. Science, 156: 821-824.

WRL, 1977. Heavy metal ion exchanger WRL/500CX. Copenhagen.

Wynn, C.S., Kirk, B.S. and McNabney, R., 1972. Pilot plant for tertiary treatment of waste water with ozone. Water, 69: 42.

Yamagata, N., 1950. J. Chem. Soc., Japan, 71: 228.

Yamagata, N., 1962. J. Radiat. Res., 3: 4.

Yamagata, N., Murata, S. and Torii, T., 1962. J. Radiat. Res., 3: 4.

Yao, J.J., Huang, Z-H. and Masten, S.J. 1998. The ozonation of pyrene: pathway and product identification. Wat. Res. Vol. 32, No. 10: 3001-3012.

Yee, W.C., 1966. Selective removal of mixed phosphate by activated alumina. J. Am. Water Works Assoc., p. 239.

Yoo, H., Ahn, K-H., Lee, H-J., Lee, K-H., Kwak, Y-J and Song, K-G. 1999. Nitrogen removal from synthetic waste water by simultaneous nitrification and denitrification (SND) via nitrite in an intermediately-aerated reactor. Wat. Res. Vol. 33, No. 1: 145-154.

Zimen, K.E., 1978. Source function for CO_2 in the atmosphere. In: Carbon Dioxide, Climate and Society. Proceedings of a IIASA Workshop, Febr. 21-24, 1978. (Jim Williams, ed.). Pergamon Press Oxford. p. 89-96.

Zimen, K.E., Offermann, P. and Hartmann, G., 1977. Source functions of CO_2 and future CO_2 burden in the atmosphere. Z. Naturforschung, 32a: 1544.

Zimmerman, F.J., 1958. Chem. Eng., 65: 17, 117.

Zuckermann, M.M. and Molof, A.H., 1970. High quality reuse water by chemical physical waste water treatment. J. Wat. Poll. Contr. Fed., p. 437. (Includes the discussion by Weber).

Æsøy, A., Ødegaard, H., Bach, K., Pujol, R. and Hamon, M. 1998. Denitrification in a packed bed biofilm reactor (biofor) - experiments with different carbon sources. Wat. Res. Vol. 32, No. 5: 1463-1470.

Index